Intercellular Communication in Cancer

Mustapha Kandouz
Editor

# Intercellular Communication in Cancer

Springer

*Editor*
Mustapha Kandouz
Department of Pathology
Wayne State University School of Medicine
Detroit, MI, USA

Karmanos Cancer Institute
Wayne State University
Detroit, MI, USA

ISBN 978-94-017-7379-9 ISBN 978-94-017-7380-5 (eBook)
DOI 10.1007/978-94-017-7380-5

Library of Congress Control Number: 2015956021

Springer Dordrecht Heidelberg New York London


Printed on acid-free paper

Springer Science+Business Media B.V. Dordrecht is part of Springer Science+Business Media (www.springer.com)

# Preface

Cells are social entities. As such, they need to interact and communicate with each other, in order to be able to build organized structures of tissues, organs, and organisms or to ensure functional and physiological coordination. Cells dedicate considerable energy and various mechanisms to ensure cell-cell communication, because it is important for their survival, behavior, and fate within their environment. For cells, communicating is vital not only because they are part of organs and tissues of which they maintain the integrity and function but also because many of their functions need to be coordinated, fine-tuned, and limited spatiotemporally. Furthermore, cells use communication to minimize the energetic and signaling burden, whereas a single minimal signal could be amplified and propagated. The diversity of the types of intercellular communications emphasizes the need for cells to communicate in different ways and for different purposes, transfer of small molecules, signaling, establishment of barriers/polarity, paracellular permeability, and transmission of cytoskeleton-generated forces, all of which play critical roles in normal physiology and pathological development such as in cancer. As in many other diseases, an essential part of tumor genesis and progression is the deregulation of the ability of cells to communicate and transmit information to and receive signaling input from neighboring cells.

Communication as a basic cellular need uses different modes. A classical mode involves growth factors and other signaling molecules released by cells to act either locally as autocrine or paracrine moieties or to be systemically transported via the body fluids to affect the biology of remote target cells. However, the subject of this book on *Intercellular Communication in Cancer* is another mode of cell-cell communication that involves a direct contact between cell membranes.

First, we review junctional modes of intercellular communication. These modes are associated with the building of elaborate protein complexes, arranged to form specialized junctions between cells. Chapter 1 by Talhouk and colleagues shows an example of how the so-called gap junctions and their building blocks the connexin proteins regulate normal and malignant processes, i.e., in the mammary gland. Connexins can also be important in the communication of cancer cells with cellular components of their microenvironment, a process discussed in Chap. 2 by Lathia and colleagues in the context of brain tumors. Gap junctions have been classically known to ensure the transfer of ions and other small moieties. More recently, they have been shown as conduits for the transfer of important nucleic acid molecules such as microRNAs. As explained by Rameshwar and colleagues in Chap. 3, this mechanism could have an important impact on cellular therapy of glioblastomas.

The next chapters will introduce examples of other types of junctional intercellular communication. In Chap. 4, we can read from Jacquier-Sarlin and colleagues about how adherens junctions are regulated to play an essential function in cell differentiation and what the impact is on tumor development. In Chap. 5, Sen and colleagues, after introducing cadherins, the major component of adherens junctions, explain different computational and mathematical approaches used to explore different aspects of cadherin-mediated cell-cell adhesions. Furthermore, Al-Moustafa and colleagues, in Chap. 6, provide an overview of the role of E-cadherin/catenin complex modulations in human oral cancer and shed light on how the high-risk human papillomaviruses (HPVs) interfere with this complex to promote oral cancer progression.

In addition to junction-based interactions, we are also witnessing a growing interest in the role of non-junctional modes of cell-cell communication. First, Alaoui-Jamali and colleagues discuss the role of small cell-derived membranous vesicles released by cancer cells called exosomes, in metastatic signaling. They describe in Chap. 7 the use of exosomes as tumor biomarkers and their therapeutic impact. Another mode of cell-cell communication, introduced by Lou and colleagues in Chap. 8, addresses another type of membranous structures called tunneling nanotubes, which are made of elongated intercellular conduits that form bridges between cells. In other respects, in Chap. 9, Parris discusses the concept of cell-cell fusion between abnormal cells with "task force" cells of the immune system and its role in cancer progression. In line with these chapters, as Kandouz and colleagues explain in Chap. 10, cell-cell communication can be achieved without necessarily involving junctional structures, via Eph receptors and ephrin ligands, which constitute the largest family of receptor tyrosine kinases.

This book would have been incomplete if no chapters were dedicated to viewing cell-cell communication in the context of a community dynamic. Komurov in Chap. 11 describes computational approaches to modeling of molecular interactions in multicellular systems, while Goldberg and colleagues describe in Chap. 12 a process called contact normalization of tumor cells by communicating with neighboring normal cells. Last but not least, in Chap. 13, Nelson and colleagues provide an overview of intercellular communication within the tumor microenvironment.

We believe that the chapters presented in this volume provide a global overview of different modes of membrane-to-membrane intercellular communication. They are intended to update researchers in the field about novel developments and also to constitute an important knowledge base for researchers contemplating to engage in this exciting and rich research area. Nevertheless, it should be noted that this book is not an exhaustive repertoire of all known types of cell-cell communication. Rather, we made admittedly subjective choices to illustrate the diversity of these modes and their instrumental role in cancer.

Detroit, USA Mustapha Kandouz

# Contents


1 **Beyond the Channel: Role of Connexins in Regulating Normal and Cancerous Processes in the Mammary Gland** ............................. 1
Dana Bazzoun, Sophie Lelièvre, and Rabih Talhouk

2 **Connexins: Bridging the Gap Between Cancer Cell Communication in Glioblastoma** .................................................. 29
Maksim Sinyuk and Justin D. Lathia

3 **MicroRNA and Gap Junctions in Glioblastoma Cells: Implications for Cellular Therapy** ............................................. 43
Jessian L. Munoz, Vivian Rodriguez-Cruz, and Pranela Rameshwar

4 **Dynamic Regulation of Adherens Junctions: Implication in Cell Differentiation and Tumor Development** ................................ 53
Marjolaine A. Pelissier-Rota, Nicolas T. Chartier, and Muriel R. Jacquier-Sarlin

5 **Cadherin-Based Cell-Cell Adhesions: Adhesion Structure, Signalling and Computational Modeling** ........................... 151
Sandeep Kumar and Shamik Sen

6 **E-Cadherin/Catenin Complex Modulations in Human Oral Cancer** ......................................................................... 169
Ala-Eddin Al Moustafa and Amber Yasmeen

7 **Tumor-Derived Exosomes in Metastasis Signaling and Implications** ...................................................................... 189
Yuan Hang Cao, Krikor Bijian, Maisa I. Alkailani, and Moulay A. Alaoui-Jamali

8 **Tunneling Nanotubes: Intercellular Conduits for Direct Cell-to-Cell Communication in Cancer** .............................. 201
Emil Lou and Subbaya Subramanian

**9 Cell-Cell Fusion, Chemotaxis and Metastasis** ..... 227
George E. Parris

**10 Role of the Family of Ephs and Ephrins in Cell-Cell Communication in Cancer** ..... 255
Mohamed Amessou and Mustapha Kandouz

**11 Computational Approaches to Modeling of Molecular Interactions in Multicellular Systems** ..... 287
Kakajan Komurov

**12 Contact Normalization or Escape from the Matrix** ..... 297
Harini Krishnan and Gary S. Goldberg

**13 Intercellular Communication, the Tumor Microenvironment, and Tumor Progression** ..... 343
Mei-Fong Pang and Celeste M. Nelson

**Index** ..... 363

# Chapter 1
# Beyond the Channel: Role of Connexins in Regulating Normal and Cancerous Processes in the Mammary Gland

**Dana Bazzoun, Sophie Lelièvre, and Rabih Talhouk**

**Abstract** The mammary epithelium possesses a well-defined architecture mediated by cell-extracellular matrix and cell-cell junctions that is essential for the coordinated continuous development of the mammary gland. The dynamic remodeling of the mammary gland is orchestrated by cellular responses to environmental cues that are relayed through the interplay between cell-cell junctions themselves including tight junctions, adherens junctions, desmosomes and gap junctions, and their interacting partners, notably polarity proteins. In this chapter, we address the molecular dynamics of gap junctions and the roles that these junctions have been ascribed in modulating normal and mammary cancer behaviors. We aim to highlight how connexins, the building blocks of gap junctions, have transcended their gap junction-dependent functions as structural cellular components and are now perceived as signaling hubs. These structural entities are able to integrate messages from the cell's surrounding and modulate cytoplasmic downstream signaling pathways that regulate cell function and often gene expression during the normal differentiation of the mammary epithelium. We also elaborate on the changes in the expression, function and localization of gap junctions and connexins and their consequences for mammary cancer progression. Finally, we present recent breast cancer therapies that target gap junction proteins.

**Keywords** Gap junction • Connexin • Mammary gland • Cell junctions • Normal development • Tumorigenesis

---

D. Bazzoun • R. Talhouk (✉)
Department of Biology, Faculty of Arts and Sciences, American University of Beirut (AUB), Beirut, Lebanon
e-mail: rtalhouk@aub.edu.lb

S. Lelièvre
Department of Basic Medical Sciences, Purdue University, West Lafayette, IN, USA

Center for Cancer Research, Purdue University, West Lafayette, IN, USA



M. Kandouz (ed.), *Intercellular Communication in Cancer*,
DOI 10.1007/978-94-017-7380-5_1

# 1.1 Introduction

Being the organ responsible for milk production and one that continues to develop during the lifetime of a female, the mammary gland is considered a distinctive feature of mammals [1]. The extensive remodeling that this organ undergoes in response to microenvironmental cues reflects a delicate and orchestrated developmental process that is determined to a large extent by cell-cell and cell-extracellular matrix (ECM) interactions characterized by a reciprocal exchange of mechanical and biochemical stimuli [2, 3]. Importantly, the formation of cell junction complexes that mediate cell adhesion and communication determines the architecture of the mammary epithelium by regulating proper cellular differentiation. In the epithelium, cellular differentiation requires interaction with adjacent cells and with the microenvironment (e.g., the stroma, hormones, and growth factors including contact with the ECM. These interactions are made through multiprotein junction complexes that play a central role in the maintenance of normalcy in the mammary gland. In the normal epithelium, cell junctions are asymmetrically organized thus, creating a basoapical axis that goes awry as soon as cancer is initiated. Well known participants in the asymmetrical organization of mammary epithelial cells are cell-cell adhesion junctions like tight junction, adherens junction and desmosomes and cell-ECM adhesion junctions, like hemidesmosomes. However, the implication of junctions like gap junction specifically in charge of cell-cell communication is not well defined in the grand scheme of the cell's interaction with its environment. Here, we focus on the implication of cell junctions, particularly gap junctions, in regulating the normal phenotype of the mammary epithelium, in addition to supporting the different stages of breast cancer development.

## *1.1.1 Normal Mammary Gland Development and Breast Cancer*

The mammary gland has been widely used as a model to study the impact of cell-cell and cell-environment interactions on normal and cancer development. The murine and the human mammary glands share functional and structural similarities, which is the reason why the mouse has long been considered as the primary animal model to study human mammary gland development. During mouse embryonic development, mammary epithelial placodes begin to branch slowly from the ectoderm into a rudimentary tree, resulting in a bilayered ductal structure with epithelial luminal cells as the inner layer and myoepithelial cells as the outer layer. Upon puberty, secondary site branching is induced by systemic hormonal changes during recurrent estrous cycles resulting in the proliferation of the ductal ends that extend into the surrounding adipose tissue (Fig. 1.1a, b) [4]. At the onset of pregnancy, full differentiation of the gland prompts massive branching, proliferation and sprouting of the lobulo-alveolar milk-secreting buds at the ducts termini. This extensive

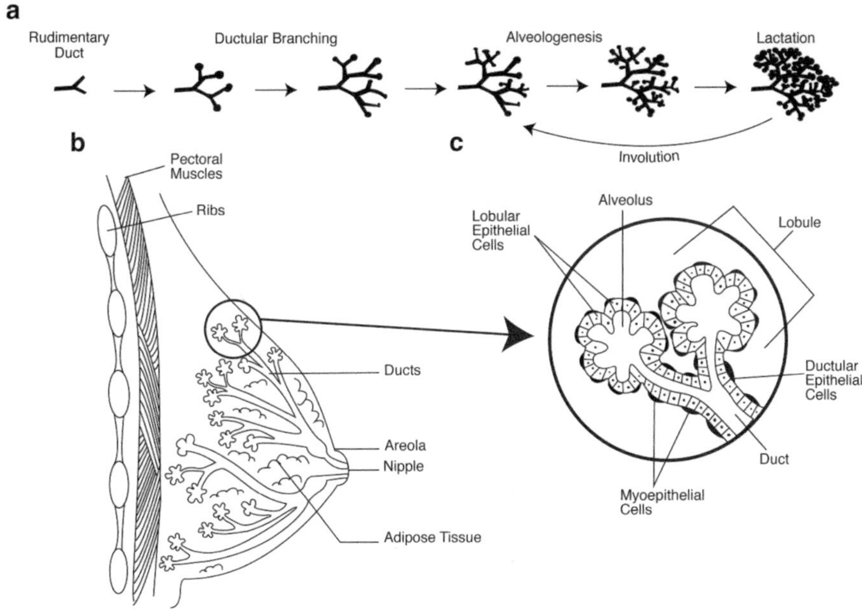


**Fig. 1.1 A schematic diagram of the different developmental stages and anatomy of the mammary gland**. (**a**) Throughout the lifetime of a female, the breast undergoes dramatic remodeling marked by ductal growth and formation of glandular structures. During fetal development, a rudimentary structure forms and remains quiescent until puberty. Ductal elongation is initiated at puberty via continuous proliferation and migration of epithelial cells. Alveologenesis is induced upon pregnancy during which lobules and milk producing luminal epithelial cells necessary for lactation are formed. At the end of lactation, the mammary epithelium undergoes extensive apoptosis resulting in involution marked by the regression of the mammary gland back almost to the pre-pregnancy state. (**b**) The mature female breast is composed of four main structures: lobules; milk ducts; adipose tissue and connective tissue. The external characteristic features of the breast include a nipple surrounded by a pigmented *areola* . Supporting the breasts are ribs that underlie the pectoral muscles. The adult mammary epithelium forms a ductal network starting at the nipple and branching into lobules. (**c**) The lobules are composed of terminal end ducts ending in glandular structures called acini or alveoli (secreting acini). These structures are lined by luminal epithelial cells, milk secreting cells, and supported basally by myoepithelial cells that are responsible for the contractility of the epithelium

remodeling supports lactation; the alveolar luminal cells apically secrete milk into the lumen; milk is then is ejected by the contraction of the myoepithelial cells that surround the alveoli (Fig. 1.1c). Post-weaning, these newly acquired structures degenerate (involution) to favor the restoration of the virgin-like architecture of the mammary gland (Fig. 1.1a) [5]. The differentiation process that accompanies the development of the mammary epithelium incorporates cell-cell adhesion and communication necessary to maintain the harmonious evolution of the epithelium. If this process is altered, breast cancer might develop. Breast cancer is the most common type of cancer in women on a global scale. It has now become associated with the highest frequency of recurrence among carcinomas and is the second most common cause of cancer-related deaths in women. Since 1999, the incidence of breast cancer *in situ* has continued to increase in women below 50 years of age and has become stable in women aged 50 and older [6]. Given the high incidence and mortality rates associated with breast cancer, a precise understanding of the molecular pathways of mammary gland differentiation is needed. Importantly, the structural and functional differentiation of the mammary epithelium relies on the tight regulation of cell–cell junction complexes and their link with polarity complexes that are specifically involved in the control of the asymmetrical organization of the cell-cell junction complexes.

### *1.1.2 Cell Junctions*

Cell-cell junctions are classified into tight junctions (TJs), adherens junctions (AJs), desmosomes, gap junctions (GJs) and the recently described tunneling nanotubes (TNT). TJs, AJs, desmosomes and GJs consist of transmembrane proteins with intracellular domains that enable interactions with partner signaling molecules, and extracellular domains that mediate communication between neighboring cells. Full development of the epithelial polarity axis requires the asymmetrical distribution of both cell junction complex and polarity complex proteins. As such, besides their structural role in adhesion and communication, cell junctions are capable of relaying signals to the nucleus through their interaction with a variety of cytoplasmic partners, and subsequently regulating the expression of genes involved in development, differentiation and oncogenesis [7].

Apically, TJs maintain tissue homeostasis by acting as selective permeable barriers to paracellular movement in secreting glands, and are major contributors to tissue polarity as they mark the apical side of membranes. They encompass transmembrane occludins, claudins and junctional adhesion molecules (JAMs) associated with cytoplasmic partner proteins such as the Zona Occludens (ZO) family members [8]. The extracellular domains of occludins regulate the paracellular diffusion of small hydrophilic molecules between cells and target occludins to tight junctions at the cell membrane. The interaction among claudins dictate TJ gate function as it has been found that changes in their expression altered TJ conductivity [9]. Claudins also recruit occludins to TJs [10]. Claudins, occludins, ZO proteins and JAMs is essential for the formation of functional TJs, and if disrupted, will disable the junction reformation even after repletion of calcium normally necessary to maintain TJs [11]. Furthermore, ZO-1 has been suggested to be a scaffold protein to which transmembrane and cytoplasmic proteins bind, and a common binding partner of TJs, AJs and GJs proteins [12, 13]. TJ proteins control signaling pathways related to tissue homeostasis. For example, JAM proteins act as regulators of epithelial cell morphology by enhancing β1-integrin expression through controlling Rap1 GTPase activity [14], consequently contributing to cell attachment to the basement membrane, a specialized form of ECM delineating the breast epithelium. By regulating TGF-β type I receptor localization, occludins induce the TGF-β-dependent TJ disassembly during epithelial-to-mesenchymal transition, a phenomenon in which cells acquire mesenchymal features enabling them to detach from the epithelium during cancer metastasis or normal tissue development [15]. Interestingly, studies have revealed a TJ-mediated role in cell fate modulation necessary for the progression of cancer. The expression of occludins and claudin-1, -4 and -6 was reported to be down-regulated in breast cancer cell lines [16, 17], while claudin-1 overexpression enhanced apoptosis in tumor nodules of MDA-MB-361 breast cancer cells produced in three-dimensional (3D) cell culture [16].

AJs reflect the same level of complexity as TJs. They mediate intercellular adhesion as well as initiate downstream signaling. They are characterized by transmembrane cadherins that mediate calcium-dependent intercellular adhesion, and regulate cell signaling and transcription. When cell-cell contacts are formed, cadherins are recruited to enhance intercellular contacts [18]. Classical cadherins such as E-cadherin, N-cadherin and P-cadherin permit firm cellular adhesion and anchorage of cytoskeletal intermediate filaments to the membrane [19]. Cadherins, namely E-cadherin, assembles with members of the catenin family of proteins (α-catenin, β-catenin and p120 catenin). E-cadherin is a negative regulator of Wnt signaling pathways that are implicated in cell proliferation, differentiation, gene transcription and cell adhesion. By binding to E-cadherin, β-catenin is recruited to the cell membrane away from the nucleus, thus preventing its pro-proliferative potential [20]. In the mammary ducts, when E-cadherin, expressed in the luminal epithelial cells, is blocked, the growth of the epithelial tissue is inhibited by interfering with downstream signaling pathways initiated at the AJ complex [19].

In addition to E-cadherin, desmogleins and desmocollins, members of the desmosomal cadherin superfamily, form desmosomes that are adhesive intercellular junctions crucial to tissues experiencing mechanical stress, such as the myocardium, bladder, gastrointestinal mucosa, breast and skin [21–23]. These proteins recruit intermediate filaments to the sites of desmosome assembly through binding to other proteins such as members of the armadillo gene family, plakoglobin and plakophilins [24]. Although desmosomes are critical for maintaining stable cell–cell adhesion, emerging evidence indicates that they are also dynamic structures that contribute to cellular processes beyond cell adhesion. It has been reported that the misexpression of desmoglein3 in transgenic mice resulted in epidermal hyperproliferation and abnormal differentiation [25]. Furthermore, expressing desmoglein3 in the upper layers of the epidermis induced a more severe phenotype, including a reduction in epidermal barrier function and early postnatal lethality due to excessive water loss [26]. Loss of desmoplakin in breast cancer is correlated with increased proliferation and tumor size [27]. Moreover, it has been reported that reduced levels of desmoplakin is pronounced in breast cancer metastasis compared to primary tumors [28]. In addition, desmosomal adhesion was enhanced in response to increased estrogen levels [29], suggesting a potentially interesting avenue of investigation in mammary gland differentiation and cancer development.

Cadherins or their associated catenins are usually altered in tumors [30]. The association between loss or down-regulation of E-cadherin and the progression of sporadic breast cancer has been fairly documented. Early studies on breast cancer cell lines have associated low E-cadherin expression levels with invasive properties of breast carcinomas [31, 32]. Moreover, levels of β-catenin are usually elevated in the nuclei of tumor cells and associated with the activation of genes involved in proliferation and cell cycle progression necessary for tumor development [33, 34].

GJs are located at cell-cell contacts and spread underneath TJs. They are unique in that they mediate intercellular signals by connecting the cytoplasms of two neighboring cells, enabling various functions such as rapid transmission of action potentials and diffusion of metabolites, nutrients, second messengers such as 1,4,5-inositol-trisphosphate (IP3), calcium ions, and cyclic nucleotides, hence participating in the modulation of gene transcription, proliferation and apoptosis [35]. GJs include clusters of tens to thousands of intercellular channels, each of which is formed by the end-to-end docking of two hemichannels, also referred to as connexons, composed of six connexin (Cx) subunits. Except for skeletal muscle cells, erythrocytes and circulating lymphocytes, most cells in normal tissues communicate via GJs, highlighting the immense roles that proper GJs' function and Cxs' expression levels play in regulating growth, differentiation, developmental signaling and cell death [36, 37]. Cxs are no longer studied as structural components of the GJ that solely confer a channel role. Their regulation via translational and post-translational mechanisms and their interaction with cytoskeletal elements, junctional proteins and polarity proteins makes them signaling nodes with impact on normal and cancer cell functions. The role of GJs in the normal development of the mammary gland and in breast cancer will be detailed in later sections. Both GJs and TNTs facilitate the transfer of intracellular mediators between the cytoplasms of neighboring cells. Specifically, TNTs facilitate and allow the exchange of small molecules, vesicles and organelles. They were reported to be expressed in a variety of cell lines including bovine and human mammary epithelial cells [38, 39]. In breast cancer, TNTs were recently demonstrated to mediate the interaction between breast cancer cells and human mesenchymal stromal cells that promotes cancer growth and progression [40], yet the role of TNTs in normal function and cancer development of different tissues is still poorly understood.

Cell junctions appear independently involved in the maintenance of tissue morphology and homeostasis; nevertheless, they display multiple interactions and overlapping localization among each other. Therefore, it is impossible to focus a review of GJs without taking into account their connections with the other cell-cell junctions. Studies have shown that several growth factors that regulate the function of GJs are also implicated in the regulation of TJ permeability. In addition, Cxs can co-immunoprecipitate with core TJ proteins, claudins and occludins, and with AJ protein, N-cadherin [12, 41]. Cell junctions also share a common pool of downstream effectors and signaling partners that in turn integrate signals transduced by those junctions to govern tissue differentiation. For example, it has been demonstrated that GJs and AJs share a common downstream effector, β-catenin that binds to both Cxs and cadherins [42, 43]. This intimate link between the different types of junctions reflects the close interaction required for their proper formation and function as will be discussed in later sections of this chapter. Cell junctions not only rely on interactions among each other to secure proper function, but also on their association with polarity proteins that are key players in epithelial differentiation.

### 1.1.3 Cell Junctions and Polarity Proteins

Understanding the relationship between cell junctions and polarity proteins is unavoidable to further clarify pathways controlling normal differentiation of epithelia. Polarity in the normal epithelium is referred to as basoapical polarity where polarity proteins and cell junctions are unevenly localized between the apex and the base (against the basement membrane) of the cell [44]. Polarity proteins are classified into three groups: Crumbs (Crumbs/PALS/PATJ) and Par (Par3/Par6/aPKC/Cdc42) complexes that are apical, and Scribble (Scrib/Dlg/Lgl) that is basolateral [45]. Initially and before cell junction formation, Par3 localizes to the apical side of the cell. This is followed by the generation of cell–cell contacts through the enrichment of E-cadherin, JAM-A, ZO-1 and nectin, a cellular adhesion molecule, at the membrane [46]. The membrane localization of the mentioned proteins activates Rac1 which, in turn, induces cell junction complexes expansion by inhibiting RhoA to [47]. Numerous studies have been initiated to identify how junction and polarity proteins affect each other with TJ being the most studied junctions in relation to these proteins. For example, TJ maintenance has been regulated by Par3/Par6/aPKC complex, one of the early polarity pathways activated during differentiation. JAM was found to directly associate with Par3 leading to the recruitment of the Par3/Par6/aPKC complex which, in turn, activates Cdc42, a Rho family GTPase that induces the maturation of TJ via the activation of aPKC kinase activity [48–50]. In addition, Lgl which regulates the disassembly of the Par3/Par6/aPKC complex and controls the maturation of TJs during epithelial tissue remodeling is inactivated by aPKC [51]. There are some insights on the role of polarity proteins in modulating AJs by regulating E-cadherin. For example, Scrib was reported to stabilize the coupling between E-cadherin and catenins, as its depletion disrupts E-cadherin mediated cell adhesion [52].

Information on the potential interaction between Cxs and polarity proteins. One report has indicated that drebrin, a developmentally regulated brain protein that mediates cell polarity and maintains the distribution of the different plasma membrane domains, directly interacts with Cx43. This interaction was found to enhance the stability of Cx43 at the membrane. Interestingly, drebrin's depletion in a kidney epithelial cell line lead to Cx43 degradation [53]. Further interactions between Cxs and polarity proteins are yet to be unveiled in order to understand the mechanisms through which Cxs might regulate the differentiation of many epithelia including the mammary epithelium [7]. In this chapter, we mainly discuss the contribution of GJs to mammary epithelial differentiation and pathophysiology not only as channels joining the cytosols of two neighboring cells but also as signaling hubs that activate downstream signaling cascades via complex formation with a range of adaptor proteins. GJs are expressed in a temporal and developmental stage-dependent manner in the mammary gland, which illustrates a paramount role in mediating normal epithelial architecture. Furthermore, GJs are dynamically modulated to support the neoplastic development of the mammary gland by being up or down-regulated throughout cancer progression to support detachment, invasion and metastasis.

These emerging roles, being channel-dependent or -independent, add to the complexity of the proteins forming GJs, making them potential candidates for breast cancertherapy as will be discussed in the last section of this chapter.

## 1.2 Gap Junctions

### *1.2.1 Connexins: Gene, Protein and Biogenesis*

Given the dynamic nature of the GJ observed in response to the developmental stages of tissues, it is crucial to apprehend its building blocks, Cxs. The family of Cxs consists of 21 genes in the human genome and 20 genes in the mouse genome. Most Cxs genes share a similar structure consisting of a 5′-untranslated exon 1 that is separated from the coding exon 2 by an intron of variable length. Interestingly, Cx genes have distinct transcription start sites and distinct 5′-untranslated region depending on the tissues, making their expression tissue specific [54].

The Cx protein is composed of nine main domains including highly conserved N-terminus, two extracellular loops, and four transmembrane domains among different isoforms. On the other hand, the C-terminus and the cytoplasmic loop domains are variable in terms of the protein length and primary aminoacid sequence which brings about the diverse Cxs and connexon types [55, 56]. Cxs orient their N-termini, cytoplasmic loops, and C-termini towards the cytosol (Fig. 1.2a) allowing for interaction with catenins, cytoskeletal proteins, and ZO proteins among others which, in turn, contributes to Cxs' and GJ's functions [57, 58]. Cxs nomenclature can either be as per their molecular weights such as Cx26, Cx32, and Cx43, corresponding to molecular weights of 26, 32 and 43 kDa respectively, or based on sequence similarity and length of the cytoplasmic domain with classification into α-, β- and γ-groups [59]. Their genes are abbreviated with Gj and numbered in chronological order of discovery. For example, Cx43 was the first α-group Cx to be identified thus the gene is dubbed, *GJA1*, and Cx32 was the first Cx of the β-group annotated as *GJB1* [60]. Like most proteins, Cxs use the usual translation machinery of the ribosomes attached to the endoplasmic reticulum (ER). After translation, Cxs are sequentially released to the ER lumen through the Sec61 or translocon [57, 61]. This step is followed by oligomerization of monomeric Cxs into connexons before reaching the plasma membrane to form GJs. It was initially thought that oligomerization occurred in the ER, yet some reports have indicated that Cxs remain as monomers in the ER and will only oligomerize while being transported from the ER to the Golgi apparatus [62]. The resulting connexons are formed of six Cx proteins that assemble into a homomeric (identical Cx subunits) or a heteromeric (different Cx subunits) hemichannels and packaged in vesicles for delivery to the membrane via a microtubule-dependent or -independent mediated transport [63–65]. The docked connexons either remain uncoupled or interact with apposing connexons via their extracellular loops to form the channels collectively termed GJs (Fig. 1.2b) [66, 67]. GJs may be homotypic or heterotypic when two identical or different

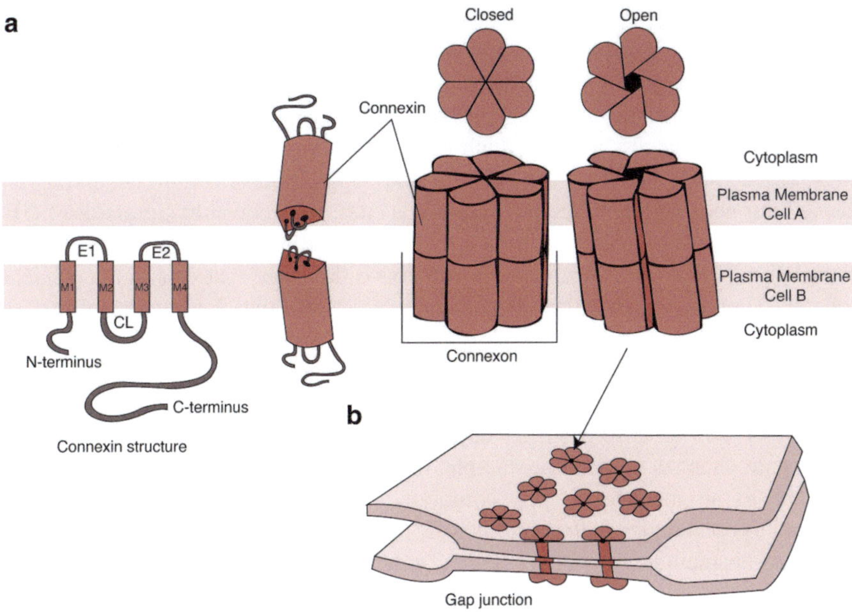


**Fig. 1.2 A schematic model illustrating Cxs assembly into GJs.** (**a**) Cxs have their four domains (m1–m4) spanning the plasma membrane and creating two extracellular loops (E1–E2), one cytoplasmic loop (CL) with of their N- and C-termini deposited towards the cytoplasm. Upon insertion into the plasma membrane, six Cxs can position themselves into a donut-like configuration called a connexon. (**b**) Connexons may either remain as hemichannels or dock with compatible connexons from neighboring cells to form structures collectively known as GJs

connexons dock, respectively on opposing membranes of two cells. Heterotypic channels have different properties compared to homotypic channels due to their variable permeability and conductance [68]. It has been suggested that the principle signal for Cxs assembly into a connexon is located at their C-terminus while the selectivity signal determining subunit compatibility is at the N-terminus. These signals represent stretches of aminoacids that differ among the Cx isoforms resulting in different surface structures of protein/protein interfaces. This organization prevents the interaction of isoforms with differently folded motifs while enabling interactions among Cxs with similarly folded motifs [61]. Thus, different signals, different structural motifs within Cxs polypeptides might regulate recognition and co-oligomerization of different connexin isoforms.

### *1.2.2 Channel: Function, Regulation and Break-Down*

The active role that GJs perform in mediating intercellular signaling and communication and in contributing to tissue homeostasis makes their regulation a complex process. GJs were shown to be involved in bone modeling, embryonic growth, alveolar differentiation, neural function in the developing central nervous system and central nervous system signaling among other systems [69–71]. Intercellular communication is mainly mediated by GJs and referred to as gap junction intercellular communication (GJIC); it enables the exchange of ions, essential metabolites and second messengers among which $Ca^{2+}$, cAMP, IP3, nucleotides and microRNAs [72]. GJs alternate between "closed" and "open" conformations depending on the cellular need and the microenvironmental cues. Consequently, the status of GJs is dependent on calcium concentration, pH, transjunctional potential, and Cx phosphorylation [73, 74]. For example, reports have indicated that calcium-dependent cellular events might modulate the function of GJs through the interaction of Cxs with calcium-affected molecules. In addition, there are regions within the intracellular loop and C-terminus of Cxs that respond to changes in pH levels. Such sites were found to be variable among different Cx proteins, indicating that the Cx composition of GJs in different cells determines their communication function and response to changes in pH. For example, in contrast to Cx26 channels that close in response to an increase in pH; homotypic Cx45 and heterotypic Cx45/Cx43 GJs tend to open under such conditions [75, 76]. Furthermore, GJ function is regulated by transmembrane voltages that vary depending on the type of Cxs forming the hemichannels. Cx43 channels were reported to be relatively insensitive to changes in transjunctional voltage compared to channels composed of Cx45 [77]. Interestingly, studies conducted on Cx26 and Cx32 channels revealed that the cytoplasmic N-terminal domain contains charged residues that determine the response of the channel to changes in voltage. For instance, Cx26, Cx30 and Cx50 close at positive voltages, and Cx31, Cx32, Cx37, Cx40, Cx43, Cx45 and Cx57 close at negative voltages [78]. Phosphorylation status of Cxs, which determines its trafficking and degradation, is another factor that regulates the function of GJ either by modulating the efficacy of passage of certain molecules or closing and opening the channel for the transfer of molecules [79]. For example, the phosphorylation of Cx50 and Cx43 in lens fibers enhanced GJ permeability and stability, respectively [80, 81].

GJs are targeted for degradation through a multistep process starting with junctional internalization. They are internalized into one of the contacting cells by double-membrane vesicles that could encompass either the whole or parts of the junction [82]. Following internalization, GJ degradation involves lysososomal and proteasomal pathways, yet the exact degradation mechanisms and process governing the choice of either pathway have not been fully elucidated [83]. It has been reported that proteasomal degradation destabilizes phosphorylated GJs at the plasma membrane of lens epithelial cells [84, 85]. On the other hand, lysosomal degradation pathway directly regulates Cx degradation before GJs are formed either during transport or after incorporation into the membrane [86]. Furthermore, degradation not only affects GJs and replenishes the plaques with newly formed channels, but also, indirectly regulates Cx stability by mediating Cx-associated proteins turnover, such as ZO-1 [86] and others.

## 1.3 Channel-Dependent and -Independent Roles of Connexins

As previously mentioned, GJs are permeable to second messengers and metabolites [72]. The permeability to the transported metabolites is in large regulated by the Cx isoforms composing the junctions. Cxs are not solely considered as structural components of GJs that mediate the selectivity, permeability and function of the GJ; they also demonstrate channel-independent roles by acting as integral components of signaling hubs that relay signals from the microenvironment to the nucleus, affecting the expression of many genes [57, 87].

### *1.3.1 Selectivity and Permeability of Gap Junctions*

Permeability of GJs is highly dependent on the Cx make-up of the GJ. For example, a study has evaluated single channel permeabilities of homotypic Cx26, Cx32, Cx37, Cx40, Cx43, Cx45 and heterotypic Cx26/Cx32, Cx37/Cx43 GJ channels in *Xenopus laevis* oocytes and revealed that the permeability of heterotypic channels was determined by the permeability of the less permeable Cxs [88]. The permeability of Cx36 channels for cationic dyes is about 10-fold higher than for anionic dyes of the same net charge and similar molecular mass [77]. When examining single channel permeability of GJs in cardiac cells, the ratio of single channel conductance to permeability was much higher for Cx30.2 GJs than for Cx40, Cx43, and Cx45 channels [68]. Homotypic channels made of Cx43 have a 100-fold higher selectivity for ATP compared to channels made of Cx32; metabolites such as glutamate, glutathione, ADP and AMP displayed a 10-fold higher permeability through Cx43 channels than through Cx32 channels [89]. These studies demonstrate that Cxs differentially regulate the intercellular transfer of natural permeants. However, the physical basis of this selectivity remains to be examined as transfer of metabolites through these channels is not simply size- or charge-dependent [90].

### *1.3.2 Channel-Independent Role of Connexins*

It has become unarguable that Cxs have channel-independent effects. They have transcended their channel-restricted roles to encompass a variety of functions, including the regulation of expression of genes involved in developmental and differentiation signaling pathways. Early reports indicated that Cx26 acts as a tumor suppressor since it inhibits migration and invasion of MDA-MB-231, highly invasive breast cancer cells, by regulating the expression levels of β1-integrin and matrix metalloproteinases [91, 92]. Similar results were obtained for Cx32 that suppressed proliferation, invasion and metastasis of renal carcinoma cells in culture by modulating Src, VEGF and other signaling molecules [93]. Data suggest that alterations in Cx expression affect genes expression indirectly by regulating their transcription factors. It has been shown that Cxs can recruit sp1 and sp3 transcription factors to the Cx-responsive elements, found in the promoter regions of responsive genes, through the ERK/PI3K pathway and regulates their expression.

Cxs can also regulate cell differentiation in a GJIC-independent manner. For example, it has been shown that the C-terminal domains of Cx43 and Cx45.6 prevent premature lens cell and neuronal differentiation during embryonic development [94, 95]. Interestingly, findings of a mitochondrial localization of Cx43 link Cxs to pathological processes such as hyperhomocysteinemia. Microarray data showed that Cx43 knock-out (KO) in Sertoli cells can affect 658 genes with 135 genes up-regulated and 523 genes down-regulated, many of which were related to mitotic and meiotic progression of spermatogenesis [96]. Furthermore, Cx45- and Cx43-rich GJs, when formed between bone marrow stromal cells, can induce the nuclear translocation of the transcription factor Sp1, which, in turn, activates the expression of immunity related genes such as *Cxcl12* [97]. Not only differentiation but also cell migration and cytoskeletal rearrangements can be regulated by Cxs. In fact, Cx43 mediates the motility of cardiac neural crest cells by associating with actin-binding proteins, namely vinculin and drebrin [98]. Interestingly, Cx43 appears to display some signaling roles while not at the membrane. Full length Cx43 or its C-terminus was reported to localize to the nucleus and inhibit cell growth in a mechanism that is yet to be determined [99]. The channel-independent roles of Cxs are mainly mediated by their interaction with associated partners such as junctional proteins.

### *1.3.3 Crosstalk between Connexins and Other Junctional Proteins*

There is an intertwined relationship between the different cell junctions since the expression and assembly of their component proteins depend on one another. Several studies have indicated that GJ and AJ formation are intimately linked. For instance, knock-down of Cx43 in fibroblasts resulted in impaired membrane localization of AJs [100]. In NIH3T3 cells, it was found that Cx43 colocalizes and co-immunoprecipitates with N-cadherin with no evidence of direct binding [100]. It was also observed that blockage of Cx33 induced a mislocalization of N-cadherin [101]. These data suggest that a direct feedback between classical adherens and GJ proteins is necessary for their proper function and the overall cell behavior. Another emerging cross-talk is that between GJs and TJs. Occludin was found to interact with Cx32 in immortalized mouse and cultured rat hepatocytes [102]. In addition, both claudin-5 and occludin were shown to colocalize with Cx40 and Cx43 in porcine blood–brain barrier endothelial cells [47]. Moreover, Cx32 transfection in mouse hepatocytes derived from Cx32-deficient mice induced TJ strands formation and enhanced occludin, claudin-1 and ZO-1 expression that led to a stronger TJ functionality [103]. Cx43 was also shown to regulate the blood-testes barrier via its effects on the reassembly of TJs [81]. Therefore, there exists an intimate spatial and temporal relationship between GJs and other cellular junctions in the normal tissue context, and the disruption of Cxs might influence the proper localization and functionality of other cell junctions

## 1.4 Roles of Connexins in the Differentiation of Mammary Epithelial Cells

Given the dynamic nature of the mammary gland, GJs provide pivotal cross-talk of mammary epithelial cells with the surrounding cells and their local microenvironment, which is critical for the gland differentiation states. In fact, GJs were found to be implicated not only in the normal differentiation of the mammary gland but also in its pathophysiology by acting as nexus joining adjacent cells and by complexing with associated proteins that relay downstream signaling cascades. As such, understanding GJ regulation, expression of different Cxs in a stage-specific manner and the association with binding partners is essential for the investigation of normal and cancerous processes in the mammary gland.

### *1.4.1 Temporal and Developmental Expression of Connexins*

Among the different Cxs, it has been reported that normal human mammary epithelial cells express both Cx26 and Cx43 and display functional GJIC among themselves and with human mammary fibroblasts [104, 105]. In particular, Cx43 was shown to be abundant in human myoepithelial cells with minimal expression in luminal cells that rather express Cx26 [106]. *In vitro*, it has been found that Cx43 and not Cx26 is expressed in nonmalignant human mammary epithelial cells, HMT-3522-S1 ([107], Lelièvre laboratory, unpublished data). In addition, Cx43 is also expressed in another non-neoplastic human mammary epithelial cell line, MCF10A [108] where it controls normal-like phenotype in 3D culture [34]. As it is difficult to temporally study the expression of Cxs during normal human mammary gland development due to limited access to representative tissues, the rodent model was extensively used as an alternative, given the high similarities between both mammary glands and the conservation of Cx genes between humans and rodents [109–111]. In the murine mammary gland, not only Cx26 and Cx43 but also Cx32 and Cx30 were reported to be expressed. Interestingly, there is a differential expression of these Cxs during pregnancy, lactation and involution. Cx26 and Cx32 were found to be expressed at the basolateral membranes of luminal cells at all developmental stages of the gland [112]. Both Cxs were reported to have an elevated expression during pregnancy that then peaked around lactation and decreased during involution. This variable expression throughout the different developmental stages of the gland suggests a potential role in regulating milk production [110]. In addition, Cx30 mRNA and protein levels were barely detected in the virgin gland of mice but were significantly elevated throughout later stages of pregnancy and during lactation, then drastically decreased during involution. Cx30 was found to localize with Cx26 [112]. At parturition, a drop in Cx30 mRNA levels was correlated with a peak in Cx32 expression [113]. The temporal expression of Cx30 protein coincided with the onset of β-casein and whey acidic protein (WAP) expression, both markers of mammary gland differentiation, suggesting the central role that Cx30 might play in inducing proper differentiation in the mammary epithelium. Cx30 expression was evident in differentiation permissive coculture conditions of SCp2 and SCg6, the epithelial and myoepithelial subclones of CID-9 mouse mammary cells, respectively, but not when differentiated SCp2 cells were cultured alone in the presence of Matrigel. The heterocellular interaction between SCp2 and SGg6 was characterized by membranous Cx30 forming a complex assembly with β-catenin, α-catenin and ZO-2 resulting in the production of β-casein. This suggests a complex GJ-mediated Cx30-dependent pathway for mammary epithelial differentiation [57, 114].

As for Cx43, while its mRNA expression decreased during mid-pregnancy and lactation, its active phosphorylated form was specifically up-regulated during lactation. More so, studies in our laboratory suggested that Cx43 may be localized to the myoepithelial-epithelial cell contact regions indicating that it is not only expressed by myoepithelial cells but also possibly in secretory alveolar epithelial cells [112, 115]. In light of these findings, it is expected that deregulation of the expression or localization of mammary Cxs might have detrimental effects on the differentiation and development of the mammary gland. For example, while the loss of Cx26 from the mammary epithelium before puberty resulted in an altered lobuloalveolar development and function during lactation, its ablation during the latter part of pregnancy in the secretory epithelial cells had no consequences on alveolar morphogenesis and function. These observations indicate that Cx26 is essential during early stages of the mammary gland development but not after complete functional differentiation. On the other hand, Cx32-null mice developed a differentiated mammary gland that was indistinguishable from that of the wild-type epithelium, and were able to nurture their pups, demonstrating that Cx32 is not essential for the functional development of the gland [116]. In the *Gja1Jrt/+* mouse model, decreased Cx43 protein levels, reduced GJIC in myoepithelial cells resulting in a delay in the mammary gland development with no changes in the morphology of the gland and the expression of other Cxs compared to wild-type mice at parturition. Most importantly, high levels of milk proteins were detected in these mice, but there was a failure in delivery of milk proteins to the ducts following oxytocin stimulation, suggesting a central role for Cx43 in the contraction of myoepithelial cells and ejection of milk [117]. Given these findings, Cxs appear to be dynamically modulated to support the different stages of the mammary gland development, whereby alterations in these proteins lead to perturbations in this delicate developmental process and can ultimately lead to the loss of proper function of the gland and/or tumorigenesis.

### *1.4.2 Gap Junctions in Normal Mammary Epithelial Differentiation*

Several studies reported a change in channel composition and permeability at different stages of the mammary gland development. In fact, at the onset of parturition, initially heteromeric Cx26/Cx32 hemichannels are formed to be later changed into homomeric Cx32 channels after parturition, allowing the passage of larger permeants. Osmolytes such as Taurine, which accumulates in mammary cells during lactation, were shown to inhibit the molecular permeability of Cx26/Cx32 channels while sparing that of homomeric Cx32, heteromeric Cx26/Cx30 and Cx30/Cx32 channels, implying a role for Cx30 and Cx32 during lactation [113, 118, 119]. Furthermore, the role of GJs in the differentiation of mammary epithelial cells was investigated in CID-9 mouse mammary strain consisting mostly of epithelial cells in addition to myoepithelial-like and fibroblast cells cultured on plastic and on a nonadhesive substratum. The stimulation of GJIC with cAMP was sufficient to drive β-casein expression. This effect was not observed in homogeneous SCp2 cultures (i.e. epithelial clones from CID-9 cells) in the absence of a basement membrane. On the other hand, culturing SCp2 cells with SCg6 (i.e. myoepithelial-like clones from CID-9 cells) enabled the establishment of GJIC between the two cell populations. As a result of GJIC between the two cell populations, β-casein milk protein was expressed, emphasizing the importance of gap junction-mediated epithelial/myoepithelial interactions in mammary gland development [109]. The differentiation of SCp2 cells in SCp2/SCg6 cells was disrupted in response to lower concentrations of GJ inhibitors in coculture compared to SCp2 cells cultured on Matrigel, which implies that heterocellular GJIC presents yet another level in the differentiation hierarchy of mammary epithelial cells [34, 114]. In addition to the β1-integrin/STAT5-mediated pathway where STAT5 is phosphorylated and translocated to the nucleus to activate β-casein expression (Watson and Burdon 1995); a β1-integrin/STAT5-independent but GJ-dependent pathway was reported to activate β-casein expression in CID-9 cells cultured on a non-adhesive PolyHEMA substratum. OCT-1, which activates β-casein expression [120], rather than STAT5 was suggested to be the GJ downstream effector of β-casein expression [34]. The above findings indicate that GJ and cell-matrix interaction-mediated signaling are essential for differentiation of mammary epithelial cells, and that such signaling is synergistic [121]. The above suggests that the differentiation of mammary epithelial cells depends partially on the assembly of GJ complexes and the interaction of Cxs within this complex.

## 1.5 Connexins and Gap Junctions in Mammary Tumorigenesis

Several studies have demonstrated that altered GJIC and aberrant expression of Cxs are associated with breast cancer. GJs, although down-regulated during early stages of tumorigenesis, are needed for the migration and invasion of tumor cells and their interaction with endothelial cells during intravasation and extravasation of cancer cells at later stages of tumorigenesis (Fig. 1.3) [122, 123]. It has been shown that Cx26 and Cx43 were down-regulated in human breast cancer cell lines. This down-regulation has been linked to alterations in cell-cell adhesion and activation of oncogenes, such as RAS and C-ERBB2 [110]. Cx43 expression is reduced in all stages of human breast cancer and carcinogen-induced breast tumors that have an increased activity of the PKC, EGF and Src pathways [124]. Furthermore, phosphorylation of Cx43 is enhanced in myoepithelial cells, transformed luminal cells of *in situ* and invasive breast carcinomas that lack Cx43 localization at the cell membrane [110]. During metastasis, Cxs were shown to play a major role in cell intravasation and extravasation, consequently enabling metastasis. Evidence indicates that Cx43 expression was up-regulated in sites of micrometastases to the lungs at the tumor cell/endothelial cell contact, which in turn enhanced GJIC, supporting the significant role of Cx43 and GJIC in metastatic stages of breast cancer [125]. Similarly, it has been demonstrated that breast cancer cells utilize Cx26 and Cx43 to initiate brain metastatic lesion formation in association with the vasculature. When Cx26 and Cx43 were silenced, extravasation was blocked and brain colonization was inhibited [126]. The dynamic changes in Cx expression in a cancer stage-dependent manner is further supported by evidence indicating that 70 % of Cx26, Cx43 and Cx32-negative primary breast tumors were able to develop Cx26, Cx43 and Cx32-positive lymph node metastases [127].

In breast carcinogenesis in particular, there are contradictory results on the role of Cxs; while Cxs have widely been known as tumor suppressor genes, recent data indicate that some invasive breast carcinomas express high levels of Cxs. These paradoxical roles depend to a great extent on the stage of tumorigenesis.

### 1.5.1 Gap Junctional Intercellular Communication-Dependent Mechanisms

As GJs mediate intercellular exchange of a wide range of regulatory molecules that affect normal and cancer cells, several studies have attempted to decipher the role of GJIC in the different stages of breast tumorigenesis. In bone metastasis, GJIC was found to form between breast cancer cells and human osteoblastic cells [128]. In the breast cancer cell lines MCF-7 and MDA-MB-435, homotypic GJIC was found to be less abundant than heterotypic GJIC between either cell lines and hFOB, a human osteoblastic cell line, which correlated with increased tumorigenesis and metastatic potential of breast cancer cells [128, 129]. Similarly, GJIC-deficient HBL100 breast cancer epithelial cells exhibited increased GJIC and diapedesis through the endothelial monolayer upon Cx43 expression [130]. Stabilizing cell adhesion and enhancing communication between cancer cells enable effective migration [131, 132]. Interestingly, invasive tumors are characterized by an enhanced expression of E-, N-cadherin and TJ proteins that form complexes at the cell membrane. Furthermore, GJs were reported to be more concentrated in tumor cells compared to their normal counterparts. For example, over-expression of Cx43 and Cx26

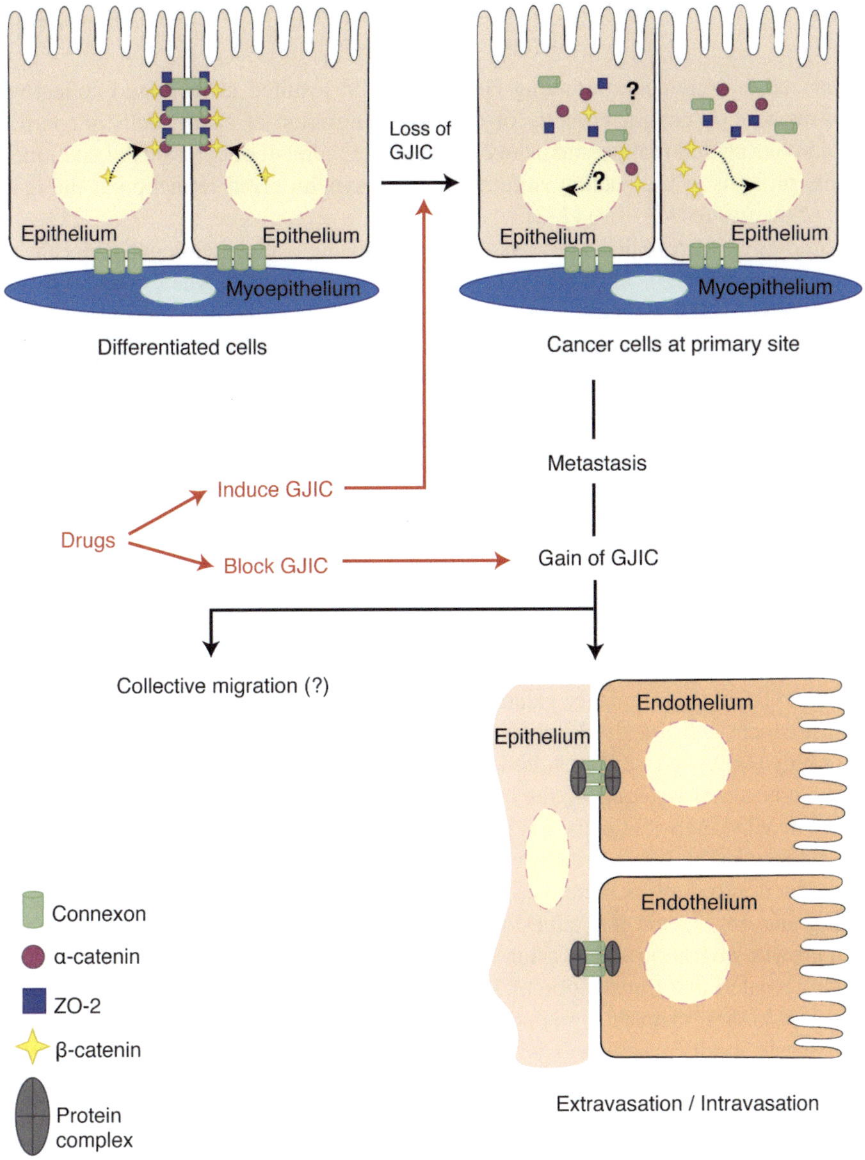


**Fig. 1.3 GJ dependent and independent role of Cxs in normal and cancerous mammary epithelial development.** In a differentiated mammary epithelium, junctions such as GJs are assembled at the membrane of epithelial cells and their corresponding myoepithelial partners. Within GJs, Cxs form a complex assembly with cytoplasmic partner proteins such as ZO-2, α- and β-catenins that are able to transduce signals initiated at the gap junctional complex towards the nucleus. These signals can ultimately bring about changes in the expression of genes involved in the differentiation of the mammary epithelium. Alterations in the expression and/or localization of Cxs may lead to the loss of GJIC where Cx-associated proteins are released resulting in the inadequate activation of signaling cascades. This phenomenon marks tumor initiation at the cancer primary site. During later stages of cancer progression such as metastasis, invasive epithelial cells reestablish GJIC to facilitate their interaction with the endothelial barrier and possibly enable collective cell migration. Upon GJIC formation, Cxs gain interactions with their proteins complexes to support a tumor-promoting rather than a tumor-suppressing role of GJs that is typically displayed in the normal context [122]. As such, depending on the stage of the cancer, drugs could be designed to either inhibit or promote GJIC as a means to effectively block tumor progression

enhanced communication among HeLa cells and resulted in increased collective invasiveness. A central property of collective migration of cancer cells is coordinated polarization that is established by the rearrangement of cytoskeletal and junctional proteins at the leading edge to facilitate synchronized retraction at the rear end of the group (Fig. 1.3) [133].

Studies have revealed that Cxs can also mediate a channel-dependent cancer suppressive role. For example, it has been documented that when induced to be expressed in MCF-7 cells, Cx43 displays a tumor suppressive role by restoring GJIC and decreasing anchorage-independent growth, migration and invasion [134]. Clearly, GJs are not only essential for a differentiated mammary epithelial phenotype, they are also central coordinators of tumorigenesis as cancer cells need communication, adhesion and adequate polarization for progression and metastasis.

### *1.5.2 Gap Junctional Intercellular Communication-Independent Mechanisms*

There is an increasing body of evidence for the tumor suppressive GJ-independent roles of Cxs in the mammary gland. The stable transfection with Cx26 and Cx43 in breast cancer cell lines lacking Cx expression (HBL100) or having defects in GJ assembly (MDA-MB-231) resulted in an evident inhibition of tumor growth *in vivo* without necessarily requiring the formation of GJs. Indeed, exogenously expressed Cx43 in MDA-MB-231 cells was retained within the cell and did not assemble into GJ plaques [135]. It has been also shown that the expression of Cx26 variants that assembled into nonfunctional GJs in MDA-MB-435 cells was capable of up-regulating anti-angiogenic and adhesion markers, such as β-integrin, hence inhibiting cell growth, migration and invasion in a GJ-independent pathway [91, 136]. Another study revealed that transfection of MDA-MB-435 with Cx43 induced a significant decrease in their metastatic potential via the down-regulation of N-cadherin expression levels and increased their sensitivity to apoptosis. Furthermore, the expression of both Cx26 and Cx43 in MDA-MB-231 cells resulted in the promotion of mesenchymal to epithelial transition and the regulation of the release of pro-angiogenic and anti-angiogenic molecules that inhibited angiogenesis*in vitro* and *in vivo* [110, 137]. While reduced Cx43 levels in Cx43 mutant mice were associated with extensive mammary gland hyperplasia and mammaglobin-positive mammary gland metastases to the lung; normal Cx43 levels inhibited mammary tumor metastasis to the lungs [138]. Also, the overexpression of Cx43 in MDA-MB-231 and MCF-7 that represent different invasive stages of breast cancer induced a partial reversion of the cancer phenotype in a culture-dependent manner. Exogenous expression of Cx43 in two-dimensional (2D) and 3D cultures of MCF-7 was associated with membrane localization of Cx43 in contrast to MDA-MB-231 that only displayed a membranous Cx43 under 3D culture conditions. Consequently, the membrane localization of Cx43 induced the recruitment of ZO-1, ZO-2 and α-catenin; in particular, β-catenin was recruited away from the nucleus in 2D cultures of MCF-7 and 3D cultures of MCF-7 and MDA-MB-231 cells. While the tumor cell behavior-suppressive effects of Cx43 were displayed in 2D and 3D cultures of MCF-7, it was only evident in 3D cultures of MDA-MB-231 cells. Thus, the Cx43 tumor suppressive effect occurs in a context-dependent manner, where GJ assembly at the membrane and the association of Cx43 with its interacting partners is implicated in reducing the proliferation rate, invasiveness, and, hence, malignant phenotype of MCF-7 cells in 2D and 3D cultures, and of MDA-MB-231 cells only in 3D cultures [34].

The channel-dependent and -independent functions of Cxs require understanding the role of Cxs and GJs in a context-dependent manner. In other words, we should reevaluate the role of Cxs in tumorigenesis as either tumor suppressors or enhancers depending on the tumor microenvironment and the stage of tumorigenesis. Moreover, as Cxs are not the sole players in the GJ complex, it is expected that alterations in the expression or localization of GJ-associated proteins contributes to the progression of breast cancers [139].

## 1.6 Gap Junctions and Connexins as Therapeutic Targets in Breast Cancer

Cellular communication influences the promotion or the prevention of a cancerous behavior, therefore cell junction proteins, such as Cxs, emerge as new potential targets in anticancer therapy (Fig. 1.3). Gamma linolenic acid, a polyunsaturated fatty acid known for its anti-proliferative and cytotoxic effects on cancer cells, was shown to reduce adhesion of breast (MDA-MB-231) and colon (HT115) cancer cells to endothelial cells by specifically improving the GJIC between endothelial cells [140]. Combinatorial therapy with tamoxifen and retinoic acid, employed for the treatment of breast cancer, was able to induce a reversion of the tumor phenotype of MCF-7 cells as a result of enhanced GJIC, colocalization of Cx26 and Cx43 at cell–cell contacts, reduced proliferation, increased cell adhesion through the up-regulation of E-cadherin expression, accumulation of cells in the G0/G1 cell cycle phase and decrease in telomerase activity [141]. Furthermore, it has been reported that the combinational treatment of T47D, human ductal breast epithelial tumor cells, with tamoxifen and PQ1, a gap junction activator, enhanced tamoxifen efficacy as PQ1 facilitated intercellular passage of tamoxifen and consequently, mediated its rapid action to inhibit cancer cell proliferation and colony growth [142]. The treatment of T47D cells by PQ7, a second-generation substituted quinoline, induced an increase in Cx43 expression levels and enhanced GJIC, leading to 50 % reduced colony growth *in vitro* and complete regression of tumor growth in xenograft models [143]. The therapeutic potential of GJs in combinational treatments can also be attributed at least partially to the "bystander effect", whereby the routes for drug delivery within a tumor mass are mediated by the restoration of GJIC between neighboring cells to ultimately induce cell death [122].

## 1.7 Conclusion

The dynamic cues from the microenvironment of the cell that are relayed from cell-cell and cell-ECM interactions are critical for the proper development and function of the mammary gland. GJs mediate the intercellular transfer of molecules that regulate normal mammary gland development and breast cancer growth. Yet, there is immense evidence on a role for Cxs beyond the GJIC, as they interact with a wide array of associated proteins that initiate and transduce downstream signals. Among those binding partners are junctional proteins, associated adapter and signaling proteins and polarity proteins, that all converge to affect the expression of genes involved in normal- and cancer-related signaling pathways in the mammary gland. Besides their tumor suppressive roles, Cxs play diverse roles in breast carcinogenesis throughout the different stages of progression. Initially, the dissociation of the complex formed by Cxs can lead to the loss of GJIC and tumor initiation; whereas at later stages such as during intravasation/extravasation this complex has to reassemble to support the interaction between breast cancer cells and endothelial cells, and also possibly during the collective migration of the leading edge of cancer cells towards their secondary target sites [133]. Given that Cxs execute their functions through down-stream effectors, a better understanding of the extent of their involvement in a stage- and context-dependent fashion is needed to decipher their role in mammary gland differentiation and tumorigenesis. One of the major limitations to accomplish such a goal is the ability to mimic the *in vivo* cellular microenvironment in culture models that recapitulate the differentiation of epithelial cells in normal tissues and that of cancer cells in specific progression stages [144]. Despite the efforts to maintain adequate growth conditions in 3D culture models, the physiological interactions that exist *in vivo* between mammary epithelial cells, myoepithelial cells and fibroblasts throughout the development of the mammary gland are yet to be readily explored in such models. There have been recent advances in organ-on-chip technology that can recapitulate the luminal portion of the branched breast epithelium with channels of variable diameters similar to that in the mammary gland [145]. In addition, the use of robotic bioprinters for the multilayered additive biofabrication of functional 3D tissues and organs [146] hold promise in providing culture models that capture the stage of development and context. We ought to integrate our knowledge of the roles GJ-dependent and GJ-independent roles of Cxs into the development of new culture models that will help generate a functional map of Cxs. This approach is necessary to design new therapeutic strategies for breast cancer that take into consideration the cancer stage and its microenvironment.

**Acknowledgments** The authors are grateful to Sabreen Fostok for critical reading of the manuscript and to Diala El-Zein for working on the schematic diagrams. We wish to acknowledge the support of the American University of Beirut Research Board, the Lebanese National Council for Scientific Research (to RT and DB); and the UNESCO-L'ORÉAL International Fellowship for Women in Science-2012 (to DB). SL and RT are members of International Breast Cancer & Nutrition (IBCN).

## References


1. Maller O, Martinson H, Schedin P (2010) Extracellular matrix composition reveals complex and dynamic stromal-epithelial interactions in the mammary gland. J Mammary Gland Biol Neoplasia 15:301–318
2. Lelièvre SA (2009) Contributions of extracellular matrix signaling and tissue architecture to nuclear mechanisms and spatial organization of gene expression control. (BBA)-Gen Subj 9:925–935
3. Talhouk R (2012) On cell–matrix interactions in mammary gland development and breast cancer. Cold Spring Harb Perspect Biol 4:8
4. Tiede B, Kang Y (2011) From milk to malignancy: the role of mammary stem cells in development, pregnancy and breast cancer. Cell Res 21:245–257
5. Hens JR, Wysolmersk JJ (2005) Key stages of mammary gland development: molecular mechanisms involved in the formation of the embryonic mammary gland. Breast Cancer Res 7:220–224
6. Breen N, Gentleman JF, Schiller JS (2011) Update on mammography trends: comparisons of rates in 2000, 2005, and 2008. Cancer 117:2209–2218
7. Bazzoun D, Lelievre SA, Talhouk RS (2013) Polarity proteins as regulators of cell junction complexes: implications for breast cancer. Pharmacol Ther 138:418–427
8. Brennan K, Offiah G, Mcsherry E, Hopkins A (2010) Tight junctions: a barrier to the initiation and progression of breast cancer? J Biomed Biotechnol 2010:460607
9. Krause G, Winkler L, Mueller SL, Haseloff RF, Piontek J, Blasig IE (2008) Structure and function of claudins. Biochim Biophys Acta 1778:631–645
10. Martin TA, Jiang WG (2009) Loss of tight junction barrier function and its role in cancer metastasis. Biochim Biophys Acta 1788:872–891
11. Nusrat A, Brown GT, Tom J, Drake A, Bui TT, Quan C et al (2005) Multiple protein interactions involving proposed extracellular loop domains of the tight junction protein occludin. Mol Biol Cell 16:1725–1734
12. Giepmans BN (2004) Gap junctions and connexin-interacting proteins. Cardiovasc Res 62:233–245
13. Utepbergenov DI, Fanning AS, Anderson JM (2006) Dimerization of the scaffolding protein ZO-1 through the second PDZ domain. J Biol Chem 281:24671–24677
14. Mandell KJ, Babbin BA, Nusrat A, Parkos CA (2005) Junctional adhesion molecule 1 regulates epithelial cell morphology through effects on β1 integrins and Rap1 activity. J Biol Chem 280:11665–11674
15. Barrios-Rodiles M, Brown KR, Ozdamar B, Bose R, Liu Z, Donovan RS et al (2005) High throughput mapping of a dynamic signaling network in mammalian cells. Science 307:1621–1625
16. Hoevel T, Macek R, Swisshelm K, Kubbies M (2004) Reexpression of the TJ protein CLDN1 induces apoptosis in breast tumor spheroids. Int J Cancer 108:374–383
17. Osanai M, Murata M, Nishikiori N, Chiba H, Sawada N (2007) Epigenetic silencing of claudine-6 promotes anchorage independent growth of breast carcinoma cells. Cancer Res 98:1557–1562
18. Ehrlich JS, Hansen MD, Nelson WJ (2002) Spatio-temporal regulation of rac1 localization and lamellipodia dynamics during epithelial cell–cell adhesion. Dev Cell 3:259–270
19. Lanigan F, McKiernan E, Brennan DJ, Hegarty S, Millikan RC, McBryan J et al (2009) Increased claudin-4 expression is associated with poor prognosis and high tumour grade in breast cancer. Int J Cancer 124:2088–2097
20. Wijnhoven BPL, Dinjens WNM, Pignatelli M (2000) E-cadherin-βcatenin cell–cell adhesion complex and human cancer. Br J Surg 87:992–1005
21. Getsios S, Amargo EV, Dusek RL, Ishii K, Sheu L, Godsel LM, Green KJ (2004) Coordinated expression of desmoglein 1 and desmocollin 1 regulates intercellular adhesion. Differentiation 72:419–433

22. Holthofer B, Windoffer R, Troyanovsky S, Leube RE (2007) Structure and function of desmosomes. Int Rev Cytol 264:65–163
23. Runswick SK, O'Hare MJ, Jones L, Streuli CH, Garrod DR (2001) Desmosomal adhesion regulates epithelial morphogenesis and cell positioning. Nat Cell Biol 3:823–830
24. Nollet F, Kools P, van Roy F (2000) Phylogenetic analysis of the cadherin superfamily allows identification of six major subfamilies besides several solitary members. J Mol Biol 299:551–572
25. Merritt AJ, Berika MY, Zhai W, Kirk SE, Ji B, Hardman MJ, Garrod DR (2002) Suprabasal desmoglein 3 expression in the epidermis of transgenic mice results in hyperproliferation and abnormal differentiation. Mol Cell Biol 22:5846–5858
26. Elias PM, Matsuyoshi N, Wu H, Lin C, Wang ZH, Brown BE, Stanley JR (2001) Desmoglein isoform distribution affects stratum corneum structure and function. J Cell Biol 153:243–249
27. Davies EL, Gee JMW, Cochrane RA, Jiang WG, Sharma AK, Nicholson RI et al (1999) The immunohistochemical expression of desmoplakin and its role in vivo in the progression and metastasis of breast cancer. Eur J Cancer 35:902–907
28. Knudsen KA, Wheelock MJ (2005) Cadherins and the mammary gland. J Cell Biochem 95:488–496
29. Maynadier M, Chambon M, Basile I, Gleizes M, Nirde P, Gary-Bobo M et al (2012) Estrogens promote cell-cell adhesion of normal and malignant mammary cells through increased desmosome formation. Mol Cell Endocrinol 25:126–133
30. Cowing P, Rowland's TM, Hat sell SJ (2005) Catherin's and catena's in breast cancer. Curr Opin Cell Biol 17:499–508
31. Charafe-Jauffret E, Tarpin C, Bardou VJ, Bertucci F, Ginestier C, Braud AC et al (2004) Immunophenotypic analysis of inflammatory breast cancers: identification of an 'inflammatory signature'. J Pathol 202:265–273
32. Kowalski PJ, Rubin MA, Kleer CG (2003) E-cadherin expression in primary carcinomas of the breast in its distant metastases. Breast Cancer Res 5:R217–R222
33. Hatsell S, Rowlands T, Hiremath M, Cowin P (2003) β-catenin and Tcfs in mammary development and cancer. J Mammary Gland Biol Neoplasia 8:145–158
34. Talhouk RS, Fares MB, Rahme GJ, Hariri HH, Rayess T, Dbouk HA et al (2013) Context dependent reversion of tumor phenotype by connexin-43 expression in MDA-MB231 cells and MCF-7 cells: Role of β-catenin/connexin43 association. Exp Cell Res 319:3065–3080
35. Guttman JA, Finlay BB (2009) Tight junctions as targets of infectious agents. Biochim Biophys Acta Biomembr 1788:832–841
36. Laird DW (2009) The gap junction proteome and its relationship to disease. Cell Press 20:92–101
37. Vinken M, Vanhaecke T, Papeleu P, Snykers S, Henkens T et al (2006) Connexins and their channels in cell growth and cell death. Cell Signal 18:592–600
38. Domhan S, Ma L, Tai A, Anaya Z, Beheshti A, Zeier M et al (2011) Intercellular communication by exchange of cytoplasmic material via tunneling nano-tube like structures in primary human renal epithelial cells. PLoS One 6:e21283
39. Wustner D (2007) Plasma membrane sterol distribution resembles the surface topography of living cells. Mol Biol Cell 18:211–228
40. Pietilä M, Lehenkari P, Kuvaja P, Kaakinen M, Kaul SC, Wadhwa R (2013) Mortalin antibody-conjugated quantum dot transfer from human mesenchymal stromal cells to breast cancer cells requires cell–cell interaction. Exp Cell Res 319:2770–2780
41. Nagasawa K, Chiba H, Fujita H, Kojima T, Saito T, Endo T et al (2006) Possible involvement of gap junctions in the barrier function of tight junctions of brain and lung endothelial cells. J Cell Physiol 208:123–132
42. Ai Z, Fischer A, Spray DC, Brown AM, Fishman GI (2000) Wnt-1 regulation of connexin43 in cardiac myocytes. J Clin Invest 105:161–171

43. Xu X, Li WE, Huang GY, Meyer R, Chen T, Luo Y et al (2001) N-cadherin and Cx43alpha1 gap junctions modulates mouse neural crest cell motility via distinct pathways. Cell Commun Adhes 8:321–324
44. Nelson WJ (2003) Adaptation of core mechanisms to generate cell polarity. Nature 422:766–774
45. Dow LE, Elsum IA, King CL, Kinross KM, Richardson HE, Humbert PO (2008) Loss of human scribble cooperates with H-Ras to promote cell invasion through deregulation of MAPK signaling. Oncogene 27:5988–6001
46. Ebnet K (2008) Organization of multiprotein complexes at cell–cell junctions. Histochem Cell Biol 130:1–20
47. Nakagawa M, Fukata M, Yamaga M, Itoh N, Kaibuchi K (2001) Recruitment and activation of Rac1 by the formation of E-cadherin-mediated cell–cell adhesion sites. J Cell Sci 114:1829–1838
48. Bradfield PF, Nourshargh S, Aurrand-Lions M, Imhof BA (2007) JAM family and related proteins in leukocyte migration (Vestweber series). Arterioscler Thromb Vasc Biol 27:2104–2112
49. Ebnet K, Suzuki A, Ohno S, Vestweber D (2004) Junctional adhesion molecules (JAMs): more molecules with dual functions? J Cell Sci 117:19–29
50. Weber C, Fraemohs L, Dejana E (2007) The role of junctional adhesion molecules in vascular inflammation. Nat Rev Immunol 7:467–477
51. Betschinger J, Mechtler K, Knoblich JA (2003) The par complex directs asymmetric cell division by phosphorylating the cytoskeletal protein Lgl. Nature 422:326–330
52. Qin Y, Capaldo C, Gumbiner BM, Macara IG (2005) The mammalian scribble polarity protein regulates epithelial cell adhesion and migration through E-cadherin. J Cell Biol 171:1061–1071
53. Butkevich E, Hülsmann S, Wenzel D, Shirao T, Duden R, Majoul I (2004) Drebrin is a novel connexin-43 binding partner that links gap junctions to the submembrane cytoskeleton. Curr Biol 14:650–658
54. Pfenniger A, Wohlwend A, Kwak BR (2011) Mutations in connexin genes and disease. Eur J Clin Invest 41:103–116
55. Evans WH, Martin PE (2002) Gap junctions: structure and function. Molec Membr Biol 19:121–136
56. Rhett JM, Gourdie RG (2012) The perinexus: a new feature of Cx43 gap junction organization. Heart Rhythm 9:619–623
57. Dbouk HA, Mroue RM, El-Sabban ME, Talhouk RS (2009) Connexins: a myriad of functions extending beyond assembly of gap junction channels. Cell Commun Signal 7:4
58. Herve JC, Bourmeyster N, Sarrouilhe D, Duffy HS (2007) Gap junctional complexes: from partners to functions. Prog Biophys Mol Biol 94:29–65
59. Rackauskas M, Neverauskas V, Skeberdis VA (2010) Diversity and properties of connexin gap junction channels. Medicina 46:1
60. Eiberger J, Degen J, Romualdi A, Deutsch U, Willecke K, Söhl G (2001) Connexin genes in the mouse and human genome. Cell Adhes Commun 8:163–165
61. Segretain D, Falk MM (2004) Regulation of connexin biosynthesis, assembly, gap junction formation, and removal. BBA-Biomembr 1662:3–21
62. Koval M (2006) Pathways and control of connexin oligomerization. Trends Cell Biol 16:159–166
63. Johnson RG, Meyer RA, Li XR, Preus DM, Tan L, Grunenwald H et al (2002) Gap junctions assemble in the presence of cytoskeletal inhibitors, but enhanced assembly requires microtubules. Exp Cell Res 275:67–80
64. Lauf U, Giepmans BN, Lopez P, Braconnot S, Chen SC, Falk MM (2002) Dynamic trafficking and delivery of connexons to the plasma membrane and accretion to gap junctions in living cells. Proc Natl Acad Sci U S A 99:10446–10451

65. Martin PE, Blundell G, Ahmad S, Errington RJ, Evans WH (2001) Multiple pathways in the trafficking and assembly of connexin 26, 32 and 43 into gap junction intercellular communication channels. J Cell Sci 114:3845–3855
66. Duffy HS, Delmar M, Spray DC (2002) Formation of the gap junction nexus: binding partners for connexins. J Physiol Paris 96:243–249
67. Evans WH, De Vuyst E, Leybaert L (2006) The gap junction cellular internet: connexin hemichannels enter the signalling limelight. Biochem J 397:1–14
68. Rackauskas M, Verselis VK, Bukauskas FF (2007) Permeability of homotypic and heterotypic gap junction channels formed of cardiac connexins mCx30. 2, Cx40, Cx43, and Cx45. Am J Physiol Heart C 293:H1729–H1736
69. Houghton FD (2005) Role of gap junctions during early embryo development. Reproduction 129:129–135
70. Yang SR, Cho SD, Ahn NS, Jung JW, Park JS, Jo EH et al (2005) Role of gap junctional intercellular communication (GJIC) through p38 and ERK1/2 pathway in the differentiation of rat neuronal stem cells. J Vet Med Sci 67:291–294
71. Yeager M, Harris AL (2007) Gap junction channel structure in the early 21st century: facts and fantasies. Curr Opin Cell Biol 19:521–528
72. Katakowski M, Buller B, Wang X, Rogers T, Chopp M (2010) Functional microRNA is transferred between glioma cells. Cancer Res 70:8259–8263
73. Alev C, Urschel S, Sonntag S, Zoidl G, Fort AG, Höher T et al (2008) The neuronal connexin36 interacts with and is phosphorylated by CaMKII in a way similar to CaMKII interaction with glutamate receptors. Proc Natl Acad Sci U S A 105:20964–20969
74. Peracchia C (2004) Chemical gating of gap junction channels; roles of calcium, pH and calmodulin. Biochim Biophys Acta 1662:61–80
75. González-Nieto D, Gómez-Hernández JM, Larrosa B, Gutiérrez C, Muñoz MD, Fasciani I et al (2008) Regulation of neuronal connexin-36 channels by pH. Proc Natl Acad Sci U S A 105:17169–17174
76. Palacios-Prado N, Briggs SW, Skeberdis VA, Pranevicius M, Bennett MV, Bukauskas FF (2010) pH-dependent modulation of voltage gating in connexin45 homotypic and connexin45/connexin43 heterotypic gap junctions. Proc Natl Acad Sci U S A 107:9897–9902
77. Paulauskas N, Pranevicius H, Mockus J, Bukauskas FF (2012) Stochastic 16-state model of voltage gating of gap-junction channels enclosing fast and slow gates. Biophys J 102:2471–2480
78. Srinivas M, Kronengold J, Bukauskas FF, Bargiello TA, Verselis VK (2005) Correlative studies of gating in Cx46 and Cx50 hemichannels and gap junction channels. Biophys J 88:1725–1739
79. Moreno AP, Lau AF (2007) Gap junction channel gating modulated through protein phosphorylation. Prog Biophys Mol Biol 94:107–119
80. Dunn WB, Broadhurst DI, Atherton HJ, Goodacre R, Griffin JL (2011) Systems level studies of mammalian metabolomes: the roles of mass spectrometry and nuclear magnetic resonance spectroscopy. Chem Soc Rev 40:387–426
81. Li MW, Mruk DD, Lee WM, Cheng CY (2010) Connexin 43 is critical to maintain the homeostasis of the blood–testis barrier via its effects on tight junction reassembly. Proc Natl Acad Sci U S A 107:17998–18003
82. Jordan K, Chodock R, Hand AR, Laird DW (2001) The origin of annular junctions: a mechanism of gap junction internalization. J Cell Sci 114:763–773
83. Leithe E, Brech A, Rivedal E (2006) Endocytic processing of connexin43 gap junctions: a morphological study. Biochem J 393:59–67
84. Girao H, Pereira P (2003) Phosphorylation of connexin 43 acts as a stimuli for proteasome-dependent degradation of the protein in lens epithelial cells. Mol Vis 9:24–30
85. Thomas MA, Zosso N, Scerri I, Demaurex N, Chanson M, Staub O (2003) A tyrosine-based sorting signal is involved in connexin43 stability and gap junction turnover. J Cell Sci 116:2213–2222

86. Laird DW (2006) Life cycle of connexins in health and disease. Biochem J 394:527–543
87. Naus CC, Matsuuchi L (2013) Gap junction proteins on the move: connexins, the cytoskeleton and migration. BBA-Biomembr 1828:94–108
88. Weber PA, Chang HC, Spaeth KE, Nitsche JM, Nicholson BJ (2004) The permeability of gap junction channels to probes of different size is dependent on connexin composition and permeant-pore affinities. Biophys J 87:958–973
89. Goldberg GS, Moreno AP, Lampe PD (2002) Gap junctions between cells expressing connexin 43 or 32 show inverse permselectivity to adenosine and ATP. J Biol Chem 27:36725–36730
90. Harris AL (2007) Connexin channel permeability to cytoplasmic molecules. Prog Biophys Mol Biol 94(1):120–143
91. Kalra J, Shao Q, Qin H, Thomas T, Alaoui-Jamali MA, Laird DW (2006) Cx26 inhibits breast MDA-MB-435 cell tumorigenic properties by a gap junctional intercellular communication-independent mechanism. Carcinogenesis 27:2528–2537
92. Lee SW, Tomasetto C, Sager R (1991) Positive selection of candidate tumor-suppressor genes by subtractive hybridization. Proc Natl Acad Sci U S A 88:2825–2829
93. Sato H, Hagiwara H, Ohde Y, Senba H, Virgona N, Yano T (2007) Regulation of renal cell carcinoma cell proliferation, invasion and metastasis by connexin 32 gene. J Membr Biol 216:17–21
94. Banks EA, Yu XS, Shi Q, Jiang JX (2007) Promotion of lens epithelial-fiber differentiation by the C-terminus of connexin 45.6 a role independent of gap junction communication. J Cell Sci 120:3602–3612
95. Cina C, Maass K, Theis M, Willecke K, Bechberger JF, Naus CC (2009) Involvement of the cytoplasmic C-terminal domain of connexin43 in neuronal migration. J Neurosci 29:2009–2021
96. Giese S, Hossain H, Markmann M, Chakraborty T, Tchatalbachev S, Guillou F et al (2012) Sertoli-cell-specific knockout of connexin 43 leads to multiple alterations in testicular gene expression in prepubertal mice. Dis Model Mech 5:895–913
97. Schajnovitz A, Itkin T, D'Uva G, Kalinkovich A, Golan K, Ludin A et al (2011) CXCL12 secretion by bone marrow stromal cells is dependent on cell contact and mediated by connexin43 and connexin45 gap junctions. Nat Immunol 12:391–398
98. Xu X, Francis R, Wei CJ, Linask KL, Lo CW (2006) Connexin 43-mediated modulation of polarized cell movement and the directional migration of cardiac neural crest cells. Development 133:3629–3639
99. Dang X, Doble BW, Kardami E (2003) The carboxy-tail of connexin-43 localizes to the nucleus and inhibits cell growth. Mol Cell Biochem 242:35–38
100. Wei CJ, Xu X, Lo CW (2004) Connexins and cell signaling in development and disease. Annu Rev Cell Dev Biol 20:811–838
101. Chung SS, Lee WM, Cheng CY (1999) Study on the formation of specialized inter-Sertoli cell junctions in vitro. J Cell Physiol 181:258–272
102. Kojima T, Kokai Y, Chiba H, Yamamoto M, Mochizuki Y, Sawada N (2001) Cx32 but not Cx26 is associated with tight junctions in primary cultures of rat hepatocytes. Exp Cell Res 263:193–201
103. Kojima T, Spray DC, Kokai Y, Chiba H, Mochizuki Y, Sawada N (2002) Cx32 formation and/or Cx32-mediated intercellular communication induces expression and function of tight junctions in hepatocytic cell line. Exp Cell Res 276:40–51
104. Lee SW, Tomasetto C, Paul D, Keyomarsi K, Sager R (1992) Transcriptional downregulation of gap-junction proteins blocks junctional communication in human mammary tumor cell lines. J Cell Biol 118:1213–1222
105. Tomasetto C, Neveu MJ, Daley J, Horan PK, Sager R (1993) Specificity of gap junction communication among human mammary cells and connexin transfectants in culture. J Cell Biol 122:157–167

106. Laird DW, Fistouris P, Batist G, Alpert L, Huynh HT, Carystinos GD, Alaoui-Jamali MA (1999) Deficiency of connexin43 gap junctions is an independent marker for breast tumors. Cancer Res 59:4104–4110
107. Park CC, Henshall-Powell RL, Erickson AC, Talhouk R, Parvin B, Bissell MJ (2003) Ionizing radiation induces heritable disruption of epithelial cell interactions. Proc Natl Acad Sci U S A 100:10728–10733
108. Rakib MA, Kim YS, Jang WJ, Choi BD, Kim JO, Kong IK et al (2010) Attenuation of 12-O-tetradecanoylphorbol-13-acetate (TPA)-induced gap junctional intercellular communication (GJIC) inhibition in MCF-10A cells by c9, t11-conjugated linoleic acid. J Agric Food Chem 58:12022–12030
109. El-Sabban ME, Abi-Mosleh LF, Talhouk RS (2003) Developmental regulation of gap junctions and their role in mammary epithelial cell differentiation. J Mammary Gland Biol Neoplasia 8:463–473
110. McLachlan E, Shao Q, Laird DW (2007) Connexins and gap junctions in mammary gland development and breast cancer progression. J Membr Biol 218:107–121
111. Sohl G, Willecke K (2004) Gap junctions and the connexin protein family. Cardiovasc Res 62:228–232
112. Talhouk RS, Elble RC, Bassam R, Daher M, Sfeir A, Mosleh LA et al (2005) Developmental expression patterns and regulation of connexins in the mouse mammary gland: Expression of connexin30 in lactogenesis. Cell Tissue Res 319:49–59
113. Locke D, Jamieson S, Stein T, Liu J, Hodgins M, Harris A et al (2007) Nature of Cx30-containing channels in the adult mouse mammary gland. Cell Tissue Res 328:97–107
114. Talhouk RS, Mroue R, Mokalled M, Abi-Mosleh L, Nehme R, Ismail A et al (2008) Heterocellular interaction enhances recruitment of α and β-catenins and ZO-2 into functional gap-junction complexes and induces gap junction-dependant differentiation of mammary epithelial cells. Exp Cell Res 314:3275–3291
115. Talhouk RS, Zeinieh MP, Mikati MA, El-Sabban ME (2008) Gap junctional intercellular communication in hypoxia-ischemia-induced neuronal injury. Prog Neurobiol 84:57–76
116. Bry C, Maass K, Miyoshi K, Willecke K, Ott T, Robinson G et al (2004) Loss of connexin 26 in mammary epithelium during early but not during late pregnancy results in unscheduled apoptosis and impaired development. Dev Biol 267:418–429
117. Plante I, Laird DW (2008) Decreased levels of connexin43 result in impaired development of the mammary gland in a mouse model of oculodentodigital dysplasia. Dev Biol 318:312–322
118. Locke D, Perusinghe N, Newman T, Jayatilake H, Evans WH, Monaghan P (2000) Developmental expression and assembly of connexins into homomeric and heteromeric gap junction hemichannels in the mouse mammary gland. J Cell Physiol 183:228–237
119. Locke D, Stein T, Davies C, Morris J, Harris A, Evans W et al (2004) Altered permeability and modulatory character of connexin channels during mammary gland development. Exp Cell Res 298:643–660
120. Dong B, Zhao F (2007) Involvement of the ubiquitous Oct-1 transcription factor in hormonal induction of β-casein gene expression. Biochem J 401:57–64
121. Streuli CH, Bailey N, Bissell MJ (1991) Control of mammary epithelial differentiation: basement membrane induces tissue-specific gene expression in the absence of cell-cell interaction and morphological polarity. J Cell Biol 115:1383–1395
122. El-Saghir JA, El-Habre ET, El-Sabban ME, Talhouk RS (2011) Connexins: a junctional crossroad to breast cancer. Int J Dev Biol 55:773–780
123. Czyz J (2008) The stage-specific function of gap junctions during tumourigenesis. Cell Mol Biol Lett 13:92–102
124. Carystinos GD, Bier A, Batist G (2001) The role of connexin-mediated cell-cell communication in breast cancer metastasis. J Mammary Gland Biol Neoplasia 6:431–440

125. Elzarrad M, Haroon A, Willecke K, Dobrowolski R, Gillespie M, Al-mehdi A (2008) Connexin-43 upregulation in micrometastases and tumor vasculature and its role in tumor cell attachment to pulmonary endothelium. BMC Med 6:20
126. Stoletov K, Strnadel J, Zardouzian E, Momiyama M, Park FD, Kelber JA et al (2013) Role of connexins in metastatic breast cancer and melanoma brain colonization. J Cell Sci 126:904–913
127. Kanczuga-Koda L, Sulkowski S, Lenczewski A, Koda M, Wincewicz A, Baltaziak M (2006) Increased expression of connexins 26 and 43 in lymph node metastases of breast cancer. J Clin Pathol 59:429–433
128. Kapoor P, Saunders MM, Li Z, Zhou Z, Sheaffer N, Kunze EL et al (2004) Breast cancer metastatic potential: correlation with increased heterotypic gap junctional intercellular communication between breast cancer cells and osteoblastic cells. Int J Cancer 111:693–697
129. Saunders MM, Seraj MJ, Li Z, Zhou Z, Winter CR, Welch DR et al (2001) Breast cancer metastatic potential correlates with a breakdown in homospecific and heterospecific gap junctional intercellular communication. Cancer Res 61:1765–1767
130. Pollmann MA, Shao Q, Laird DW, Sandig M (2005) Connexin 43 mediated gap junctional communication enhances breast tumor cell diapedesis in culture. Breast Cancer Res 7:R522–R534
131. Carmona-Fontaine C, Matthews CK, Kuriyama S, Moreno M, Dunn GA, Parsons M et al (2008) Contact inhibition of locomotion in vivo controls neural crest directional migration. Nature 456:957–961
132. Friedl P (2004) Prespecification and plasticity: shifting mechanisms of cell migration. Curr Opin Cell Biol 16:14–23
133. Friedl P, Gilmour D (2009) Collective cell migration in morphogenesis, regeneration and cancer. Nat Rev Mol Cell Biol 10:445–457
134. Momiyama M, Omori Y, Ishizaki Y, Nishikawa Y, Tokairin T, Ogawa J et al (2003) Connexin26-mediated gap junctional communication reverses the malignant phenotype of MCF-7 breast cancer cells. Cancer Sci 94:501–507
135. Qin H, Shao Q, Curtis H, Galipeau J, Belliveau DJ, Wang T et al (2002) Retroviral delivery of connexin genes to human breast tumor cells inhibits in vivo tumor growth by a mechanism that is independent of significant gap junctional intercellular communication. J Biol Chem 277:29132–29138
136. Qin H, Shao Q, Igdoura SA, Alaoui-Jamali MA, Laird DW (2003) Lysosomal and proteasomal degradation play distinct roles in the life cycle of Cx43 in gap junctional intercellular communication deficient and -competent breast tumor cells. J Biol Chem 278:30005–30014
137. Shao Q, Wang H, McLachlan E, Veitch GI, Laird DW (2005) Down-regulation of Cx43 by retroviral delivery of small interfering RNA promotes an aggressive breast cancer cell phenotype. Cancer Res 65:2705–2711
138. Plante I, Stewart MKG, Barr K, Allan AL, Laird DW (2010) Cx43 suppresses mammary tumor metastasis to the lung in a Cx43 mutant mouse model of human disease. Oncogene 30:1681–1692
139. Mroue R, El-Sabban M, Talhouk R (2011) Connexins and the gap in context. Integr Biol (Camb) 3:255–266
140. Jiang W, Bryce R, Mansel R (1997) Gamma linolenic acid regulates gap junction communication in endothelial cells and their interaction with tumour cells. Prostaglandins Leukot Essent Fatty Acids 56:307–316
141. Saez CG, Velasquez L, Montoya M, Eugenin E, Alvarez MG (2003) Increased gap junctional intercellular communication is directly related to the anti-tumor effect of all-trans-retinoic acid plus tamoxifen in a human mammary cancer cell line. J Cell Biochem 89:450–461
142. Gakhar G, Hua DH, Nguyen TA (2010) Combinational treatment of gap junctional activator and tamoxifen in breast cancer cells. Anticancer Drugs 21:77–88

143. Heiniger B, Gakhar G, Prasain K, Hua D, Nguyen T (2010) Second-generation substituted quinolines as anticancer drugs for breast cancer. Anticancer Res 30:3927–3932
144. Vidi PA, Bissell MJ, Lelièvre SA (2013) Three-dimensional culture of human breast epithelial cells: The how and the why. In: Epithelial cell culture protocols (2nd edn). Springer Publisher, Humana Press, pp 193–219.
145. Grafton MM, Wang L, Vidi PA, Leary J, Lelievre SA (2011) Breast on-a-chip: mimicry of the channeling system of the breast for development of theranostics. Integr Biol 3:451–459
146. Mironov V, Vladimir K, Markwald R (2011) Organ printing: from bioprinter to organ biofabrication line. Curr Opin Biotech 22(5):667–673

# Chapter 2
# Connexins: Bridging the Gap Between Cancer Cell Communication in Glioblastoma

**Maksim Sinyuk and Justin D. Lathia**

**Abstract** Despite concerted clinical and research efforts, glioblastoma (GBM), the most prevalent primary malignant brain tumor, remains uniformly lethal. Like other advanced cancers, GBM is characterized by extensive cellular heterogeneity and is organized in a hierarchy with self-renewing, therapeutically resistant cancer stem cells (CSCs) at the apex. While communication between GBM tumor cells and their surrounding stroma supports tumor survival and expansion, the mechanisms behind direct cell-cell communication and its contribution to tumor growth have yet to be fully elucidated. In particular, the biological importance of intercellular communication between GBM tumor cells, including CSCs and non-stem tumor cells (NSTCs) has yet to be determined. Gap junctions (GJs) are specialized structures, composed of connexin proteins, allowing for the diffusion of small molecules and ions directly between the cytoplasm of adjacent cells, enabling them to respond to each other and external stimuli rapidly and coordinately. Connexins have been found to help promote tumor cell growth, invasiveness, and tumorigenicity, making them attractive anti-tumor targets. However a complete understanding of the function of connexins and GJs in GBM remains an area of active investigation. Here we discuss recent advances in connexin function as they relate to our understanding of cellular communication and malignancy in GBM.

**Keywords** Glioblastoma • Gap junctions • Connexins • Cancer stem cells

M. Sinyuk
Department of Cellular and Molecular Medicine, Lerner Research Institute, Cleveland Clinic, Cleveland, OH, USA

Department of Biological, Geological, and Environmental Sciences, Cleveland State University, Cleveland, OH, USA

J.D. Lathia (✉)
Department of Cellular and Molecular Medicine, Lerner Research Institute, Cleveland Clinic, Cleveland, OH, USA

Department of Biological, Geological, and Environmental Sciences, Cleveland State University, Cleveland, OH, USA

Department of Molecular Medicine, Lerner College of Medicine, Case Western University, Cleveland, OH, USA

Case Comprehensive Cancer Center, Case Western Reserve University, Cleveland, OH, USA
e-mail: lathiaj@ccf.org



M. Kandouz (ed.), *Intercellular Communication in Cancer*,
DOI 10.1007/978-94-017-7380-5_2

Over the past century, major medical innovations have improved the standard of living for much of the developed world. Likewise, the average life expectancy has increased in many countries as a direct result of our continued efforts to understand the biology of disease and apply this knowledge toward the development of more efficacious therapeutic interventions. However, while mortality rates have decreased for many illnesses such as heart and cerebrovascular disease, little progress has been made in the fight against cancer, as death rates due to malignant neoplasms have remained steady over the past 50 years [1]. Primary central nervous system (CNS) tumors, in particular, are among the most dangerous malignancies, comprising only 2 % of all cancer diagnoses but accounting for a disproportionate rate of morbidity and mortality.

The most common aggressive of all primary CNS tumors is glioblastoma (GBM), accounting for 20 % of all intracranial tumors in the United States [2]. GBM is classified by the World Health Organization (WHO) as a grade IV glioma and is thought to arise from glial cells: non-neuronal cells responsible for maintaining homeostasis, forming myelin, and providing protection for neurons in the brain and peripheral nervous system [3]. Most GBM tumors (~90 %) appear to rapidly expand de novo mainly in elderly patients, without evidence of a precursor lesion, and are classified as primary GBM. However, a smaller fraction (~10 %) of secondary GBM tumors are thought to progress from low-grade astrocytomas, mainly in younger patients. Primary and secondary GBM tumors are largely indistinguishable histologically but differ in their genetic and epigenetic profiles. Additionally, secondary GBMs carry a vastly favorable prognosis compared to primary GBM diagnoses [4]. The histological criteria for GBM diagnosis include cellular pleomorphism, nuclear atypia, vascular thrombosis, microvascular proliferation, and necrosis with lesions displaying both intra- and inter-tumoral heterogeneity [4]. Currently, GBM is uniformly fatal, and treatment is only palliative, consisting mainly of maximal safe surgical resection, chemotherapy, and radiation. Independent prognostic factors include patient age, performance status, number of lesions, and extent of resection [5]. Despite aggressive therapy, median survival time after diagnosis is 12–18 months, while the 5-year survival rate remains at 5 % [6].

The high degree of invasiveness characterizing GBM is a major impediment for treatment, as surgical resection is often unable to remove the entirety of the tumor, leaving behind a population of infiltrating cancerous cells that egress away from the primary site [7]. Further surgical intervention is made difficult or impossible as the remaining tumor cells continue to migrate along myelinated axons, vascular basement membranes, or the subependyma to infiltrate anatomically critical structures in the brain and escape the reach of current therapeutics [8]. Likewise, due to the heterogeneous nature of GBM, the administration of chemotherapy and radiation may have deleterious consequences for patient survival. According to the stochastic model of tumor formation, all tumor cells are thought to possess the ability to propagate a tumor, with genetic cues dictating which cells drive tumor progression. Accordingly, a small number of tumor cells, through clonal selection, are thought to be capable of randomly developing resistance to current therapeutics. As such, treatment may inadvertently select for more aggressive tumor clones, helping explain why recurrence is almost always inevitable following initial tumor debulking [9]. Recent clinical trials interrogating the efficacy of several antiangiogenic agents in conjunction with radiation and chemotherapy appear to reinforce the concept of clonal selection in GBM. Antiangiogenic therapy resulted in short-term tumor burden control and improved progression-free survival. However, overall survival was not impacted [10, 11], indicating the rise of a more aggressive, recurrent population of tumor cells following initial therapy.

The stochastic model is an attractive intellectual concept regarding the formation and maintenance of GBM. It proposes that tumors arise from a single clone, allowing for the sequential selection of progressively more malignant cancer cells. As tumor cells acquire additional mutations, some variants are destroyed while others are imparted with growth advantages, permitting clonal expansion. These clones become the predominant subpopulation until a more favored variant appears. Eventually, the acquired genetic instability and associated selection process results in a heterogeneous population of cancer cells making up a tumor [12]. However, this model does not take into account environmental factors and epigenetic variables that influence cell behavior independently of genomic control. As such, it is an incomplete and simplistic view of tumor biology, necessitating additional models to better represent the complex nature of GBM.

A surrogate hypothesis is the hierarchical model of cancer, which posits the existence of a cellular hierarchy within a tumor. A small sub-population of cells, referred to as cancer stem cells (CSCs), exist at the top of the hierarchy and are capable of propagating tumor heterogeneity. The hierarchical model accommodates the possibility that CSCs are capable of retaining responsiveness to external environmental cues, eliciting their genomically determined potential for self-renewal and recapitulation of the cellular diversity composing the bulk of the tumor [13]. However, the two models should not be considered to be mutually exclusive. The CSC hypothesis does not exclude stochastic selection or the acquisition of resistance by tumor cells, and instead, both models should be viewed complimentarily to address the complexity of tumorigenesis [14].

Indeed, recent evidence has demonstrated the existence of CSCs crucial for GBM initiation and maintenance [15]. Unlike their rapidly proliferative non-stem tumor cell (NSTC) counterparts, CSCs are resistant to chemotherapy and radiation as a result of an increased DNA repair capacity [16, 17]. The exclusive ability of CSCs to self-renew and differentiate into multiple lineages is a major factor in GBM tumorigenesis and recurrence [14]. In light of their unique phenotype, CSCs have since been recognized as attractive targets for the development of novel, combinatorial GBM therapeutics aimed at eradicating both the bulk of the tumor as well as the resistant CSC population. However, prior to successful clinical translation, several challenges remain in the integration of CSC-specific interventions alongside current standard-of-care modalities. Among them are complications in the categorization of cellular differentiation states in GBM lesions, given that stemness is a dynamic property within a tumor. As such, CSCs have the capacity to differentiate into NSTCs while still retaining the ability to revert back to a CSC state in response to microenvironmental cues such as hypoxia [18], pH [19], and metabolism [20].

Additionally, CSCs share common gene-expression signatures as well as cell signaling pathways with neural progenitor cells (NPCs) [21], hampering the development of agents capable of destroying the former while preserving the latter.

While multiple cell types likely contribute to GBM growth, cancer cells display a remarkable ability to tailor and influence their microenvironment for the promotion of tumor growth, maintenance, and migration [22]. Physically, tumor cells are capable of remodeling their surrounding extracellular matrix (ECM), both through the production of matrix metalloproteinases (MMPs) and the degradation or synthesis of collagen ligands to facilitate invasion into the brain parenchyma. In addition, glioma cells apply physical stresses on the surrounding collagen matrix, as a result of traction forces exerted by individual cells and compressive forces generated by the expansion of the tumor bulk, to influence tumor cell proliferation and malignant outcome [23]. In addition, it is becoming more appreciated that GBM is not simply composed of small numbers of CSCs and their NSTC progeny but rather contains a proportion of host cells and tissue. The capability of normal tissue to directly influence tumor biology and vice versa should not be taken lightly as mutual interactions between neoplastic and non-neoplastic cells produces a local milieu, favoring tumor cell growth and immune escape. Tumor-associated cells in the GBM microenvironment, such as microglia, vascular cells, peripheral immune cells, and NPCs, also play important roles in the pathology of GBM, often exerting pro-tumorigenic effects [22].

Likewise, the host immune system is capable of interacting with GBM tumor cells. GBM CSCs are capable of driving tumor growth by actively attenuating immunosurveillance through the secretion or expression of immunosuppressive factors or by the recruitment of accessory cells, which locally suppress the immune response until tumors reach a size at which they surpass immune pressure, resulting in progression and malignancy. In addition, GBM CSCs are capable of recruiting multiple cell types with tumor-supportive phenotypes. In vitro, CSC-conditioned medium was found to increase monocyte migration compared to cell suspensions generated from GBM NSTCs [24]. Likewise, CSCs have been shown to secrete soluble colony stimulating factor-1 (sCSF-1), C-C motif ligand 2 (CCL2), and macrophage inhibitory cytokine 1 (MIC-1) to enhance monocyte infiltration into the tumor [24]. However, upon recruitment of peripheral monocytes into the tumor, the secreted sCSF-1 and CCL2 polarize them toward the immunosuppressive M2 macrophage phenotype, while MIC-1 simultaneously inhibits their phagocytic ability [24]. Moreover, recent work has revealed that CSCs are similarly capable of recruiting tumor-associated macrophages (TAMs) into the tumor microenvironment by secreting periostin, a protein normally thought to support the adhesion and migration of epithelial cells, through its receptor, integrin $\alpha_V\beta_3$ [25].

Based on these observations, GBM should not be thought of as one distinct entity residing in normal brain tissue but rather as an aberrant organ. Like normal organs, GBM tumors are composed of multiple cellular and stromal aspects working in concert for proper function, under physiological conditions, or malignancy, under neoplastic conditions. Additionally, both normal and cancerous cells must be able to interact and communicate with various, surrounding cell types to execute biological functions at the tissue level that could not otherwise be accomplished [26]. The information exchanged between cells may involve direct cell-to-cell contact or the release of soluble mediators capable of acting in an autocrine or paracrine manner, depending on the nature of the signaling pathway. However GBM tumors do not frequently metastasize to other organs of the body and remain confined to the brain parenchyma, which is itself enclosed by the blood-brain-barrier, limiting cellular cross-talk across peripheral circulation. As a result, GBM tumors often histologically manifest as dense hypercellular masses with little room between individual cells. The close confines of the GBM microenvironment necessitates rapid tumor cell communication both between other tumor cells as well as with the surrounding stroma in order to coordinately respond to chemical and physical stimuli. Both the spatial limitations of the brain and the temporal need to quickly adapt to an ever-changing environmental milieu make it likely that GBM tumor cells rely on autocrine and paracrine signaling pathways through direct cell-cell contact as a means of communication. Gap junctions (GJs) represent a well-documented means of intercellular communication in various tissues. GJ-mediated communication has been demonstrated to be essential in normal embryonic development [27], electric coupling in cardiac muscle [28] and neurons [29], as well as in normal hematopoiesis [30]. Additionally, connexin expression in non-excitable tissues has key roles in organ development [31], skeletal development [32], and growth control [33].

GJs are aggregates of intercellular channels composed of a family of 24 proteins, termed connexins, that allow the direct transport of cytoplasmic contents from cell to cell. Six co-oligomerized connexin subunits form a connexon, also known as a hemichannel. The connexin subunits making up a connexon can either be identical (homomeric) or disimilar (heteromeric), although not all connexin subunits are capable of forming a functional hemichannel [34]. Two hemichannels on different cells are then able to dock and form a homotypic or heterotypic GJ channel, depending on the connexin isotype. A functional GJ channel allows for the diffusion of small molecules up to 1 kDa in size between the cytoplasm of adjoining cells. GJs favor the intercellular exchange of metabolites such as adenosine diphosphate (ADP), glucose, glutamate and glutathione [35], as well as second messengers such as calcium ions ($Ca^{2+}$), cyclic adenosine monophosphate (cAMP), and inositol triphosphate ($IP_3$) [36].

The expression of connexin proteins is both tissue specific and developmentally regulated, making the number of combinations of possible intercellular channels broad. The variability of connexin signaling also plays an important role in the physiological properties of the various gap junctionhemichannels, including conductance and permeability [37]. Traditionally, connexin function has been linked to the formation of gap junction channels although it is becoming more appreciated that connexin hemichannels are capable of serving as aqueous pores permeable to ions and small molecules [38, 39], that link the intra- and extracellular compartments. In addition, dysregulated connexin expression has also been linked to at least ten distinct diseases, such as X-linked Charcot-Marie-Tooth disease [40], keratitis-ichthyosis-deafness syndrome [41], and oculodentodigital dysplasia [42]. The importance of regulating connexin function in normal development may also underscore the role that aberrant expression plays in tumor formation and growth.

Historically, connexins have been thought to function as tumor suppressors in several animal models of cancer, including hepatoma [43] and thyroid tumors [44], as well as human carcinoma of the stomach [45], which was evidenced by a lack of electrical coupling between tumor cells. However, recent evidence has suggested that connexins may also promote tumorigenesis. Forcing connexin expression in both non-metastatic and metastatic tumor cells with no prior functional connexin activity was shown to decrease proliferation and cell migration [46], and promote a mesenchymal to epithelial transition [47]. Likewise, in several advanced cancers, GJ function was associated with invasion [48], intravasation [49], extravasation [50], and metastasis of tumor cells [51], facilitating late-stage disease progression. Gap junctions also participate in the "bystander effect" following radiation therapy in which cells that are not directly exposed to radiation but are in the vicinity likewise respond to the exposure and display significant levels of genetic change and lethality.

In the CNS, abundant connexin expression has been demonstrated in multiple cell lineages, including neurons, astrocytes, and microglia. Under physiological conditions, connexins are thought to be important in normal neurogenesis as well as neuronal electrical signaling [35]. Connexin subunit expression was detected at the very early stages of neural development, with connexin 43 (Cx43) and connexin 45 (Cx45) robustly expressed and essential for rat NPC proliferation and survival [52]. Follow-up studies demonstrated that embryonic NPCs possessed active gap junctions as, confirmed by dye-coupling studies, and that in the absence of essential growth factors, Cx43 overexpression was sufficient to preserve NPC self-renewal, which was otherwise compromised in differentiation-inducing conditions [53]. Along with preserving self-renewal, connexins also impact lineage commitment, with connexin 36 (Cx36) being important in the modulation of NPC differentiation into neurons and glia [54]. Reduced Cx36 expression decreased neuronal commitment, and overexpression restored neuronal differentiation along with oligodendrocyte commitment [54]. It has recently been appreciated that connexins may also function to impact invasion as well as cellular signaling programs through interaction with scaffolding proteins via their cytoplasmic tails [55]. There is evidence for each of these aforementioned functions in the developing brain. NPCs utilize Cx43 and connexin 26 (Cx26) for tangential migration of newly-born neurons [56, 57]. While Cx43 reduction has a profound impact on rat and mouse NPCs, it was dispensable for human NPC function, and Wnt/β-catenin signaling was activated in Cx43-reduced conditions [58], suggesting that the Cx43 may serve to suppress Wnt/β-catenin signaling. These results also demonstrate the species difference that may exist for connexins and highlight the need for additional studies in multiple systems. The existence of connexin signaling in NPC maintenance is also important in the adult mouse brain, with connexin 30 (Cx30) and Cx43 found to mediate intercellular coupling between radial glial cells in the dentate gyrus. It was found that mice lacking Cx43 and Cx30 in radial glial cells showed complete inhibition of cell proliferation in the subgranular zone of the adult dentate gyrus [59]. In addition,

GJ-mediated communication was found to be crucial for several brain processes, including neuronal energy supply, electrical and chemical synapses, calcium waves, spatial buffering of $K^+$ and glutamate, maintenance of myelin and blood–brain barrier integrity [35].

While there is an established literature for the function of connexins during neural development, the role that connexins play in GBM and CSCs is only beginning to be investigated. Some studies have found that Cx43 is decreased in high-grade brain tumors [60], while others demonstrate that Cx43 is capable of conferring chemotherapeutic resistance to human glioma cells [61] through the upregulation of key pathways including the epidermal growth factor receptor [62]. Overall, this is reflected in the lack of a consensus for the pro- or anti-tumorigenic role for connexins in GBM, which has thus far mainly focused on data surrounding Cx43 [63]. Efforts have begun to interrogate the expression and function of connexins in CSCs. Similar to the early work in GBM, pro- and anti-tumorigenic roles have emerged and these may be model- and connexin subunit-specific. Overexpression of Cx43 was found to inhibit CSC self-renewal, invasiveness, and tumorigenicity via E-cadherin, which in turn influenced Wnt/β-catenin signaling, increasing the latency of GBM tumors [64]. However, others have demonstrated that CSCs predominantly express Cx46, while their NSTCs express Cx43. As CSCs were differentiated, Cx46 was reduced while Cx43 increased, and targeting Cx46 rather than Cx43 was found to compromise CSC maintenance [65]. Along with serving as a functional regulator of CSC maintenance and possible driver of tumor progression, modulating connexins may be an adjuvant therapeutic approach. The rationale for this is the "bystander effect," whereby damage generated from irradiating one cell may be passed to another via gap junctions. Recent work in a mouse medulloblastoma model used genetic approaches to downregulate Cx43 and demonstrated that gap junction-mediated communication is crucial for the transmission of radiation. Upregulation of Cx43 was found to cause tumor regression in the distal CNS, the area not exposed to direct radiation therapy, further supporting its anti-tumor role. Surprisingly, Cx43 was also found to be upregulated in non-targeted tissue following irradiation, which may allow for the transduction of potentially oncogenic signals to remote tissue through this "bystander effect" [66]. As evidenced, additional work is necessary to completely unravel the function of connexins in GBM, especially in the context of the "bystander effect" and therapeutic resistance. However, evidence strongly suggests that connexins are key regulators of GBM phenotypes and are emerging as attractive targets for potential therapeutic modalities aimed at reducing GBM invasiveness, proliferation, and lethality.

To better understand connexin biology and its role in disease, two non-specific pan-GJ inhibitors are currently being investigated in pre-clinical trials, Carbenoxolone (CBX) and 1-Octanol. CBX is currently approved in the clinical treatment of esophageal and mouth ulcers in the United Kingdom [67], while 1-Octanol is currently being interrogated for the treatment of essential tremors [68]. CBX, in particular, has been investigated in several advanced cancers, including thyroid [69], leukemia [70], and GBM [65] due to its minimal cytotoxic nature. Combinatorial treatment of primary human glioma isolates with CBX and mesenchymal stem cells (MSCs) engineered to express tumor necrosis factor-related apoptosis-inducing ligand (TRAIL) has been shown to enhance glioma cell death through the upregulation of death receptor 5. Importantly, dual therapy utilizing TRAIL and CBX prolonged mouse survival by ~27 % compared with control animals, suggesting a favorable clinical translation [71]. Likewise, both 1-Octanol and CBX were shown to prolong animal survival in intracranial and subcutaneous xenograft models of GBM. Interestingly, when GJ inhibition strategies were combined with temozolomide (TMZ), an additive effect on survival was seen, suggesting that the use of GJ inhibition strategies alongside current therapeutic modalities could positively impact patient outcome [65].

Targeting GJ-mediated communication in GBM and other cancers is emerging as an exciting prospective strategy with potentially translatable results. Specifically, the additive survival advantage that GJ inhibition, alongside chemotherapy, confers in animal models of GBM is particularly promising. However, several caveats remain to be addressed regarding both CBX and 1-Octanol before clinical trials are implemented. Both agents demonstrate remarkable efficacy for inhibiting GJs and tumor cell growth in vitro and in vivo. However, their mechanism of action is poorly understood. In particular it should be noted that they do not specifically block individual connexin subunits or GJs. Rather, they are pan-inhibitors, ostensibly blocking all connexin function and making it difficult to study which particular connexin subunits are involved in tumor biology. It is also important to note that blocking all GJs may have unintended off-target effects that need to be addressed before considering clinical trials. Additionally, the exact methods by which the agents inhibit connexin function is an ongoing area of investigation. It has been hypothesized that both CBX and 1-Octanol act on cell membranes to alter fluidity and disrupt the transmembrane domains of connexin proteins, rendering them inert. However, this explanation has yet to be fully investigated and remains speculative. The last, and possibly most important, point to consider regarding GJ inhibition is the exact mechanism behind tumor cell death after treatment with CBX or 1-Octanol. Several likely explanations for this phenomenon have therefore been proposed. As previously mentioned, GBM tumor cells exist in a closely packed microenvironment and communicate predominantly through cell-cell contact mediated by GJs. As such, tumor cells are better able to respond to external stimuli and escape damage from sources such as chemotherapeutics and radiation by exchanging information and rendering themselves less susceptible to perturbation. In addition, GJs may allow for the release of potentially lethal intercellular components, such as reactive oxygen species (ROS), generated in response to cell damage. Conversely, GJ hemichannels may also facilitate the uptake of molecules necessary to protect tumor cells from ROS-induced DNA damage. Recent work in normal hematopoietic stem cells (HSCs) has supported this concept, as Cx43 deficient HSCs displayed decreased survival and increased senescence as a direct result of their inability to transfer ROS to the hematopoietic microenvironment following myeloablation, demonstrating that Cx43 is able to play a protective role during stressful conditions such as hematopoietic recovery [72].

Even though the exact molecular mechanisms behind connexin signaling are only now beginning to be elucidated, the potential to disrupt GJs, and consequently tumor cells, by pharmacologically targeting connexins remains an attractive strategy in a field that has had limited clinical success over the past decades. However it is prudent to consider that connexins may have additional functions which have yet to be fully described. To this end, cytoplasmic partners have been thought to be capable of interacting with the intracellular domains of connexin proteins, providing a potential means of specifically targeting individual subunits. The ablation of one universal connexin may have unintended secondary effects or no effects at all, as compensatory mechanisms likely exist among various connexin proteins. Rather, GJ inhibition strategies should be contextualized in light of the overall tumor or, even more effectively, in light of the cell-of-origin of the tumor to target the root of the malignancy rather than the branches. Of paramount importance is the development of novel mimetic peptides or agents capable of disrupting individual connexin subunits to minimize the harm done to normal tissue in the course of treatment. Cancer therapy as a whole is moving away from a "one-size-fits-all" paradigm and towards a more individualized model. Targeting specific connexin subunits, depending on tumor subtype, is therefore complementary to the emerging trends regarding cancer care and should be considered for further attention. Additional work is also necessary to tease out the direct molecular mechanisms responsible for connexin signaling, but efforts are slowly beginning to concentrate on this line of inquiry. With careful methodology and proper animal models, elucidating connexin signaling has the potential to make a transformative impact for the development of therapies capable of improving the outcome of patients diagnosed with not only GBM but also other neoplasias for which little hope currently exists.

**Conflict of Interest** The authors wish to disclose that they do not have any relevant affiliations or financial involvement with any organization or entity with a financial interest in or financial conflict with the subject matter or materials discussed in the manuscript.

## References

1. Hoyert DL, Xu JQ (2011) Deaths: preliminary data for 2011, vol 61, 6th edn, National vital statistics reports. National Center for Health Statistics, Hyattsville
2. Buckner JC, Brown PD, O'Neill BP et al (2007) Central nervous system tumors. Mayo Clin Proc 82:1271–1286. doi:10.4065/82.10.1271
3. Gourine AV, Kasymov V, Marina N et al (2010) Astrocytes control breathing through pH-dependent release of ATP. Science 329:571–575. doi:10.1126/science.1190721
4. Ohgaki H, Kleihues P (2013) The definition of primary and secondary glioblastoma. Clin Cancer Res 19:764–772
5. Okita Y, Narita Y, Miyakita Y et al (2012) Pathological findings and prognostic factors in recurrent glioblastomas. Brain Tumor Pathol 29:192–200. doi:10.1007/s10014-012-0084-2
6. Grossman SA, Ye X, Piantadosi S et al (2010) Survival of patients with newly diagnosed glioblastoma treated with radiation and temozolomide in research studies in the United States. Clin Cancer Res 16:2443–2449

7. Wilson TA, Karajannis MA, Harter DH (2014) Glioblastoma multiforme: state of the art and future therapeutics. Surg Neurol Int 5:64. doi:10.4103/2152-7806.132138
8. Hoelzinger DB, Demuth T, Berens ME (2007) Autocrine factors that sustain glioma invasion and paracrine biology in the brain microenvironment. J Natl Cancer Inst 99:1583–1593
9. Bonavia R, Inda M-M, Cavenee WK, Furnari FB (2011) Heterogeneity maintenance in glioblastoma: a social network. Cancer Res 71:4055–4060
10. Gilbert MR, Dignam JJ, Armstrong TS et al (2014) A randomized trial of bevacizumab for newly diagnosed glioblastoma. N Engl J Med 370:699–708. doi:10.1056/NEJMoa1308573
11. Khasraw M, Grant R, Wheeler H, Pavlakis N (2014) Antiangiogenic therapy for high-grade glioma. Cochrane Database Syst Rev 9, CD008218
12. Vogelstein B, Papadopoulos N, Velculescu VE et al (2013) Cancer genome landscapes. Science 339:1546–1558
13. Nguyen LV, Vanner R, Dirks P, Eaves CJ (2012) Cancer stem cells: an evolving concept. Nat Rev Cancer 12:133–143. doi:10.1038/nrc3184
14. Visvader JE, Lindeman GJ (2012) Cancer stem cells: current status and evolving complexities. Cell Stem Cell 10:717–728. doi:10.1016/j.stem.2012.05.007
15. Singh SK, Hawkins C, Clarke ID et al (2004) Identification of human brain tumour initiating cells. Nature 432:396–401. doi:10.1038/nature03128
16. Bao S, Wu Q, McLendon RE et al (2006) Glioma stem cells promote radioresistance by preferential activation of the DNA damage response. Nature 444:756–760. doi:10.1038/nature05236
17. Chen J, Li Y, Yu T-S et al (2012) A restricted cell population propagates glioblastoma growth following chemotherapy. Nature 488:522–526. doi:10.1038/nature11287
18. Heddleston JM, Li Z, McLendon RE et al (2009) The hypoxic microenvironment maintains glioblastoma stem cells and promotes reprogramming towards a cancer stem cell phenotype. Cell Cycle 8:3274–3284. doi:10.4161/cc.8.20.9701
19. Hjelmeland AB, Wu Q, Heddleston JM et al (2011) Acidic stress promotes a glioma stem cell phenotype. Cell Death Differ 18:829–840
20. Flavahan WA, Wu Q, Hitomi M et al (2013) Brain tumor initiating cells adapt to restricted nutrition through preferential glucose uptake. Nat Neurosci 16:1373–1382. doi:10.1038/nn.3510
21. Lee J, Kotliarova S, Kotliarov Y et al (2006) Tumor stem cells derived from glioblastomas cultured in bFGF and EGF more closely mirror the phenotype and genotype of primary tumors than do serum-cultured cell lines. Cancer Cell 9:391–403. doi:10.1016/j.ccr.2006.03.030
22. Charles NA, Holland EC, Gilbertson R et al (2011) The brain tumor microenvironment. Glia 59:1169–1180. doi:10.1002/glia.21136
23. Payne LS, Huang PH (2013) The pathobiology of collagens in glioma. Mol Cancer Res 11:1129–1140
24. Wu A, Wei J, Kong L-Y et al (2010) Glioma cancer stem cells induce immunosuppressive macrophages/microglia. Neuro Oncol 12:1113–1125
25. Baril P, Gangeswaran R, Mahon PC et al (2006) Periostin promotes invasiveness and resistance of pancreatic cancer cells to hypoxia-induced cell death: role of the [beta]4 integrin and the PI3k pathway. Oncogene 26:2082–2094
26. Salido-Guadarrama I, Romero-Cordoba S, Peralta-Zaragoza O et al (2014) MicroRNAs transported by exosomes in body fluids as mediators of intercellular communication in cancer. Onco Targets Ther 7:1327–1338. doi:10.2147/OTT.S61562
27. Warner AE, Guthrie SC, Gilula NB (1984) Antibodies to gap-junctional protein selectively disrupt junctional communication in the early amphibian embryo. Nature 311:127–131. doi:10.1038/311127a0
28. Beauchamp P, Desplantez T, McCain ML et al (2012) Electrical coupling and propagation in engineered ventricular myocardium with heterogeneous expression of connexin43. Circ Res 110:1445–1453

29. Allison DW, Ohran AJ, Stobbs SH et al (2006) Connexin-36 gap junctions mediate electrical coupling between ventral tegmental area GABA neurons. Synapse 60:20–31. doi:10.1002/syn.20272
30. Gonzalez-Nieto D, Li L, Kohler A et al (2012) Connexin-43 in the osteogenic BM niche regulates its cellular composition and the bidirectional traffic of hematopoietic stem cells and progenitors. Blood 119:5144–5154. doi:10.1182/blood-2011-07-368506
31. Kamiya K, Yum SW, Kurebayashi N et al (2014) Assembly of the cochlear gap junction macromolecular complex requires connexin 26. J Clin Invest 124:1598–1607. doi:10.1172/JCI67621
32. Stains JP, Civitelli R (2005) Gap junctions in skeletal development and function. Biochim Biophys Acta 1719:69–81. doi:10.1016/j.bbamem.2005.10.012
33. Kihara AH, Santos TO, Osuna-Melo EJ et al (2010) Connexin-mediated communication controls cell proliferation and is essential in retinal histogenesis. Int J Dev Neurosci 28:39–52. doi:10.1016/j.ijdevneu.2009.09.006
34. Hervé J-C, Derangeon M (2013) Gap-junction-mediated cell-to-cell communication. Cell Tissue Res 352:21–31. doi:10.1007/s00441-012-1485-6
35. Orellana JA, Martinez AD, Retamal MA (2013) Gap junction channels and hemichannels in the CNS: regulation by signaling molecules. Neuropharmacology 75:567–582. doi:10.1016/j.neuropharm.2013.02.020
36. Goldberg GS, Lampe PD, Nicholson BJ (1999) Selective transfer of endogenous metabolites through gap junctions composed of different connexins. Nat Cell Biol 1:457–459. doi:10.1038/15693
37. Maeda S, Tsukihara T (2011) Structure of the gap junction channel and its implications for its biological functions. Cell Mol Life Sci 68:1115–1129. doi:10.1007/s00018-010-0551-z
38. Bruzzone S, Guida L, Zocchi E et al (2000) Connexin 43 hemichannels mediate Ca2+ -regulated transmembrane NAD+ fluxes in intact cells. FASEB J 15:10–12. doi:10.1096/fj.00-0566fje.
39. Cherian PP, Siller-Jackson AJ, Gu S et al (2005) Mechanical strain opens connexin 43 hemichannels in osteocytes: a novel mechanism for the release of prostaglandin. Mol Biol Cell 16:3100–3106
40. Karadima G, Koutsis G, Raftopoulou M et al (2014) Four novel connexin 32 mutations in X-linked Charcot–Marie–Tooth disease. Phenotypic variability and central nervous system involvement. J Neurol Sci 341:158–161. doi:10.1016/j.jns.2014.04.007
41. Arita K, Akiyama M, Aizawa T et al (2006) A novel N14Y mutation in Connexin26 in keratitis-ichthyosis-deafness syndrome: analyses of altered Gap junctional communication and molecular structure of N terminus of mutated connexin26. Am J Pathol 169:416–423. doi:10.2353/ajpath.2006.051242
42. Paznekas WA, Boyadjiev SA, Shapiro RE et al (2003) Connexin 43 (GJA1) mutations cause the pleiotropic phenotype of oculodentodigital dysplasia. Am J Hum Genet 72:408–418. doi:10.1086/346090
43. Loewenstein WR, Kanno Y (1967) Intercellular communication and tissue growth: I. Cancerous growth. J Cell Biol 33:225–234
44. Jamakosmanović A, Loewenstein WR (1968) Intercellular communication and tissue growth: III. Thyroid cancer. J Cell Biol 38:556–561
45. Kanno Y, Matsui Y (1968) Cellular uncoupling in cancerous stomach epithelium. Nature 218:775–776. doi:10.1038/218775b0
46. Hirschi K, Xu C, Tsukamoto T, Sager R (1996) Gap junction genes Cx26 and Cx43 individually suppress the cancer phenotype of human mammary carcinoma cells and restore differentiation potential. Cell Growth Differ 7:861–870
47. Kalluri R, Weinberg RA (2009) The basics of epithelial-mesenchymal transition. J Clin Invest 119:1420–1428. doi:10.1172/JCI39104
48. Ezumi K, Yamamoto H, Murata K et al (2008) Aberrant expression of connexin 26 is associated with lung metastasis of colorectal cancer. Clin Cancer Res 14:677–684

49. Ito A, Katoh F, Kataoka TR et al (2000) A role for heterologous gap junctions between melanoma and endothelial cells in metastasis. J Clin Invest 105:1189–1197. doi:10.1172/JCI8257
50. Elzarrad MK, Haroon A, Willecke K et al (2008) Connexin-43 upregulation in micrometastases and tumor vasculature and its role in tumor cell attachment to pulmonary endothelium. BMC Med 6:20. doi:10.1186/1741-7015-6-20
51. Lin JH-C, Takano T, Cotrina ML et al (2002) Connexin 43 enhances the adhesivity and mediates the invasion of malignant glioma cells. J Neurosci 22:4302–4311
52. Cai J, Cheng A, Luo Y et al (2004) Membrane properties of rat embryonic multipotent neural stem cells. J Neurochem 88:212–226. doi:10.1046/j.1471-4159.2003.02184.x
53. Cheng A, Tang H, Cai J et al (2004) Gap junctional communication is required to maintain mouse cortical neural progenitor cells in a proliferative state. Dev Biol 272:203–216. doi:10.1016/j.ydbio.2004.04.031
54. Hartfield EM, Rinaldi F, Glover CP et al (2011) Connexin 36 expression regulates neuronal differentiation from neural progenitor cells. PLoS One 6, e14746. doi:10.1371/journal.pone.0014746
55. Zhou JZ, Jiang JX (2014) Gap junction and hemichannel-independent actions of connexins on cell and tissue functions – an update. FEBS Lett 588:1186–1192. doi:10.1016/j.febslet.2014.01.001
56. Elias LAB, Wang DD, Kriegstein AR (2007) Gap junction adhesion is necessary for radial migration in the neocortex. Nature 448:901–907. doi:10.1038/nature06063
57. Elias LAB, Turmaine M, Parnavelas JG, Kriegstein AR (2010) Connexin 43 mediates the tangential to radial migratory switch in ventrally derived cortical interneurons. J Neurosci 30:7072–7077
58. Rinaldi F, Hartfield EM, Crompton LA et al (2014) Cross-regulation of connexin43 and [beta]-catenin influences differentiation of human neural progenitor cells. Cell Death Dis 5:e1017
59. Kunze A, Congreso MR, Hartmann C et al (2009) Connexin expression by radial glia-like cells is required for neurogenesis in the adult dentate gyrus. Proc Natl Acad Sci 106:11336–11341
60. Soroceanu L, Manning TJ, Sontheimer H (2001) Reduced expression of connexin-43 and functional gap junction coupling in human gliomas. Glia 33:107–117. doi:10.1002/1098-1136(200102)33:2<107::AID-GLIA1010>3.0.CO;2-4
61. Gielen PR, Aftab Q, Ma N et al (2013) Connexin43 confers temozolomide resistance in human glioma cells by modulating the mitochondrial apoptosis pathway. Neuropharmacology 75:539–548. doi:10.1016/j.neuropharm.2013.05.002
62. Munoz JL, Rodriguez-Cruz V, Greco SJ et al (2014) Temozolomide resistance in glioblastoma cells occurs partly through epidermal growth factor receptor-mediated induction of connexin 43. Cell Death Dis 5, e1145
63. Sin W-C, Crespin S, Mesnil M (2012) Opposing roles of connexin43 in glioma progression. Biochim Biophys Acta 1818:2058–2067. doi:10.1016/j.bbamem.2011.10.022
64. Yu S-C, Xiao H-L, Jiang X-F et al (2012) Connexin 43 reverses malignant phenotypes of glioma stem cells by modulating E-cadherin. Stem Cells 30:108–120. doi:10.1002/stem.1685
65. Hitomi M, Deleyrolle LP, Mulkearns-Hubert E, Jarrar A, Li M, Sinyuk M, Otvos B, Brunet S, Flavahan WA, Hubert CG, Goan W, Hale JS, Alvarado AG, Zhang A, Rohaus M, Oli M, Vedam-Mai V, Fortin JM, Futch HS, Griffith B, Wu Q, Xia C, Gong X, Ahluwalia MS, Rich JN, Reynolds BA, Lathia JD (2015) Differential connexin function enhances self-renewal in glioblastoma. Cell Rep 11(7):1031–1042
66. Mancuso M, Pasquali E, Leonardi S et al (2011) Role of connexin43 and ATP in long-range bystander radiation damage and oncogenesis in vivo. Oncogene 30:4601–4608
67. Doll R, Langman MJ, Shawdon HH (1968) Treatment of gastric ulcer with carbenoxolone: antagonistic effect of spironolactone. Gut 9:42–45
68. Nahab F, Wittevrongel L, Ippolito D et al (2011) An open-label, single-dose, crossover study of the pharmacokinetics and metabolism of two oral formulations of 1-octanol in patients with essential tremor. Neurotherapeutics 8:753–762. doi:10.1007/s13311-011-0045-1

69. Jensen K, Patel A, Klubo-Gwiezdzinska J et al (2011) Inhibition of gap junction transfer sensitizes thyroid cancer cells to anoikis. Endocr Relat Cancer 18:613–626
70. Moosavi M, Moasses Ghafary S, Asvadi-Kermani I et al (2011) Carbenoxolone induces apoptosis and inhibits survivin and survivin-ΔEx3 genes expression in human leukemia K562 cells. DARU J Pharm Sci 19:455–461
71. Yulyana Y, Endaya BB, Ng WH et al (2013) Carbenoxolone enhances TRAIL-induced apoptosis through the upregulation of death receptor 5 and inhibition of Gap junction intercellular communication in human glioma. Stem Cells Dev 22:1870–1882. doi:10.1089/scd.2012.0529
72. Taniguchi Ishikawa E, Gonzalez-Nieto D, Ghiaur G et al (2012) Connexin-43 prevents hematopoietic stem cell senescence through transfer of reactive oxygen species to bone marrow stromal cells. Proc Natl Acad Sci 109:9071–9076

# Chapter 3
# MicroRNA and Gap Junctions in Glioblastoma Cells: Implications for Cellular Therapy

**Jessian L. Munoz, Vivian Rodriguez-Cruz, and Pranela Rameshwar**

**Abstract** Glioblastoma (GBM), the most common and aggressive adult human primary brain tumor, has poor prognosis with the median survival of 11 months. Several major cellular pathways linked to cell growth and survival, have been characterized in GBM. The frontline therapy for GBM, Temozolomide (TMZ), induces its own resistance, partly through miRNA-9. The increase in miRNA-9 decreases the translation of Patch receptor, resulting in enhanced hedgehog signaling and induced expression of the multiple drug resistance gene. CD133+ GBM cells, which have been thought to be the early cell type, expressed high levels of miRNA-9, indicating that this miR can be a target. The TMZ-resistant GBM cells can transfer miRNA-9 among cells through gap junctions and through the release of small vesicles, in particular exosomes. Here we discussed how these findings can be applied to develop new treatments for a disease with poor prognosis.

**Keywords** Glioblastoma • miRNA • Connexin • Temozolomide • Exosomes

## 3.1 Glioblastoma Multiforme (GBM)

GBM is the most common and aggressive adult human primary brain tumor, with approximately 10,000 new cases per year in the United States. Median survival is 11 months, with a 5-year survival rate of 5 % despite current surgery, with concomitant radiotherapy and Temozolomide chemotherapy or bevacizumab (Avastin®) monoclonal antibody therapy [1]. However, Avastin was only approved for cases that

J.L. Munoz • P. Rameshwar (✉)
Department of Medicine, Hematology-Oncology, New Jersey Medical School, Rutgers School of Biomedical Health Sciences, 185 South Orange Avenue, MSB E-585, Newark, NJ 07103, USA
e-mail: rameshwa@njms.rutgers.edu

V. Rodriguez-Cruz
Chemistry Department, University of Puerto Rico, Cayey, Puerto Rico



M. Kandouz (ed.), *Intercellular Communication in Cancer*,
DOI 10.1007/978-94-017-7380-5_3

continue to progress after the standard care. Malignant gliomas include oligodendrogliomas and oligoastrocytomas. Patients with both subtypes of GBM respond to temozolomide, but with little effect during recurrence [2].

Symptoms of GBM include seizures, nausea, hemiparesis and vomiting as well as, memory loss and neurological deficiencies due to temporal and frontal lobe involvement [3]. Histologically, GBM characteristically are presented with a small area of necrosis surrounded by anaplastic cells and hyperplastic vascularization, collectively termed "pseudopalisading necrosis". Although metastasis beyond the central nervous system is uncommon, about 50 % of GBMs occupy more than one lobe or are bilateral, a rare phenomenon known as a "butterfly" glioma in which the tumor infiltrates and crosses the corpus collosum [4].

A number of cellular pathways have been characterized in GBM, among them Phosphoinositol-3-phosphate Kinase (PI3K), Mitogen Activated Protein Kinase (MAPK), and Hedgehog (HH), all of which promote cellular growth and survival [5]. In addition, the Notch receptor and its pathway have also been shown to mediate proliferation, differentiation and apoptosis. Forced over-expression of Notch-1 in glioma cells leads to an increase in proliferation and formation of Nestin$^+$, neurosphere-forming stem cells [6]. Thus, this data suggests that in the case of gliomas, neurodevelopmental pathways such as Notch signaling, might be involved in avoiding apoptosis and promoting proliferation and the development of a "stem-cell" phenotype.

## 3.2 MicroRNA (miRNA) Profiles of Glioblastoma

MiRNAs are small, non-coding RNAs that constitute ~1–2 % of mammalian genes. Since their discovery in 2000, miRNAs have been implicated in a number of biological processes including neural patterning, neurodevelopment and oncogenesis. Despite their relatively small size, miRNAs are stable in the normal circulation, thus enabling their utilization as a biomarker in oncology [7]. MiRs interact with naïve complementary RNA molecules at the 3′ untranslated region (UTR) to suppress protein translation. Initially transcribed as longer precursor molecules (primary-miRNA), this molecule is cleaved by the nuclear RNase Drosha, creating pre-miRNA. Pre-miRNA is then shuttled out of the nucleus into the cytoplasm by Ran-GTP dependent Exportin 5. Cytoplasmic Pre-miRNA is further processed by Dicer, a class III RNase. This mature miRNA forms a multimeric protein complex, RISC, which together bind to and regulate the translational expression of cellular target mRNA.

In general, miRNAs are considered as regulators of genes, post-transcriptionally. This occurs through interactions with the 3′ UTR of transcripts (mRNA) to suppress translation [8]. Interestingly, miRNAs and their targets are generally conserved, suggesting their critical functions in development. Interestingly, a single miRNA has been reported to suppress target genes. In most instances, outcomes of a single miRNA are generally modest. The efficiency of miRNAs in regulating gene expression appears to require one or more miRNAs at clusters at multiple sites within the transcript. Although the role of miRNAs in development has been exhaustively studied, it is yet to be determined if the miRNAs are involved in 'fine-tuning' the developmental processes or if they mediate a central role in development.

The polycystronic mir-17-92 cluster located on chromosome 13q32-33, was the first example of miR involvement in human malignancy. This chromosomal segment is known to undergo amplification via c-Myc, a well-studied proto-oncogene [9].

MiRNA expression profiles have been shown to correlate with the development, maintenance, proliferation and migration of various cancers. Thus, miRNAs have been classified as both tumor suppressor and oncogenes. In addition, 50 % of miRNAs are located in sites of chromosomal instability associated with cancer. As an example (the human homolog of *C. elegans lin-4*), is located at the chromosomal 11q24 site [10]. 11q24 is commonly deleted in breast and lung cancers. On the other hand, *Sonoki et al* reported a patient case in which the patient developed B-cell ALL with an insertion of the mir-125 gene into the gamma heavy chain gene [11]. These sets of preliminary data suggest miRNA can be tumor suppressors, oncogenes or both depending perhaps on the stage and histology of the cancer.

MiRNAs can be potential targets to treat GBM. One type of miRNA can damage different mRNA. Since miRNA have been associated with tumor suppressive or oncogenic roles in GBM, identifying and understanding miRNA's on GBM can help on diagnosing and treatment of patients with GBM. These can also be used as biomarkers [12].

Studies have been performed to identify the miRNA expression on GBM by microarray or RT-PCR. The most predominant miRNA present in GBM are miRNA-21, 181, 221, 137, 124, 128 and 451 [13]. The miRNA-21, 221 and 451 are upregulated on GBM tumors compared with normal brain tissue. On the other hand the miRNA-128b, 181a, 181b and 181c are downregulated on GBM. Studies performed with tissue from patients treated with radiotherapy and Temozolomide shown that the miR-181b and 181c are downregulated miRNA, which can be used to shown the response of patients for this treatment [14]. The miRNA 21 and 181b are the only ones expressed on all types of tumors while others are only present on a specific grade of tumor. MiRNA can be used to differentiate grade III anaplastic astrocytoma (AA) from GBM. Approximately 29 miRNA can be used to identify AA especially miRNA-7, 124 and 137 [15].

There is a difference in expression of miRNA on non-stem cell GBM ($CD133^-$) and stem cell GBM ($CD133^+$). On the $CD133^-$ population the miRNA-451, 486, 16, 107 and 185 are upregulated [16]. These miRNA can be linked to the loss of stem cell pluripotency in cells, this is why if these miRs are upregulated in $CD133^+$ populations, these cells loss stem cell properties.

By identifying the molecular targets that miR-21, it was determined that miR-21 acts like an oncogene by blocking genes that enable the cell to undergo apoptosis [17]. MiR-21 does this by specifically targeting p53, transforming growth factor (TGF)-β and mitochondrial apoptosis tumor suppressor genes in GBM cells. The invasion and migration of the GBM cell is related to the miRNA-10b [18].

MiRNA-196 is associated to target HOX clusters and Annexin A1 (ANXA1), a mediator of apoptosis and an inhibitor of cell proliferation [19]. Another miRNA that targets the HOX cluster, specifically negative regulate HOXD10, is miRNA-23a. HOXD10 act as a tumor suppressor in gliomas. The miRNA-221/222 can be associated to cellular growth, inhibition $p27^{kip1}$ that regulates the cellular grow leads to enhanced cellular growth, and inhibition of apoptosis [20].

Several miRNAs can influence the growth of glioma cells. The upregulation of miRNA-181a, b, and c can cause growth inhibition, apoptosis and suppressed invasion. They can also sensitize GBM cells to radiotherapy [21]. MiRNA-7 suppresses the EGFR and Akt pathways, both are related to growth and survival of GBM cells [22].

## 3.3 Gap Junctions in the CNS and Neuro-Oncology

Gap Junctional Intercellular Communication (GJIC) allows for the transfer of small molecules across continuous cytoplasms of neighboring cells. GJIC is established by the union of two hemichannels, formed by members of the connexin protein family, on adjacent membranes. Astrocytes highly express connexin-43 (Cx43) and astrocytomas further upregulate Cx43 [23]. The belief that connexin mediated cellular communication restricts cell growth and proliferation was initially accepted and supported by many experiments and animal models, but has recently been challenged with more recent data confirming the importance of gap junctions in therapy resistance, progression, and migration (metastasis) [24].

Cx43 is currently considered both a tumor suppressor gene and an oncogene or a conditional tumor suppressor [25]. During late stage epithelial-to-mesenchymal transitions (EMT) and tumor infiltration, GJIC is enhanced, allowing for cell migration. The importance of GJIC in early stage acquired resistance to chemotherapy has not been explored.

Glioma cells resistant to therapy are highly invasive and cannot be removed surgically. GBM is the only human tumor, which can initiate in one hemisphere and migrate to the opposite hemisphere resulting in a phenomenon known as "butterfly tumor" invasion. During GBM migration, the cells interact with and migrate along endogenous astrocytes ensheathing the microvasculature [26]. This interaction is bilateral and accompanied by astrocytic migration to the tumor core forming the peritumoral astroglial hurdle (Fig. 3.1).

In a series of publications, Baklaushev, Yusubalieva and colleagues developed a Connexin-43 targeting monoclonal antibody with specificity for the second extracellular loop [27, 28]. In vitro data showed that MAbE2Cx43 could inhibit Gap Junction formation and thus halt dye transfer among cells. When delivered intravenously, the MAbE2Cx43 antibody effectively targeted glioma xenografts and periglioma tissue. Intravenous delivery of MAbE2Cx43 to C6 rat glioma xenotransplanted mice resulted in increase animal survival, 20 % of animals were progression free for 6 months, compared to untreated mice which succumbed 4–5 weeks after

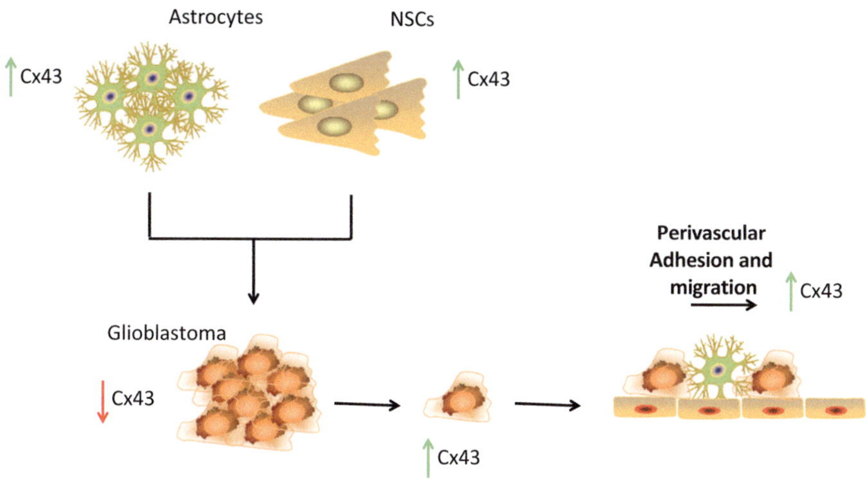


**Fig. 3.1** Shown are astrocytes at the perivascular region surrounding the tumor. Although NSCs express high level of Cx43 there is reduced Cx43 in the bulk cells. At the lining, Cx43 is increased in the astrocytes

implantation. Histological analysis confirmed reduced parenchymal and perivascular migration of glioma cells. MRI analysis showed reduced tumor growth and the formation of cysts or cavities.

## 3.4 GJIC Transport and Regulation by MicroRNA

The direct transport of miRNA through gap junctions was shown initially in breast cancer cells and later in GBM cells. In the studies involving GBM cells, no specific miRNA of interest was shown to be transferred but rather a proof-of-principle was established [29]. To do this, human U87 and rat L9 cells were engineered to express cel-mir-67, a *C. elegan* miRNA with no known homology to human or rat cells. Then cells were co-cultured with others containing a Luciferase vector tagged in the 3′ region by tandem cel-mir-67 binding sites. In situ hybridization of these cells confirmed functional transfer of this miRNA and this effect was inhibited by using carbenoxolone, a known gap junction-uncoupling compound.

Research has recently focused on exploiting the ability of miRNA to be transported via gap junctions as a novel method of targeted delivery. Lee et al explored the usage of four different sources of Mesenchymal Stem Cells (MSCs): adipose tissue, bone marrow, placenta and umbilical cord blood [30]. As previously mentioned, mir-124 and -145 are expressed at low levels in GBM cells, in addition these miRNA were found to be expressed at low levels in MSCs [31, 32]. Thus, MSCs were transfected with florescent-tagged mir-124 and 145 mimicking molecules and these were confirmed to be transferred from MSCs to GBM cells. The source of the

MSCs did not affect the efficiency of miRNA transfer. This model was confirmed using a GBM intracranial xenografts were established under stereotaxic guidance, MSCs were able to transfer the labeled microRNA into these GBM cells.

In addition to transporting miRNA, Gap Junctions composed of connexins can also be regulated by specific miRNAs. MiR-125b, a microRNA enriched within the central nervous system has also been shown to be enriched in GBM [33]. Jin et al. showed miR-125b expression promoted GBM proliferation and inhibited apoptosis. GJA1 was identified as a functional target of miR-125b, overexpression of Cx43 reversed the oncogenic effects of miR-125b.

MiRNA-221/222 has also been shown to be upregulated in GBM cells, playing an important role in the regulation of a number of cellular processes [34]. Bioinformatic analyses suggested miR-221/222 may regulate Cx43 expression. Hao and colleagues confirmed roles for miR-221/222 regulation of CX43, which was mediated through the GJA1 3′ UTR [25]. In cells lacking miR-221/222 cell proliferation and migration was enhanced and increased apoptosis was observed. Thus, in GBM, miR-221/222 is considered an oncogenic miRNA and this effect is at least partially mediated through regulation of Cx43.

## 3.5 Contact-Independent Transport of MiRNA

MiRNA-targeting therapeutics in GBM is an area of extensive research [35]. Yet, functional delivery of these targeted treatments without the use of viral vectors has yet to be successful. Cell-base delivery mechanisms have become an attractive mechanism for miRNA/anti-miRNA treatments, among cell types, stem cells show innate tropism to GBM cells [36]. Among those currently being studied are Neural Stem Cells (NSCs) and MSCs. Although NSCs have shown to be effective at drug delivery, difficulty harvesting and culturing as well as the immunogenic nature of these cells has limited their usage in humans [37]. In contrast, adult bone marrow-derived MSCs have shown to be not only as efficient in cancer targeting but also non-immunogenic, thus allowing for allogeneic transplantation and 'off-the-shelf' usage [38]. MSCs have been shown to communicate with cancer cells through Gap Junctions as well as through secreted exosomes, the contributive importance of each has not been determined.

We showed functional transport of anti-miR-9 from human MSCs to GBM cell lines and also, in low passaged GBM cells [39]. The transfer of anti-miR-9 was determined to be caused by contract independent mechanisms, through exosomes. The shuttled anti-miR-9 reversed ATP-dependent drug efflux and enhanced TMZ-induced cell death. Lee and colleagues used MSC exosomes to deliver synthetic mir-124 and mir-145 mimicking molecules to GBM cells [30]. Target genes SCP1 and Sox2 were identified and experiments were validated using an intracranial xenograft model. Taken together, these two studies suggest that MSC-derived

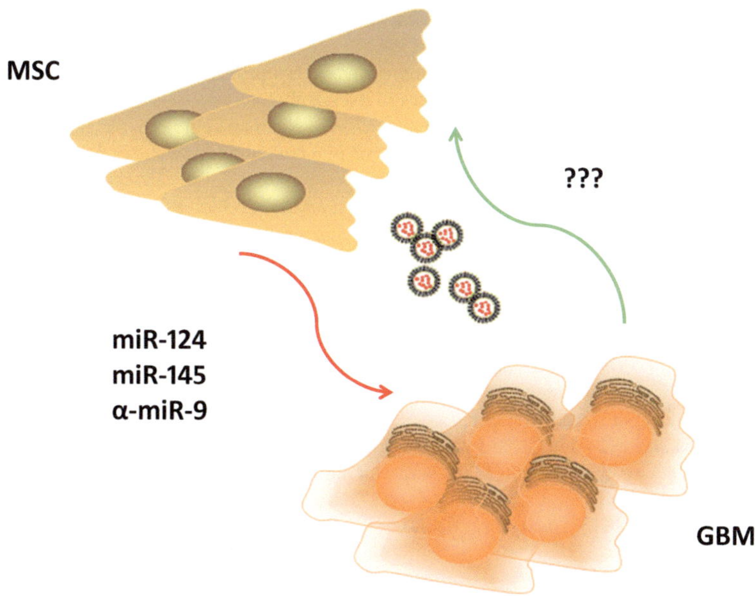


**Fig. 3.2** MSCs are shown to be a cellular method to deliver miRNAs to brain tumor. The miRNA-loaded MSCs deliver the miRNA through exosomes

exosomes may transport both functional microRNA and anti-microRNA to GBM, cells in vitro and in vivo, resulting in cellular changes and enhanced cell death (Fig. 3.2).

## References

1. Lassen U et al (2013) Phase II study of bevacizumab and temsirolimus combination therapy for recurrent glioblastoma multiforme. Anticancer Res 33:1657–1660
2. Norden AD et al (2013) Phase 2 study of dose-intense temozolomide in recurrent glioblastoma. Neuro Oncol 15:930–935
3. Prakash O et al (2012) Gliomas and seizures. Med Hypotheses 79:622–626
4. Dziurzynski K et al (2012) Butterfly glioblastomas: a retrospective review and qualitative assessment of outcomes. J Neurooncol 109:555–563
5. Drier Y, Sheffer M, Domany E (2013) Pathway-based personalized analysis of cancer. Proc Natl Acad Sci 110:6388–6393
6. Gursel DB, Berry N, Boockvar JA (2012) The contribution of notch signaling to glioblastoma via activation of cancer stem cell self-renewal: the role of the endothelial network. Neurosurgery 70:N19–N21
7. Luningschror P et al (1833) MicroRNAs in pluripotency, reprogramming and cell fate induction. Biochim Biophys Acta 2013:1894–1903
8. Ji F, Lv X, Jiao J (2013) The role of microRNAs in neural stem cells and neurogenesis. J Genet Genomics 40:61–66

9. Olive V, Jiang I, He L (2010) Mir-17-92, a cluster of miRNAs in the midst of the cancer network. Int J Biochem Cell Biol 42:1348–1354
10. Mar-Aguilar F et al (2013) Serum circulating microRNA profiling for identification of potential breast cancer biomarkers. Dis Markers 34:163–169
11. Sonoki T et al (2005) Insertion of microRNA-125b-1, a human homologue of lin-4, into a rearranged immunoglobulin heavy chain gene locus in a patient with precursor B-cell acute lymphoblastic leukemia. Leukemia 19:2009–2010
12. Hayes J, Peruzzi PP, Lawler S (2014) MicroRNAs in cancer: biomarkers, functions and therapy. Trends Mol Med: 20:460–469
13. Asadi-Moghaddam K, Chiocca EA, Lawler SE (2010) Potential role of miRNAs and their inhibitors in glioma treatment. Expert Rev Anticancer Ther 10:1753–1762
14. Slaby O et al (2010) MicroRNA-181 family predicts response to concomitant chemoradiotherapy with temozolomide in glioblastoma patients. Neoplasma 57:264–269
15. Conti A et al (2009) miR-21 and 221 upregulation and miR-181b downregulation in human grade II-IV astrocytic tumors. J Neurooncol 93:325–332
16. Shi L et al (2012) Functional differences of miR-125b on the invasion of primary glioblastoma CD133-negative cells and CD133-positive cells. Neuromolecular Med 14:303–316
17. Calore F, Lovat F, Garofalo M (2013) Non-coding RNAs and cancer. Int J Mol Sci 14:17085–17110
18. Guessous F et al (2013) Oncogenic effects of miR-10b in glioblastoma stem cells. J Neurooncol 112:153–163
19. Guan Y et al (2010) MiRNA-196 is upregulated in glioblastoma but not in anaplastic astrocytoma and has prognostic significance. Clin Cancer Res 16:4289–4297
20. Zhang J et al (2010) miR-221/222 promote malignant progression of glioma through activation of the Akt pathway. Int J Oncol 36:913–920
21. Gwak HS et al (2012) Silencing of microRNA-21 confers radio-sensitivity through inhibition of the PI3K/AKT pathway and enhancing autophagy in malignant glioma cell lines. PLoS One 7, e47449
22. Kefas B et al (2008) microRNA-7 inhibits the epidermal growth factor receptor and the Akt pathway and is down-regulated in glioblastoma. Cancer Res 68:3566–3572
23. Wang X et al (2013) The role of connexin 43 and hemichannels correlated with the astrocytic death following ischemia/reperfusion insult. Cell Mol Neurobiol 33:401–410
24. Matsuuchi L, Naus CC (2013) Gap junction proteins on the move: connexins, the cytoskeleton and migration. Biochim Biophys Acta 1828:94–108
25. Hao J et al (2012) miR-221/222 is the regulator of Cx43 expression in human glioblastoma cells. Oncol Rep 27:1504–1510
26. Aubert M et al (2008) A model for glioma cell migration on collagen and astrocytes. J R Soc Interface 5:75–83
27. Baklaushev VP et al (2009) Isolation of extracellular recombinant fragment of rat connexin-43. Bull Exp Biol Med 148:389–393
28. Baklaushev VP et al (2011) Visualization of connexin 43-positive cells of glioma and the periglioma zone by means of intravenously injected monoclonal antibodies. Drug Deliv 18:331–337
29. Katakowski M et al (2010) Functional microRNA is transferred between glioma cells. Cancer Res 70:8259–8263
30. Lee HK et al (2013) Mesenchymal stem cells deliver synthetic microRNA mimics to glioma cells and glioma stem cells and inhibit their cell migration and self-renewal. Oncotarget 4:346–361
31. Xia H et al (2012) Loss of brain-enriched miR-124 microRNA enhances stem-like traits and invasiveness of glioma cells. J Biol Chem 287:9962–9971
32. Rani SB et al (2013) MiR-145 functions as a tumor-suppressive RNA by targeting Sox9 and adducin 3 in human glioma cells. Neuro Oncol 15:1302–1316

33. Wu N et al (2013) MiR-125b acts as an oncogene in glioblastoma cells and inhibits cell apoptosis through p53 and p38MAPK-independent pathways. Br J Cancer 109:2853–2863
34. Quintavalle C et al (2012) miR-221/222 overexpression in human glioblastoma increases invasiveness by targeting the protein phosphate PTPmu. Oncogene 31:858–868
35. Moller HG et al (2013) A systematic review of microRNA in glioblastoma multiforme: micro-modulators in the mesenchymal mode of migration and invasion. Mol Neurobiol 47:131–144
36. Doucette T et al (2011) Mesenchymal stem cells display tumor-specific tropism in an RCAS/Ntv-a glioma model. Neoplasia 13:716–725
37. Reekmans K et al (2012) Current challenges for the advancement of neural stem cell biology and transplantation research. Stem Cell Rev 8:262–278
38. Rameshwar P (2009) Casting doubt on the safety of "off-the-shelf" mesenchymal stem cells for cell therapy. Mol Ther 17:216–218
39. Munoz JL et al (2013) Delivery of functional anti-miR-9 by mesenchymal stem cell-derived exosomes to glioblastoma multiforme cells conferred chemosensitivity. Mol Ther Nucleic Acids 2, e126

# Chapter 4
# Dynamic Regulation of Adherens Junctions: Implication in Cell Differentiation and Tumor Development

**Marjolaine A. Pelissier-Rota, Nicolas T. Chartier, and Muriel R. Jacquier-Sarlin**

**Abstract** Adherens junctions (AJ) are fundamental cellular subcompartments required for the integrity of animal tissues and organs. The core complex of AJ is formed of transmembrane cell-cell adhesion molecules, cadherins, and adaptor, the catenins that link the complex to cytoskeletal and regulatory networks in the cell. Clustering of these molecules at junctions induces signaling pathways that regulate cellular responses, with crucial effects on the physiology and pathology of the epithelium, from cell differentiation to tumor development. Loss of E-cadherin expression is a hallmark of epithelial-to-mesenchymal transition (EMT), the process allowing the dissemination and the invasion of metastatic tumors. In this chapter, we focus on intestinal epithelium to highlight how cell-cell junction proteins, especially AJ proteins, interact together in normal and cancerous contexts to regulate major cellular processes such as cell differentiation and migration. For instance, imbalance in the signaling properties of catenins results in cell cycle deregulation during cancer and metastasis. We will also describe signaling and trafficking machineries that drive dynamic regulation of AJ and how these mechanisms are co-opted in cancer. In particular, the crosstalk between AJ, polarity proteins, extracellular matrix and microenvironment will be discussed. Recent evidences indicate that cell-cell junctions modulate chromatin function and mutually, we will also discuss epigenetic mechanisms that regulate AJ. Finally, we will examine more recent hypotheses about AJ implications in cancer stem cells and give an overview of cancer therapies that target cell-cell junction proteins.

**Keywords** Adherens junctions • Colon cancer therapy • Cell signalling • Tissue homeostasis • Transcriptional and post-transcriptional regulation

---

M.A. Pelissier-Rota • M.R. Jacquier-Sarlin (✉)
Centre de Recherche Inserm U836, Institute of Neurosciences, Grenoble, France

University Joseph Fourier of Grenoble, Grenoble, France
e-mail: muriel.jacquier-sarlin@ujf-grenoble.fr

N.T. Chartier
BIOTEC, TU Dresden, 47/49 Tatzberg, Dresden 01307, Germany



M. Kandouz (ed.), *Intercellular Communication in Cancer*,
DOI 10.1007/978-94-017-7380-5_4

## Abbreviations

| | |
|---|---|
| Ach | Acetylcholine |
| ADAM | A disintegrin and metalloprotease |
| AJ | Adherens junction |
| AOM | Azoxymethane |
| APC | Adenomatous polyposis coli |
| ARM | Armadillo |
| CAC | Colitis associated cancer |
| CAM | Cell-cell adhesion |
| CBD | C-terminal catenin |
| cdc42cdc42 | Cell division control protein 42 homolog |
| CD | Crohn'disease |
| Cdx2 | Caudal-type homeobox transcription factor |
| CH | Cadherin |
| CRC | Colorectal cancer |
| CREB | Cyclic AMP response element binding protein |
| CRF2 | Corticotropin releasing factor receptor 2 |
| CtBP | C-terminal binding protein |
| DNMT | DNA methyl transferase |
| DSS | Dextran sodium sulfate |
| EC | Extracellular repeats domain |
| EGF | Epidermal growth factor |
| ERK | Extracellular signalkinase |
| EpiSC | Epiblast-like stem cell |
| ECM | Extracellular matrix |
| EMT | Epithelial to mesenchymal transition |
| Eph | Erythropoietin producing hepathocytes |
| ESC | Embryonic stem cell |
| FAK | Focal adhesion kinase |
| FGF | Fibroblast growth factor |
| Gastrointestinal | GI |
| GEF | GTPase-specific nucleotide exchange factor |
| GF | Growth factor |
| GLIS2 | Gli- similar protein 2 |
| GSK3 | Glycogen synthase kinase |
| GTP | Guanosine tri phosphate |
| GDI | Guanine nucleotide dissociation inhibitor |
| HDACS | Histone deacetylase |
| HGF | Hepatocyte growth factor |
| IBD | Inflammatory bowel disease |
| IEC | Intestinal epithelial cell |
| ILK | Integrin linked kinase |
| INFα | Interferon alpha |
| iPSC | Pluripotent stem cell |

| | |
|---|---|
| JMD | Juxtamembrane domain |
| Klk6 | Kallikrein 6,LI-cadherin |
| mAChR | Muscarinic acetylcholine receptor |
| MeCP2 | Methyl-CpG-binding protein |
| MET | Mesenchymal to epithelial transition |
| miRNA | MicroRNA |
| MLC | Myosin light chain |
| MMP | Matrix metalloproteinase |
| mTOR | Mammalian target of rapamycin |
| MD | Maternal deprivation |
| NLS | Nuclear localization sequence |
| PAR | PARtitionning defective mutant |
| PI3K | Phosphoinositide-3-kinase |
| PJS | Peutz Jehgers syndrome |
| p120ctnp120ctn | p120 catenin |
| PRC2 | Polycomb repressive complex 2 |
| PSEN | Presenilin |
| PSP | Serine/threonine phosphatase |
| PTEN | Phosphatase and tensin homolog |
| PTP | Protein tyrosine phosphatase |
| Rac | Ras-related C3 botulinum toxin substrate |
| ROK | Rho-associated protein kinase |
| sE-cadh | Soluble E-cadherin |
| SAHA | Suberoylanilide hydoxamic acids |
| Src | Homologue of the Rous sarcoma virus oncogene protein |
| sN-cad | Soluble N-cadherin |
| SPARC | Secreted protein acidic and rich in cysteine |
| TCF/LEF | T cell factor/Lymphoid enhancer factor |
| TJ | Tight junction |
| TGF-β | Transforming growth factor |
| UC | Ulcerative colitis |
| Ucn | Urocortin |
| ZEB | Zinc finger E-box-binding homeoboxe |

## 4.1 Introduction

Intestinal epithelial homeostasis and function depend on various cell populations distributed on the different layers that constitute the intestinal mucosa. These cells participate in the control of intestinal secretory, motor and immune functions as well as epithelial permeability. The main function of the small intestine is dedicated to nutrient absorption during digestion, whereas the colon mainly functions to resorb water. This functional difference is reflected by the structure and composition of the epithelium. The small intestinal epithelium is organized in a glandular structure with crypts and finger like structures called villi that maximize the exchange surface. Villi are composed of various differentiated cells: absorptive enterocytes, mucus-secreting goblet cells, hormone-producing enteroendocrine cells. The colon is only constituted of crypts. This epithelium also constitutes a barrier against numerous antigenic stimulations, which separates immune cells from the potentially harmful luminal content such as solutes, microorganisms, and antigens. The importance of an intact mucosal layer is highlighted by pathological disorders such as inflammatory bowel disease (IBD), in which disruption of the epithelial barrier leads to severe inflammation of the submucosal tissue compartments [1]. Enterocytes are specialized cells characterized by structural features including polarized morphology and a complex set of cell–cell and cell-extracellular matrix junctions. They represent 80 % of the cells forming the epithelium monolayers. They are derived from stem cells located in the crypts of Lieberkühn and differentiate as they migrate within 3–5 days toward the tip of the villus region where they are shed [2]. During this directional cell migration, the epithelial integrity of the intestine has to be ensured. This requires the adhesive contacts to rearrange and adapt to different conditions in a highly dynamic way. These intercellular junctions, which consist of a plethora of membrane-associated and transmembrane proteins organized in spatially restricted complexes, provide mechanical strength by anchoring cell-cell contacts to the cytoskeleton. Adhesion complexes are also preferential sites for signal transduction controlling multiple aspects of the cellular behavior, mainly proliferation and differentiation [3–5]. Three types of junctional complexes mainly execute cell-cell adhesion: tight junction types (TJs), adherens junctions (AJs) and desmosomes. Although each type of junction has its particular mechanism of formation, regulation and function, extensive communication between them mutually influences their dynamics and signaling properties. On the basis of pioneering experiments on E-cadherin blocking, it was thought for a long time that E-cadherin-dependent adhesion is a prerequisite for the assembly of other specialized cell-cell junctions [6].

In this chapter, we will focus on the structural and regulatory components of AJ, and the interactions between them, collectively referred as the cadherin adhesome, or cadhesome [7] (Figs. 4.1 and 4.2). As the assembly and regulation of cohesive intercellular junctions is central to morphogenesis and tissue homeostasis, we will describe the complex crosstalk between AJ and polarity proteins, extracellular matrix proteins or molecules of the surrounding environment and their roles in the regulation of theses junctions. Finally, we will describe how AJ are altered in pathogenesis and discuss the strategies developed for the treatment of intestinal disorders associated with deregulation of adhesive systems.

## 4.2 Structure and Organization of Adherens Junctions

Cadherins constitute a superfamily of Ca+-dependent cell-cell adhesion molecules, which comprises over 100 members; [11]. Typically, cadherin are able to bind a cadherin of the same type on an adjacent cell (homophilic interactions), but they can also interact with other types of cadherin or transmembrane receptors (heterophilic

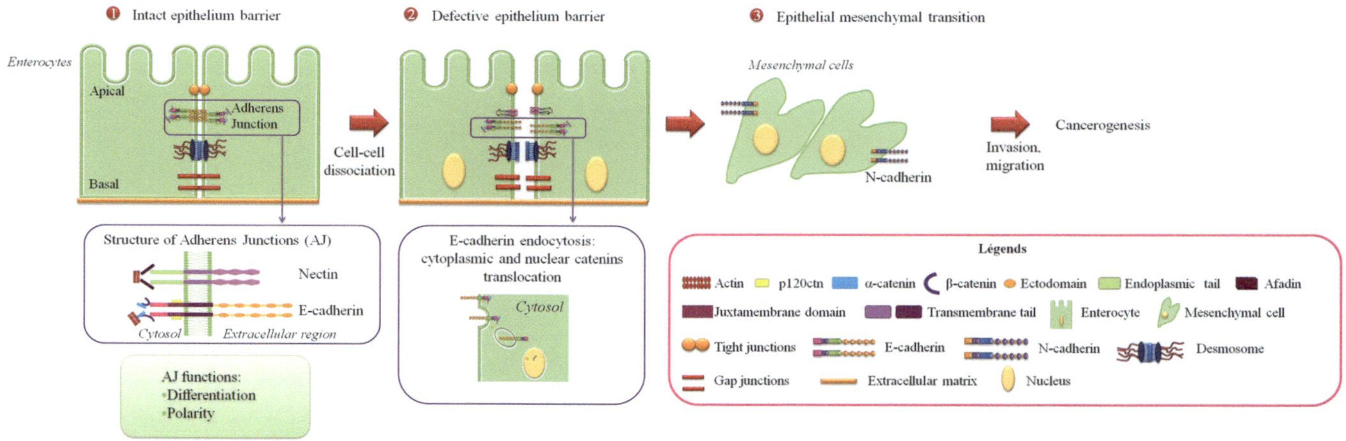


**Fig. 4.1** Adherens junctions from epithelial differentiation to cancer development. (1) In normal differentiated adult epithelium, epithelial cells constitute a single layer of polarized cells maintained by various adhesive complexes such as tight junctions, gap junctions and adherens junctions. Most of these junctions are connected to the actin cytoskeleton. The adhesive core of AJ is composed by two transmembrane proteins: cadherins and nectins. Cadherin possess: (i) an extracellular domain, which is subdivided in 5 ectodomains (EC1–EC5), each subset contains calcium-binding site; (ii) A short transmembrane domain; (iii) A cytoplasmic tail divided in two domains: the juxtamembrane domain (JMD) which is adjacent to the cadherin transmembrane segment and the C-terminal catenin binding domain (CBD). p120ctn binding to JMD stabilizes classical cadherin. β-catenin binds to Cadherin CBD and α-catenin, which connects the junction to the actin cytoskeleton. Nectin also contains an extracellular region, a transmembrane domain and a cytoplasmic tail, which binds. Afadin linked to actin filaments. (2) p120 catenin (p120ctn) is a gatekeeper of the stabilization of cadherin-catenin complexes to the cell surface. Under stress conditions, the epithelium is altered. This step is characterized by cell-cell dissociation resulting from an uncoupling between p120ctn and E-cadherin, which stimulates E-cadherin endocytosis and p120ctn and β-catenin translocation into the cytoplasm or the nucleus. (3) Cell-cell dissociation is the initial step of the epithelial to mesenchymal transition (EMT). Mesenchymal cells have a fusiform morphology and interact with each other only through focal points. During this process there is a shift in cadherin expression: epithelial cells express high levels of E-cadherin, whereas mesenchymal cells rather express N-cadherin. Thus, EMT entails invasion and cellular proliferation, two processes implicated in cancerogenesis

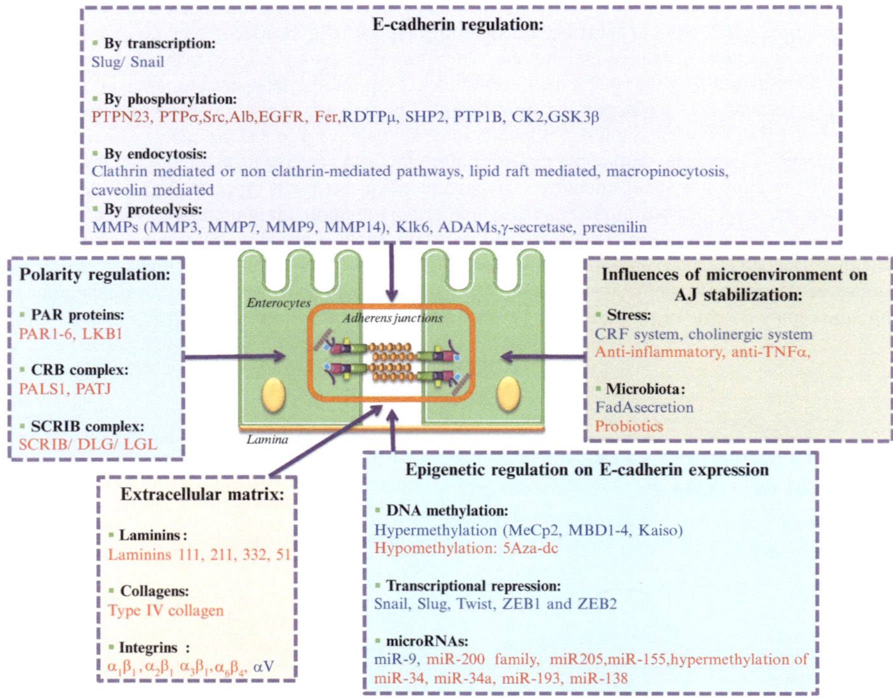


**Fig. 4.2** Modulators of epithelial AJ dynamic. Overview of the different processes and actors involved in the regulation of AJ dynamic in epithelial cells. "Positive" regulators that favor the expression of AJ proteins and their stabilization are presented in red. On the opposite, "negative" regulators of AJ are shown in blue

interactions). The binding between cadherin extracellular domains is relatively weak, but cell-cell adhesion may be strengthened by their lateral clustering, a process mediated by protein linkages between the cadherin cytoplasmic domain and the actin cytoskeleton [12]. Cadherins are classified within six subset: classical/type I cadherins, atypical/type II cadherins, desmocollins, desmogleins, protocadherins, Flamingo cadherins, all of which interact with various cadherin–related proteins [13]. Several species of classical cadherins are named after the tissue in which they are predominantly expressed, such as E-cadherins (epithelial cadherins), N-cadherins (neural-cadherins), P-cadherins (placenta cadherins), VE-cadherins (vacular endithelial cadherins) [14]. Thus, cadherins are very diverse in both structure and function, and their mutual interactions influence biological responses in complex and versatile ways. Their roles in suppression versus initiation or progression of multiple tumor types have been discussed in an excellent review from Frans Van Roy [15].

Mammalian AJ contains classical cadherins (approximately 20 members) that share a common domain organization (for review [16]): they possess an extracellular N-terminal region or ectodomain, a single pass trans-membrane segment and a highly conserved short cytoplasmic tail that contains catenin-binding sites [17]. The cytoplasmic tail interacts with β-catenin, which in turn binds α-catenin, allowing an adjustable linkage between the cadherin-catenin complex and the actin cytoskeleton [18]. Adhesion is mediated by the extracellular fragment, whereas the cytoplasmic sequence regulates signaling, turn-over and modulates actin assembly [19, 20].

### *4.2.1 Interaction between Neighboring Cadherins: The Pivotal Role of Ectodomain*

The extracellular domain of E-cadherin principally forms homophilic bonds with E-cadherins on adjacent cells [21, 22]. The ectodomain is divided into five repetitive domains of approximately 110 amino acids length, called extracellular repeats (EC) domains [23]. EC are ranging from the membrane-most distal EC1 domain to the membrane-most proximal EC5 domain. Each subdomain contains a highly conserved calcium-binding site that controls the adhesive function of AJ by modifying the ectodomain conformation [24]. In 2010, Hong and colleagues demonstrated that cadherin mutants lacking the cytoplasmic domain are able to drive AJ formation, indicating that the structural determinants of this process reside in the ectodomain [25].

Cell-cell adhesion is a dynamic process where cadherins adjust their binding kinetics in order to resist to mechanical perturbations [26]. This process is mediated by cooperative formation of distinct cadherin–cadherin interfaces in *cis*: between cadherins in a parallel orientation on the same cell and in *trans*: between opposing cadherins, on two adjacent cells. Classical cadherin ectodomains form *trans* adhesive homodimers through their membrane-distal EC1 domains, bridging the intermembrane space between opposing cells. Atomic resolution structures of classical cadherin ectodomain and electron micrographs of isolated natural and recombinant E-cadherin extracellular domains indicate that the extracellular domain of classical cadherins generally has a stable curved structure [17, 24, 27].

All classical cadherins share a common binding mechanism in which the dimerization is swapping of the amino-terminal β-strand between interacting EC1 domains. Mutational, structural and cellular experiments revealed that this strand swapping is stabilized by three interactions. One interaction is the insertion of conserved tryptophan at position 2 (Trp 2) residue into a hydrophobic pocket of the interacting partner; [17, 28, 29]. The second interaction is the formation of salt bridges between N-terminal amino acid ($NH3^{+}$) and the carboxyl group of Glu 89 residue [30]. An additional interaction is hydrogen bonds between Asp1 and Asn27 and between Trp2 and Asp90. In the calcium bound form, ectodomains are rigid and molecules from the opposing plasma membrane have favorable orientation for the *trans* swap [24, 31]. When *trans*-dimer are formed, the flexibility is markedly reduced, which lowers the entropic penalty associated with *cis*-dimer formation [32]. Micropipette manipulations show that the binding between two opposite cells occurs in two steps: an initial rapid stage inducing *trans*-dimer, a second slower stage ascribed to *cis*-adhesion [33]. While the *trans*-dimerization is well established, the existence and functional roles of *cis*-dimers are not clearly resolved. Cadherin adhesion is enhanced by lateral assembly as a dimer (*cis*-dimer) on surface cell that interacts with *cis*-dimer on opposite cell to form *trans*-adhesion. X-ray observations of several classical cadherins suggested that interactions between the apex of EC1 and the base of EC2 and EC3 of neighboring cadherins induce dimerization in a *cis* orientation [17, 26]. However, nuclear magnetic resonance measurements and single-molecule fluorescence resonance energy transfer experiments could not detect *cis*-dimerization formation [28, 34]. Recent theoretical studies predict that *cis*-dimer formation requires prior *trans*-assembly [32].

### *4.2.2 Interactions with the Cytoplasmic Domain of E-cadherin*

The cytoplasmic domain is not essential for cadherin dimerization during cell-cell contact but it plays a pivotal role in sensing and transducing adhesion related signal [25, 26]. The cytoplasmic fragment can be divided into two subdomains: the juxtamembrane domain (JMD), which is adjacent to the cadherin transmembrane segment and the C-terminal binding domain (CBD), each of which has a sequence motif of about 30–35 residues, respectively called CH2 and CH3 (CH, cadherin homology region) [35]. The JDM supports many aspects essential for cadherins fuctions including clustering, adhesive strengthening and suppression of cell motility; [36, 37]. The deletion of the CBD, including CH3, abrogated $Ca^{2+}$-dependent cell aggregation [38]. This region was narrowed down to a core region of 30 amino acids that included up to eight well-conserved Ser residues that were highly phosphorylated and shown to be important for catenin binding and cell-cell adhesion. Interestingly, several experiments indicate that the JMD has both positive and negative functions in cadherin clustering and cell–cell adhesion. The clustering potential of the JMD appears to be activated by the cadherin *trans*-dimerization and is not observed in the absence of *trans*-adhesion [39]. Thus, cadherin clustering in vivo might be secondary to the formation of adhesive cadherin dimers, and requires the JMD but not the CBD.

These domains interact with various armadillo family proteins: the JMD binds to p120 catenin (p120ctn) and the CBD interacts with β-catenin and plakoglobin [40, 41].

#### 4.2.2.1 β-catenin

About 20 years ago, β-catenin was independently discovered twice, on the basis of its different functions: structural and signaling [42]. Togeher with its structural as a binging partner of E-cadherin, β-catenin plays a pivotal role as a transcriptional co-activator of Wnt signaling [43] (for review see [44, 45]). When Wnt signal is absent, the majority of β-catenin is localized in the cortical region as a component of cadherin adhesion complex. β-catenin interacts with α-catenin via the distal part of its N-terminal domain and binds to the CBD of cadherin through an armadillo domain. β-catenin can interact with the cytoplasmic domain of all classical cadherin, however the most studied are the consequences of β-catenin binding to E-cadherin. This association occurs early after E-cadherin synthesis, in the endoplasmic reticulum, where it supports the transport of E-cadherin to the membrane [46]. More detailed roles of cortical, cytoplasmic and nuclear β-catenin roles are discussed in Sect. 4.3.1.

#### 4.2.2.2 p120ctn

p120ctn is a member of a subfamily of Armadillo (ARM) repeat containing proteins [47]. ARM domain repeats are found in a large variety of proteins and facilitate protein–protein interactions [48]. p120ctn is encoded by the gene *CTNND1*, that consists of 21 exons encoding 968 amino acids [49]. Cloning and analysis of *CTNND1* revealed four alternatively spliced exons and four transcriptional start sites at amino acid residues 1, 55, 102 and 324, that can initiate expression of isoforms 1, 2, 3 and 4. Together they can potentially produce 64 different isoforms of p120ctn, named after the transcriptional start site used (1–4), and the alternatively spliced exons they express (A–D) [49, 50]. This variety of isoforms might reflect tissue-or context specific expression. Indeed, isoforms 1 and 3 are the most widely and abundantly found, with isoform 1 being preferentially expressed in motile, mesenchymal-type cells, whereas isoform 3 characterizes epithelial cells. The isoform 1 contains an N-terminal coiled-coil domain, which is absent from others. A regulatory region harboring most of the tyrosine and serine/threonine phosphorylation sites in p120ctn follows the N-terminal domain [51]. The C-terminal tail of p120ctn also contains several serine/threonine phosphorylation sites as well as the alternatively spliced exon A and exon B. The presence of the alternatively spliced exon C results in the disruption of a nuclear localization signal (NLS), the inhibition of binding to RhoA and decreased endothelial growth factor (EGF)-induced phosphorylation [49, 52–55]. p120ctn regulates many functions both upstream and downstream of E-cadherin, which are developed in details in more specific Sect. 4.3.2.

Heterotrimeric G proteins of the G12 subfamily mediate cellular signals leading to events such as cytoskeletal rearrangements, cell proliferation, and oncogenic transformation [56–60]. The G12 subfamily of G proteins is comprised of the ubiquitously expressed members Gα12 and Gα13, which share 67 % amino acid identity [61]. Several recent studies have revealed direct effector proteins through which G12 subfamily members may transmit signals leading to various cellular responses. In particular, Gα12/13 have been found to interact with the cytoplasmic tail of E- cadherin in a region different from β-catenin and p120ctn, and trigger the release of β-catenin from cadherin complexes, which localizes to the cytoplasm and nucleus where it participates in transcriptional activation, thereby disrupting cadherin-mediated cell-cell adhesion [59, 62, 63]. Alternatively, the activation of RhoA by G12 subfamily proteins can counteract the p120ctn-mediated inhibition of RhoA. The inhibitory activity of Gα12 on p120ctn-mediated RhoA regulation does not require Gα12 binding to p120ctn while both members of the G12 subfamily can interact with p120ctn in the presence or absence of E-cadherin [64]. In fact, Gα12 binds to a region overlapping with the phosphorylation domain in the N-terminus of p120ctn and Gα12 down-regulates the overall tyrosine phosphorylation on p120ctn, including phosphorylation of residue Y228 induced by Src and EGF [65].

#### 4.2.2.3 α-catenin

α-catenin is a crucial component of AJ: cells devoid of α-catenin do not form intercellular contacts while its expression provokes AJ assembly and favors apico-basal cell polarity establishment [66]. Unlike other catenins, α-catenin binds indirectly to cadherins through the amino acids 118–149 of β-catenin [40, 67]. α-catenin binds simultaneously to the β-catenin-cadherin complex and actin either directly or indirectly through actin-binding-proteins. β-catenin and actin binding to α-catenin is mutually exclusive and the binding site overlaps with α-catenin's homodimerization region, thus α-catenin binds to AJ only in its monomeric form to give a β – catenin/α-catenin heterodimer [68]. In fact, α-catenin monomer has higher affinity for β – catenin than the homodimer and, on the opposite, actin pelleting experiments showed that α-catenin homodimer has a higher affinity for actin filament than the monomer. One of the direct consequence is that the binding of α-catenin to AJ complex negatively regulates its activity on actin polymerization [69].

## 4.3 Catenin-Dependent Signaling to and from AJ: Roles in Intestinal Epithelium

Cadherins engaged in AJ can modulate many cellular processes: intercellular adhesion of course, but also epithelial differentiation, cytoskeleton modulation and even cell proliferation through contact inhibition. The majority of these regulations depend on catenins and their associated proteins. The next section summarizes the roles of β-catenin and p120ctn on cell adhesion and proliferation as well as on cytoskeleton regulation, and their effect on intestinal epithelium function.

### *4.3.1 β-catenin and the Wnt Pathway*

β-catenin regulates cell-to-cell adhesion and gene transcription through its interactions with cadherins and Wnt signaling. In the AJ, the main function of β-catenin is to provide a link with the cortical actomyosin cytoskeleton through α-catenin [70], Vinculin [71] and Eplin/Lima 2 [72]. E-cadherin mutants enable to bind β-catenin (mutants with deleted CBD) fail to accumulate at the AJ and are recognized by an ubiquitin ligase, Hakai that marks E-cadherin for lysosomal degradation. Overall, β-catenin enhances the cadherin adhesion via two major mechanisms: (1) by preventing the cadherin degradation and regulating the cadherin trafficking from the endoplasmic reticulum to the plasma membrane; (2) by increasing the cadherin stability at plasma membrane through α-catenin recruitment [73, 74]. However, the affinity of β-catenin interaction with E-cadherin is constantly controlled: this association is increased upon serine phosphorylation events occurring in the CBD of cadherin, while on the opposite, growth factors (GF) dependent phosphorylation of β-catenin on Tyrosine 654 is responsible for the release of β-catenin from the cadherin CBD [40, 75]. When non-associated to E-cadherin, β-catenin works as a rheostat and a transducer of cellular environment signaling: under normal conditions and in the absence of Wnt signaling, cytoplasmic β-catenin is recruited and phosphorylated by the formed by the adenomatous polyposis coli protein with Axin and Glycogen synthase kinase 3 (APC/Axin/GSK3) complex. This phosphorylation favors β-catenin ubiquitination and its degradation by the proteasome [76]. Conversely, in condition where the binding of Wnt ligand to its transmembrane receptor Frizzled activates the Wnt pathway, the destruction complex gets inactivated and β-catenin accumulates in the cytoplasm and the nucleus [43]. In the nucleus, β-catenin interacts with the T cell factor/Lymphoid enhancer factor (TCF/LEF) transcription factors and potentiates the transcriptional activity of the complex [77], leading to the expression of proliferation genes such as Myc and cyclin D1 [78, 79]. Therefore, in this case β-catenin plays a positive role towards cell proliferation. Thus a balance between membranes, cytosolic and nuclear β-catenin exists and exerts a primordial role on epithelial cells function.

In the physiologic intestinal epithelium, cells in the crypt compartment receive a strong Wntsignal and E-cadherin expression is reduced, possibly because of Snail-dependent repression [80] (see Sect. 4.5). This leads to an accumulation of nuclear β-catenin in the crypt compartment that favors the proliferation of precursor cells [81]. Later, during epithelial differentiation, when epithelial progenitors migrate from the crypt toward the villi, strengthening of intercellular contacts implicating E-cadherin induces β-catenin sequestration at the cortex, leading to a reduction of the cytoplasmic β-catenin pool that interacts with TCF/LEF thus resulting in a decrease in cell proliferation [4, 82].

## *4.3.2 p120ctn: The Multitasking Partner*

When localized at the AJ, p120ctn is a gatekeeper of AJ formation, stability, turnover, and also provide a connection to the microtubule cytoskeleton. When not associated to E-cadherin, p120ctn is present in the cytoplasm but to date, no mechanism of environmentally controlled cytoplasmic degradation equivalent to APC/Axin/GSK3 complex for β-catenin was described, although calpain-mediated degradation of p120ctn has been showed [83, 84]. However, in another parallel with β-catenin, p120ctn functions in the cytoplasm and in the nucleus multiple pathways.

### 4.3.2.1 p120ctn at the Junction

At the plasma membrane, the p120ctn central ARM domain associates with the E-cadherin JMD core region (JDM $^{core}$), resulting in stabilization and maturation of the AJ [54, 85]. To better understand how the interaction between p120ctn and JMD modulates the stability of cadherin, Ishiyama and al. elaborated variants of p120ctn to co-crystallize with E-cadherin JMD domain. They showed that the binding of p120ctn to E-cadherin involves electrostatic interactions, hydrogen bonds and salt bridges between the acidic region of the JDM $^{core}$ and the basic p120ctn region. A second region includes mainly hydrophobic interactions between C-terminal of JMD $^{core}$ and the N-terminal part of p120ctn [41]. Several studies showed that a reduction of p120ctn expression (by genetic approaches or an uncoupling between cadherin and p120ctn) provokes the exposition of E-cadherin dileucine motif (LL 743–744), responsible for cadherin internalization and therefore reduces the amount of cadherin available for cell-cell adhesion. Miranda et al., demonstrated that this motif is required for the proper sorting of newly synthetized E-cadherin to the basolateral membrane of polarized epithelial cells [86]. However, E-cadherin mutants unable to bind p120ctn are properly targeted to the plasma membrane, suggesting that p120ctn acts as a “cap” on the cadherin cytoplasmic tail to mask an endocytic signal [87]. Based on structural evidences, it appears that p120ctn binding physically impedes the interaction of cadherin molecules with E3 ligases. Indeed, the association of the Cbl-like protein Hakai to E-cadherin subsequently ubiquitinilates E-cadherin either at the membrane or rapidly after endocytosis. This process determines whether E-cadherin is recycled to the membrane or targeted for lysosomal degradation [88–90]. Alternatively, p120ctn can recruit presenilin 1 (PSEN1) to the AJ [91–93]. This protease-mediated cleavage of E-cadherin results in the disassembly of AJs and the release of a C-terminal fragments of E-cadherin in the cytoplasm, that is capable of transporting p120ctn to the nucleus, thus affecting Wnt signaling, as discussed in more detail below (Sect. 4.2.2.3). An alternative mechanism of AJ degradation is facilitated by NUMB, which can interact with the C-terminal region of p120ctn resulting in stimulation of clathrin-dependent endocytosis of AJ complex components [94–96].

#### 4.3.2.2 p120ctn and Small GTPases

Rho family GTPases such as Rho, Rac, and cdc42 (Cell division control protein 42 homolog) play important roles in the regulation of cytoskeletal organization and actin dynamics. Their localized activity is required to induce proper cell polarity, migration and AJ formation and maturation [97]. Interestingly, p120ctn regulates the activity of these small monomeric GTPases: it directly interacts with and inhibits RhoA, acting as a Rho-GDI [98] while it activates Rac and cdc42 through Vav2, a guanine nucleotide exchange factor (GEF) for these GTPases [99]. The initiation of cadherin-mediated adhesion was found to inhibit RhoA activity but to stimulate Rac1 activity [100], and this activation is dependent on binding of p120ctn to the cadherin [101]. However when p120ctn binds to cadherin, it loses the ability to inhibit RhoA activity [102]. Interestingly, Izumi and colleagues found that clathrin-dependent endocytosis of E-cadherin was inhibited by activation of Rac and cdc42 through IQGAP1 and actin filaments, suggesting that p120ctn may regulate indirectly cadherin endocytosis via small GTPases [103]. Thus, p120ctn can modulate the dynamic and morphology of the actin cytoskeleton, and for instance, its overexpression in fibroblasts leads to a characteristic "dendritic phenotype", presenting long membranous extensions and filopodias. The direct consequence of this regulation is that p120ctn can affect cell motility depending on the level and the type of cadherin expressed at the plasma membrane [104, 105].

#### 4.3.2.3 p120ctn and KAISO in the Nucleus

In addition to its membrane-related and cytosolic functions, the presence of an NLS and NES sequences, and ARM-domain render p120ctn capable of shuttling in and out of the nucleus [55, 106, 107]. Alternatively, p120ctn might travel to the nuclear compartment through its association and subsequent translocation with Gli-similar protein Glis2 [108, 109] or a C- terminal fragment of E-cadherin [110]. Thus p120ctn can modulate gene expression through its direct interaction with the transcriptional repressor Kaiso (ZTBT33) and the transcription factors Glis2 [108, 111]. Kaiso is a member of the POZ-ZF or BTB family of transcription factors that can bind to methylated DNA and/or a specific Kaiso binding sequence (TCCTGCNA) [112–115]. The fixation of Kaiso to its target genes represses their transcription, and p120ctn association to Kaiso releases this repression [113]. Interestingly, some of these genes, such as the ones encoding Wnt11, Cyclin D1 and matrilysin are also targets activated by the β-catenin/TCF/LEF complex, which have been linked to non canonical and canonical Wnt-dependent developmental processes [113, 116–118]. Therefore by interacting with Kaiso, p120ctn potentiates the proliferative effect of β-catenin, highlighting the convergence of the two catenins-dependent pathways in the regulation of cell proliferation [118–120]. Nuclear Kaiso might also indirectly inhibit canonical Wnt signaling by sequestering β–catenin and TCF/LEF transcription factors that act downstream from canonical Wnt signaling [121, 122]. Overall, it appears that p120ctn is an important regulator of Kaiso-dependent repression of transcription. Furthermore, in a similar manner to β-catenin, p120ctn can control post-translational processes in the cytosol and regulate transcriptional processes in the nucleus.

#### 4.3.2.4 p120ctn and Microtubules

Lastly, p120ctn influences cell cycle by interacting with microtubules and centrosomes during mitosis [123–125]. These interactions have several consequences: (i) they can support p120ctn trafficking along microtubules through kinesin; (ii) they favor microtubules bundling and stability [126] and (iii) they can regulate proper centrosome duplication, possibly through the stabilization of Cyclin E levels [123, 127]. p120ctn also interacts with CLASP proteins, microtubules plus-end binding proteins, that allow the association and the stabilization of microtubules at the AJ, which seems to play a role in the determination of epidermal cells progenitors [128].

## 4.4 Dynamic Regulation of Adherens Junctions

### *4.4.1 Regulation by Polarity Proteins*

The intestinal epithelium supports the two main functions of the intestine: allow optimal nutriment absorption and act as a barrier toward external medium. Both functions important point of reference as well as a docking platform for signalization enabling enterocyte polarization. Indeed determination of AJ location on the baso-apical axis is an important step in epithelial differentiation and allows the strong maintenance of epithelial polarity, with adhesion complexes being at the border between basal and apical domains. The next section will discuss the interconnections between AJ components and polarity establishment, with a focus on intestinal epithelium.

#### 4.4.1.1 PAR Proteins

Intrinsic polarity regulators were well described in the last 20 years thanks to genetic studies in model organisms such as the fruit fly *D. melanogaster* and the nematode *C. elegans*. Originally identified in *C. elegans*, the PAR protein family (for PARtitionning defective mutants) comprises seven proteins (PAR-1 to 6 and αPKC), each of them being involved in intrinsic antero-posterior polarity establishment and/or maintenance in early *C. elegans* embryo [129]. With the exception of PAR-2, these proteins have homologues in mammals and have an effect on epithelial baso-apical polarity, acting to maintain the basolateral and apical domains depending on their localization. In mammalian epithelial cells, PAR proteins function by a mutual exclusion mechanism, where PAR-1 (or MARK), for Microtubule Affinity Regulating Kinase, is localized at the basolateral domain and maintains the integrity of this domain by excluding the apical protein PAR-3 by phosphorylation [130, 131]. Phosphorylated PAR-3 is then trapped in the cytoplasm by PAR-5 (14-3-3 in mammals) so that it is excluded from the basolateral domain [132]. In turn, the kinase αPKC, associated with the apical proteins PAR-3 and PAR-6 phosphorylates PAR-1 to exclude it from the apical cortex [133, 134]. PAR-6 also interacts with the small GTPase cdc42 in the apical domain [135]. Such mutual inhibition mechanism allows robust polarity establishment and maintenance. During the early stages of intercellular adhesion formation, the interaction of nectin and afadin with PAR-3 allows the subsequent assembly of AJ and TJ [136]. Then PAR-3 exclusion is required to separate TJ components from AJ components, this is allowed by PAR-3 binding to a Rac1 GEF, TIAM-1 and the subsequent activation of Rac1 GTPase [137]. Then apical PAR-3 is kept excluded from tight junction by αPKC phosphorylation [138]. In vitro, the inhibition of PAR-1, αPKC or PAR-3 affects the capacity of MDCK epithelial cells to generate a proper apical lumen [139, 140].

#### 4.4.1.2 CRUMB and SCRIB Complexes

While PAR proteins are general regulators of cell polarity and act both to polarize epithelial cells, migrating cells and the cells undergoing asymmetric divisions, some protein complexes regulate polarity more specifically in epithelia. These proteins consist of an apical complex formed by the transmembrane protein CRUMB and its cytosolic interactors PALS1 and PATJ, and a basolateral complex composed of the cytoplasmic proteins SCRIB, Disc Large (DLG) and LGL (for review, see [141]). Originally identified in *D. melanogaster*, they modulate apical and basolateral domains identity. For instance, overexpression of CRUMB increases the size of the apical domain [142]. The complexes are closely associated to TJ but also to AJ, and it was shown that depleting PALS1 results in the blocking of E-cadherin transport to the membrane [143]. The basal SCRIB/DLG/LGL complex functions by inhibiting the expansion of the apical domain and it was proposed to do so by competing with PAR3 for the binding to PAR-6/αPKC [144]. Interestingly, SCRIB localization at the intercellular junctions depends on E-cadherin [145] and its depletion disrupts E-cadherin-dependent contacts [146]. Consistently, SCRIB and its associated partners are found mislocalized or down-regulated in many cancers including colon cancers [147, 148].

Mammalian epithelial cells, including intestinal epithelial cells (IEC), use partially redundant and therefore strong signaling pathways to ensure proper polarity establishment. In the first steps of intercellular contact formation, tight and adherens junctions depend on cell polarity signaling to be localized at the apical/lateral side of epithelial cells. Then, in a second step of maintenance of polarity, these junctions serve as docking plateform for the different complexes and favor a positive feedback loop to amplify and strengthen epithelial polarity.

### *4.4.2 Regulation by the Extracellular Matrix*

There are two major types of cell adhesion: the cell-matrix and the intercellular adhesion. Epithelial cells adhere to a specialized extracellular matrix (ECM) called basal lamina through protein complexes called hemidesmosomes and focal adhesions, localized at the basal pole of the differentiated epithelial cells. These complexes mainly involve the transmembrane adhesion receptors integrins, associated to diverse structural and regulatory intracellular adaptators. The capacity of these adhesion systems to modulate intracellular signaling in response to surrounding environmental conditions combined with their ability to modulate adhesion strength during epithelial differentiation support the notion that they are important for intestinal epithelium functions [149–152]. In the next section, we will present the different actors of cell to matrix adhesion in IEC and their connection with the AJ proteins.

Basal lamina is the frontier between epithelial cells and conjunctive tissue. It is composed of macromolecules being synthetized both by epithelial cells themselves and the conjunctive tissue. These macromolecules include collagens, laminins, fibronectines and proteoglycans associated in a multidimensional network [153, 154]. In intestinal epithelium basal lamina, the principal component is type IV collagen, that builds a strong foundation for other macromolecules and can interact with epithelial cells through $\alpha 1\beta 1$, $\alpha 2\beta 1$ and $\alpha 3\beta 1$ integrins [155, 156]. Type VII collagen was also observed in adult human small intestine [157, 158]. Intestinal ECM is also enriched in laminins, glycoproteins made of three chains, $\alpha$, $\beta$ and $\gamma$ interacting through disulfide bounds [159, 160]. From the 19 laminin isoforms identified to date [161], 4 are expressed in the intestinal epithelium: laminins 111, 211, 332 and 511 [162, 163]. Depending on the isoforms, laminins also interact with integrin receptors at the basal pole of the epithelial cells. Other MEC components are present in a more broad range of basal lamina, such as fibronectin, secreted by fibroblasts from the conjunctive environment, interacting on a concensus RGD motif of integrins. Long proteoglycans as perlecan are part of the ECM too, allowing the establishment of a hydrophilic gel prone to molecules diffusions and protein-

protein interactions. Interestingly, immunohistochemistry experiments have shown that the repartition of basal lamina components along the crypt-villus axis is not uniform. Indeed, as it is known that laminin isoforms change during embryonic development [154, 164], there is also a spatial variation of these isoforms in adult crypts. Laminins 111 and 211are essentially located in the crypts while laminins 332 and 511are enriched in villi [165]. Therefore, laminin 332 forms a gradient from the base of villi to the apex, correlating with enterocytic differentiation. Such specific repartition suggests that the ECM, depending on its composition, can activate different signalization pathways that regulate proliferation and differentiation of intestinal epithelial cells [163].

ECM signals to epithelial cells through integrins, whose functions are to anchor cells to the MEC, to stabilize the cytoskeleton and allow a bi-directional signaling. Each integrin heterodimer can recognize many targets proteins in the ECM with different affinities; it is thus possible to classify sub-family depending on their β subunits, the α subunit providing specificity towards ligand. β1 integrins are the most represented, both in term of heterodimer number and in their tissue distribution. They recognize most of ECM proteins: fibronectin, vitronectin as well as laminin and collagens. In the intestinal epithelium, the laminin-specific α7β1 integrin but also α2β1 and α3β1 are expressed. Surprisingly, however, conditional deletion of β1 integrin in mouse does not modify adherence capacities to ECM, but induces an hyperproliferation due to a misregulation of the Hedgehog pathway [166]. These results suggest that contrarily to other tissues such as epidermis or mammary gland epithelium, β1 integrins have a preferential role in signal transduction than in adhesion in the intestinal epithelium. Thus, other heterodimers regulate at least partially IEC adhesion to the ECM. This is the case of α6β4, found in complexes called hemidesmosomes, allowing the anchorage to laminin 332 [167]. Consistently, deletion of β4 subunit cytoplasmic domain induces adhesion and proliferative defects in mice [168]. Work from JF Beaulieu also revealed the presence in the intestine of a variant called β4Actd- devoid of its carboxy-terminal extremity [165]. This variant can interact with the β6 subunit but the association to laminin 332 is lost. It is expressed more predominantly in the crypts while the untruncated form is expressed in the villi, when cells are fully differentiated, concomitantly with the expression profile of laminins. Such a precise regulation of cell to ECM interaction during the proliferation and differentiation process of IEC suggest a close relationship between integrin signaling and subsequent differentiation programs implicating intercellular junctions proteins as E-cadherin or the catenins.

Once engaged to their ECM substrate, integrins transduce an intracellular signal through kinases such as the focal adhesion kinase (FAK) or the integrin linked kinase (ILK) (for review, see [169]), some of which have a direct impact on intercellular adhesions. For instance, integrin β1-dependent activation of FAK in colon cancer cells leads to Src activation and E-cadherin delocalization [170], a process that seems to depend on the control of peripheral phospho-myosin by Src [171, 172]. The exact signaling pathways are still not totally clear and seem to be dependent on the cell type and tissue investigated. For instance, it was proposed that TGF-β signaling through the phosphoinositide-3 kinase (PI3K) and the phosphatase and tensin homolog (PTEN) can down-regulate E-cadherin-dependent intercellular contacts in pancreatic carcinoma cells [173] while another study indicates that FAK activation upon TGF-β treatment induces E-cadherin expression in colon carcinoma cells [174]. However in the latter publication, the localization of overexpressed E-cadherin is not discussed and it could be trapped in cytoplasmic vesicles. More recently, the α3β1 integrin has also been shown to regulate the β-catenin pool through interactions with E-cadherin [175]. Immune fluorescent studies demonstrated that α3β1 integrin co-localizes with E-cadherin. Biochemical studies in alveolar epithelial cells also have demonstrated that α3β1 integrin physically associates with E-cadherin and the TGF-β receptor. This tri-molecular complex of α3β1 integrin, E-cadherin and the TGF-β receptor results in the tyrosine phosphorylation of β-catenin at position Y654, which subsequently associates with Smad2 (a TGF-β signaling molecule) and then translocates to the nucleus where it can regulate transcription.

This apparently negative effect of integrin-based adhesion on AJ is however balanced by a stabilizing effect of α6β4 activation. Indeed, both in keratinocytes and in human colon carcinoma cells, laminin 332 signals through PI3K and small GTPases (cdc42 and Rac1b respectively) to stabilize AJ [176, 177]. For instance, it was shown that αV integrin, E-cadherin and IGF-1R are able to interact in human colon carcinoma cell lines [178]. In this context, the binding of IGF1 to IGF1-R disrupts the complex and favors the relocation of αV integrin to focal contacts, thus favoring migration.

As IEC differentiate and migrate toward the apex of villi, they secrete their own ECM proteins and model an ECM gradient for which they adapt their integrin receptors [179]. The crosstalk between integrin and cadherin signaling described above therefore allow differentiating cells to sense their position on the crypt-villus axis and adapt their intercellular contacts and thus their degree of differentiation consequently.

### *4.4.3 Roles of Nectins*

Nectins, which were originally identified as virus receptors, are another family of cell-cell adhesion molecule (CAM) [180, 181]. They are $Ca^{2+}$-independent immunoglobulin-like cell adhesion molecules and the nectin family comprises four members, nectin-1 to nectin-4. Nectins are able to interact in *trans* both homophilically and heterophilically, with heterophilic trans-interactions that are much stronger than their homophilic *trans*-interactions. Compared with the stronger *trans*-interaction between E-cadherin molecules, which can facilitate long-term cell–cell adhesion, the weak *trans*-interaction between nectins is more adapted to the formation of transient cell–cell adhesions with a fast turnover. Studies performed in cultured cell lines reveal that nectins initiate the formation of AJ before that cadherins start to form cell–cell adhesions [182]. Once the initial cell-cell contacts are formed between two neighbouring cells by nectins, cadherins are recruited at these contact sites, resulting in the formation of strong cell-cell adhesion. The nectin and cadherin systems are then physically associated to establish AJ. Many potential physical links between the nectin–cadherin adhesion systems have been identified; the most direct one being those between afadin and α-catenin or p120ctn [183]. In fact disruption of afadin by knockout or knockdown techniques inhibits the formation of E-cadherin-based AJ [136, 184].

Either cadherin or nectin engagement can control the activity of the Rho family GTPases, which is a crucial regulator of actin dynamics. Activated forms of these GTPases affect cadherin activity and junction stability suggesting a close reciprocal relationship [185]. One model suggests that initial engagement of cells occurs via nectins-mediated activation of Rac and cdc42, which then stimulate the formation of lamellipodial protrusions and cadherin binding [186]. In addition, Rho GTPases may regulate actomyosin contractions, an important process for junctional rearrangements during morphogenetic movements [187]. Local regulation of small GTPase activity is most likely mediated by the recruitment of Rho family GTPase-specific GEFs and GTPase-activating proteins to AJ, several of which interact with core components. AJ may regulate Rho small GTPases via their upstream regulators, the Rap small GTPases. Indeed, RapGEFs directly interact with E-cadherin and engagement of E-cadherin activates Rap. Also, Afadin is an effector of Rap activity, suggesting the possibility of a positive loop reinforcing intercellular adhesion (reviewed in Ref. [188]).

### *4.4.4 The Dynamic Actin Cytoskeleton*

While much research is focused on identifying the molecular links that bridge the core cadherin-catenin complex to the actin cytoskeleton, less attention is directed toward understanding how actin is initially assembled and organized at AJ. However understanding junctional actin assembly is of importance as it is constantly turning over to maintain AJ steady state [69]. Indeed, new actin assembly must accompany the formation of new cadherin contacts that occurs during morphogenetic cell movements and dynamic actin networks can play a key role when cadherin is co-opted during disease, such as during inflammation and cancer. Excellent reviews from the Yap Lab's highlight the diverse and dynamic cytoskeletons that contribute to cadherin function and describe how cadherins cooperate with multiple dynamic cytoskeletal networks at AJ [189, 190]. Actin dynamics at intercellular adhesive contact sites are very different to those of the cadherin complex, suggesting the absence of a stable interaction. Regardless, both genetic and cell biological data strongly indicate that regulation of actin polymerization does take place at or in close vicinity of the AJ. This is at least partially dependent on α-catenin (reviewed in Refs. [191, 192]). In nascent adhesions, α-cateninis presumed to be primarily bound to cadherin, but as the density of cadherin rises over a certain threshold, a subset of α-catenin can dimerize and bind to F-actin in a way that inhibits Arp2/3 from binding [68, 193], thus promoting the shift from a protrusive activity to contractility.

Cadherin ligation and clustering is a sequential process that needs the small GTPases Rac1 and cdc42, which promote lamellipodial and filopodial protrusions [100, 194]. These protrusions help to increase the contact area with the neighbouring cell and thus positively feedback on cadherin ligation and clustering. Following the recruitment of RacGAPs and RhoGEFs [195, 196], there appears to be a switch from Rac1 and cdc42 to RhoA activity [197]. Such a Rho-Rac antagonism observed during AJ assembly could be mediated by the activation of the p190RhoGAP. Indeed, Rac1 activity causes translocation of p190RhoGAP to AJ, where it couples to cadherin complex via p120ctn and induces a local inhibition of Rho [198]. In IEC this interaction mainly occurs in lipid raft since we found that p190RhoGAP is recruited in these sites together with E-cadherin and p120ctn and that Rac1 is necessary for this step [199]. As the protrusive activity subsides, RhoA activates the formin mDia1 and Rho Kinase (RhoK), whose joint activities lead to the formation of contractile actomyosin bundles [200, 201]. Additionally, Myosin II that concentrates with cadherins in AJ of epithelial cells participates in actin contractility [187, 197, 200, 202, 203]. There is also emerging evidence that formins and Arp2/3 may cooperate at AJ [204, 205]. The following model has been proposed: Arp2/3-mediated actin nucleation reactions and actin branching activity tend to build highly dynamic, protrusive actin structures such as lamellipodia while formin-dependent linear actin filament polymerization, important for fillopodia formation, tends to be associated with contractile actin arrays [201, 206]. Then, the physical tension generated locally appears to recruit additional actin polymerization factors or actin binding proteins such as vinculin, spectrin, cortactin, zyxin, Ena Ena/Vasp, Wave, and Wasp; theses molecules, which can associate with AJ may be responsible for local regulatory activity (reviewed [7, 191, 192]). N-WASP and WAVE localize to AJ and may activate Arp2/3 to trigger actin assembly [207–210]. In addition, the Arp2/3 complex tightly associates with cadherin-enriched membranes, but it largely remains inactive unless α-actinin also binds to the membrane [211]. In the contractile phase, α-catenin that remains bound to cadherin may expose a cryptic binding site for vinculin that reinforce the link between cadherin and F-actin [71, 212] Other proteins, such as myosin VI and eplin must be specifically recruited to AJ under tension, facilitating a positive feedback between contractility and maturation of AJ [213, 214].

### *4.4.5 Transcriptional Regulation*

Zinc finger proteins of the Slug/Snail family are repressors of E-cadherin gene transcription [215, 216], and their expression is activated by signaling from the fibroblast growth factor FGF-R type I or TGF-β and ErbB1 and ErbB2 [216–218]. Slug/Snail expression results in the loss of cell-cell adhesion and increased cell migration. In parallel, Wnt signaling also regulates E-cadherin expression regardless of or synergistically with Slug/Snail [217]. Thus, the repression of cadherin expression by Slug/Snail or TCF/β-catenin complex reduces cell-cell adhesion, and works to amplify and/or sustain Wnt signaling. In the nucleus, TGF-β signaling and its downstream effectors SMAD, also coordinate β-catenin-mediated activation of specific gene transcription. Genes activated by Wnt signaling alone, or by both TGF-β and Wnt, are different [219, 220]. Together, these results show that a variety of GF receptor pathways intersect with the Wnt pathway by regulating the availability of signaling β-catenin, either by disrupting the cadherin-catenin complex or by repressing cadherin expression.

## 4.4.6 Regulation by Phosphorylation

Junction dynamics are regulated by rapid and transient shifts in protein phosphorylation or dephosphorylation under the control of specific kinases and phosphatases (for review [221]) (Figs. 4.2 and 4.3). This process is directly implicated in the regulation of cadherin adhesiveness and functions. So far, 12 serine/threonine kinases (S/T) and nine tyrosine kinases and one S/T and 12 tyrosine phosphatases have been implicated in regulating phospho-residues of AJ proteins. They are all cytoplasmic and the exact mechanism of their recruitment into AJ is largely unknown. If most likely every kinase and phosphatase can recognize at least one docking site within the AJ, it is not currently known which are permanent residents of AJ and which are transient components. The following section will sum up the key phosphatases and kinases implicated in AJ regulation.

### 4.4.6.1 Receptors with Kinase Activity

Receptor tyrosine kinases are commonly activated by an external ligand, such as a GF or cytokines that induce dimerization, *cis*-phosphorylation or autophosphorylation and subsequent activation of the catalytic domain [222, 223]. Cortical or cytoplasmic S/T kinases can be activated by the binding of Rho GTPases, for instance ROCK1 is activated by RhoA and PAK1 is activated by Rac1 and cdc42 [224, 225]. S/T kinases are also regulated by tyrosine phosphorylation and tyrosine kinases and phosphatases are themselves regulated by S/T phosphorylation, in a complex feedback network. Cadherin ligation and clustering may act as an activation signal for some kinases. Most notably, Src and Fer were shown to be recruited to the membrane upon cadherin binding and EGFR signaling was shown to be stimulated by AJ formation independently of EGF ligand [226–228]. Furthermore, cadherin clustering has been found to induce indirectly the activation of RhoGTPases, which in turn could activate S/T kinases [229].

The structural integrity and the functionality of the cadherin-catenin complex itself are regulated by many different kinases. The numerous kinases involved and the complexity of their networks makes it difficult to identify a global effect of phosphorylation on AJ. For instance, experiments using different phosphatase inhibitors leaded to the conclusion that phosphorylation is a negative regulator of

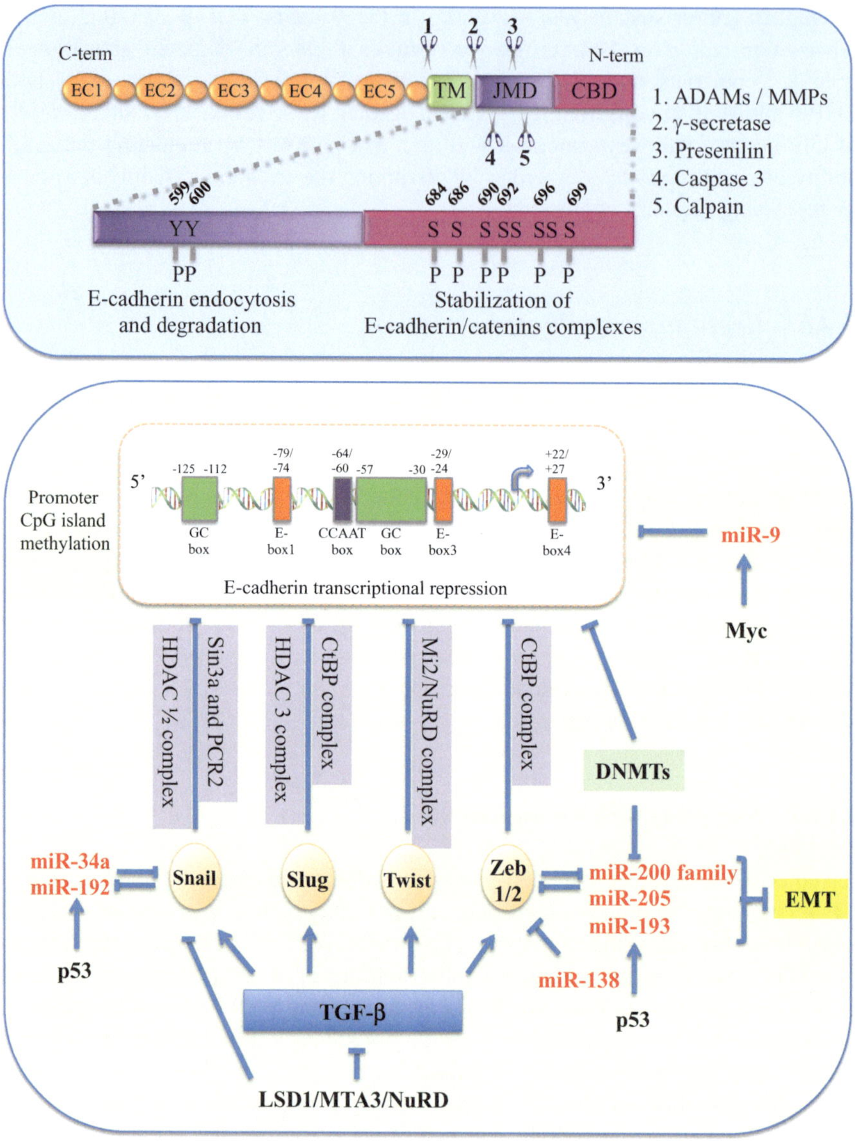


**Fig. 4.3** Transcriptional and post-transcriptional regulation of E-cadherin. Transcriptional regulation (lower panel): E-cadherin promoter with negative regulators such as Snail, Slug, Twist and Zeb1/2 transcription factors. Epigenetic silencing has been associated with CpG methylation in the promoter region and binding of specific transcriptional repressors to E-box sequences in the promoter. Upon TGF-β stimulation, the transcription factors could recruit various DNA/Histones modification complexes to repress E-cadherin and initiate EMT. LSD1/MAT3/NurD represses TGF-β and Snail. miR-200 family negatively regulates Zeb1 and 2, while it is negatively regulated by DNMTs mediated epigenetic silencing. miR-34a and miR-192 activated by p53 negatively regulate Snail, while miR-138 repress Zeb2.miR-9 activated by myc directly represses E-cadherin. CtBP, C-terminal binding protein; CCAAT/enhancer binding protein; DNMTs, DNA methyltrans-

AJ. Indeed these inhibitors induce a dramatic increase in phospho-tyrosine signals at AJ, followed by the disassembly of AJ [230, 231]. Consistent with the notion that excessive tyrosine phosphorylation causes AJ disassembly, cells expressing constitutively active Src kinase lost intercellular contacts, and inhibition of tyrosine kinase activity by the drug tyrphostin was able to restore AJ in the Src-transformed cells [231]. On the opposite, more recent experiments based on loss of function of specific kinases are in favor of a positive role of phosphorylation in AJ assembly. For example, Src and Fyn were found to be essential for the formation of AJ in mouse keratinocytes [232]. Src activity is important for the recruitment of PI3K to AJ and promotes the expansion of nascent cadherin-adhesive contacts [233]. Along the same lines, Abl1 tyrosine kinase activity was shown to be important for the maintenance of AJ in epithelial cells, and S/T phosphorylation of E-cadherin by protein kinase D1 (PRKD1) was found to be associated with increased cellular adhesion and decreased cellular motility in prostate cancer [234, 235].

Concerning E-cadherin, phosphorylation of various S/T sites was described to increase its binding activity towards β-catenin. This is the case of phosphorylation of residues $Ser^{684}$, $Ser^{686}$ and $Ser^{692}$ by CK2 and GSK3β kinases or residues $Ser^{840}$, $Ser^{851}$ and $Ser^{853}$ in the C-terminus of human E-cadherin by CSNK1E or PRKD1kinases. In contrast, phosphorylation of $Ser^{846}$ inhibits this interaction [236, 237]. Phosphorylation by Src of three tyrosines in position 753–755 on human E-cadherin creates a docking site for the E3-ligase Hakai, which leads to internalization of E-cadherin facilitating disassembly of the AJ [88, 238]. Catenins are also regulated by phosphorylation. For instance, phosphorylation of β-catenin at $Tyr^{489}$ or $Tyr^{654}$ disrupts binding to cadherin, and at $Tyr^{142}$ disrupts binding to $\alpha$-catenin [239]. Src phosphorylates β-catenin at $Tyr^{654}$, Abl at $Tyr^{489}$, EGFFR at $Tyr^{654}$ and Fer at $Tyr^{142}$ [240–242]. The subcellular distribution of non junctional β-catenin also depends on GSK3 and Casein Kinase I phosphorylations in the "destruction complex" [19]. Finally, p120ctn is regulated by phosphorylation too. Eight tyrosine residues in the N-terminus of p120ctn can be phosphorylated by Src [243]. Upon phosphorylation, these sites serve as docking sites for proteins carrying SH2 domains, such as the tyrosine phosphatase SHP-1 [52]. Under certain conditions tyrosine phosphorylation of p120ctn was shown to increase its affinity to cadherin, but this does not seem to be a general rule (reviewed in Ref. [51]).

**Fig. 4.3** (continued) ferases; HDAC, histone deacetylase, sin3A, histone deacetylase subunit; miR, microRNA. Post-transcriptional regulation involved proteolysis of E-cadherin and phosphorylation (upper panel). Cleavage sites of the proteases are shown by the scissors symbol. Ectodomain shedding occurs when matrix metalloproteases (MMPs) or disintegrin and metalloproteinases (ADAMs) molecules cleave a stable 80 kDa soluble fragment. Proteolysis of the cytoplasmic domain is mediated by γ-secretase, caspase 3, presenilin 1 and calpain. Regulation by phosphorylation involves tyrosine phosphorylation by receptor tyrosine kinase or the homologue of the Rous sarcoma virus oncogene protein (c-Src). These phosphorylations may lead to endocytosis and degradation of E-cadherin and are counteracted by protein tyrosine phosphatases (PTPμ). Serine/threonine phosphorylation by glycogen synthase kinase-3B, caseine kinase II or protein kinase D1 stabilizes the E-cadherin/catenin complex

In conclusion, the effect phosphorylation on AJ dynamics has to be examined residue-by-residue since some phosphorylation events lead to the switching "on" of a protein or interaction whereas other, even on the same protein, serve to switch it "off".

#### 4.4.6.2 Phosphatase Regulation

Phosphatases are categorized into two major classes, protein tyrosine phosphatases (PTPs) and protein serine/threonine phosphatases (PSPs). Receptor tyrosine phosphatases may be activated by homophilic association with their counterparts on neighboring cells [244], as well as by tyrosine phosphorylation [245]. Several S/T kinases implicated at cell-cell junctions are activated by PTPs (PTPμ, PTP1B, PTB-PEST and others) after removing inhibitory phosphates (for review [246, 247]). The effects of the principal phosphatases on AJ are summarized on Table 4.1.

**PP2A** PP2A appears to play an important role in the dynamic of AJ and TJ assembly. Studies in MDCK cells showed that sequestration of PP2A in AJ due to hypoglycosylation of E-cadherin reduces the binding of PP2A to TJ proteins as ZO-1 and claudin-1and thus favors their phosphorylation on Ser residues [248].

**SHP2** The Src homology-2 (SH2) domain-containing phosphatases (SHPs) comprise a sub-family of non-receptor protein-tyrosine phosphatases with two human members, SHP1 (encoded by *PTPN6*) and SHP2 (encoded by *PTPN11*). SHP2 is composed of two SH2 domains (N-SH2, C-SH2), a PTP domain, a C-terminal tail

**Table 4.1** Phosphorylation regulators of AJ. Junction dynamics are regulated by transient shifts in protein phosphorylation or dephosphorylation under the control of specific phosphatases. This table sums up the key phosphatases implicated in AJ regulation

| | Substrate | Effect |
|---|---|---|
| **PP2A** | E-cadherin | Sequesters PP2A at AJ to facilitate TJ assembly |
| | VE-cadherin | Prevents Ser phosphorylation and vascular junction disassembly |
| **SHP2** | β-catenin, VE-cadherin | Prevents β-cateninphosphorylation and disruption of VE-cadherin endothelial AJ complexes |
| | P190RhoGAP | Modulates RhoA activation |
| | Src, FAK | Indirectly up-regulates ERK1/2 activity, decreased FAKphosphorylation, decreased AJ and TJ proteins internalized |
| **PTP1B** | β-catenin,E-cadherin | Dephosphorylates and stabilizes E-cadherin:β-catenin complex at AJ |
| | VEGFR | VEGFR dephosphorylation stabilizes cell-cell contacts |
| **PTBxPEST** | RhoA- Rac1 | Promotes RhoA and limits Rac1 activity to enhance AJ assembly; Prevents aberrant epithelial cell motility |
| **RDTPμ** | Cadherins | Dephosphorylates cadherins; stabilizes intercellular contacts |
| **PTPσ** | β-catenin,E-cadherin | Dephosphorylation, prevents their removal from AJ |

with tyrosine phosphorylation sites (Y542 and Y580) and a proline-rich motif [249]. Its role in the regulation of junctional complexes and barrier function is most established in endothelial cells where it is a key regulator of VE-cadherin [250]. Numerous studies support a link between increased protein tyrosine phosphorylation and vascular permeability. The model is that SHP2 associates with β-catenin in VE-cadherin complexes. When SHP2 dissociates from the complex, β-catenin, γ-catenin and p120ctn phosphorylation increase impairing their association with α-catenin and thus leading to cell junction disassembly and intercellular gap [251–253]. SHP2 inhibition in pulmonary endothelial cells attenuates the phosphorylation-dependent activity of p190RhoGAP and induces a concomitant increase in RhoA activation [254]. Therefore, SHP2 appears to play a major role in supporting basal pulmonary endothelial AJ complex integrity and barrier function by coordinating the tyrosine phosphorylation levels of VE-cadherin, β-catenin, as well as p190RhoGAP and the activity of RhoA.

**PTP1B** Encoded by the *PTPN1* gene, the non-receptor protein tyrosine phosphatase PTP1B (protein tyrosine phosphatase 1B) contains an N-terminal PTP domain, followed by two tandem proline-rich motifs that may allow interaction with SH3 domain-containing proteins. PTP1B contains a C-terminal hydrophobic domain that targets the enzyme to its intracellular location on the cytoplasmic surface of the endoplasmic reticulum. PTP1B regulation of junctional proteins was initially identified in studies on N-cadherin in neural cells [255]. PTP1B was also shown to play an important role in the regulation of AJ in the colon cell line Caco-2 where it regulates cadherin-based adhesion by binding directly to the cadherin cytoplasmic tail and dephosphorylating β-catenin at $Tyr^{654}$ [256]. Phosphorylation of PTP1B at $Tyr^{152}$ by the tyrosine kinase Fer is required for the interaction with cadherin [240]. Indeed, Fer binding to 120ctn promotes the interaction between PTP1B and cadherin and favors the subsequent dephosphorylation of β-catenin at $Tyr^{654}$.

**PTB-PEST** Encoded by the *PTPN12* gene, PTP-PEST is a 88 kDa cytosolic PTP belonging to the PEST-domain-enriched tyrosine phosphatase (PEP) subfamily of non-receptor PTPases. PTP-PEST functions as a suppressor of colonic carcinoma epithelial cell motility by controlling Rho GTPase activity and the assembly of AJ. In the KM12C colonic epithelial cell line, PTP-PEST localizes to AJ and its knockdown caused a disruption of cell-cell contacts due to a defect in AJ assembly, but with no effect on E-cadherin expression [257].

**RDTPμ** Receptor protein tyrosine phosphatase mu (RPTPμ) is a member of the type IIB receptor protein tyrosine phosphatases (RPTPs). This sub-family of RPTPs can be considered as bi-functional cell surface molecules since their ectodomains mediate stable homophilic cell-adhesive interactions, whereas their intracellular catalytic regions can modulate the phosphorylation state of cadherin complexes. RPTPμ interacts with the scaffolding protein RACK1 (receptor for activated protein C kinase) [258]. RPTPμ can interact with several members of the cadherin family of cell-adhesion molecules (N, E, R and VE cadherins) [250]. Functionally, it is thought that such an interaction localizes the phosphatase activity of RPTPμ near the cadherin–catenin complexes, maintaining them in a dephosphorylated state and thus stabilizing the intercellular contacts. Sui et al. showed that RPTPμ can bind directly to VE-cadherin and that overexpression of wild-type RPTPμ decreased tyrosine phosphorylation of VE-cadherin [259]

**PTPσ** PTPsigma (PTPσ) belongs to the type II a subfamily of RPTPs that includes the phosphatases LAR, PTPσ, and PTPδ [260, 261]. The extracellular region of PTPσ comprises three immunoglobulin-like domains and four to nine fibronectin type III domains [262]. This region has structural similarity to cell adhesion molecules. A role for PTPσ in cell junction regulation was identified by Murchie and colleagues who showed that mice lacking the gene encoding PTPσ develop mild colitis [263]. PTPσ is a positive regulator of intestinal epithelial barrier, which mediates its effects by modulating epithelial cell adhesion through targeting of apical junction complex-associated proteins (including ezrin), a process impaired in IBD. The AJ proteins, E-cadherin and β-catenin are substrates for PTPσ and E-cadherin is hyper-phosphorylated on tyrosine in the colons of the *PtprS*-KO mice relative to wild-type. Phosphorylation of E-cadherin leads to its cellular redistribution away from the AJ and cell dissociation [264, 265].

**PTPN23** PTPN23 directly induces E-cadherin internalization and is associated with increased invasion. Src, E-cadherin and β-catenin appear to be direct substrates of PTPN23. An increase in Src activity with the subsequent loss of PTPN23 is responsible for the expression of mesenchymal markers and the observed increase invasion in mammary cell lines [266].

**PTEN** PTEN acts as a tumor suppressor, which is inactivated in many sporadic cancers (for review see [267]). PTEN antagonizes the activity of PI3K by dephosphorylating phosphoinositides. PTEN prevents junction instability and is consequently implicated in cell migration and invasion, extracellular matrix interaction and cellular polarity [268].

### *4.4.7 Regulation by Endocytosis*

Cadherin molecules are not always stably exposed at the cell surface but are constantly turning over in a high dynamic way through exocytic and endocytic events [269, 270] (Fig. 4.2). Endocytic pathways are often misregulated in cancers and a shift in the balance between recycling and degradation can result, for example, in the degradation of E-cadherin and increased cell migration. Endocytic signaling pathways enclose a vast number of mechanisms (see for review [271–273]). E-cadherinendocytosis can be mediated by clathrin-mediated vesicles or non clathrin pathways (for review [274]).

#### 4.4.7.1 Clathrin-Mediated Endocytosis Pathway

Degradation or recycling of E-cadherin is dependent on its phosphorylation state. In fact, internalization of E-cadherin from AJ is initiated by the Src-mediated tyrosine phosphorylation of E-cadherin [275, 276]. This posttranslational modification induces the dissociation of p120ctn from E-cadherin, and the binding of the ubiquitin-ligases, Hakai or Mdm2 which result in the ubiquitination of E-cadherin and internalization within clathrin-coated endosomes and subsequent degradation by the proteasome complex if recycling does not occur [88, 90, 238].

#### 4.4.7.2 Clathrin-Independent Endocytic Pathways

Cadherin turnover is also associated with clathrin-independent endocytic pathways. Studies suggested that cadherinendocytosis may occur through both caveolin-mediated and macropinocytosis-like pathways. Akhtar and al. found that a dominant-active form of the small GTPase Rac1 could disrupt cell-cell adhesion through the endocytosis of E-cadherin. This pathway appears to be distinct from the uptake of transferrin, which is clathrin-mediated, and through structures that co-localized with caveolin [269]. Lu et al., provides further evidence for caveolin-mediated cadherin endocytosis; they demonstrated that EGF signaling could disrupt cell-cell adhesion by triggering the caveolin-mediated internalization of E-cadherin, a mechanism that may be relevant to EMT in cancers [277]. In contrast, Bryant and colleagues characterized the EGF-induced internalization of E-cadherin in a breast carcinoma cell line, in which E-cadherin is internalized along with the cadherin-binding proteins p120ctn and β-catenin, as Rac1-modulated macropinocytosis, rather than caveolin-mediated [278]. Paterson and colleagues observed E-cadherin endocytosis that is both clathrin- and caveolin-independent, but dynamin-dependent. This pathway, which they identify as similar to macropinocytosis, appears to affect cadherin that is not engaged in trans-interactions in an AJ [279]. Lastly, the desmosomal cadherin desmoglein 3 undergoes lipid-raft-mediated endocytosis, though it is unclear if this pathway is available to classical cadherins as well [274]. Though some of the specific details of the clathrin-independent pathways remain unclear, it appears that both clathrin-dependent and clathrin-independent endocytic pathways play a role in cadherin turnover and degradation (Table 4.2).

### *4.4.8 Regulation by Proteolysis*

In colorectal tumorigenesis, E-cadherin may play a tumor suppressor role or an oncogenic role depending on its degree of proteolysis, which is determined by tumor environment. Various proteases can mediate the cleavage of either the ectodomain or the cytoplasmic domain of E-cadherin, a post-translational process that substantially changes the functional properties of the protein (for review [280]) (Fig. 4.3).

**Table 4.2** Endocytic pathways for AJ. The table resumes the different endocytic pathways for E-cadherin and the protein actors involved

| Internalization route | Cadherin | Adaptators and regulators |
|---|---|---|
| **Clathrin mediated** | VE-cadherin | p120ctn, β-arrestin, VEGF, Ang1, AP-2 |
| | E-cadherin | p120ctn, Dab2, β-catenin, PIPKIγ, Rac1/IQGAP1, FGF, HGF |
| | N-cadherin | NMDAR, β-catenin |
| **Caveolea mediated** | E-cadherin | Caveolin-1, Rac-1 |
| **Lipid raf mediated** | Dsg3 | Tyr phosphorylation |
| | E-cadherin | Arf6 |
| **Macropinocytosis** | E-cadherin | EGF,Rac-1 |
| | B-catenin | B-catenin, Rac-1/IQGAP1 |

#### 4.4.8.1 Cadherin Ectodomain Shedding

The family of membrane-anchored metalloproteases that are known as 'a disintegrin and metalloprotease' (ADAM) proteins are key components in protein ectodomain shedding [281]. Several proteases like matrix metalloproteinase MMPs involving stromelysin-1 (MMP3), matrilysin (MMP7), MMP9, and MT1-MMP (MMP14), and serine proteases such as plasmin kallikreins 6 and 7, and ADAM10 and 15, can cleave mature E-cadherin (120 kDa) in vitro and shed the E-cadherin ectodomain (soluble E-cadherin, sE-CAD) [282–289]. Several groups have shown that sE-CAD present in cell culture medium stimulates epithelial cancer cell dispersion and promotes cell-cell junction disruption [287, 289–291]. The cleaved fragment may still keep functional activity either as paracrine/autocrine competitor of cadherin function or as an anchor for migration when bound to the extracellular matrix. For example, the sE-CAD produced by overexpressed kallikrein-7 in pancreatic adenocarcinomas may function as pseudo ligands to block normal E-cadherin interactions and promote invasion [284]. Growth factor, such as HGF, can also stimulate E-cadherin ectodomain shedding [292]. Indeed, ADAM15-generated sE-CAD appeared to stabilize heterodimerization of the HER2 (ErbB2) receptor tyrosine kinase with HER3, thus leading to Erk signaling, which stimulates both cell proliferation and migration [293]. Alternatively, heterotypic interactions between ECM-anchored cadherin fragments and other receptors may occur. Interestingly, collagen-binding integrins initiate a signaling cascade that promotes MMP9-dependent E-cadherin ectodomain shedding [293]. Furthermore, HT1080 fibrosarcoma cells adhere in an α2β1 integrin dependent way to pentameric E-cadherin [294]. The ectodomain of E-cadherin is also a substrate of secreted cathepsins B, L, and S, often produced by various cells in the tumor microenvironment [295].

Proteases-mediated cleavage of ectodmain also occurs on N-cadherin, whose gain in expression is associated with loss of E-cadherin expression in cancer cells [296]. Mature N-cadherin (130 kDa) is cleaved by several enzymes: plasmin, ADAM10, MT1-MMP (membrane type 1-matrix metalloproteinase), MT5-MMP that shed a 90 kDa fragment in the environment. In contrast, presenilin 1 (PSEN1)/

γ-secretase cleaves N-cadherin in the cytoplasmic part releasing a free cytoplasmic 35 kDa fragment [297–299]. Proteolysis of N-cadherin is a sequential process: (1) ADAM10 associates with N-cadherin and cleaves the extracellular domain at the level of the EC5 domain, resulting in the release of an N-terminal fragment of 90 kDa (soluble N-cadherin, sN-CAD); (2) Then the C-terminal, membrane-bound fragment becomes a substrate for a PSEN1/γ-secretase complex that cleaves it at the membrane-cytosol interface, thus generating a cytoplasmic fragment of 35 kDa [300]. This fragment can then regulate transcription by (1) binding to the transcriptional co-activator CBP (CREB binding protein) (CREB = cyclic AMP response element binding protein) and promoting its proteasomal degradation; (2) entering the nucleus to repress N-cadherin transcription and promote β-catenin dependent transcription [301]. sN-CAD conserves its activity as an adhesive ligand when attached directly with collagen or other components of the basement membrane such as fibronectin or laminin, providing a signal for the stimulation of neurite outgrowth or invasion of cancer cells and angiogenesis [36, 302, 303].

#### 4.4.8.2 Proteolysis of the Cytoplasmic Domain of E-cadherin: Gain of Oncogenic Function

Nuclear E-cadherin expression has been reported in a few human colorectal primary tumors and hepatic metastases, however its functional implications remain unexplored [304–307]. To study the effect of E-cadherin in the formation of metastasis, Céspedes et al., established a colorectal cancer metastasis model by microinjecting SW480-ADH human colorectal cancer cells, transfected with E-cadherin, in the ceca of nude mice [308]. They found that peritoneal tumor foci developed by E-cadherin overexpressing cells presented an increase in E-cadherin proteolysis and nuclear translocation, and enhanced expression of proteolytically active PSEN1, which was linked to increased tumor growth and shortened mouse survival. These foci showed aggressive growth, suggesting that nuclear E-cadherin translocation may represent an oncogenic event. It is also consistent with the ability of nuclear E-cadherin to activate a non-canonical Wnt pathway [110]. This hypothesis agrees with the competition in vitro between E-cadherin cytosolic tail and LEF1 for the binding to β-catenin, leading to the suppression of LEF1-mediated transactivation in SW480 cells [309, 310]. In this model, the ratio between full-length and proteolyzed cytosolic E-cadherin changes with the environment. Thus, depending on the organ where the tumor grew, the dramatic oncogenic effect can be observed or not. This effect is probably related to the amount of active PSEN1 present in a given environment and with the total cellular levels of E-cadherin. An oncogenic effect in colorectal carcinoma has also been recently demonstrated for EpCAM, another epithelial-specific cell adhesion molecule that is overexpressed, proteolysed, and translocated to the nucleus [311, 312]. Like E-cadherin EpCAM can be proteolyzed following sequential cleavage by TNFα converting enzyme (TACE/ADAM17) and a γ-secretase complex containing presenilin 2 (PSEN-2), thus releasing the extracellular domain into the culture medium, and the cytoplasmic domain into the cytoplasm where it takes part to the large nuclear complex containing transcriptional regulators β-catenin/TCF/LEF complex [313]. Finally, other enzymes like caspases and calpain cleave E-cadherin in its cytoplasmic part releasing an intracellular 37 kDa c-terminal fragment [314–316]. This fragment disturbs β-catenin subcellular localization and stimulates β-catenin dependent cyclin D1 expression in the human keratinocyte HaCaT cell line, stimulating cell proliferation and migration rather than cell-cell adhesion [286].

In conclusion, the role of E-cadherin in colorectal cancer will depend on the relative abundance of full-length and proteolyzed E-cadherin species within the tumor cell. Membranous full-length E-cadherin promotes differentiation, intercellular adhesion that restricts invasion and migration and inhibits tumorigenesis. Consistently, its deletion or down-regulation in animal models or in humans exerts opposed effects favoring tumorigenesis by activating the Wnt/β-catenin and other signaling pathways [19, 317–319]. Accordingly, an association exists between increased membranous E-cadherin expression in human primary colorectal tumors and favorable outcome [320]. If proteolyzed E-cadherin fragments predominate over full-length E-cadherin, these fragments would induce cell-cell dissociation, and migration or translocate to the nucleus and activate an oncogenic program. This is in agreement with the reported changes in the ratio of membranous and cytosolic E-cadherin depending on the metastatic site of colon tumors, and its association with tumor recurrence in colorectal cancer patients [320].

## 4.5 Transcriptional and Epigenetic Regulation of Adherens Junctions

Development of cancer is a consequence of both genetic and epigenetic lesions. The former result from mutations or gain or loss of DNA sequence, whereas the latter result from hereditable changes in chromatin, such as DNA-associated proteins post-translational modifications as methylation, phosphorylation or acetylation. Inactivation of E-cadherin is thought to be an important step in the development of most epithelial derived tumor types. Mutations within the E-cadherin gene have been identified as the cause of familial gastric cancer and loss of expression of E-cadherin has been found to be widespread in sporadically occurring epithelial tumors. Inactivating mutations, such as found in gastric and breast cancer are absent in colon cancers [321], thus, the mechanisms of E-cadherin down-regulation in colon cancer mainly involves epigenetic regulation (Fig. 4.3). Epigenetic regulation encompasses three types of changes: DNA methylation, histone modifications, and microRNAs, each of which has been shown to play a key role in controlling EMT and cancer metastasis (for review [322]).

### 4.5.1 Promoter Methylation

The silencing of E-cadherin expression by hypermethylation is a common event in cancer [323, 324]. In colorectal carcinoma, Darwanto et al., showed that methyl-CpG-binding protein (MeCP2) and 5'CpG island hypermethylation cooperatively regulate E-cadherin gene [325]. DNMTs are recruited to DNA by interaction with chromatin remodeling complexes, and gene silencing is subsequently conferred by the recruitment of methyl-DNA-binding proteins, such as MeCP2, MBD1–4 and Kaiso [113, 326]. The methylated CpG island of the repressed E-cadherin promoter is bound by MeCP2 and MBD2. Binding of these nuclear factors results in the recruitment of histone deacetylase (HDACs) to the methylated E-cadherin promoter area, leading to histone H3 deacetylation, which is essential for the silencing of the methylated E-cadherin gene [327]. Down regulation of E-cadherin expression by promoter methylation is reversible and fits with the hypothesis that at the metastatic sites, disseminated mesenchyme cancer cells must regain some of their epithelial functions, undergoing a mesenchymal-epithelial transition (MET) [328]. The methylation process can be inhibited through the use of 5′-aza-2′-deoxycytidineb (5Aza-dC), a nucleoside analog that forms adducts with DNMTs, leading to hypo-methylation and increased gene expression [329–331]. Indeed reactivation of E-cadherin expression has been observed in certain cancer cell lines upon treatment with this demethylating agent [332, 333].

### 4.5.2 Transcriptional Repression

The E-box-binding factors Snail, Slug, Twist and zinc finger E-box-binding homeobox 1 and 2 (ZEB1 and ZEB2, respectively) were originally characterized as transcriptional repressors of cell adhesion genes that encode components of AJ such as E-cadherin but also desmosomes and TJs [334, 335]. Two of these transcription factors, Snail and Twist1, have been considered to be master regulators of EMT [334, 336, 337]. Snail is not expressed in normal colon, whereas cancers exhibit a high expression of the protein [338]. Repression of E-cadherin by Snail/Twist involves the recruitment of repressor complexes that inhibit gene expression through chromatin remodeling such as HDAC and polycomb proteins. Deacetylation of histones by HDAC prevents the opening of chromatin and the binding of the transcription machinery whereas polycomb proteins act as a platform to recruit DNMT, suggesting that the two epigenetic mechanisms could intersect [339]. Snail1 interacts with histone deacetylase (HDAC) 1 and 2, SIN3a and components of the polycomb repressive complex 2 (PRC2) [80, 340, 341]. Snail2/Slug recruits co-repressors such HDAC1/3 and C-terminal binding protein (CtBP) [342–344]. Studies in several cancer cell lines showed that CtBP and Smad proteins co-repressor complexes might also collaborate in the repression of CDH1 with ZEB factors [345–347]. Importantly, CtBP seems to be crucial in controlling CDH1 expression in colorectal tumors as they can regulate ZEB1 activity [348]. The mechanism underlying Twist-mediated repression of E-cadherin is less known but seems to involve several other components such as the Mi2/nucleosome remodeling and deacetylases (Mi2/NuRD) complex [349].

### 4.5.3 MicroRNAs

MicroRNAS (miRNAs) are highly conserved, small noncoding RNA molecules of 19–24 nucleotides which bind to the complementary sequences within target mRNA molecules. They post-transcriptionally govern gene expression, thus affecting a broad range of physiological processes. However miRNA have been shown to be aberrantly expressed and to play causal roles in a variety of pathologies including cancer [350, 351]. During the last years, the implication of a growing number of miRNAs in the regulation of EMT transcription factors or EMT-related pathways has been observed in several human tumor contexts (for review [351]). Interestingly, poorly differentiated carcinomas present an overall decrease of total miRNA expression, suggesting a role for the miRNAs in the regulation of different mechanisms governing epithelial plasticity [352] (Fig. 4.3). A few miRNAs have been involved in the direct regulation of E-cadherin expression but it was shown that miR9 directly targets the E-cadherin gene in breast cancer cells [353]. miR-9-mediated E-cadherin down-regulation results in the activation of β-catenin signaling, which contributes to up-regulated expression of mesenchymal and angiogenic genes required for metastatic outgrowth. Several miRNAs have been identified that directly target key transcription factors involved in EMT, such as Snail and Zeb, while others modulate various signaling pathways that promote EMT. The best example is the miR-200 family (miR-200a, miR-200b, miR-200c, miR-141 and miR-429). The miR-200 family reduces the expression of the transcription factors Zeb-1 and 2 and, conversely, Zeb-1 and 2 repress the expression of the miR-200 gene cluster, thereby generating a regulatory loop that can reinforce the ability of Zeb proteins to maintain a stable mesenchymal phenotype, as observed at the invasive front of colon carcinoma cells [354, 355]. According to modulation of signaling pathways that promote EMT, Smad dependent-TGFβ-signaling induced miR-155 expression leading to reduce RhoA expression and the dissolution of cell TJs [356]. p53 also induces the expression of miR-34a and miR-192, which then repress Snail1 and Zeb2, respectively, to block EMT [357, 358]. miR-138 can also repress Zeb2 to suppress EMT [359].

Recently it has been shown that specific miRNA repression could be achieved by CpG miRNA promoter hypermethylation. This novel mechanism increases the mesenchymal characteristics of tumor cells. Epigenetic regulation of the miR-200 family and its relationship with aggressiveness and chemoresistant behavior has been established in different cancer cell types including colon and breast carcinomas

[360–362]. Hypermethylation of the miR-34 promoter is found in prostate tumors (80 %) and in melanoma (63 %); it is also present in many cell lines derived from breast, colon, lung, and other types of tumors.

## 4.6 Role of AJ in Intestine Physiopathology: From Development to Colorectal Cancer

### *4.6.1 Role in Stem Cell Biology and Embryology*

Development, organization and homeostasis of most of animal tissues and organs depend on the formation of specific cell-cell junctions (Fig. 4.4). The spatiotemporal expression pattern of cadherins during embryogenesis is critical for cell migration, celldifferentiation and tissue morphogenesis. In this line, E-cadherin knockout embryos fail to undergo proper embryogenesis and do not form the trophectoderm epithelium [363]. In the postnatal environment, cadherins play a role in the maintenance of tissue architecture and physiological homeostasis of mature tissues [364, 365]. Recently, various studies focused on the roles of cell–cell adhesion complexes in stem cell self-renewal and differentiation, and their effect on stem cell fate and somatic cell reprogramming (for review [366]). The major characteristic of stem cells is their self-renewal capacities, correlated with signals that retain cells in an undifferentiated state, which is fundamental for their pluripotency and limitless proliferation. Key signals in this process are provided by components of the Wnt pathway, LIF/STAT3 and c-Myc, and by transcription factors Nanog, Oct3/4, Klf4, and Sox2 [367]. Embryonic stem cells (ESC) form compact colonies due to intercellular adhesion dependent of E-cadherin-catenin complexes. These interactions constitute a driving force in survival, self-renewal and pluripotency maintenance of naïve ESC [368]. E-cadherin also participates in the differentiation of stem cells in well-differentiated teratomas or in chimeric animals by interacting with key components of the naive stemness pathway. In addition, depleting E-cadherin converts naive mouse ESC into primed epiblast-like stem cells (EpiSC) [369]. Mouse ESC exclusively expresses E-cadherin at the cell membrane, whereas mouse EpiSC express both E- and N-cadherin. In line with this, a mesenchymal-to-epithelial transition (MET) occurs during reprogramming of somatic cells towards induced pluripotent stem cells (iPSC), leading to down regulation of N-cadherin and acquisition of high E-cadherin levels [370]. E-cadherin up-regulation considerably increases reprogramming efficiency and the established iPS cells possess E-cadherin-mediated cell–cell contact, morphologically indistinguishable from ESC [371]. Consistently, abrogation of cell-cell contact by an inhibitory peptide or a neutralizing antibody directed against the extracellular domain of E-cadherin compromises iPSC generation. Thus, MET is essential for iPSC formation and can be induced by Klf4 (a transcription factor involved in the epithelial cell differentiation) or BMP signaling [372, 373]. Klf4 competes with the transcription factor Zeb2 to bind and activate the E-cadherin promoter [374, 375].

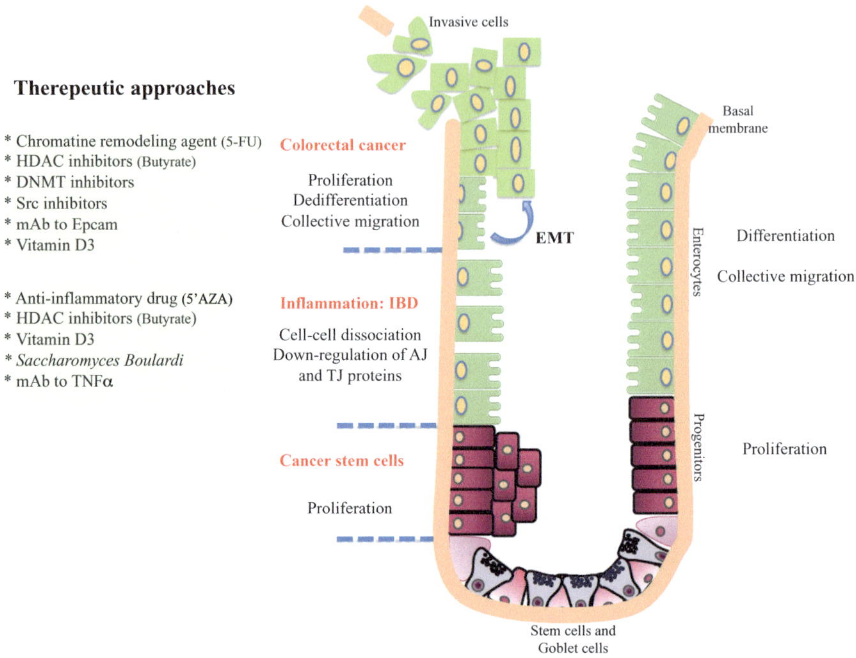


**Fig. 4.4** Pathologies associated with the intestinal epithelium. (1) Embryonic stem cells (ESC) form compact colonies due to intercellular adhesion dependent of E-cadherin-catenin complexes. These interactions constitute a driving force in survival, self-renewal and pluripotency maintenance of naïve ESC. 3) However they can transform into cancer stem cells and then proliferate. (2) Inflammatory bowel diseases (IBD) develop following alteration of the epithelial barrier, which increases intestinal permeability. Default in epithelial barrier results from cell-cell dissociation and down-regulation of junctional proteins. (3) Long-standing IBD favors the development of colitis-associated colorectal cancers (CAC), where normal colonic epithelium becomes hyper-proliferative and forms adenomatous polyps that progress to malignant diseases. In sporadic colorectal cancer, E-cadherin downregulation and p120ctn delocalization have been reported to associate with dedifferentiation, progression, and metastasis

Together with E-cadherin, β-catenin exerts a dual function in stem cell biology: first, its interaction with α-catenin at cell junctions in ESC is responsible for the adhesive function of β-catenin, which might be crucial in the formation and maintenance of ESC colonies as well as for maintaining their naive state [376]; second, β-catenin participates in Wnt signaling and within the nucleus it stimulates transcription of genes important for stemness [377, 378].

Evidence is rising that other junctional proteins may play a role in stem cell biology. This is particularly true for the Epithelial cell adhesion molecule (EpCAM), which has been described to participate in stem cell signaling (for review [312]). EpCAM is a 30/40 kDa type I glycosylated membrane protein located at intercellular junction spaces that was first described as a dominant antigen in human colon carcinoma tissue [379]. EpCAM is expressed in a variety of human epithelial tissues, and strongly expressed in cancers, progenitors and stem cells [312, 380]. Structurally, EpCAM comprises an extracellular domain with epidermal growth factor (EGF)- and thyroglobulin repeat-like domains, a single transmembrane domain, and a short 26-amino acid intracellular domain called EpICD. Additionally, EpCAM is connected to the cytoskeleton upon interactions of EpICD with α-actinin [381]. Like cadherins, EpCAM can mediate cell-cell contacts but also transmits signals from the plasma membrane into the nucleus in order to regulate gene transcription after proteolysis. Indeed EpICD fragment contains a binding sequence to LEF-1 DNA consensus sites [313]. The resulting nuclear complex up-regulates target genes including (i) *c-myc* and *cyclins*, and additional genes involved in cell growth and proliferation, cell cycle, and cell death [382, 383]; (ii) key factors of the conversion of somatic cells into pluripotent stem (iPS) such as Nanog, Klf4, Sox2 and Oct4 [384–386]. The Wntsignal transduction pathways, which are used for both EpCAM signaling and regulation of the *epcam* gene is crucial for normal and malignant stem cells [387, 388]. This is supported by experiments of EpCAM knockdown in murine ESC using siRNA, and by its forced overexpression. EpCAM reduction resulted in diminished proliferation and expression of stem cell markers, whereas forced expression allowed the stabilization of Oct3/4 expression even under differentiating conditions [381]. In contrast to normal cells, in colon carcinoma cell lines, EpICD translocates into the nucleus, where it enhances c-myc transcription and cell proliferation [311, 313]. In conclusion, cancer-initiating and normal stem cells, which express high-level of EpCAM molecules may benefit from activated EpCAM for proliferation, self-renewal, and anchorage-independent growth and invasiveness.

### 4.6.2 Role in Epithelial Cell Differentiation

Cadherins have functions that extend beyond cell-cell adhesion. These proteins are linked to multiple intracellular signaling pathways, including Wnt, PI3K/Akt and FGF pathways [390–392]. Such signaling pathways are particularly important in intestinal cell proliferation and differentiation. However we found that only "mature" AJ are able to induce epithelial cell differentiation [393]. The maturation of AJ results in part from a crosstalk between integrins and cadherins and takes place in cholesterol -enriched regions called lipid raft where p120ctn is recruited to bind E-cadherin [176, 199]. Involvement of ECM proteins in the reinforcement of E-cadherin/actin complexes has been previously described in Caco-2 cells [394]. In this work, Caco-2 cells cultured on complex ECM, were induced to express apolipoprotein A IV, a marker of IEC differentiation, via the recruitment of integrins. Among basement membrane molecules, laminins present the highest variability in their spatial and temporal expression, either during intestinal development or in the adult [154, 165, 395, 396]. Laminin 111 isoform is able to induce the expression of the differentiation marker sucrase-isomaltase in Caco-2 cells, together with the nuclear shuttling of nucleolin, a transcriptional regulator [397]. The differentiation process induced by laminin 111 requires the activation of the p38/MAPK. Consistent with these findings, the p38 cascade is known to be crucial in intestinal epithelial differentiation, as this pathway acts on the key intestinal transcription factor, Cdx2 [398]. Similarly Laprise et al., found that PI3K/AKT pathway activated by cadherins in turn stimulates the p38/MAPK pathway [3].

### 4.6.3 Role in Inflammation: Example of IBD

Inflammatory bowel diseases (IBD), which include Chron's disease (CD) and ulcerative colitis (UC), consist of a measurable over-inflammatory response leading to gut damage [399]. The etiology of IBD is unclear but it is proposed that genetic susceptibility and environmental factors favor dysfunctional immune response and increased intestinal permeability that participate to the development of IBD (reviewed by [400, 401]). Permeability defects could be due to the pronounced apoptosis that occurs during inflammation processes, however, it has been shown that IEC apoptosis alone is not sufficient for the entire deficit. Several studies have provided evidences for the perturbation of AJ or TJ in IBD, but the question of whether any altered expression of junctional molecules is a primary event in IBD mucosa, or a phenomenon secondary to the inflammatory process has yet to be clarified. It has been hypothesized that permeability defects might represent a primary disorder in CD, since intestinal permeability alterations have been observed not only in inflamed gut tissues but also in areas lacking any sign of macroscopic injury [402, 403]. In UC, perturbations in permeability seem to be limited to the inflamed intestinal segment [404, 405]. Various junctional molecules are affected by the actively inflamed status in IBD, in particular the expression of ZO-1, occludin, E-cadherin and desmoglein-2 [405]. Transgenic animal models revealed the importance of E-cadherin in maintaining the epithelial barrier by showing that AJ proteins contributed to IBD-like processes (for review [406, 407]). Jankowski and co-workers have demonstrated that deregulation of E- and P- cadherin correlates with the progression of human colitis [408]. E-cadherin mRNA transcripts were clearly expressed in actively inflamed mucosa of CD and UC, whereas the protein is less detected, suggesting a post-transcriptional regulation of barrier integrity as it was observed with cytokines and several GF (TGF, HGF, TNF) [409]. Using non-invasive techniques, various studies demonstrated an increased intestinal permeability in CD, which is most likely attributed to the disruption of intercellular junctions by Th1 cytokines (TNFα and INFγ) that are characteristic of this disease [410–413]. IFNγ can prime intestinal epithelial monolayers to respond to TNFα by disrupting TJ morphology and barrier function via myosin light chain (MLC) kinase up-regulation and MLC phosphorylation [414]. The mechanism by which IFNγ induces permeability changes is incompletely understood but it has been associated to endocytosis of occludin, JAM-A and claudin1 following the activation of Rho GTPases [415].

Differentiation of IEC is likewise affected during chronic colitis; indeed gland atrophy and mucin depletion have been suggested as diagnostic markers for UC

[416, 417]. In the Dextran Sulfate Sodium (DSS)-induced colitis murine model, that affects primarily the distal colon, the production of markers of goblet cell is strongly induced in crypts in mildly inflamed mucosa, whereas there is a marked reduction in strongly affected areas in the descending colon [418]. Because mucins protect the epithelial barrier from luminal pathogens, it seems likely that the epithelium responds to crypt epithelial damage and ulceration during severe inflammation, primarily by increasing proliferation and thus, reduced differentiation [419]. Paneth cells at the small intestinal crypt base secrete defensins into the gut lumen that act as potent bactericidal agents [420]. A loss of defensin expression has been linked to the development of IBD [421]. Additionally, a decrease in the number of Paneth cells may be seen in active ileitis, whereas increased Paneth cell proliferation and ectopic differentiation in the colon (Paneth cell metaplasia) are observed in regenerating epithelium in chronic colitis [422]. Therefore, improper differentiation of IEC not only may cause intestinal inflammation, but could also perpetuate impaired epithelial homeostasis and accelerate intestinal neoplasia [423]. Long-standing IBD favors the development of colitis-associated colorectal cancers (CAC), which are believed to result from oxidative DNA damage in epithelial progenitor cells [424]. In this model, enhanced crypt turnover during inflammation contributes to intestinal tumorigenesis by increasing the likelihood of acquiring mutations in known oncogenes such as APC, whose defect causes familial adenomatous polyposis; this is an early event in the development of a majority of colonic adenocarcinomas.

The chronic inflammatory states may lead to environments that foster genomic lesions and tumors initiation. Furthermore, disorganization of inter-epithelial junctions could participate in the infection process as well as in the cellular dedifferentiation process that precedes carcinogenesis [425]. These epithelial alterations are more pronounced in UC in which the development of malignancies is more frequent than in CD tissues, suggesting that disturbances of junction associated molecules are likely to be involved in carcinogenesis from IBD patients.

Interestingly, recent studies have shown that the Wnt/β-catenin pathway can be activated by inflammatory cytokines during intestinal inflammation, through PI3K/AKT signaling [426–428]. Exposure of IEC to inflammatory mediators results in the persistent transcriptional activation of β-catenin mediated by Akt phosphorylation, and this process leads to increased cell proliferation at the onset of colitis. However, sustained cytokine treatment result in the induction of the β-catenin/TCF target gene Dkk1 (Dickkopf-related protein 1), a potent secreted Wnt antagonist. Consequently, extended intestinal inflammation is associated with a reduction of Wnt signaling in the inflamed mucosa and with excessive enterocyte apoptosis induced by INF-β and mediated by Dkk1. The Barrett laboratory reported that activation of T cells in the intestinal mucosa result in PI3K-mediated phosphorylation of Akt, stabilization of β-catenin, and increase cell proliferation in the crypts. Importantly, inhibition of PI3K was sufficient to prevent the progression from chronic colitis to mucosal dysplasia by inhibiting β-catenin transcriptional activity and thus epithelial hyper-proliferation [427]. In accordance with these findings, loss of PTEN, a negative regulator of Akt, leads to intestinal polyposis by increasing the activity of β-catenin through Akt [429].

### 4.6.4 Role in Colorectal Cancer

Colorectal cancer (CRC) is the third most common malignancy worldwide [430]. The pathogenesis is characterized by clinical and histopathological changes known as the adenoma-carcinoma sequence, where normal colonic epithelium becomes hyper-proliferative and forms adenomatous polyps that progress to malignant disease [431]. Several studies have reported mutations of the tumor suppressor APC gene in CRC [432–434]. These mutations are amongst the earliest genetic events found in the majority (up to 80 %) of sporadic cases of CRC [435–437]. The main tumor suppressor function of APC has been shown to reside in its ability to mediate the proteasomal degradation of cytoplasmic β-catenin, a key member of the Wnt signaling cascade [438]. In CRC, mutations of both APC and β-catenin cause deregulation of intracellular β-catenin levels which leads to the nuclear accumulation of β-catenin [438]. This causes aberrant and constitutive expression of Wnt target genes, and thus the development of CRC [439].

#### 4.6.4.1 Cadherins in Cancer

Loss of intercellular adhesions by mutations or reduced expression of E-cadherin is a hallmark of epithelial cancer and tumor progression [15, 365]. Interestingly, while the gene encoding E-cadherin, *CDH-1*, is found mutated in numerous gastric cancers [440], mutations are rare in colon tumors, where E-cadherin function is more favorably shut down at the transcriptional or post-transcriptional level (see Sect. 4.4.5) [441]. Overall, the transformation process is characterized by a down-regulation of E-cadherin and an up-regulation of mesenchymal cadherins such as N-cadherin and cadherin-11, which render epithelial cells more invasive [391, 442]. The switch from E-cadherin expression to P-cadherin also results in the translocation of p120ctn to the cytosol, which in turn induces cell migration by activating Rac1 and cdc42 [443]. E-cadherin is described as a tumor suppressor that maintains epithelial integrity and inhibits migration and invasion [19]. Consistently, forced expression of E-cadherin suppresses tumorigenesis, whereas its loss is associated with adenoma to carcinoma transition in experimental models [318]. In sporadic colorectal cancer, E-cadherin down-regulation has been reported be associated with dedifferentiation and metastasis, whereas its membranous expression confers better prognosis [317, 320]. Intriguingly, E-cadherin is expressed in a subset of metastatic lesions in colorectal cancer [444, 445]. Because p120ctn participates in stabilizing the cadherin-catenin complex, many cancer types are characterized by a loss or a delocalization of p120ctn (reviewed in Ref. [446]. However, reduction in E-cadherin expression impacts on catenin levels and localization, resulting in an acceleration of the tumorigenic development. It is well recognized that inactivation of E-cadherin leads to cytosolic translocation of p120ctn in colon, breast, bladder, lung, pancreas, prostate and stomach tumors, which has been associated with tumor malignancy (reviewed by [447]). Indeed, studies in breast and colon cancer indicated that cytosolic p120ctn controls the invasive phenotype of E-cadherin-deficient tumor cells [448, 449]. Recently it was shown that p120ctn interacts with hNanos1, the human ortholog of the Drosophila zinc-finger protein, Nanos [450]. Transcription of hNanos1 mRNA is often suppressed after E-cadherin expression. Conditional expression of hNanos1 in human colon DLD1 cancer cells induces cytoplasmic translocation of p120ctn, up-regulates expression of MMP1 at the mRNA and protein levels, and increases migratory and invasive abilities [450, 451].

Several mechanisms of E-cadherin repression have been reported in CRC, including gene mutations and promoter hypermethylation. However, transcriptional repression of E-cadherin and associated up-regulation of Snail are also considered to play a role in the progression of CRC [452–455]. Analysis of Snail and Slug expressions in human CRC indicated that the two transcription factors were respectively overexpressed in 78 % and 37 % of the tumor samples examined [453, 454].

#### 4.6.4.2 Catenins in Cancer

β-catenin and p120ctn act as true molecular switches that favor intercellular adhesion through E-cadherin stabilization in healthy and differentiated epithelium, while they can activate migration and cell proliferation when intercellular contacts are lost. This is a perfect mechanism to regulate homeostasis in the rapidly and constantly renewing intestinal epithelium but it also represents an Achilles' heel when it comes to tumorigenesis. Signaling through catenins is affected in many type of cancers but this chapter will focus on GI tumors (for review [456]). For instance, nuclear accumulation of β-catenin is observed in 20–90 % of GI cancers [457] and p120ctn delocalization is observed in more than 50 % of colorectal cancers, correlating with tumor aggressiveness [448]. p120ctn expression is also down regulated in gastric cancers (56 % of analyzed tumors) [458], 34 % of esophageal carcinomas [459] and mislocalized in all esophageal cancers [460]. Consistently, mouse models bearing tissue-specific p120ctn deletion develop neoplastic lesions in the oral cavity and esophagus as well as in the intestine [461].

Deleterious effects of catenin expression are also highlighted by the numerous mutations identified in their genes in GI cancers. For instance, mutations in the genes encoding β-catenin or APC are early events in the neoplastic transformation of CRC (for review see [462]). β-catenin is mutated in 20–25 % of CRC depending on the tumorigenic stage [463, 464]. Mouse models with APC deletions or β-catenin overexpression in the intestine are prone to tumor development [465]. While in the case of β-catenin, pro-oncogenic effects are directly related to its transcriptional activity with TCF/LEF and the resulting increase in the expression of proliferation and invasion related proteins, the case of p120ctn might be subtler. Indeed, p120ctn exhibits both pro-oncogenic and tumor suppressor properties (reviewed in Ref. [466]), depending on the isoform and the tissue being investigated. Building the map of p120ctn multiple functions in tumorigenesis is one of the important challenges for the coming years.

#### 4.6.4.3 Polarity and Cancer

Consistently with PAR's roles in polarity establishment, these genes are found affected in cancer: PAR-3 is down-regulated in oesophageal carcinoma [467], PAR-6 is overexpressed in breast and lung cancer [468–470] and αPKC is also overexpressed in hepatocellular carcinoma, bladder, pancreatic, head and neck and breast cancers [141].

One particular PAR protein has been extensively studied in the context of tumorigenesis, the serine/threonine kinase PAR-4, known as LKB1 in mammals. Indeed, mutations in LKB1 are responsible for Peutz-Jehgers syndrome (PJS), a condition in which patients display a high incidence (93 %) of cancer in various epithelial tissues and develop intestinal polyps [471, 472] making of this kinase a tumor suppressor. LKB1 was described to act in different pathways such as the regulation of cell growth, energy metabolism and cell polarity. Through the activation of AMPK and the downstream regulation of mTOR pathway, LKB1 can modulate cell growth and proliferation depending on the metabolic status of the cell. In human epithelial cells, LKB1 is localized at the AJ in an E-Cadherin-dependent manner [473]. Also, LKB1 activity is required, in conjunction with the MST4 kinase, to form the apical brush border [474]. Surprisingly, it was shown that LKB1 activation by its associated pseudo kinase STRAD is able to polarize a single epithelial cell [475], suggesting that it can also act totally independently of intercellular contacts. Apart of Peutz-Jeghers syndrome, loss of heterozygosity of LKB1 is found in sporadic cancers, among which breasts and colon cancers [476]. Interestingly, mice model depleted for LKB1 resumed the clinical observation of GII polyps that remain benign and rarely become malignant [477]. Specific depletion of LKB1 in the small intestine leads to an alteration in Paneth and Goblet cells differentiation [478].

#### 4.6.4.4 Role of EPH Receptor in Cancer

The role for erythropoietin-producing hepatoma (EPH) receptors in cancer has been recently reviewed by Li et al., and will be treated as a chapter in this book [479]. In this section, we will focus on the mutual regulation between EPH receptors and E-cadherin-based cell-cell adhesions. For example E-cadherin co-localizes with EPHA2 in cell-cell contacts along the lateral membranes of normal epithelial cells and the loss of E-cadherin results in the associated loss of EPHA2. Consistently, the expression of E-cadherin is required for EPHA2 phosphorylation and localization showing that E-cadherin can regulate the expression and subcellular localization of EPHA2 [480, 481]. Conversely, the ligand-mediated activation of EPHA2 enhances E-cadherin–based cell-cell contacts, as well as the apical-basal polarization of epithelial cells [482]. Moreover, although EPHA2 is overexpressed in metastatic breast cancer cells, as well as in gastric and colorectal cancers exhibiting liver metastasis, its phosphorylation and E-cadherin expression level are decreased suggesting that EPHA2 activation in metastasis is independent of E-cadherin-induced phosphorylation [481, 483–485]. Such phosphorylation of EPHA2 could occur through various

GF signaling involved in the EMT process [486]. Furthermore, overexpression of EPHA2 promotes the destabilization of AJ and the migration of breast cancer cell by a mechanism involving RhoGTPase regulation [487, 488]. In a same manner, overexpression of EPHA2 resulted in up-regulation of the EMT molecular markers N-cadherin and Snail and down-regulation of E-cadherin through canonical Wnt signaling in gastric cancer cells [489]. These results indicate that E-cadherin and ligand-independent mechanisms represent a predominant approach to activate EPH receptors in human malignancies.

While the expression of EPHB receptors is essential for correct positioning of IEC along the crypt/villus axis via the Wnt signaling pathway, the activation of Wnt signaling also induces EPHB expression in initial compartmentalized intestinal tumor cells [490, 491]. Later on, the loss of EPHB receptors accelerates the progression of colorectal cancer [492]. Studies in SW480 colon cancer cells have shown that EPHB2 and EPHB4 are two TCF/β-catenin-regulated genes. Moreover, the preferential binding of β-catenin to CBP rather than p300, contributes to cancer progression and metastasis by increasing the expression of EPHB4 and decreasing the expression of EPHB2 [493]. In contrast, EPHB3 can act as a tumor suppressor by two ways: (i) it might affect the redistribution of cellular E-cadherin to the plasma membrane, thus enhancing E-cadherin–mediated cell-cell adhesion through the Rac1 pathway [494]; (ii) by promoting compartmentalization, it restricts the spreading of EPHB-expressing tumor cells into EPHRIN B1-positive regions [495]. In addition, EPHB3 overexpression induces an up-regulation of E-cadherin, as well as a decrease in nuclear β-catenin. Moreover, TCF/β-catenin activity was found to be reduced in EPHB3-overexpressing colon cancer cells [496]. These results indicate that EPHB3 receptor acts as cancer suppressor through the inhibition of the Wnt signaling pathway.

### *4.6.5 The Intriguing Liver-Intestine Cadherin*

Liver–intestine (LI)-cadherin (also referred as CDH17) is a member of the cadherin superfamily, although it contains seven molecular domains rather than five [496, 497]. Moreover, the cytoplasmic domains of this cadherin is truncated and lacks the two armadillo protein-binding domains, which are typical of classical cadherins. LI-cadherin is located solely on the basolateral surface of enterocytes and Goblet cells in the intestine and colon but is absent of the apical plasma membrane and so excluded from AJ [496]. In contrast to E-cadherin, LI-cadherin is exclusively localized in cholesterol-rich microdomains and its expression is strictly dependent on cholesterol levels. LI-cadherin is one of the transcriptional targets of Cdx2, which proved to be important during early differentiation and maintenance of intestinal epithelium [498]. During mouse embryogenesis LI-cadherin expression in the intestinal epithelium begins at embryonic day 12.5 [499]. The expression onset coincides with the formation of intestinal villi. Excessive tissue remodeling, growth, and differentiation characterize this developmental stage, however the exact biological function of LI-cadherin is currently not known (for review [500]). It has been speculated that due to the high lateral mobility, LI-cadherin could be responsible for fast and flexible initial formation of adhesive contacts between cells in development and during tissue regeneration [497, 501, 502]. Thanks to its heterophilic interaction with E-cadherin it might also regulate E-cadherin mediated adhesion [503]. In pathological conditions, the distribution of LI-cadherin expression can be changed as up or down-regulation of LI-cadherin expression might be correlated with tumor generation and metastasis. A study of human colorectal cancer showed that reduced LI-cadherin expression is associated with high tumor grade, lymphatic invasion and lymph node metastasis [504]. This is in principal concordance with the function of other cadherins, which inhibit tumor growth. However, LI-cadherin up-regulation was also described in advanced stage of gastric, hepatocellular or colorectal carcinomas [505–507]. It is presently not understood if the up-regulation of LI-cadherin in cells normally lacking it is a cause for the tumor progression or if LI-cadherin is up-regulated to compensate the impaired E-cadherin mediated adhesion in the corresponding tumor cells. Activation of $\alpha2\beta1$ integrin by LI-cadherin was recently implicated in increasing proliferation of highly metastatic colorectal cancer cells and in increasing adhesion of tumor cells to collagen type IV [505]. LI-cadherin silencing in such cells suppressed tumor growth in vivo as well as liver metastasis, and this was associated with reduced FAK activation and RAS signaling.

### *4.6.6 EMT in Physiological and Pathological Adherens Junctions-Dependent Processes*

EMT can be induced by a large variety of stimuli, ranging from GF signal transduction pathways to the loss of E-cadherin function by protein degradation or gene mutation. In many carcinomas, the tumor-associated stroma produces a variety of growth factors (HGF, EGF, FGF, IGF) and cytokines including different members of the TGFβ superfamily, Wnt and Notch (for review [508–510]); all of which are able to induce the expression of various transcriptional repressors of E-cadherin gene expression, including Snail-1 and 2, Zeb1 and 2, and Twist.

#### 4.6.6.1 Modification of Adhesion Process that Occurs During EMT

During EMT, cells lose their epithelial polarity and dissolve their adherent and tight junctions, favoring a high plasticity in cell-cell adhesion and communication with the ECM through focal adhesions. The Armadillo-catenin proteins are often lost from the cadherin-mediated cell-cell contacts and shuttle to the nucleus to potentiate EMT signaling events [511]. A Rho-mediated network of stress fibers, at the tip of which ECM adhesion molecules localize, replaces circumferential F-actin fibers [512–514]. These changes are sufficient for the cells to separate, lose their apico-basal polarity and gain a more fusiform shape, all facilitating cellular migration [515]. However, cells produced by EMT are not always motile and invasive. The expression of intermediate filaments is also changing during EMT, and vimentin, a typical filament of mesenchymal cells, replaces for instance cytokeratin being characteristic of epithelial cells. Matrix metalloproteases such as MMP-1, -2, -3, -7 and -14 are frequently up-regulated, potentially enabling cells to detach from each other (via cadherin ectodomain shedding) and to penetrate the basement membrane. ECM synthesis changes from basal lamina proteins to interstitial forms including collagen type I, fibronectin, secreted protein acidic and rich in cysteine (SPARC) and tenascin C [516].

#### 4.6.6.2 EMT and CRC

Invasion by colorectal carcinomas is characterized by an EMT-like "dedifferentiation" of the tumor cells. However a re-differentiation towards an epithelial phenotype, resembling a MET is detectable in metastases. This indicates that malignant progression is based on dynamic processes, which cannot be explained solely by irreversible genetic alterations, but must be additionally regulated by the tumor environment. The main oncoprotein in CRC is β-catenin, which is overexpressed due to mutations in the APC tumor suppressor in most cases. EMT of the tumor cells is associated with a nuclear accumulation β-catenin, which is reversed in metastases. Nuclear β-catenin is involved in two fundamental processes in embryonic development: EMT and stem cell maintenance. Accumulating data demonstrate that aberrant nuclear expression of β-catenin can confer these two abilities to tumor cells in the adult. The unusual combination of EMT with stem cell competence might result in a migrating tumor stem cell, which drives tumor invasion and metastasis. Although EMT processes are documented in many cancer cell models in vitro, the significance of EMT during cancer progression and even its relevance in human cancer tissues has remained a matter of debate. This resistance was mainly due to the lack of convincing evidence of EMT in clinical samples. The description of cords or small aggregates of tumor cells extending or detaching from the tumor mass into the adjacent stroma have recently provided morphological evidence of EMT at invasive fronts of human tumors [517]. Similarly, in colon carcinoma, EMT occurs at the invasive front and produces single migratory cells that lose E-cadherin expression. This is concomitant with deregulation of the Wnt pathway and a selective loss of the basement membrane [518].

### *4.6.7 Adherens Junctions in Collective Cell Migration*

Two modes of cell migration collaborate to assume tissue remodeling events that include embryonic morphogenesis, immunity, wound healing or diseases such as chronic inflammations and cancer metastasis: single cell migration and collective cell migration [334, 519]. They both include a crosstalk between different cellular adhesion complexes: cell migration at the single cell level as observed during EMT, requires disassembly of cell-cell contacts that occurs concomitantly to the enrichment of cell-ECM interactions, whereas collective migration is a directional cell movement characterized by the maintenance of intercellular contacts that occurs without EMT or with incomplete EMT (for review [520, 521]). In such migration, cells move as clusters of various sizes and forms (sheets, strands, ducts) rather than individually, and use similar actin- and myosin-mediated protrusions and guidance by environmental factors. Furthermore, moving group of cells structurally modify the tissue along the migration path, either by clearing the way or by causing ECM modifications, including the deposition of a basement membrane. Initially restricted to epithelial cell types, which are able to maintain stable cell–cell junctions, the field of collective cell migration now includes cooperative behavior of mesenchymal cells [522].

As discuss previously, EMT process has been clearly demonstrated in vitro to increase cancer cell motility and invasion, however the infrequent identification of single invading cells during clinical examination of tumor sections, possibly due to the detection-limit difficulties, has led to a debate about the role of EMT in these processes [523, 524]. In vivo, events of EMT can be suggested by the observation of mesenchymal markers increase and a corresponding down-regulation or loss of E-cadherin, as visualized after immunohistochemistry labeling. In both high- and low-grade epithelial cancers, such as breast, colon and prostate carcinoma, distant metastases are typically similar in histology to the primary tumor, with few signs of EMT, suggesting that predominantly collective cell invasion occurs in these cancers [525, 526]. Several hypotheses have been advanced to take in account these observations (for review [527]): (i) invasion and metastasis can take place without EMT; (ii) routine pathological examination has difficulty in identifying single cell migration; (iii) only small numbers of cells may partially or transiently undergo EMT for invasion.

Evidences for collective invasion of tumor cells came from the analysis of human cancer lesions, in which neoplastic multicellular clusters have crossed the tissue boundaries and have extended into the adjacent stroma while retaining intact cell–cell junctions and intact cadherin expression [528]. In vitro, Friedl lab showed that implantation of human tumor samples into 3D collagen matrix results in multicellular clusters that actively migrate from tissue into the surrounding collagen [529]. However, they also provide direct in vivo evidence for collective invasion of cancer cells using high-resolution multi-photon imaging together with dorsal skinfold windows, after injecting HT-1080 fibrosarcoma cells into the deep dermis of mice [529]. Similar observations were done by real-time imaging of MTLn3E breast tumors [530]. These two contributions also reported that tumor microenvironment is a key determinant for the choice between singe cell invasive migration and collective invasion. This can occur by various mechanisms involving: (i) chemokines and GF that might be either freely diffusing or tethered to the ECM macromolecules [531]; (ii) autocrine or paracrine release of soluble factors by cells in the group or stromal cells, such as stromal cell-derived factor 1, FGF and TGF-β [532–538];

(iii) specific expression of integrins in the leader cells [538–540]. Together with the differential expression of GF receptors in cells located at the front and rear of the cluster, these factors contribute to maintain the polarity of cells within a cluster [519, 534]. Cadherins and cell-cell adhesions may also be a key determinant of invasion mode. On good example is that down-regulation of E-cadherin, (transcriptional or post-transcriptional) or blockade of β1 integrin during collective migration, leads to transitions in cancer cell invasion modes, rather than stopping the movement [540, 541].

Cellular and molecular mechanisms underlying collective migration are still a « work in progress », however several reports highlight the particularly role of active dynamic AJs between motile cells (for review [520, 542]). Because cancer is a slow and long-term process that is not readily amenable to direct microscopic observation, most of the data concerning the mechanistic of collective migration came from models based on morphogenesis (*Drosophila ovary* or zebrafish models) or regeneration (astrocytes migration in a wound-healing assays). However, it was hypothesized that collective movements observed during development require cadherin-mediated AJ, which are also involved in the collective invasion of cancer cells and that they share similar mechanistic. Finally, collective cell migration involves diverse processes rather than a uniform process, which depend on both the cell type from which the group originates and the tissue that is transmigrated [519, 543]. High-resolution confocal microscopy coupled with photoswitchable fluorescent proteins (GFP-tagged E-cadherin), indicate that Src/FAK signaling regulates collective cell migration of the human A431 squamous carcinoma cell line in vivo (reviewed by [544]). Indeed, treatment of tumors with the kinase inhibitors PF-562,271 and dasatinib, results in a shorter recovery half-time of GFP-E-cadherin at sites of cell-cell adhesion and inhibition of cell migration within the primary tumor [545]. Further in vitro investigation revealed that these dynamic changes in E-cadherin mobility correlate with inhibition of E-cadherin endocytosis, inhibition of collective invasion into a 3D collagen matrix and strengthening of cell-cell adhesion. These imaging studies provide new insights into the existence of regulatory crosstalk between components of AJ and integrin-ECM adhesions in vivo [546]. Using the Drosophila ovary as a model, Cai et al. demonstrated that interactions between border and nurse cells function in a positive feedback loop with Rac and actin assembly to stabilize forward-directed protrusion and directionally persistent movement thus contributing to the directed collective migration of border cells between the nurse cells [547]. Moreover, using the FRET based traction sensor, the authors have assessed the forces exerted through cadherin-mediated contacts and showed that the tension is higher at cells at the front than at the rear of the cluster. In the leader cells, this tension participates in a positive feedback loop, which locally increases GF-induced signaling. E-cadherin-mediated adhesion between individual border cells communicates direction from the lead cell to the followers, whereas E-cadherin-mediated adhesion between motile cells and polar cells holds the cluster together and polarizes each individual cell.

Dynamic role of AJ between motile cells have been studied in astrocytes migrating in a wound-healing assay using fluorescently labeled N-cadherin [548].

They found that N-cadherin engaged in AJ undergoes a continuous treadmilling resulting from an actin-coupled retrograde flow along the lateral sides of adjacent leading cells sustained by a polarized recycling of cadherins. At the leading edge, N-cadherin complexes accumulate in a p120ctn-dependent manner providing material for the continuous formation of new AJ at the front of adjacent migrating cells. GSK3-dependent phosphorylation of p120ctn at $Thr^{310}$ is essential for the regulation of AJ from the front to the rear of migrating cells dynamics during collective cell migration; it induces the local dissociation of p120ctn from AJ and clathrin-coated pits mediated endocytosis of N-cadherin. In p120ctn-depleted cells, N-cadherin endocytosis at the rear is increased, leading to the progressive detachment of the leading cells with an increase cell speed, suggesting that AJ serve as a break for the migration of the leading cells while preventing their separation from the rest of the cluster. In carcinoma cells, loss of E-cadherin expression, together with up-regulation of N-cadherin and neural cell adhesion molecules, may provide a collective migration in which cell-cell junctions are retained similarly to that previously described. High-resolution in vivo imaging in the pre-clinical setting will aid understanding of tumor/host interplay and predicting responses to therapeutic intervention.

## 4.7 Stress as an Example of Environmental-Mediated Alteration of AJ

Intestinal homeostasis depends on the complex network formed by the various cell populations distributed on the different layers of the GI tract. Among these cells, enteric nervous system, immunocytes, and enterochromaffin cells constitute a microenvironment for IEC, capable of regulating their organization as well as their function through cell contact or mediator signals. On one hand, stress modulates the activity of neuroendocrine and immune systems altering the release of cytokines or neuroendocrine factors that may disturb barrier integrity [549–553]. On the other hand, neuronal activation following stress may regulate expression of stress-ligand and receptors as well as their function in IEC, directly according to the close proximity between enteric neurons and these cells or indirectly after stimulation of the mucosal immune cells [554, 555]. In this part, we will focus on the role of stress neuromediators are key regulators of intestinal barrier (for review [556]).

### *4.7.1 Stress as an Inducer of Cell Permeability*

In the past 10 years the influence of psychological and environmental stress on pathogenesis received increased awareness [557]. In particular, stress is recognized to participate in the development and/or aggravation of GI disorders such as IBD [553, 558–561]. Stress effects are mediated through the secretion of specific stress neuromediators, such as corticotropin releasing factor (CRF) or its analogs Urocortin 2 and 3 (Ucn2/3) [553]. These peptides act through the corticotropin releasing factor receptors 1 and 2 (CRF1/CRF2), two class II G protein coupled receptors (GPCR) with different affinities [559]. CRF receptors are primarily coupled to Gαs and trigger cAMP formation via adenylyl cyclase activation [558]. This signaling pathway has been described to be responsible for the intercellular dissociation of IEC [562]. In addition to their canonical pathways, GPCRs as well as CRF receptors are able to activate the Src kinase by promoting its auto-phosphorylation on Y-418 [563]. Thus, by modulating the phosphorylation status of intercellular junction proteins, Src activation could lead to the opening of the barrier [21]. Using HT-29 and SW620 cells, two CRC cell lines differing in their abilities to perform cell-cell contacts, we found that CRF2 signals through Src/ERK pathway to induce the alteration of cell-cell junctions. This occurs through an increased endocytosis of E-cadherin and the shuttle of p120ctn and Kaiso into the nucleus [564]. Furthermore, both immobilization stress of rats and intravenous/intracerebral injection of CRF in conscious non stressed rats, lead to increased colonic mucosal levels of cyclooxygenase-2 (COX-2) mRNA and prostaglandin E2 (PGE2) secretion [565, 566]. RhoA-dependent COX-2 signaling has been shown to disrupt formation of AJ in HCA-7 cells, whereas PGE2 signaling mediates TJ disassembly through a mechanism that involves PKC and claudin1 in Caco-2 cells [567].

If CRF system is well documented in the GI tract for its expression and regulation by stress and inflammation, its implication in CRC is poorly investigated [399, 568]. Ligands and receptors are expressed and secreted by various normal and cancer cells (for review [569]). Not only the level of expression but also the cellular distributions of CRF system component have to be taken in account for their role in cancer. Indeed, CRF2 cytoplasmic pattern was associated with diseases in a more advanced stage [570, 571]. Accordingly, our results describe aberrant expression of CRF2 and ligands in both CRC tumors and cell lines, according to their grade and/or differentiation status [564]. CRF2 expression and cytoplasmic distribution increase in de-differentiated or metastatic tumors. This suggests a role for CRF2 in the loss of cellular organization and tumor progression. In HT-29 cells, this signaling pathway also leads to the remodeling of cell adhesion by the phosphorylation of FAK and a modification of actin cytoskeleton and focal adhesion complexes. In CRC, the activation of the Src kinase and the associated disturbance of E-cadherin expression and/or localization is correlated to histological grading and is of worse prognosis [572]. Cell-cell dissociation provokes a remodeling of cell adhesion and an increase in migration and invasion, two major processes in CRC development. The CRF system has been proposed to affect migration and invasion of tumors, thus supporting tumor progression, in part by modulating RhoGTPases activity [573, 574]. The CRF induces migration and invasion of B16F10 murine melanoma cells, a process that also depends on a transient ERK activation via the CRF1 [575]. It participates to actin polymerization and FAK phosphorylation, which lead to MCF-7 motility [576]. Furthermore, CRF signaling is implicated in carcinogenesis-related process such as proliferation, apoptosis and angiogenesis [577–581]. Using the 4T1 breast cancer cell line, Arranz et al. demonstrated that peripheral CRF modifies the expression of SMAD2 and β-catenin, induces cell proliferation and increases the TGF-β action on proliferation, confirming its impact on TGF-β and Wnt signaling [582].

In conclusion, CRF system mediated-signaling controls cellular organization and may promote metastatic potential of human CRC cells through an EMT-like process. Autocrine/paracrine actions of the CRF system have been suggested to be involved in the microenvironment control of the tumor and neighboring cells [576, 583]. In the tumor microenvironment, CRF is released by endothelial and immune cells and by the local neuronal innervations [584]. The non-tumoral cells could also be source of CRF ligands, which is influenced by stress and inflammation.

### *4.7.2 Crosstalk Between CRF and Cholinergic Systems*

Pioneer experiments showing a crosstalk between stress and muscarinic receptors came from experiments in Wistar-Kyoto rats. Comparative analysis of stress-induced cell permeability between Wistar Kyoto and the parent Wistar strain indicated that increase of permeability was higher in Wistar Kyoto rats compared to the parents. An important difference between the two strains is their acetylcholinesterase activity, which is much lower in the Wistar Kyoto rats, suggesting that acethylcholine (ACh) plays a role in these rats prior to restraint stress exposure [585]. This, and the fact that atropine, considered to act only in the periphery, can prevent the breakdown of the barrier, suggests the involvement of peripheral cholinergic muscarinic mechanism in the epithelial tightness [586, 587]. More recently, it has been proposed that maternal deprivation (MD)-induced barrier dysfunction was dependent on the release of Ach by enteric nerves following CRF-mediated activation of CRF2 [588]. Acute stress also activates cholinergic pathways to trigger exocrine pancreatic secretion. Trypsin released in these conditions may be responsible for colonic barrier alterations through activation of PAR2 [589].

ACh signals through either muscarinic (G-protein-coupled) receptors (mAChR) or nicotinic (ligand-gated ion channels receptors (nAChR), which are expressed in neuronal and non neuronal cells as well as in cancer cells [590, 591]). Both receptors were identified as important regulators of cancer cells [592–594]. ACh production results in neurocrine, paracrine and autocrine promotion of cell proliferation, survival and migration. In GI tissue, the major mAChR subunits are M1 mAChR, M2 mAChR and M3 AChR, but colon epithelial cells mainly express M1 AChR and M3 AChR [595–597]. Their expression is increased up to eight-fold in cancer compared to normal tissue [598]. This overexpression, which is observed in 60 % of colon cancer examined, is associated with a drop of degradation and an increase in ACh synthesis [599].

Colon epithelial cells express muscarinic receptor subtypes 1 and 3 (M1 mAChR and M3 mAChR). In colon cancer cells, activation of M3 mAChR stimulates cell proliferation induced by ACh or bile acid [600, 601]. These agonists binding to M3 mAChR results in MMP7 activation that releases EGF receptor (EGFR) ligand named HBEGF [602, 603]. Co-expression of M3 mAChR and EGFR in many colon cancer cell lines, associated with over-expression of these receptors in the majority of colon cancer, suggest that the functional interaction between M3 mAChR and EGFR is important for colon cancer regulation. The transactivation of EGFR also mediates the effect of M3 mAChR on cell migration and invasion. The downstream signaling involves ERK and PI3K-dependent RhoA activation as well as FAK activation and alterations in the actin cytoskeleton [604, 605]. In HT-29 and Caco-2, we found that ACh-activated M3 mAChR also results in cell dissociation together with changes in expression and localization of intestinal TJ and AJ proteins, a process associated to epithelial barrier failure [606]. This correlates with the observation that mAChR (in particular M3 AChR) increases epithelial cell permeability [607]. Furthermore, we showed that M3 AChR not only induces modulation of cell-cell junctions but also regulates cell-MEC interactions that correlate with the acquisition of invasive potential [606]. In our work, the effect of M3 mAChR is in part mediated by CRF signaling since M3 mAChR stimulation stimulates Ucn3 production and CRF2 activation (Pelissier-Rota et al., manuscript in preparation) (Fig. 4.5).

Using different in vivo models, Raufman et al. showed that M3 mAChR gene ablation decreases both colon tumor number and size and the degree of dysplasia [597]. Indeed, only low-grade adenomas are detected from $Apc^{min/+}$ M3 $mAChR^{-/-}$ mice while both low and high-grade adenomas are observed in $Apc^{min/+}$ M3 $mAChR^{+/+}$ mice [608]. Furthermore, M3 mAChR selective inhibitors have been proven to reduce the size of small cell lung cancer xenografts in nude mice [609]. Together, these findings indicate that M3 mAChR activation plays a strong role in intestinal tumor promotion by modulating key processes of carcinogenesis. This feature is particularly important in UC patients in which the expression of M3 mAChR is increased.

### *4.7.3 Intestinal Microbiota*

The intestinal microbiota represents more than 1150 bacterial species that, in the vast majority, are able to influence IBD development. Each individual hosts approximately 160 species that are settled in the first 2 years of life and remain remarkably stable. The most represented phyla are: *Firmicutes*, 30.6–83 % (*Clostridium*, *Ruminococcus*, *Eubacterium*, *Dorea*, *Peptostreptococcus*, *Peptococcus*, *Lactobacillus*); *Bacteroidetes*, 8–48 % (*Bacteroides*); *Actinobacteria*, 0.7–16.7 % (*Bifidobacterium*) and *Proteobacteria*, 0.1–26.6 % (*Enterobacteriaceae*) [610]. Analysis of inflamed bowel segments from UC and CD patients indicates a significant reduction in the biodiversity of fecal microbiota characterized by the absence of *Firmicutes* and *Bacteroidetes* and a large representation of *Enterobacteria* (CD) and E. *coli* (CD and UC) compared to healthy people.

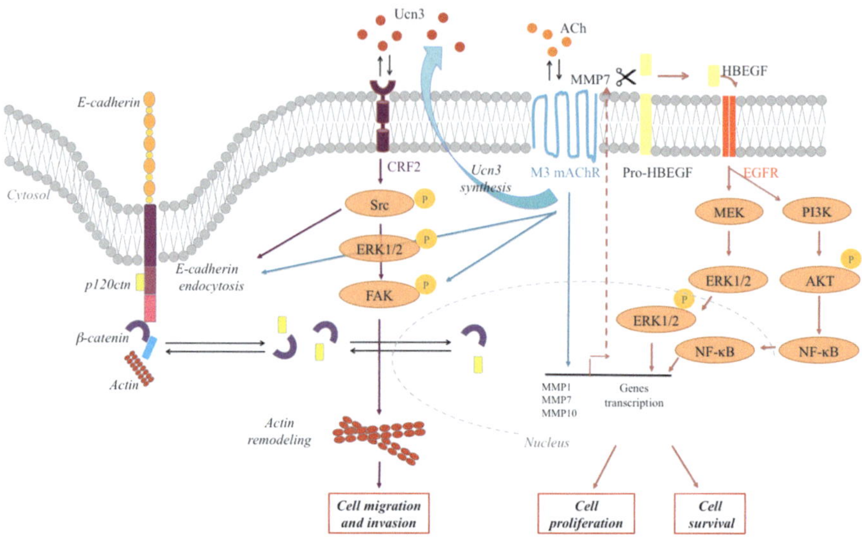


**Fig. 4.5** Model of CRF2 and mAChR crosstalk. On one hand, ACh signals through muscarinic receptors (mAChR), which are express in "normal" and cancerous epithelial cells to induce MMP7 expression. Thus MMP7 catalyze the release of HBEGF from pro-HBEGF inducing the signaling of EGFR: The MAPK/ERK pathway regulates cell proliferation whereas the PI3K/Akt pathway regulates cell survival (Raufman's team findings). On the other hand, Activation of mAChR also induces Ucn3 release. This ligand binding to CRF2 leads to Src/FAK activation following by cell dissociation, E-cadherinendocytosis, migration and invasion

#### 4.7.3.1 Probiotics

The intestinal flora is a potential target for treatments aiming at decreasing the IBD inflammation and promote digestive repair [611, 612]. For example, the non-pathogenic yeast *Saccharomyces boulardii* (*Sb*) is considered as an effective probiotic due to its resistance to the acidic pH found in the digestive tract [613, 614]. It has been shown that *Sb* added to 5-ASA, an anti-inflammatory drug classically used in IBD treatment, allows remission of crohn patients [615]. Indeed, it is suggested that *Sb* may modulate the inflammatory response of the host cell by secreting factors that facilitate the restoration of the intestinal epithelium [616]. In a model of 2,4,6-Trinitrobenzenesulfonic acid (TNBS)-induced acute colitis in mice, *Sb* protects the intestinal mucosa through inhibition of NFkB, leading to the decrease of pro-inflammatory cytokines [617]. *Sb* secretes an anti-inflammatory factor called the anti-inflammatory factor Saccharomyces (SAIF), which inhibits the production of IL-8 and reduces the response of the intestinal cells to pro-inflammatory cytokines [618, 619]. In addition to this anti-inflammatory effect, *Sb* may contribute to IEC differentiation. In vivo studies performed in rats demonstrated that the expression of *Sb* increases the expression of sucrose-isomaltase, an epithelial marker of IEC differentiation [620]. Finally, the supernatant of *Sb* contains "motogenic" factors that accelerate the intestinal restitution by increasing migration. This step results from an *Sb*-mediated increase α2β5 integrin affinity towards collagen type I and a competition with αvβ5 integrins to inhibit its interaction with vitronectin [616, 621].

#### 4.7.3.2 Microbiota and Cancer

Structure and characteristics of microbiota are altered in the CRC and in precancerous lesions; however, it is not clear whether these modifications are causal in tumor development or whether they are some of the consequences of tumorigenesis induction [622, 623]. In CRC patients, microbiota modifications result in a decrease of *Bifidibacterium*, particularly *Bifidibacterium longum* and an increase of *Proteobacteria, Dorea spp. Coprococcus*. Surprisingly, the CRC is associated with an over-representation of comensal bacteria such as *Fusobacterium nucleatum* or *Coriobacteridae*. These are Gram$^-$ bacteria, which promote the development of the CRC or the acceleration of tumorigenesis in existing adenocarcinomas by releasing a virulent factor: FadA. FadA is an adhesion molecule that facilitates the attachment and invasion of microbes to epithelial cells [624, 625]. It increases the paracellular transport by inducing the dissociation of intercellular junctions, thus facilitating the penetration of bacteria through the epithelium. This step leads to the activation of the mucosal immune system [626]. On the other hand, FadA activates the β-catenin signaling and regulates the expression of genes such as transcription factors, stem cell markers and factors stimulating cell proliferation [627]. It has been demonstrated that germ free mice develop fewer colitis than conventional animal. In addition, while the IL10$^{-/-}$ mice treated with AOM develop CAC in the presence of *Bacteroides Vulgaris*, the germ free animals do not develop tumors. Also, microbiota may also contribute to CRC via dysbiosis, a change in the distribution of metabolic activity [622, 623].

## 4.8 Adherens Junctions as Potential Targets for Colon Cancer Therapy

The lifetime risk of developing colorectal cancer in both men and women is about 1 in 20 (5.1 %). Screening for colon cancer can be done by colonoscopy to find polyps, which may develop into cancer according to the duration of their persistence in the colon. Recurrence of colon cancer is common, with an estimated 40 % of cases returning within 3–5 years of treatment. Since the use of 5-fluorouracil (5-FU) in 1962, the treatment of metastatic colorectal cancer has undergone major advances yielding significant improvements in survival over the last decade. These advances have evolved due to the benefits of routine utilization of combination chemotherapy with biologic therapy (for review [628]). In this section we summarize some aspect of cancer therapy that focus on AJ (Fig. 4.4).

### 4.8.1 Soluble Cadherins as Cancer Biomarkers

To provide definitive information about the potential of a tumor to spread and relapse, and thus develop appropriate therapy, we need clinical variables such as the prognostic value of tumor size, involvement of lymph nodes, and metastatic spread (TNM) stage. However, additional molecular markers are needed because these clinical variables often fail. Circulating cancer biomarkers are useful in oncology because there are easy to collect, but there are few candidates according to the sensitivity and specificity of the molecules in the blood of cancer patients. Their source can be either the cancer cells themselves or the associated stroma. Large molecules residing at the apical pole of epithelial cells constitute the currently used tumor markers and their presence in the circulation merely indicates the local loss of diffusion barriers, such as cell junctions and the basement membrane. Their plasmatic concentration reflects not only the tumor volume but also the extent of the barrier defect. The switch between E-cadherin and N-cadherin during the development of tumors cells towards invasive cells, associated to the remodeling of basal membrane in favor of invasion soluble N-cadherin (sN-Cad), is a promising tumor marker candidate [280]. Recently, the presence of sN-CAD was detected in the serum of both healthy individuals and cancer patients, using a direct ELISA and immunoprecipitation assays. Furthermore a correlation was found between prostate-specific antigen (PSA) and sN-CAD [629].

### 4.8.2 Chromatin Modulating Agents as Target

Transcriptional repressors of E-cadherin may be considered as useful therapeutic targets for the prevention of local invasion and distant metastasis in GI malignancies. However, these transcriptional repressors of E-cadherin also have other cellular functions. Both Snail and Twist have been associated with anti-apoptotic functions and Snail has also been implicated in cell adhesion and migration. Furthermore, according to the grade and location of cancer cells, levels of E-cadherin can be very different. This parameter must be considered before starting any pharmacological treatment targeting regulators of E-cadherin expression or EMT. Chemotherapeutic compounds currently being used for the treatment of colorectal cancer include 5-FU, in combination with cisplatin analogs such as oxaliplatin, leucovorin or Irinotecan hydrochloride or drug combinations FOLFOX or FOLFIRI [628]. Only 25 % of patients respond to 5-FU treatment; the addition of oxaliplatin can increase responses to 50 %. Due to its rapid degradation and inconsistent GI absorption, 5-FU can only be administered intravenously. These inadequacies have favored the search of analogs, which could be taken orally. While the addition of oxaliplatin to adjuvant 5-FU has resulted in 3-year disease free survival rates of 76–78 %, recurrent disease is still encountered in many patients such that new treatment options are critically needed [630, 631].

#### 4.8.2.1 DNA Methyl Transferase (DNMT) and Colon Cancer

The intricate interactions of EMT-inducing transcription factors and chromatin remodeling complexes broaden the therapeutic potentials of epigenetic drugs via DNA demethylating and histone deacetylase/demethylase inhibitors, which can act upon metastasis-related genes. The discovery of tumor suppressor genes methylation dates back for almost 30 years, however it is only recently that we have gained enough insight into this DNA modification to use it to our advantage. A great variety of nucleoside analogs such as 5-azacitidine, that require incorporation into RNA or DNA in order to inhibit DNMT has been developed as well as non-nucleoside analogs, which have demethylation capabilities and act by directly impairing the action of DNMT. RG108 is a small molecule with a significantly longer half-life in aqueous solution than 5Aza-dC, which has been designed to the active site of DNMT1 and has concentration-dependent demethylating capability in vitro without cytoxicity [632, 633]. These drugs have been tested in various conditions of administration and frequency (intermittent versus continuous therapy) and were shown to have beneficial effects in leukemia (phase I, II, and III trials). In contrast, treatment with azacitidine in patients with solid tumors (prostate cancer) at the doses and schedules initially investigated did not demonstrate significant anti-tumor activity thus limiting its used in epithelial cancer such as colon cancer (for review [331, 634]). However, experiments aiming at modifying the level of DNMT1 were used as tools to study the effect of DNA hypomethylation on intestinal tumorigenesis using the $APC^{Min/+}$ mouse model of intestinal tumors [635–637]. Treatment of $APC^{Min/+}$ mice with the 5Aza-dC, significantly reduces tumor formation in the intestine, suggesting that DNA methylation may play an important role during tumorigenesis [636]. Moreover, overexpression of DNMT3B1 in the $APC^{Min/+}$ model enhanced colorectal carcinogenesis and caused tumor suppressor gene methylation [638]. Other studies have also shown that crossing $APC^{Min/+}$ mice with DNMT1 hypomorphic mice results in complete suppression of macroscopic intestinal neoplasia [639]. At the molecular level, DNMT inhibitor 5Aza-dC was able to restore E-cadherin expression and revert cancer cells to the epithelial phenotype in various cancer cell lines [333, 640, 641]. In an in vivo model of breast cancer using cell-line MDA-MB-435S, this inhibitor restores E-cadherin expression and suppresses metastasis formation as well as primary growth, possibly through E-cadherin up-regulation [642].

Recently, epigallocatechin-3-gallate (EGCG), the principal polyphenol in green tea has been describe to inhibit DNMT [643]. Compared to 5′-aza-2′-deoxycytidine, this molecule has a lesser effect on DNA methylation and gene reactivation in various cell lines however, thanks to its tolerability and mild activity it may be used in longer-term therapy [644].

Even if therapy based on DNMT inhibitors are not used in colon cancer treatment yet, alterations such as methylation changes are found in fecal DNA of patients with colorectal cancer [645]. In fact, of the three genes studied, over 96 % of patients with cancer and 81 % of patients with precancerous lesions had at least one methylated gene. In the normal individuals, only 1 patient of the 24 studied had a methylated gene. Hypermethylated p16 in circulating DNA of patients with colorectal cancer has been associated with advanced stage and has been found in patients with recurrent disease [646, 647]. Epigenetic changes in the serum of patients already diagnosed with colorectal cancer provide a potential means of prognostication before and during therapy.

The association between aging and increased incidence of cancer has long been noted. Since alterations in the epigenome have been observed as a result of aging, the accumulation of genetic mutations is not only evoked as a precursor of the disease. As seen in patients already diagnosed with cancer, aging brings about a global hypomethylation with hypermethylation of specific gene promoters. A pertinent example is found in sporadic colorectal tumors with microsatellite instability [648].

#### 4.8.2.2 HDAC and Colon Cancer

According to the anti-inflammatory effect in the models of experimental colitis as well as the generally known anti-proliferative capacity of HDAC inhibitors, it has been suggested that HDAC inhibition would assume a strong tumor-preventing capacity in particular in cancer-associated-colitis (CAC), which differs from "non-inflammatory" colorectal cancer [649, 650]. It has been demonstrated that the binding of 5-methyl-cytosine protein causes the recruitment of HDAC, suggesting a synergy between methylation and histone acetylation to regulate gene transcription [651, 652]. These observations have stimulated clinical trials investigating the combination of a demethylating agent with an HDAC inhibitor. First trials, performed in myelodysplastic syndromes or acute myeloid leukemia, showed an improvement in clinical activity with sequential administration of these two molecules [653]. A number of clinical trials with other HDAC inhibitors have been tested and are also ongoing.

The anti-inflammatory potency of HDAC inhibitors was evaluated in DSS- and TNBS-induced colitis [654]. Oral administration of HDAC inhibitors of different classes (the suberoylanilide hydoxamic acids SAHA and ITF2357 as well as the short-chain fatty acid valproic acid) suppressed the inflammatory parameters in both models. This effect resulted in a suppression of the production of pro-inflammatory cytokines at the site of inflammation and a reduced histological inflammation score. SAHA is also capable of stimulating E-cadherin expression in endometrial carcinoma cell lines, which may represent another mechanism for the regulation of inflammation [655]. This approach was also developed in two models of inflammation-associated cancerogenesis, mediated by chronic colitis, where tumors develop at the site of inflammation. The continuous treatment with either SAHA or ITF2357 attenuated both tumor growth and development in the AOM/DSS model, which is based on a single injection of the carcinogen AOM followed by induction of a chronic DSS colitis [656–658]. In the model of IL10 knockout mice, animals spontaneously developed colitis, which progress to dysplasia and to infiltrating carcinomas with ageing. Again, treatment with HDAC inhibitors (ITF2357) was associated with an inhibition of tumorigenesis even after a previous treatment with celecoxib, an inhibitor of COX-2 which accelerates the progression of tumor development in this model [657]. These findings demonstrate the efficacy of ITF2357 in the therapeutic model of inflammation-mediated carcinogenesis. Recent human data provide evidence that ITF2357 can be administered in doses sufficient to exert anti-inflammatory capacity suggesting that these compounds might in fact provide a future pharmacological strategy for patients suffering from IBD [659]. The majority of currently available HDAC inhibitors are so-called pan-HDAC inhibitors, which inhibit the activity of a broad spectrum of HDACs [660]. One future goal will be to define the relevance of these inhibitors within the process of colorectal cancer.

IEC are integral players of intestinal homeostasis and barrier maintenance is under the influence of microbiome. In some circontances, the relationship between IEC and microbiota becomes pathological and contributes to diseases such as IBD. The recent metagenomic analyses of microbiome changes have advanced our understanding of the modified microbiome in IBD, with principally reduced representation of short chain fatty acids (acetate, propionate and butyrate, SCFAs) producing strains, which correlates with a decreased SCFA levels in feces of children with IBD. SCFAs are microbial metabolites and dietary components. Among SCFAs, butyrate induces the highest acetylation and has been proposed to inhibit class I and class II HDACs [661]. One possibility is that butyrate inhibits deacetylation, but not HDAC activity directly [662]. The link between butyrate and E-cadherin up-regulation was first discovered in two breast cancer cell-lines, in which it induces cell-cycle arrest and increases cell–cell adhesion which are reversible upon addition of E-cadherin antibodies [663]. Subsequently, butyrate was found to up-regulate E-cadherin expression in colon cancer cells and in endometrial carcinoma cells [655, 664, 665]. In Caco-2 cells, butyrate provides an enhancement of barrier function by reducing paracellular permeability and facilitating TJ assembly [666, 667]. These effects may result from the potential of SCFAs to alter epigenetic marks in epithelial cells [668]. Hyperacetylation linked to the use of HDAC inhibitors is associated with amelioration of experimental colitis in mice [654]. Increased intake of SCFAs has the capacity to improve colitis disease parameters, possibly via the impact on HDAC activity [669]. A clear impact of HDAC inhibitors is described on the generation of regulatory T cells, an effect that may contribute to the efficacy of HDAC inhibitors to colitis [670].

Butyrate can also regulate epithelial function such as cell proliferation, survival, and apoptosis [668, 671]. In particular butyrate represses p21, a potent cyclin-dependent kinase inhibitor that functions as a regulator of cell cycle progression at G1; an effect that can be overcome by overexpression of HDAC3 [672, 673]. Additionally, HDAC3 protein level was significantly up-regulated in a panel of human colon tumors compared with adjacent normal mucosa and in small intestinal adenomas derived from APC mutant mice epithelia, establishing a link between HDAC3 expression and intestinal cell transformation.

### *4.8.3 EMT as a Potential Therapeutic Target*

EMT signatures can be exploited to enhance classical chemotherapy treatments. Furthermore, pharmacological therapies designed to inhibit these proteins will aim to prevent local and distant tumor invasion.

#### 4.8.3.1 Cell Adhesion Molecules

Gain of N-cadherin expression is a typical example of the EMT signature and it has been linked with drug resistance [674]. According to the inhibitory effect of cyclic HAVD peptide on N-cadherin function in neuronal cells, Adherex (Durham, NC, US) developed a cyclic pentapeptide (ADH-1 or Exherin) as N-cadherin antagonist for use as an intravenous administered anticancer agent [675, 676]. ADH-1 has entered clinical testing and successfully passed phase I studies in patients with advanced solid tumors such as esophageal, non-small lung, renal cell and hepatocellular carcinomas expressing N-cadherin [677]. Shingosine-1Phosphate (SP1) regulates N-cadherin dependent angiogenesis via the correct trafficking of N-cadherin to the plasma membrane [678]. A monoclonal antibody to SP1 substantially reduced tumor progression in murine models via anti-angiogenic and anti-tumorigenic effects. Indeed the antibody blocks endothelial cell migration due to decreased N-cadherin shedding but also controls cancer cells proliferation through the down-regulation of β-catenin nuclear signaling.

EpCAM is also intensely used as a therapeutic target for antibody-based approaches in cancer with the strongest negative impact on prognosis. EpCAM signaling itself may be a target for therapeutic intervention. Whereas inhibitors of presenilins will not be specific for EpCAM, their use for cancer therapy may receive an additional development rationale by inhibiting EpCAM signaling. The protease inhibitors may increase levels of intact EpCAM, which serves as the extracellular anchor for anti-EpCAM antibodies with immunological activity. Another interest of EpCAM as therapeutic target lies in its expression in cancer stem cells, an approach that is currently being assessed in vitro, in animal models.

#### 4.8.3.2 Growth Factors and Tyrosine Kinase

Selective tyrosine kinase inhibitors are promising to treat cancers driven by activated tyrosine kinases such as Abl in chronic myelogenous leukemia, c-Kit in GI stromal tumors and EGFR in non-small cell lung cancer (NSCLC). It has been showed that EC-4 repeat of N-cadherin contains a sequence that associates with the HAV motif present in the FGFR leading to cell signaling [391, 679]. Indeed, synthetic peptides containing the IDPVNGQ sequence and an antibody directed against the EC-4 domain of N-cadherin inhibited both N-cadherin and FGF-induced neurite outgrowth and cancer cell migration [679, 680].

The development of targeted therapies has provided further gains on recent advances in chemotherapy approaches in colon carcinogenesis. Though the activities of these agents used alone were translated into substantial responses, their use in combination with chemotherapy resulted in the improvement of response and survival rates. For the treatment of metastatic colorectal cancer, the Food and Drug Administration (FDA) has approved three targeted agents including bevacizumab, cetuximab, and panitumumab, however the treatment only concern patients with advanced colorectal cancer whose tumors express EGFR. In subsequent studies, no correlations were found between EGPR copy number and response to biotherapy [681–683]. In addition, there is no difference in benefit for patients whose tumors express low versus high levels of the receptor [684, 685]. Even patients with negative staining for EGFR seem to demonstrate a response to the agent [684]. The lack of reliably defined EGFR patterns, as well as the variability in results combined with the cost of the treatment have limited the routine applicability of this approach, even if this treatment results in important survival gains [686, 687]. Furthermore, colorectal patients with a mutated KRAS do not respond to such inhibition of EGFR signaling [688]. Indeed, in approximately 30–50 % of colorectal cancers, KRAS can be constitutively activated such that inhibition of EGFR signaling will have no effect on KRAS activity, resulting in continued cancer cell growth [689–692]. Based on these results, the European Committee for Medicinal Products for Human Use made the decision to require confirmation of wild-type KRAS status prior to initiation of panitumumab therapy [693].

Src activity is increased in 80 % of colon cancer patients, and this increase correlates with the progression from benign polyps to carcinoma and then to metastatic cancer [694–697]. Development of metastasis constitutes the main cause of death in colon cancer; thus, their inhibition could reduce mortality. Src plays a key role in all steps of colon cancer metastasis via its downstream targets and its overexpression in rat colon epithelial cells elicits invasive behavior by increasing cell adhesion, invasion, and migration [698]. However in these conditions, Src does not modulate cell proliferation, suggesting that it may participate in late stage of carcinogenesis. Therefore, the inhibition of Src may facilitate the treatment of metastatic colon cancer to increase the survival rate. Increased activity of Src can not be attributed to an activating mutation of the Src gene only, since such mutation is found in 12 % of patients only [699, 700]. Src activity has also been associated to Src-induced drug resistance that occurs via the activation of p-glycoprotein (p-gp), a member of the ABC transporter family implicated in the transport of the drug [701, 702]. Tumor cells use p-gp to escape chemotherapeutic drug-induced death since an increase in the amount of p-gp leads to an increased efflux of drug and decreased bioavailability. An excellent recent review focuses on the role of Src in the metastasis of colon cancer and the problems of using Src inhibitors in a clinical setting [703]. Many compounds have been developed to use as Src inhibitors, and their effectiveness for the treatment of colon cancer has been tested in cultured colon cancer cells and in xenograft model in nude mice [704]. Some of them are in phase I and II clinical trials, such as dasatinib, AZD0530, and bosutinib and they are used alone or combined with other anticancer drugs, a condition that increases the effectiveness of Src inhibitors [705–707]. Curcumin, a natural product derived from a plant in the ginger family *Zingiberaceae*, is also an effective inhibitor of Src and FAK [708]. Curcumin, which increases dasatinib-induced colon cancer apoptosis, may be used for the prevention or co-treatment of colon cancer. Recently, Src-targeted therapy has been proven to weaken immune responses that are essential for the eradication of cancer cells, a parameter that may explain the unsatisfactory outcomes of clinical trials of Src inhibitors [709, 710]. Overcoming the problem of inhibiting Src in cancer cells while retaining immune system efficacy is the key to the successful application of Src-inhibition therapy.

TGF-β is a key inducer of EMT and stimulates N-cadherin dependent invasion of colon tumor-associated myofibroblasts [711, 712]. Pharmaceutical companies have avoided TGF-β antagonists, partly because they feared to affect its homeostatic roles in unaffected tissue. However, recent work highlights the clinically significant action of soluble TGF-β receptor antagonists, which bind and inactivate extracellular TGF-β, in cancer [713–715]. A specific, small organic TGF-βRI inhibitor (SB-431 542) attenuates cancer cell proliferation and motility in in vitro models [716–719].

### *4.8.4 Stem Cell as Therapeutic Target*

Cancer tissues are composed of several heterogeneous cancer cells among which a small part is supposed to have higher tumor-initiating ability. These higher tumorigenic populations are named "cancer stem cells (CSCs)" or "cancer initiating cells (CICs)" [720, 721]. More recently, the cancer stem-cell concept is gaining importance, because it suggests new approaches to anti-cancer therapies [722, 723]. CSCs/CICs are defined as small population of cancer cells which have: (i) higher tumor-initiating ability, (ii) self-renewal, and (ii) differentiation [724]. CSCs/CICs have been reported to be resistant to the majority of therapy used in colorectal cancer [725]; thus, new treatments targeting specifically CSC/CIC must be developed. Potential markers of colorectal cancer stem cells have been proposed as therapeutic target for colon cancer, including CD133, CD166, CD24, CD44, ALDH1, LGR5, and DCLK1 [724, 726, 727]. In this way, Nakanishi et al., demonstrated that doublecortin and CaM kinase-like-1 (DCLK1) distinguishes between tumor and normal stem cells in the intestine and that specific ablation of $DCLK1^{+}$stem cells resulted in a marked regression of polyps without apparent damage to the normal intestine [727, 728].

### *4.8.5 Vitamin D and Colorectal Cancer*

The most active metabolite of vitamin D3, 1α25 dihydroxyvitamin D3 ($1{,}25(OH)_2D_3$, calcitriol) is a pleotropic hormone with wide regulatory actions, which exerts its action by binding to and activating the vitamin D receptor (VDR), a member of the nuclear steroid-thyroid-retinoid-receptor super-family of transcription factors and regulates gene expression in a ligand-dependent manner [729, 730]. According to its classical role as a regulator of calcium and phosphate homeostasis and bone mineralization, this hormone has been first shown to induce differentiation in myeloid leukemia cells and inhibit the growth of melanoma cells, thus stimulating the interest of cancer research [731, 732]. It is proved now that calcitriol, the active form of vitamin D, regulates multiple signaling pathways involved in proliferation, apoptosis, differentiation, inflammation, invasion, angiogenesis and metastasis. Pioneer study for the role of vitamin D in CRC comes from Frank and Cedric Garland who proposed an inverse correlation between solar radiation and CRC mortality probably due to vitamin D synthesis [733]. To date, numerous epidemiologic and experimental studies support that CRC mortality was inversely related to calcitriol and its presence in serum [734–739]. Recent reviews describe in more details the pleotropic roles of vitamin D in cancers [740, 741]. In this section we just summarize the role of vitamin D3 on AJ proteins in relation to CRC.

According to the differentiation process, calcitriol increases the expression and/or activity of several brush border enzymes, such as alkaline phosphatase and sucrose-isomaltase, and enhances the formation of microvilli [742–745]. In addition, it increases the expression of several components of cell adhesion structures, actin cytoskeleton and intermediate filaments that are essential for the maintenance of the epithelial phenotype: (i) occludin, zonula occludens (ZO)-1 and -2, and claudin and -12 in TJ [746, 747]; (ii) E-cadherin in AJ and sequestration of β-catenin at the membrane [747]; vinculin, filamin A and keratin-13 [747, 748]. Finally calcitriol regulates several genes related to protein degradation machinery, such as proteases and their inhibitors and component of the ubiquitin-proteasome system, all may control the stability of E-cadherin at the membrane [749].

Calcitriol actions in colon cancer cells inhibit β-catenin transcriptional activity, thereby countering aberrant activation of Wnt/β-catenin signaling, which is the most common alteration in sporadic CRC. Indeed, calcitriol can inhibit β-catenin signaling pathway by: (i) increasing the amount of VDR bound to β-catenin, thus reducing its interaction with TCF/Lef transcription factor and leading to the repression of its target genes [747, 750, 751]; (ii) inducing β-catenin export from the nucleus linked to the accumulation of E-cadherin at the membrane [747]; (iii) inducing DKK1, an extracellular Wnt inhibitor [752]; (iv) regulating RhoA-ROK-p38 MAPK-MSK pathway which also increases E-cadherin expression [753, 754].

CRC responsiveness to calcitriol depends on the expression of VDR. VDR expression by normal and certain tumor epithelial cells is associated with high degree of cell differentiation whereas its expression decreases in advanced stages of tumors [755–759]. This downregulation is a cause of failure of the therapy using calcitriol or its analogs. Furthermore, Snails transcription factor represses VDR expression thereby reducing the anti-cancer effect of calcitriol [760, 761]. Increase expression of Snail in human colon tumors (60 %) is associated with a loss of responsiveness to calciriol [762]. Post-transcriptional events participate in the regulation of VDR expression. It has been shown that miR-27b and miR-298 reduce levels of VDR in LS-180 colon cancer cells [763]. miR-125b, which is overexpressed in colon cancer metastases, also down-regulates VDR expression [764]. In contrast, miR-22, which is a target of calcitriol in vivo, is required for the anti-proliferative and anti-migratory effect of calcitriol in colon cancer cells [765]. miR-122 expression is lower in tumor than in matched normal tissues in 78 % of CRC patients and correlate with VDR RNA expression. Therefore the over-expression of miRNAs that target VDR may constitute of novel strategy for CRC. Finally, co-treatment with HDAC inhibitors and DNMT reverses the resistance to the anti-proliferative effect of calcitriol in CRC [766]. A phenomenon probably due to butyrate-induced VDR over-expression [767, 768].

Multiple cell culture and animals models support a protective and therapeutic action of Vitamin D against CRC, however data from human trials are not totally clear yet. Indeed, several studies have shown that calcitriol or analogs reduce the growth of colorectal xenografts (for review [729, 769, 770]). They also reduce tumor load in $APC^{min/+}$ mice and increase the expression of E-cadherin, inhibit that of myc and reduce β-catenin nuclear levels in the small intestine and colon of $APC^{min/+}$ mice [771].

## 4.9 Conclusion

Despite their primordial structural role in stabilizing intercellular cohesion and overall epithelial tissue architecture, AJ are far from being static features. The several inputs responsible for the regulation of AJ presented in this chapter highlight their rather dynamic nature, which is of high importance, both during morphogenesis, physiological epithelial renewal in adults, as well as in pathological inflammatory or cancerous diseases. This explains why AJ proteins studies are central to the field of epithelium biology and biomedecine. However, increasing our knowledge of the cadhesome complex structure, regulation and functions is still a “work in progress”. From the “core” cadherin-catenin complexes to the myriad of cytoplasmic interactors and regulators, the complexity and the diversity of AJ signaling makes it a fascinating and rich domain of research. This complexity offers a great variety of therapeutic strategies and no doubt that the field will take advantage of the fantastic development of large scales “omics” approaches combined with more precisely aimed structural, biophysical and classical animal biology technics.

## References

1. Clayburgh DR, Shen L, Turner JR (2004) A porous defense: the leaky epithelial barrier in intestinal disease. Lab Invest 84:282–291. doi:10.1038/labinvest.3700050
2. Leblond CP, Stevens CE (1948) The constant renewal of the intestinal epithelium in the albino rat. Anat Rec 100:357–377
3. Laprise P, Chailler P, Houde M et al (2002) Phosphatidylinositol 3-kinase controls human intestinal epithelial cell differentiation by promoting adherens junction assembly and p38 MAPK activation. J Biol Chem 277:8226–8234. doi:10.1074/jbc.M110235200

4. Mariadason JM, Bordonaro M, Aslam F et al (2001) Down-regulation of beta-catenin TCF signaling is linked to colonic epithelial cell differentiation. Cancer Res 61:3465–3471
5. Peignon G, Thenet S, Schreider C et al (2006) E-cadherin-dependent transcriptional control of apolipoprotein A-IV gene expression in intestinal epithelial cells: a role for the hepatic nuclear factor 4. J Biol Chem 281:3560–3568. doi:10.1074/jbc.M506360200
6. Gumbiner B, Stevenson B, Grimaldi A (1988) The role of the cell adhesion molecule uvomorulin in the formation and maintenance of the epithelial junctional complex. J Cell Biol 107:1575–1587
7. Zaidel-Bar R (2013) Cadherin adhesome at a glance. J Cell Sci 126:373–378. doi:10.1242/jcs.111559
8. Hulpiau P, van Roy F (2009) Molecular evolution of the cadherin superfamily. Int J Biochem Cell Biol 41:349–369. doi:10.1016/j.biocel.2008.09.027
9. Takeichi M (1990) Cadherins: a molecular family important in selective cell-cell adhesion. Annu Rev Biochem 59:237–252. doi:10.1146/annurev.bi.59.070190.001321
10. Wheelock MJ, Johnson KR (2003) Cadherins as modulators of cellular phenotype. Annu Rev Cell Dev Biol 19:207–235. doi:10.1146/annurev.cellbio.19.011102.111135
11. Goodwin M, Yap AS (2004) Classical cadherin adhesion molecules: coordinating cell adhesion, signaling and the cytoskeleton. J Mol Histol 35:839–844. doi:10.1007/s10735-004-1833-2
12. Jamora C, Fuchs E (2002) Intercellular adhesion, signalling and the cytoskeleton. Nature Cell Biol 4(4):E101–8. http://doi.org/10.1038/ncb0402-e101
13. Nollet F, Kools P, van Roy F (2000) Phylogenetic analysis of the cadherin superfamily allows identification of six major subfamilies besides several solitary members. J Mol Biol 299:551–572. doi:10.1006/jmbi.2000.3777
14. Takeichi M (1988) The cadherins: cell-cell adhesion molecules controlling animal morphogenesis. Development 102:639–655
15. Van Roy F (2014) Beyond E-cadherin: roles of other cadherin superfamily members in cancer. Nat Rev Cancer 14:121–134. doi:10.1038/nrc3647
16. Niessen CM (2007) Tight junctions/adherens junctions: basic structure and function. J Invest Dermatol 127:2525–2532. doi:10.1038/sj.jid.5700865
17. Boggon TJ, Murray J, Chappuis-Flament S et al (2002) C-cadherin ectodomain structure and implications for cell adhesion mechanisms. Science 296:1308–1313. doi:10.1126/science.1071559
18. Horikawa K, Radice G, Takeichi M, Chisaka O (1999) Adhesive subdivisions intrinsic to the epithelial somites. Dev Biol 215: 182–189. doi:10.1006/dbio.1999.9463
19. Nelson WJ, Nusse R (2004) Convergence of Wnt, beta-catenin, and cadherin pathways. Science 303:1483–1487. doi:10.1126/science.1094291
20. Takeichi M (2007) The cadherin superfamily in neuronal connections and interactions. Nat Rev Neurosci 8:11–20. doi:10.1038/nrn2043
21. Gumbiner BM (2000) Regulation of cadherin adhesive activity. J Cell Biol 148:399–404
22. Foty RA, Steinberg MS (2005) The differential adhesion hypothesis: a direct evaluation. Dev Biol 278:255–263
23. Overduin M, Harvey TS, Bagby S et al (1995) Solution structure of the epithelial cadherin domain responsible for selective cell adhesion. Science 267:386–389
24. Pokutta S, Herrenknecht K, Kemler R, Engel J (1994) Conformational changes of the recombinant extracellular domain of E-cadherin upon calcium binding. Eur J Biochem 223:1019–1026
25. Hong S, Troyanovsky RB, Troyanovsky SM (2010) Spontaneous assembly and active disassembly balance adherens junction homeostasis. Proc Natl Acad Sci U S A 107:3528–3533. doi:10.1073/pnas.0911027107
26. Harrison OJ, Jin X, Hong S et al (2011) The extracellular architecture of adherens junctions revealed by crystal structures of type I cadherins. Structure 19:244–256. doi:10.1016/j.str.2010.11.016
27. Tomschy A, Fauser C, Landwehr R, Engel J (1996) Homophilic adhesion of E-cadherin occurs by a co-operative two-step interaction of N-terminal domains. EMBO J 15:3507–3514

28. Häussinger D, Ahrens T, Sass H-J et al (2002) Calcium-dependent homoassociation of E-cadherin by NMR spectroscopy: changes in mobility, conformation and mapping of contact regions. J Mol Biol 324:823–839. doi:10.1016/S0022-2836(02)01137-3
29. Parisini E, Higgins JMG, Liu J et al (2007) The crystal structure of human E-cadherin domains 1 and 2, and comparison with other cadherins in the context of adhesion mechanism. J Mol Biol 373:401–411. doi:10.1016/j.jmb.2007.08.011
30. Harrison OJ, Corps EM, Kilshaw PJ (2005) Cadherin adhesion depends on a salt bridge at the N-terminus. J Cell Sci 118:4123–4130. doi:10.1242/jcs.02539
31. Chitaev NA, Averbakh AZ, Troyanovsky RB, Troyanovsky SM (1998) Molecular organization of the desmoglein-plakoglobin complex. J Cell Sci 111(Pt 1):1941–1949
32. Wu Y, Vendome J, Shapiro L et al (2011) Transforming binding affinities from three dimensions to two with application to cadherin clustering. Nature 475:510–513. doi:10.1038/nature10183
33. Chien Y-H, Jiang N, Li F et al (2008) Two stage cadherin kinetics require multiple extracellular domains but not the cytoplasmic region. J Biol Chem 283:1848–1856. doi:10.1074/jbc.M708044200
34. Zhang Y, Sivasankar S, Nelson WJ, Chu S (2009) Resolving cadherin interactions and binding cooperativity at the single-molecule level. Proc Natl Acad Sci U S A 106:109–114. doi:10.1073/pnas.0811350106
35. Rimm DL, Kebriaei P, Morrow JS (1994) Molecular cloning reveals alternative splice forms of human alpha(E)-catenin. Biochem Biophys Res Commun 203:1691–1699. doi:10.1006/bbrc.1994.2381
36. Chen H, Paradies NE, Fedor-Chaiken M, Brackenbury R (1997) E-cadherin mediates adhesion and suppresses cell motility via distinct mechanisms. J Cell Sci 110(Pt 3):345–356
37. Yap AS, Niessen CM, Gumbiner BM (1998) The juxtamembrane region of the cadherin cytoplasmic tail supports lateral clustering, adhesive strengthening, and interaction with p120ctn. J Cell Biol 141:779–789
38. Nagafuchi A, Takeichi M (1988) Cell binding function of E-cadherin is regulated by the cytoplasmic domain. EMBO J 7:3679–3684
39. Anastasiadis PZ, Reynolds AB (2000) The p120 catenin family: complex roles in adhesion, signaling and cancer. J Cell Sci 113(Pt 8):1319–1334
40. Huber AH, Weis WI (2001) The structure of the beta-catenin/E-cadherin complex and the molecular basis of diverse ligand recognition by beta-catenin. Cell 105(3):391–402. http://doi.org/10.1016/S0092-8674(01)00330-0
41. Ishiyama N, Lee S-H, Liu S et al (2010) Dynamic and static interactions between p120 catenin and E-cadherin regulate the stability of cell-cell adhesion. Cell 141:117–128. doi:10.1016/j.cell.2010.01.017
42. McCrea PD, Turck CW, Gumbiner B (1991) A homolog of the armadillo protein in Drosophila (plakoglobin) associated with E-cadherin. Science 254:1359–1361
43. Behrens J, von Kries JP, Kühl M et al (1996) Functional interaction of beta-catenin with the transcription factor LEF-1. Nature 382:638–642. doi:10.1038/382638a0
44. Cadigan KM, Peifer M (2009) Wnt signaling from development to disease: insights from model systems. Cold Spring Harb Perspect Biol 1:a002881. doi:10.1101/cshperspect.a002881
45. Clevers H, Nusse R (2012) Wnt/β-catenin signaling and disease. Cell 149:1192–1205. doi:10.1016/j.cell.2012.05.012
46. Chen YT, Stewart DB, Nelson WJ (1999) Coupling assembly of the E-cadherin/beta-catenin complex to efficient endoplasmic reticulum exit and basal-lateral membrane targeting of E-cadherin in polarized MDCK cells. J Cell Biol 144:687–699
47. Riggleman B, Schedl P, Wieschaus E (1990) Spatial expression of the Drosophila segment polarity gene armadillo is posttranscriptionally regulated by wingless. Cell 63:549–560. doi:10.1016/0092-8674(90)90451-J
48. Hatzfeld M (1999) The armadillo family of structural proteins. Int Rev Cytol 186:179–224
49. Keirsebilck A, Bonne S, Staes K et al (1998) Molecular cloning of the human p120 ctn catenin gene (CTNND1): expression of multiple alternatively spliced isoforms. Genomics 146:129–146

50. Aho S, Rothenberger K, Uitto J (1999) Human p120 ctn catenin: tissue-specific expression of isoforms and molecular interactions with BP180/type XVII collagen. J Cell Biochem 399:390–399
51. Alemà S, Salvatore AM (2007) p120 catenin and phosphorylation: mechanisms and traits of an unresolved issue. Biochim Biophys Acta 1773:47–58. doi:10.1016/j.bbamcr.2006.06.001
52. Keilhack H, Hellman U, van Hengel J et al (2000) The protein-tyrosine phosphatase SHP-1 binds to and dephosphorylates p120 catenin. J Biol Chem 275:26376–26384. doi:10.1074/jbc.M001315200
53. Mo YY, Reynolds AB (1996) Identification of murine p120 isoforms and heterogeneous expression of p120cas isoforms in human tumor cell lines. Cancer Res 56:2633–2640
54. Thoreson MA, Anastasiadis PZ, Daniel JM et al (2000) Selective uncoupling of p120(ctn) from E-cadherin disrupts strong adhesion. J Cell Biol 148:189–202
55. Van Hengel J, Vanhoenacker P, Staes K, van Roy F (1999) Nuclear localization of the p120ctn Armadillo-like catenin is counteracted by a nuclear export signal and by E-cadherin expression. Proc Natl Acad Sci 96:7980–7985. doi:10.1073/pnas.96.14.7980
56. Buhl AM, Johnson NL, Dhanasekaran N, Johnson GL (1995) G alpha 12 and G alpha 13 stimulate Rho-dependent stress fiber formation and focal adhesion assembly. J Biol Chem 270:24631–24634
57. Goulimari P, Kitzing TM, Knieling H et al (2005) Galpha12/13 is essential for directed cell migration and localized Rho-Dia1 function. J Biol Chem 280:42242–42251. doi:10.1074/jbc.M508690200
58. Jiang H, Wu D, Simon MI (1993) The transforming activity of activated G alpha 12. FEBS Lett 330:319–322
59. Meigs TE, Fields TA, McKee DD, Casey PJ (2001) Interaction of Galpha 12 and Galpha 13 with the cytoplasmic domain of cadherin provides a mechanism for beta -catenin release. Proc Natl Acad Sci U S A 98:519–524. doi:10.1073/pnas.021350998
60. Rieken S, Sassmann A, Herroeder S et al (2006) G12/G13 family G proteins regulate marginal zone B cell maturation, migration, and polarization. J Immunol 177:2985–2993
61. Strathmann MP, Simon MI (1991) G alpha 12 and G alpha 13 subunits define a fourth class of G protein alpha subunits. Proc Natl Acad Sci U S A 88:5582–5586
62. Kaplan DD, Meigs TE, Casey PJ (2001) Distinct regions of the cadherin cytoplasmic domain are essential for functional interaction with Galpha 12 and beta-catenin. J Biol Chem 276:44037–44043. doi:10.1074/jbc.M106121200
63. Meigs TE, Fedor-Chaiken M, Kaplan DD et al (2002) Galpha12 and Galpha13 negatively regulate the adhesive functions of cadherin. J Biol Chem 277:24594–24600. doi:10.1074/jbc.M201984200
64. Krakstad BF, Ardawatia VV, Aragay AM (2004) A role for Galpha12/Galpha13 in p120ctn regulation. Proc Natl Acad Sci U S A 101:10314–10319. doi:10.1073/pnas.0401366101
65. Ardawatia VV, Masià-Balagué M, Krakstad BF et al (2011) Gα(12) binds to the N-terminal regulatory domain of p120(ctn), and downregulates p120(ctn) tyrosine phosphorylation induced by Src family kinases via a RhoA independent mechanism. Exp Cell Res 317:293–306. doi:10.1016/j.yexcr.2010.10.017
66. Hirano S, Kimoto N, Shimoyama Y et al (1992) Identification of a neural alpha-catenin as a key regulator of cadherin function and multicellular organization. Cell 70:293–301
67. Aberle H, Schwartz H, Hoschuetzky H, Kemler R (1996) Single amino acid substitutions in proteins of the armadillo gene family abolish their binding to α-catenin. J Biol Chem 271:1520–1526
68. Drees F, Pokutta S, Yamada S et al (2005) Alpha-catenin is a molecular switch that binds E-cadherin-beta-catenin and regulates actin-filament assembly. Cell 123:903–915. doi:10.1016/j.cell.2005.09.021
69. Yamada S, Pokutta S, Drees F et al (2005) Deconstructing the cadherin-catenin-actin complex. Cell 123:889–901. doi:10.1016/j.cell.2005.09.020
70. Aberle H, Butz S, Stappert J et al (1994) Assembly of the cadherin-catenin complex in vitro with recombinant proteins. J Cell Sci 107(Pt 1):3655–3663

71. Le Duc Q, Shi Q, Blonk I et al (2010) Vinculin potentiates E-cadherin mechanosensing and is recruited to actin-anchored sites within adherens junctions in a myosin II-dependent manner. J Cell Biol 189:1107–1115. doi:10.1083/jcb.201001149
72. Abe K, Takeichi M (2008) EPLIN mediates linkage of the cadherin catenin complex to F-actin and stabilizes the circumferential actin belt. Proc Natl Acad Sci U S A 105:13–19. doi:10.1073/pnas.0710504105
73. Miyashita Y, Ozawa M (2007) A dileucine motif in its cytoplasmic domain directs beta-catenin-uncoupled E-cadherin to the lysosome. J Cell Sci 120:4395–4406. doi:10.1242/jcs.03489
74. Troyanovsky RB, Klingelhöfer J, Troyanovsky SM (2011) α-Catenin contributes to the strength of E-cadherin-p120 interactions. Mol Biol Cell 22:4247–4255. doi:10.1091/mbc.E11-03-0250
75. Roura S, Miravet S, Piedra J et al (1999) Regulation of E-cadherin/catenin association by tyrosine phosphorylation. J Biol Chem 274:36734–36740
76. Aberle H, Bauer A, Stappert J et al (1997) beta-catenin is a target for the ubiquitin-proteasome pathway. EMBO J 16:3797–3804. doi:10.1093/emboj/16.13.3797
77. Van de Wetering M, Cavallo R, Dooijes D et al (1997) Armadillo coactivates transcription driven by the product of the Drosophila segment polarity gene dTCF. Cell 88:789–799
78. He TC, Sparks AB, Rago C et al (1998) Identification of c-MYC as a target of the APC pathway. Science 281:1509–1512
79. Tetsu O, McCormick F (1999) Beta-catenin regulates expression of cyclin D1 in colon carcinoma cells. Nature 398:422–426. doi:10.1038/18884
80. Batlle E, Sancho E, Francí C et al (2000) The transcription factor snail is a repressor of E-cadherin gene expression in epithelial tumour cells. Nat Cell Biol 2:84–89. doi:10.1038/35000034
81. Van de Wetering M, Sancho E, Verweij C et al (2002) The beta-catenin/TCF-4 complex imposes a crypt progenitor phenotype on colorectal cancer cells. Cell 111:241–250
82. Gottardi CJ, Wong E, Gumbiner BM (2001) E-cadherin suppresses cellular transformation by inhibiting beta-catenin signaling in an adhesion-independent manner. J Cell Biol 153:1049–1060
83. Ohno H, Uemura K, Shintani-Ishida K et al (2007) Ischemia promotes calpain-mediated degradation of p120-catenin in SH-SY5Y cells. Biochem Biophys Res Commun 353:547–552. doi:10.1016/j.bbrc.2006.12.061
84. Wang Y, Minshall RD, Schwartz DE, Hu G (2011) Cyclic stretch induces alveolar epithelial barrier dysfunction via calpain-mediated degradation of p120-catenin. Am J Physiol Lung Cell Mol Physiol 301:L197–L206. doi:10.1152/ajplung.00048.2011
85. Daniel JM, Reynolds aB (1995) The tyrosine kinase substrate p120cas binds directly to E-cadherin but not to the adenomatous polyposis coli protein or alpha-catenin. Mol Cell Biol 15:4819–4824
86. Miranda KC, Khromykh T, Christy P et al (2001) A dileucine motif targets E-cadherin to the basolateral cell surface in Madin-Darby canine kidney and LLC-PK1 epithelial cells. J Biol Chem 276:22565–22572. doi:10.1074/jbc.M101907200
87. Nanes BA, Chiasson-MacKenzie C, Lowery AM et al (2012) p120-catenin binding masks an endocytic signal conserved in classical cadherins. J Cell Biol 199:365–380. doi:10.1083/jcb.201205029
88. Fujita Y, Krause G, Scheffner M et al (2002) Hakai, a c-Cbl-like protein, ubiquitinates and induces endocytosis of the E-cadherin complex. Nat Cell Biol 4:222–231. doi:10.1038/ncb758
89. Hartsock A, Nelson WJ (2012) Competitive regulation of E-cadherin juxtamembrane domain degradation by p120-catenin binding and Hakai-mediated ubiquitination. PLoS One 7, e37476. doi:10.1371/journal.pone.0037476
90. Palacios F, Tushir JS, Fujita Y, D'Souza-Schorey C (2005) Lysosomal targeting of E-cadherin: a unique mechanism for the down-regulation of cell-cell adhesion during epithelial to mesenchymal transitions. Mol Cell Biol 25:389–402. doi:10.1128/MCB.25.1.389-402.2005
91. Baki L, Marambaud P, Efthimiopoulos S et al (2001) Presenilin-1 binds cytoplasmic epithelial cadherin, inhibits cadherin/p120 association, and regulates stability and function of the

cadherin/catenin adhesion complex. Proc Natl Acad Sci U S A 98:2381–2386. doi:10.1073/pnas.041603398
92. Marambaud P, Shioi J, Serban G et al (2002) A presenilin-1/gamma-secretase cleavage releases the E-cadherin intracellular domain and regulates disassembly of adherens junctions. EMBO J 21:1948–1956. doi:10.1093/emboj/21.8.1948
93. Spasic D, Annaert W (2008) Building gamma-secretase: the bits and pieces. J Cell Sci 121:413–420. doi:10.1242/jcs.015255
94. Sato K, Watanabe T, Wang S et al (2011) Numb controls E-cadherin endocytosis through p120 catenin with aPKC. Mol Biol Cell 22:3103–3119. doi:10.1091/mbc.E11-03-0274
95. Wang Z, Li SS-C (2010) Numb: a new player in EMT. Cell Adh Migr 4:176–179
96. Wang Z, Sandiford S, Wu C, Li SS-C (2009) Numb regulates cell-cell adhesion and polarity in response to tyrosine kinase signalling. EMBO J 28:2360–2373. doi:10.1038/emboj.2009.190
97. Etienne-Manneville S, Hall A (2002) Rho GTPases in cell biology. Nature 420:629–635. doi:10.1038/nature01148
98. Castaño J, Solanas G, Casagolda D et al (2007) Specific phosphorylation of p120-catenin regulatory domain differently modulates its binding to RhoA. Mol Cell Biol 27:1745–1757. doi:10.1128/MCB.01974-06
99. Noren NK, Liu BP, Burridge K, Kreft B (2000) p120 catenin regulates the actin cytoskeleton via Rho family GTPases. J Cell Biol 150:567–580
100. Kovacs EM, Ali RG, McCormack AJ, Yap AS (2002) E-cadherin homophilic ligation directly signals through Rac and phosphatidylinositol 3-kinase to regulate adhesive contacts. J Biol Chem 277:6708–6718. doi:10.1074/jbc.M109640200
101. Goodwin M, Kovacs EM, Thoreson MA et al (2003) Minimal mutation of the cytoplasmic tail inhibits the ability of E-cadherin to activate Rac but not phosphatidylinositol 3-kinase: direct evidence of a role for cadherin-activated Rac signaling in adhesion and contact formation. J Biol Chem 278:20533–20539. doi:10.1074/jbc.M213171200
102. Anastasiadis PZ, Moon SY, Thoreson MA et al (2000) Inhibition of RhoA by p120 catenin. Nat Cell Biol 2:637–644. doi:10.1038/35023588
103. Izumi G, Sakisaka T, Baba T et al (2004) Endocytosis of E-cadherin regulated by Rac and Cdc42 small G proteins through IQGAP1 and actin filaments. J Cell Biol 166:237–248. doi:10.1083/jcb.200401078
104. Grosheva I, Shtutman M, Elbaum M, Bershadsky AD (2001) p120 catenin affects cell motility via modulation of activity of Rho-family GTPases: a link between cell-cell contact formation and regulation of cell locomotion. J Cell Sci 114:695–707
105. Yanagisawa M, Huveldt D, Kreinest P et al (2008) A p120 catenin isoform switch affects Rho activity, induces tumor cell invasion, and predicts metastatic disease. J Biol Chem 283:18344–18354. doi:10.1074/jbc.M801192200
106. Kelly KF, Spring CM, Otchere AA, Daniel JM (2004) NLS-dependent nuclear localization of p120ctn is necessary to relieve Kaiso-mediated transcriptional repression. J Cell Sci 117:2675–2686. doi:10.1242/jcs.01101
107. Roczniak-Ferguson A, Reynolds AB (2003) Regulation of p120-catenin nucleocytoplasmic shuttling activity. J Cell Sci 116:4201–4212. doi:10.1242/jcs.00724
108. Hosking CR, Ulloa F, Hogan C et al (2007) The transcriptional repressor Glis2 is a novel binding partner for p120 catenin. Mol Biol Cell 18:1918–1927. doi:10.1091/mbc.E06-10-0941
109. Vasanth S, ZeRuth G, Kang HS, Jetten AM (2011) Identification of nuclear localization, DNA binding, and transactivating mechanisms of Kruppel-like zinc finger protein Gli-similar 2 (Glis2). J Biol Chem 286:4749–4759. doi:10.1074/jbc.M110.165951
110. Ferber EC, Kajita M, Wadlow A et al (2008) A role for the cleaved cytoplasmic domain of E-cadherin in the nucleus. J Biol Chem 283:12691–12700. doi:10.1074/jbc.M708887200
111. Daniel JM, Reynolds AB (1999) The catenin p120(ctn) interacts with Kaiso, a novel BTB/POZ domain zinc finger transcription factor. Mol Cell Biol 19:3614–3623
112. Daniel JM (2007) Dancing in and out of the nucleus: p120(ctn) and the transcription factor Kaiso. Biochim Biophys Acta 1773:59–68. doi:10.1016/j.bbamcr.2006.08.052

113. Daniel JM, Spring CM, Crawford HC et al (2002) The p120(ctn)-binding partner Kaiso is a bi-modal DNA-binding protein that recognizes both a sequence-specific consensus and methylated CpG dinucleotides. Nucleic Acids Res 30:2911–2919
114. Donaldson NS, Pierre CC, Anstey MI et al (2012) Kaiso represses the cell cycle gene cyclin D1 via sequence-specific and methyl-CpG-dependent mechanisms. PLoS One 7, e50398. doi:10.1371/journal.pone.0050398
115. Prokhortchouk A, Hendrich B, Jørgensen H et al (2001) The p120 catenin partner Kaiso is a DNA methylation-dependent transcriptional repressor. Genes Dev 15:1613–1618. doi:10.1101/gad.198501
116. Hong JY, Park J-I, Cho K et al (2010) Shared molecular mechanisms regulate multiple catenin proteins: canonical Wnt signals and components modulate p120-catenin isoform-1 and additional p120 subfamily members. J Cell Sci 123:4351–4365. doi:10.1242/jcs.067199
117. Kim SW, Park J-I, Spring CM et al (2004) Non-canonical Wnt signals are modulated by the Kaiso transcriptional repressor and p120-catenin. Nat Cell Biol 6:1212–1220. doi:10.1038/ncb1191
118. Park J, Ji H, Jun S et al (2006) Frodo links dishevelled to the p120-catenin/Kaiso pathway: distinct catenin subfamilies promote Wnt signals. Dev Cell 11:683–695. doi:10.1016/j.devcel.2006.09.022
119. Park J-I, Kim SW, Lyons JP et al (2005) Kaiso/p120-catenin and TCF/beta-catenin complexes coordinately regulate canonical Wnt gene targets. Dev Cell 8:843–854. doi:10.1016/j.devcel.2005.04.010
120. Spring CM, Kelly KF, O'Kelly I et al (2005) The catenin p120ctn inhibits Kaiso-mediated transcriptional repression of the beta-catenin/TCF target gene matrilysin. Exp Cell Res 305:253–265. doi:10.1016/j.yexcr.2005.01.007
121. Del Valle-Pérez B, Casagolda D, Lugilde E et al (2011) Wnt controls the transcriptional activity of Kaiso through CK1ε-dependent phosphorylation of p120-catenin. J Cell Sci 124:2298–2309. doi:10.1242/jcs.082693
122. Ruzov A, Hackett JA, Prokhortchouk A et al (2009) The interaction of xKaiso with xTcf3: a revised model for integration of epigenetic and Wnt signalling pathways. Development 136:723–727. doi:10.1242/dev.025577
123. Chartier NT, Oddou CI, Lainé MG et al (2007) Cyclin-dependent kinase 2/cyclin E complex is involved in p120 catenin (p120ctn)-dependent cell growth control: a new role for p120ctn in cancer. Cancer Res 67:9781–9790. doi:10.1158/0008-5472.CAN-07-0233
124. Franz CM, Ridley AJ (2004) p120 catenin associates with microtubules: inverse relationship between microtubule binding and Rho GTPase regulation. J Biol Chem 279:6588–6594. doi:10.1074/jbc.M312812200
125. Yanagisawa M, Kaverina IN, Wang A et al (2004) A novel interaction between kinesin and p120 modulates p120 localization and function. J Biol Chem 279:9512–9521. doi:10.1074/jbc.M310895200
126. Ichii T, Takeichi M (2007) p120-catenin regulates microtubule dynamics and cell migration in acadherin-independentmanner.GenesCells12:827–839.doi:10.1111/j.1365-2443.2007.01095.x
127. Jiang G, Wang Y, Dai S et al (2012) P120-catenin isoforms 1 and 3 regulate proliferation and cell cycle of lung cancer cells via β-catenin and Kaiso respectively. PLoS One 7, e30303. doi:10.1371/journal.pone.0030303
128. Shahbazi MN, Megias D, Epifano C et al (2013) CLASP2 interacts with p120-catenin and governs microtubule dynamics at adherens junctions. J Cell Biol 203:1043–1061. doi:10.1083/jcb.201306019
129. Kemphues KJ, Priess JR, Morton DG, Cheng NS (1988) Identification of genes required for cytoplasmic localization in early C. elegans embryos. Cell 52:311–320, doi: S0092-8674(88)80024-2 [pii]
130. Böhm H, Brinkmann V, Drab M et al (1997) Mammalian homologues of C. elegans PAR-1 are asymmetrically localized in epithelial cells and may influence their polarity. Curr Biol 7:603–606

131. Drewes G, Ebneth A, Preuss U et al (1997) MARK, a novel family of protein kinases that phosphorylate microtubule-associated proteins and trigger microtubule disruption. Cell 89:297–308
132. Hurd TW, Fan S, Liu CJ et al (2003) Phosphorylation-dependent binding of 14-3-3 to the polarity protein Par3 regulates cell polarity in mammalian epithelia. Curr Biol 13:2082–2090
133. Hurov JB, Watkins JL, Piwnica-Worms H (2004) Atypical PKC phosphorylates PAR-1 kinases to regulate localization and activity. Curr Biol 14:736–741. doi:10.1016/j.cub.2004.04.007
134. Suzuki A, Hirata M, Kamimura K et al (2004) aPKC acts upstream of PAR-1b in both the establishment and maintenance of mammalian epithelial polarity. Curr Biol 14:1425–1435. doi:10.1016/j.cub.2004.08.021
135. Lin D, Edwards AS, Fawcett JP et al (2000) A mammalian PAR-3-PAR-6 complex implicated in Cdc42/Rac1 and aPKC signalling and cell polarity. Nat Cell Biol 2:540–547. doi:10.1038/35019582
136. Ooshio T, Fujita N, Yamada A et al (2007) Cooperative roles of Par-3 and afadin in the formation of adherens and tight junctions. J Cell Sci 120:2352–2365. doi:10.1242/jcs.03470
137. Chen X, Macara IG (2005) Par-3 controls tight junction assembly through the Rac exchange factor Tiam1. Nat Cell Biol 7:262–269. doi:10.1038/ncb1226
138. Horikoshi Y, Suzuki A, Yamanaka T et al (2009) Interaction between PAR-3 and the aPKC-PAR-6 complex is indispensable for apical domain development of epithelial cells. J Cell Sci 122:1595–1606. doi:10.1242/jcs.043174
139. Cohen D, Brennwald PJ, Rodriguez-Boulan E, Müsch A (2004) Mammalian PAR-1 determines epithelial lumen polarity by organizing the microtubule cytoskeleton. J Cell Biol 164:717–727. doi:10.1083/jcb.200308104
140. Hao Y, Du Q, Chen X et al (2010) Par3 controls epithelial spindle orientation by aPKC-mediated phosphorylation of apical Pins. Curr Biol 20:1809–1818. doi:10.1016/j.cub.2010.09.032
141. Martin-Belmonte F, Perez-Moreno M (2012) Epithelial cell polarity, stem cells and cancer. Nat Rev Cancer 12:23–38. doi:10.1038/nrc3169
142. Wodarz A, Hinz U, Engelbert M, Knust E (1995) Expression of crumbs confers apical character on plasma membrane domains of ectodermal epithelia of Drosophila. Cell 82:67–76
143. Wang Q, Chen X-W, Margolis B (2007) PALS1 regulates E-cadherin trafficking in mammalian epithelial cells. Mol Biol Cell 18:874–885. doi:10.1091/mbc.E06-07-0651
144. Yamanaka T, Horikoshi Y, Sugiyama Y et al (2003) Mammalian Lgl forms a protein complex with PAR-6 and aPKC independently of PAR-3 to regulate epithelial cell polarity. Curr Biol 13:734–743
145. Navarro C, Nola S, Audebert S et al (2005) Junctional recruitment of mammalian Scribble relies on E-cadherin engagement. Oncogene 24:4330–4339. doi:10.1038/sj.onc.1208632
146. Qin Y, Capaldo C, Gumbiner BM, Macara IG (2005) The mammalian Scribble polarity protein regulates epithelial cell adhesion and migration through E-cadherin. J Cell Biol 171:1061–1071. doi:10.1083/jcb.200506094
147. Gardiol D, Zacchi A, Petrera F et al (2006) Human discs large and scrib are localized at the same regions in colon mucosa and changes in their expression patterns are correlated with loss of tissue architecture during malignant progression. Int J Cancer 119:1285–1290. doi:10.1002/ijc.21982
148. Schimanski CC, Schmitz G, Kashyap A et al (2005) Reduced expression of Hugl-1, the human homologue of Drosophila tumour suppressor gene lgl, contributes to progression of colorectal cancer. Oncogene 24:3100–3109. doi:10.1038/sj.onc.1208520
149. Beaulieu JF (1999) Integrins and human intestinal cell functions. Front Biosci 4:D310–D321
150. Carroll KM, Wong TT, Drabik DL, Chang EB (1988) Differentiation of rat small intestinal epithelial cells by extracellular matrix. Am J Physiol 254:G355–G360
151. Kedinger M, Lefebvre O, Duluc I et al (1998) Cellular and molecular partners involved in gut morphogenesis and differentiation. Philos Trans R Soc Lond B Biol Sci 353:847–856. doi:10.1098/rstb.1998.0249

152. Stutzmann J, Bellissent-Waydelich A, Fontao L et al (2000) Adhesion complexes implicated in intestinal epithelial cell-matrix interactions. Microsc Res Tech 51:179–190. doi:10.1002/1097-0029(20001015)51:2<179::AID-JEMT9>3.0.CO;2-4
153. Aumailley M, Krieg T (1996) Laminins: a family of diverse multifunctional molecules of basement membranes. J Invest Dermatol 106:209–214
154. Simon-Assmann P, Kedinger M, De Arcangelis A et al (1995) Extracellular matrix components in intestinal development. Experientia 51:883–900
155. Gout SP, Jacquier-Sarlin MR, Rouard-Talbot L, Rousselle P, Block MR (2001) RhoA-dependent switch between alpha2beta1 and alpha3beta1 integrins is induced by laminin-5 during early stage of HT-29 cell differentiation. Mol Biol Cell 12(10):3268–3281
156. Vandenberg P, Kern A, Ries A et al (1991) Characterization of a type IV collagen major cell binding site with affinity to the alpha 1 beta 1 and the alpha 2 beta 1 integrins. J Cell Biol 113:1475–1483
157. Leivo I, Tani T, Laitinen L et al (1996) Anchoring complex components laminin-5 and type VII collagen in intestine: association with migrating and differentiating enterocytes. J Histochem Cytochem 44:1267–1277
158. Lohi J, Leivo I, Tani T et al (1996) Laminins, tenascin and type VII collagen in colorectal mucosa. Histochem J 28:431–440
159. Burgeson RE, Chiquet M, Deutzmann R et al (1994) A new nomenclature for the laminins. Matrix Biol 14:209–211
160. Mercurio AM (1995) Laminin receptors: achieving specificity through cooperation. Trends Cell Biol 5:419–423
161. Durbeej M (2010) Laminins. Cell Tissue Res 339:259–268. doi:10.1007/s00441-009-0838-2
162. Teller IC, Beaulieu J-F (2004) Interactions between laminin and epithelial cells in intestinal health and disease. Expert Rev Mol Med 3:1–18. doi:10.1017/S1462399401003623
163. Turck N, Gross I, Gendry P et al (2005) Laminin isoforms: biological roles and effects on the intracellular distribution of nuclear proteins in intestinal epithelial cells. Exp Cell Res 303:494–503. doi:10.1016/j.yexcr.2004.10.025
164. Simo P, Simon-Assmann P, Bouziges F et al (1991) Changes in the expression of laminin during intestinal development. Development 112:477–487
165. Basora N, Herring-Gillam FE, Boudreau F et al (1999) Expression of functionally distinct variants of the beta(4)A integrin subunit in relation to the differentiation state in human intestinal cells. J Biol Chem 274:29819–29825
166. Jones RG, Li X, Gray PD et al (2006) Conditional deletion of beta1 integrins in the intestinal epithelium causes a loss of Hedgehog expression, intestinal hyperplasia, and early postnatal lethality. J Cell Biol 175:505–514. doi:10.1083/jcb.200602160
167. Giancotti FG (1996) Signal transduction by the alpha 6 beta 4 integrin: charting the path between laminin binding and nuclear events. J Cell Sci 109(Pt 6):1165–1172
168. Murgia C, Blaikie P, Kim N et al (1998) Cell cycle and adhesion defects in mice carrying a targeted deletion of the integrin beta4 cytoplasmic domain. EMBO J 17:3940–3951. doi:10.1093/emboj/17.14.3940
169. Gilcrease MZ (2007) Integrin signaling in epithelial cells. Cancer Lett 247:1–25. doi:10.1016/j.canlet.2006.03.031
170. Avizienyte E, Wyke AW, Jones RJ et al (2002) Src-induced de-regulation of E-cadherin in colon cancer cells requires integrin signalling. Nat Cell Biol 4:632–638. doi:10.1038/ncb829
171. Avizienyte E, Fincham VJ, Brunton VG, Frame MC (2004) Src SH3/2 domain-mediated peripheral accumulation of Src and phospho-myosin is linked to deregulation of E-cadherin and the epithelial-mesenchymal transition. Mol Biol Cell 15:2794–2803. doi:10.1091/mbc.E03-12-0879
172. Martinez-Rico C, Pincet F, Thiery J-P, Dufour S (2010) Integrins stimulate E-cadherin-mediated intercellular adhesion by regulating Src-kinase activation and actomyosin contractility. J Cell Sci 123:712–722. doi:10.1242/jcs.047878
173. Vogelmann R, Nguyen-Tat M-D, Giehl K et al (2005) TGFbeta-induced downregulation of E-cadherin-based cell-cell adhesion depends on PI3-kinase and PTEN. J Cell Sci 118:4901–4912. doi:10.1242/jcs.02594

174. Wang H, Radjendirane V, Wary KK, Chakrabarty S (2004) Transforming growth factor beta regulates cell-cell adhesion through extracellular matrix remodeling and activation of focal adhesion kinase in human colon carcinoma Moser cells. Oncogene 23:5558–5561. doi:10.1038/sj.onc.1207701
175. Kim Y, Kugler MC, Wei Y et al (2009) Integrin alpha3beta1-dependent beta-catenin phosphorylation links epithelial Smad signaling to cell contacts. J Cell Biol 184:309–322. doi:10.1083/jcb.200806067
176. Chartier NT, Lainé M, Gout S et al (2006) Laminin-5-integrin interaction signals through PI 3-kinase and Rac1b to promote assembly of adherens junctions in HT-29 cells. J Cell Sci 119:31–46. doi:10.1242/jcs.02698
177. Hintermann E, Yang N, O'Sullivan D et al (2005) Integrin alpha6beta4-erbB2 complex inhibits haptotaxis by up-regulating E-cadherin cell-cell junctions in keratinocytes. J Biol Chem 280:8004–8015. doi:10.1074/jbc.M406301200
178. Canonici A, Steelant W, Rigot V et al (2008) Insulin-like growth factor-I receptor, E-cadherin and alpha v integrin form a dynamic complex under the control of alpha-catenin. Int J Cancer 122:572–582. doi:10.1002/ijc.23164
179. Gout SP, Jacquier-Sarlin MR, Rouard-Talbot L et al (2001) RhoA-dependent switch between alpha2beta1 and alpha3beta1 integrins is induced by laminin-5 during early stage of HT-29 cell differentiation. Mol Biol Cell 12:3268–3281
180. Takai Y, Ikeda W, Ogita H, Rikitake Y (2008) The immunoglobulin-like cell adhesion molecule nectin and its associated protein afadin. Annu Rev Cell Dev Biol 24:309–342. doi:10.1146/annurev.cellbio.24.110707.175339
181. Takai Y, Miyoshi J, Ikeda W, Ogita H (2008) Nectins and nectin-like molecules: roles in contact inhibition of cell movement and proliferation. Nat Rev Mol Cell Biol 9:603–615. doi:10.1038/nrm2457
182. Takai Y, Irie K, Shimizu K et al (2003) Nectins and nectin-like molecules: roles in cell adhesion, migration, and polarization. Cancer Sci 94:655–667
183. Hoshino T, Sakisaka T, Baba T et al (2005) Regulation of E-cadherin endocytosis by nectin through afadin, Rap1, and p120ctn. J Biol Chem 280:24095–24103. doi:10.1074/jbc.M414447200
184. Ikeda W, Nakanishi H, Miyoshi J et al (1999) Afadin: a key molecule essential for structural organization of cell-cell junctions of polarized epithelia during embryogenesis. J Cell Biol 146:1117–1132
185. Braga VM, Yap AS (2005) The challenges of abundance: epithelial junctions and small GTPase signalling. Curr Opin Cell Biol 17:466–474. doi:10.1016/j.ceb.2005.08.012
186. Irie K, Shimizu K, Sakisaka T et al (2004) Roles and modes of action of nectins in cell-cell adhesion. Semin Cell Dev Biol 15:643–656. doi:10.1016/j.semcdb.2004.09.002
187. Bertet C, Sulak L, Lecuit T (2004) Myosin-dependent junction remodelling controls planar cell intercalation and axis elongation. Nature 429:667–671. doi:10.1038/nature02590
188. Kooistra MRH, Dubé N, Bos JL (2007) Rap1: a key regulator in cell-cell junction formation. J Cell Sci 120:17–22. doi:10.1242/jcs.03306
189. Brieher WM, Yap AS (2013) Cadherin junctions and their cytoskeleton(s). Curr Opin Cell Biol 25:39–46. doi:10.1016/j.ceb.2012.10.010
190. Michael M, Yap AS (2013) The regulation and functional impact of actin assembly at cadherin cell-cell adhesions. Semin Cell Dev Biol 24:298–307. doi:10.1016/j.semcdb.2012.12.004
191. Gates J, Peifer M (2005) Can 1000 reviews be wrong? Actin, alpha-catenin, and adherens junctions. Cell 123:769–772. doi:10.1016/j.cell.2005.11.009
192. Scott JA, Yap AS (2006) Cinderella no longer: alpha-catenin steps out of cadherin's shadow. J Cell Sci 119:4599–4605. doi:10.1242/jcs.03267
193. Benjamin JM, Kwiatkowski AV, Yang C et al (2010) AlphaE-catenin regulates actin dynamics independently of cadherin-mediated cell-cell adhesion. J Cell Biol 189:339–352. doi:10.1083/jcb.200910041
194. Kim SH, Li Z, Sacks DB (2000) E-cadherin-mediated cell-cell attachment activates Cdc42. J Biol Chem 275:36999–37005. doi:10.1074/jbc.M003430200

195. Elbediwy A, Zihni C, Terry SJ et al (2012) Epithelial junction formation requires confinement of Cdc42 activity by a novel SH3BP1 complex. J Cell Biol 198:677–693. doi:10.1083/jcb.201202094
196. Ratheesh A, Gomez GA, Priya R et al (2012) Centralspindlin and α-catenin regulate Rho signalling at the epithelial zonula adherens. Nat Cell Biol 14:818–828. doi:10.1038/ncb2532
197. Yamada S, Nelson WJ (2007) Localized zones of Rho and Rac activities drive initiation and expansion of epithelial cell-cell adhesion. J Cell Biol 178:517–527. doi:10.1083/jcb.200701058
198. Wildenberg GA, Dohn MR, Carnahan RH et al (2006) p120-catenin and p190RhoGAP regulate cell-cell adhesion by coordinating antagonism between Rac and Rho. Cell 127:1027–1039. doi:10.1016/j.cell.2006.09.046
199. Chartier NT, Lainé MG, Ducarouge B et al (2011) Enterocytic differentiation is modulated by lipid rafts-dependent assembly of adherens junctions. Exp Cell Res 317:1422–1436. doi:10.1016/j.yexcr.2011.03.009
200. Smutny M, Cox HL, Leerberg JM et al (2010) Myosin II isoforms identify distinct functional modules that support integrity of the epithelial zonula adherens. Nat Cell Biol 12:696–702. doi:10.1038/ncb2072
201. Watanabe N, Kato T, Fujita A et al (1999) Cooperation between mDia1 and ROCK in Rho-induced actin reorganization. Nat Cell Biol 1:136–143. doi:10.1038/11056
202. Gomez GA, McLachlan RW, Yap AS (2011) Productive tension: force-sensing and homeostasis of cell-cell junctions. Trends Cell Biol 21:499–505. doi:10.1016/j.tcb.2011.05.006
203. Shewan AM, Maddugoda M, Kraemer A et al (2005) Myosin 2 is a key Rho kinase target necessary for the local concentration of E-cadherin at cell-cell contacts. Mol Biol Cell 16:4531–4542. doi:10.1091/mbc.E05-04-0330
204. Blanchoin L, Michelot A (2012) Actin cytoskeleton: a team effort during actin assembly. Curr Biol 22:R643–R645. doi:10.1016/j.cub.2012.07.026
205. Block J, Breitsprecher D, Kühn S et al (2012) FMNL2 drives actin-based protrusion and migration downstream of Cdc42. Curr Biol 22:1005–1012. doi:10.1016/j.cub.2012.03.064
206. Kobielak A, Pasolli HA, Fuchs E (2004) Mammalian formin-1 participates in adherens junctions and polymerization of linear actin cables. Nat Cell Biol 6:21–30. doi:10.1038/ncb1075
207. Campellone KG, Welch MD (2010) A nucleator arms race: cellular control of actin assembly. Nat Rev Mol Cell Biol 11:237–251. doi:10.1038/nrm2867
208. Ivanov AI, Hunt D, Utech M et al (2005) Differential roles for actin polymerization and a myosin II motor in assembly of the epithelial apical junctional complex. Mol Biol Cell 16:2636–2650. doi:10.1091/mbc.E05-01-0043
209. Kovacs EM, Verma S, Ali RG et al (2011) N-WASP regulates the epithelial junctional actin cytoskeleton through a non-canonical post-nucleation pathway. Nat Cell Biol 13:934–943. doi:10.1038/ncb2290
210. Yamazaki D, Oikawa T, Takenawa T (2007) Rac-WAVE-mediated actin reorganization is required for organization and maintenance of cell-cell adhesion. J Cell Sci 120:86–100. doi:10.1242/jcs.03311
211. Tang VW, Brieher WM (2012) α-Actinin-4/FSGS1 is required for Arp2/3-dependent actin assembly at the adherens junction. J Cell Biol 196:115–130. doi:10.1083/jcb.201103116
212. Yonemura S, Wada Y, Watanabe T et al (2010) alpha-Catenin as a tension transducer that induces adherens junction development. Nat Cell Biol 12:533–542. doi:10.1038/ncb2055
213. Maddugoda MP, Crampton MS, Shewan AM, Yap AS (2007) Myosin VI and vinculin cooperate during the morphogenesis of cadherin cell cell contacts in mammalian epithelial cells. J Cell Biol 178:529–540. doi:10.1083/jcb.200612042
214. Taguchi K, Ishiuchi T, Takeichi M (2011) Mechanosensitive EPLIN-dependent remodeling of adherens junctions regulates epithelial reshaping. J Cell Biol 194:643–656. doi:10.1083/jcb.201104124
215. Cano A, Pérez-moreno MA, Rodrigo I et al (2000) The transcription factor Snail controls epithelial – mesenchymal transitions by repressing E-cadherin expression. Nat Cell Biol 2:76–83

216. Conacci-sorrell M, Simcha I, Ben-yedidia T et al (2003) Autoregulation of E-cadherin expression by cadherin – cadherin interactions: the roles. J Cell Biol 163:847–857. doi:10.1083/jcb.200308162
217. Chem JB, Peinado H, Quintanilla M, Cano A (2003) Molecular basis of cell and developmental biology : transforming growth factor β -1 induces snail transcription factor in epithelial cell lines : mechanisms for transitions Hector Peinado, Miguel Quintanilla and transforming growth factor -1 Induces Sn. doi: 10.1074/jbc.M211304200
218. Ciruna B, Rossant J (2001) FGF signaling regulates mesoderm cell fate specification and morphogenetic movement at the primitive streak. Dev Cell 1:37–49
219. Nishita M, Hashimoto MK, Ogata S, Laurent MN, Ueno N, Shibuya H, Cho KWY (2000) Interaction between Wnt and TGF- beta signalling pathways during formation of Spemann's organizer. Nature 357(403):781–785. doi:10.1038/35001602
220. Riese J, Yu X, Munnerlyn A, Eresh S, Hsu S, Grosschedl R, Bienz M (1997) LEF-1, a nuclear factor coordinating signaling inputs from wingless and decapentaplegic. Cell 88(6):777–787
221. Bertocchi C, Vaman Rao M, Zaidel-Bar R (2012) Regulation of adherens junction dynamics by phosphorylation switches. J Signal Transduct 2012:125295. doi:10.1155/2012/125295
222. Hubbard SR, Till JH (2000) Protein tyrosine kinase structure and function. Annu Rev Biochem 69:373–398. doi:10.1146/annurev.biochem.69.1.373
223. Stoker AW (2005) Protein tyrosine phosphatases and signalling. J Endocrinol 185:19–33. doi:10.1677/joe.1.06069
224. Fujisawa K, Fujita A, Ishizaki T et al (1996) Identification of the Rho-binding domain of p160ROCK, a Rho-associated coiled-coil containing protein kinase. J Biol Chem 271:23022–23028
225. Lu W, Mayer BJ (1999) Mechanism of activation of Pak1 kinase by membrane localization. Oncogene 18:797–806. doi:10.1038/sj.onc.1202361
226. El Sayegh TY, Arora PD, Fan L et al (2005) Phosphorylation of N-cadherin-associated cortactin by Fer kinase regulates N-cadherin mobility and intercellular adhesion strength. Mol Biol Cell 16:5514–5527. doi:10.1091/mbc.E05-05-0410
227. Joo C-K, Kim H-S, Park J-Y et al (2008) Ligand release-independent transactivation of epidermal growth factor receptor by transforming growth factor-beta involves multiple signaling pathways. Oncogene 27:614–628. doi:10.1038/sj.onc.1210649
228. Zandy NL, Pendergast AM (2008) Abl tyrosine kinases modulate cadherin-dependent adhesion upstream and downstream of Rho family GTPases. Cell Cycle 7:444–448
229. Watanabe T, Sato K, Kaibuchi K (2009) Cadherin-mediated intercellular adhesion and signaling cascades involving small GTPases. Cold Spring Harb Perspect Biol 1:a003020. doi:10.1101/cshperspect.a003020
230. Serres M, Grangeasse C, Haftek M et al (1997) Hyperphosphorylation of beta-catenin on serine-threonine residues and loss of cell-cell contacts induced by calyculin A and okadaic acid in human epidermal cells. Exp Cell Res 231:163–172. doi:10.1006/excr.1996.3443
231. Volberg T, Zick Y, Dror R et al (1992) The effect of tyrosine-specific protein phosphorylation on the assembly of adherens-type junctions. EMBO J 11:1733–1742
232. Calautti E, Cabodi S, Stein PL et al (1998) Tyrosine phosphorylation and src family kinases control keratinocyte cell-cell adhesion. J Cell Biol 141:1449–1465
233. Pang J-H, Kraemer A, Stehbens SJ et al (2005) Recruitment of phosphoinositide 3-kinase defines a positive contribution of tyrosine kinase signaling to E-cadherin function. J Biol Chem 280:3043–3050. doi:10.1074/jbc.M412148200
234. Jaggi M, Rao PS, Smith DJ et al (2005) E-cadherin phosphorylation by protein kinase D1/protein kinase C{mu} is associated with altered cellular aggregation and motility in prostate cancer. Cancer Res 65:483–492
235. Zandy NL, Playford M, Pendergast AM (2007) Abl tyrosine kinases regulate cell-cell adhesion through Rho GTPases. Proc Natl Acad Sci U S A 104:17686–17691. doi:10.1073/pnas.0703077104

236. Dupre-Crochet S, Figueroa A, Hogan C et al (2007) Casein kinase 1 is a novel negative regulator of E-cadherin-based cell-cell contacts. Mol Cell Biol 27:3804–3816. doi:10.1128/MCB.01590-06
237. Lilien J, Balsamo J, Arregui C, Xu G (2002) Turn-off, drop-out: functional state switching of cadherins. Dev Dyn 224:18–29. doi:10.1002/dvdy.10087
238. Pece S, Gutkind JS (2002) E-cadherin and Hakai: signalling, remodeling or destruction? Nat Cell Biol 4:E72–E74. doi:10.1038/ncb0402-e72
239. Hellberg CB, Burden-Gulley SM, Pietz GE, Brady-Kalnay SM (2002) Expression of the receptor protein-tyrosine phosphatase, PTPmu, restores E-cadherin-dependent adhesion in human prostate carcinoma cells. J Biol Chem 277:11165–11173. doi:10.1074/jbc.M112157200
240. Grazia Lampugnani M, Zanetti A, Corada M et al (2003) Contact inhibition of VEGF-induced proliferation requires vascular endothelial cadherin, beta-catenin, and the phosphatase DEP-1/CD148. J Cell Biol 161:793–804. doi:10.1083/jcb.200209019
241. Nawroth R, Poell G, Ranft A et al (2002) VE-PTP and VE-cadherin ectodomains interact to facilitate regulation of phosphorylation and cell contacts. EMBO J 21:4885–4895
242. Ohsugi M, Butz S, Kemler R (1999) Beta-catenin is a major tyrosine-phosphorylated protein during mouse oocyte maturation and preimplantation development. Dev Dyn 216:168–176. doi:10.1002/(SICI)1097-0177(199910)216:2<168::AID-DVDY7>3.0.CO;2-R
243. Mariner DJ, Anastasiadis P, Keilhack H et al (2001) Identification of Src phosphorylation sites in the catenin p120ctn. J Biol Chem 276:28006–28013. doi:10.1074/jbc.M102443200
244. Zondag GC, Moolenaar WH (1997) Receptor protein tyrosine phosphatases: involvement in cell-cell interaction and signaling. Biochimie 79:477–483
245. Dadke S, Kusari A, Kusari J (2001) Phosphorylation and activation of protein tyrosine phosphatase (PTP) 1B by insulin receptor. Mol Cell Biochem 221:147–154
246. Alonso A, Sasin J, Bottini N et al (2004) Protein tyrosine phosphatases in the human genome. Cell 117:699–711. doi:10.1016/j.cell.2004.05.018
247. Sallee JL, Wittchen ES, Burridge K (2006) Regulation of cell adhesion by protein-tyrosine phosphatases: II. Cell-cell adhesion. J Biol Chem 281:16189–16192. doi:10.1074/jbc.R600003200
248. Nita-Lazar M, Rebustini I, Walker J, Kukuruzinska MA (2010) Hypoglycosylated E-cadherin promotes the assembly of tight junctions through the recruitment of PP2A to adherens junctions. Exp Cell Res 316:1871–1884. doi:10.1016/j.yexcr.2010.02.008
249. Chan G, Kalaitzidis D, Neel BG (2008) The tyrosine phosphatase Shp2 (PTPN11) in cancer. Cancer Metastasis Rev 27:179–192. doi:10.1007/s10555-008-9126-y
250. Brady-Kalnay SM, Mourton T, Nixon JP et al (1998) Dynamic interaction of PTPmu with multiple cadherins in vivo. J Cell Biol 141:287–296
251. Ozawa M, Kemler R (1998) Altered cell adhesion activity by pervanadate due to the dissociation of alpha-catenin from the E-cadherin.catenin complex. J Biol Chem 273:6166–6170
252. Tsukatani Y, Suzuki K, Takahashi K (1997) Loss of density-dependent growth inhibition and dissociation of alpha-catenin from E-cadherin. J Cell Physiol 173:54–63. doi:10.1002/(SICI)1097-4652(199710)173:1<54::AID-JCP7>3.0.CO;2-I
253. Ukropec JA, Hollinger MK, Salva SM, Woolkalis MJ (2000) SHP2 association with VE-cadherin complexes in human endothelial cells is regulated by thrombin. J Biol Chem 275:5983–5986
254. Grinnell KL, Casserly B, Harrington EO (2010) Role of protein tyrosine phosphatase SHP2 in barrier function of pulmonary endothelium. Am J Physiol Lung Cell Mol Physiol 298:L361–L370. doi:10.1152/ajplung.00374.2009
255. Balsamo J, Leung T, Ernst H et al (1996) Regulated binding of PTP1B-like phosphatase to N-cadherin: control of cadherin-mediated adhesion by dephosphorylation of beta-catenin. J Cell Biol 134:801–813
256. Danilkovitch-Miagkova A, Miagkov A, Skeel A et al (2001) Oncogenic mutants of RON and MET receptor tyrosine kinases cause activation of the beta-catenin pathway. Mol Cell Biol 21:5857–5868

257. Espejo R, Rengifo-Cam W, Schaller MD et al (2010) PTP-PEST controls motility, adherens junction assembly, and Rho GTPase activity in colon cancer cells. Am J Physiol Cell Physiol 299:C454–C463. doi:10.1152/ajpcell.00148.2010
258. Mourton T, Hellberg CB, Burden-Gulley SM et al (2001) The PTPmu protein-tyrosine phosphatase binds and recruits the scaffolding protein RACK1 to cell-cell contacts. J Biol Chem 276:14896–14901. doi:10.1074/jbc.M010823200
259. Sui XF, Kiser TD, Hyun SW et al (2005) Receptor protein tyrosine phosphatase micro regulates the paracellular pathway in human lung microvascular endothelia. Am J Pathol 166:1247–1258
260. Chagnon MJ, Uetani N, Tremblay ML (2004) Functional significance of the LAR receptor protein tyrosine phosphatase family in development and diseases. Biochem Cell Biol 82:664–675. doi:10.1139/o04-120
261. Tonks NK (2006) Protein tyrosine phosphatases: from genes, to function, to disease. Nat Rev Mol Cell Biol 7:833–846. doi:10.1038/nrm2039
262. Pulido R, Serra-Pagès C, Tang M, Streuli M (1995) The LAR/PTP delta/PTP sigma subfamily of transmembrane protein-tyrosine-phosphatases: multiple human LAR, PTP delta, and PTP sigma isoforms are expressed in a tissue-specific manner and associate with the LAR-interacting protein LIP.1. Proc Natl Acad Sci U S A 92:11686–11690
263. Murchie R, Guo C-H, Persaud A et al (2014) Protein tyrosine phosphatase σ targets apical junction complex proteins in the intestine and regulates epithelial permeability. Proc Natl Acad Sci U S A 111:693–698. doi:10.1073/pnas.1315017111
264. Rioux JD, Silverberg MS, Daly MJ et al (2000) Genomewide search in Canadian families with inflammatory bowel disease reveals two novel susceptibility loci. Am J Hum Genet 66:1863–1870. doi:10.1086/302913
265. Wagner J, Gordon LA, Heng HH et al (1996) Physical mapping of receptor type protein tyrosine phosphatase sigma (PTPRS) to human chromosome 19p13.3. Genomics 38:76–78. doi:10.1006/geno.1996.0594
266. Le Bras GF, Taubenslag KJ, Andl CD (2012) The regulation of cell-cell adhesion during epithelial-mesenchymal transition, motility and tumor progression. Cell Adh Migr 6:365–373. doi:10.4161/cam.21326
267. Chalhoub N, Baker SJ (2009) PTEN and the PI3-kinase pathway in cancer. Annu Rev Pathol 4:127–150. doi:10.1146/annurev.pathol.4.110807.092311
268. Lin G-P, Li X-Y, Huang J-W et al (2007) Inhibitory effects of tumor suppressor gene PTEN on proliferation and metastasis of breast cancer ZR-75-1 cells. Ai Zheng 26:1069–1073
269. Akhtar N, Hotchin NA (2001) RAC1 regulates adherens junctions through endocytosis of E-cadherin. Mol Biol Cell 12:847–862
270. Xiao K, Garner J, Buckley KM et al (2005) p120-Catenin regulates clathrin-dependent endocytosis of VE-cadherin. Mol Biol Cell 16:5141–5151. doi:10.1091/mbc.E05-05-0440
271. De Beco S, Amblard F, Coscoy S (2012) New insights into the regulation of E-cadherin distribution by endocytosis. Int Rev Cell Mol Biol 295:63–108. doi:10.1016/B978-0-12-394306-4.00008-3
272. Platta HW, Stenmark H (2011) Endocytosis and signaling. Curr Opin Cell Biol 23:393–403. doi:10.1016/j.ceb.2011.03.008
273. Sigismund S, Confalonieri S, Ciliberto A et al (2012) Endocytosis and signaling: cell logistics shape the eukaryotic cell plan. Physiol Rev 92:273–366. doi:10.1152/physrev.00005.2011
274. Delva E, Kowalczyk AP (2009) Regulation of cadherin trafficking. Traffic 10:259–267. doi:10.1111/j.1600-0854.2008.00862.x
275. McLachlan RW, Kraemer A, Helwani FM et al (2007) E-cadherin adhesion activates c-Src signaling at cell-cell contacts. Mol Biol Cell 18:3214–3223. doi:10.1091/mbc.E06-12-1154
276. Papkoff J (1997) Regulation of complexed and free catenin pools by distinct mechanisms. Differential effects of Wnt-1 and v-Src. J Biol Chem 272:4536–4543
277. Lu Z, Ghosh S, Wang Z, Hunter T (2003) Downregulation of caveolin-1 function by EGF leads to the loss of E-cadherin, increased transcriptional activity of beta-catenin, and enhanced tumor cell invasion. Cancer Cell 4:499–515

278. Bryant DM, Kerr MC, Hammond LA et al (2007) EGF induces macropinocytosis and SNX1-modulated recycling of E-cadherin. J Cell Sci 120:1818–1828. doi:10.1242/jcs.000653
279. Paterson AD, Parton RG, Ferguson C et al (2003) Characterization of E-cadherin endocytosis in isolated MCF-7 and chinese hamster ovary cells: the initial fate of unbound E-cadherin. J Biol Chem 278:21050–21057. doi:10.1074/jbc.M300082200
280. De Wever O, Derycke ÆL, Hendrix ÆA et al (2007) Soluble cadherins as cancer biomarkers. 685–697. doi:10.1007/s10585-007-9104-8
281. Seals DF, Courtneidge SA (2003) The ADAMs family of metalloproteases: multidomain proteins with multiple functions. Genes Dev 17:7–30. doi:10.1101/gad.1039703
282. Davies G, Jiang WG, Mason MD (2001) Matrilysin mediates extracellular cleavage of E-cadherin from prostate cancer cells: a key mechanism in hepatocyte growth factor/scatter factor-induced cell-cell dissociation and in vitro invasion. Clin Cancer Res 7:3289–3297
283. Hayashido Y, Hamana T, Yoshioka Y et al (2005) Plasminogen activator/plasmin system suppresses cell-cell adhesion of oral squamous cell carcinoma cells via proteolysis of E-cadherin. Int J Oncol 27:693–698
284. Johnson SK, Ramani VC, Hennings L, Haun RS (2007) Kallikrein 7 enhances pancreatic cancer cell invasion by shedding E-cadherin. Cancer 109:1811–1820. doi:10.1002/cncr.22606
285. Lochter A, Galosy S, Muschler J et al (1997) Matrix metalloproteinase stromelysin-1 triggers a cascade of molecular alterations that leads to stable epithelial-to-mesenchymal conversion and a premalignant phenotype in mammary epithelial cells. J Cell Biol 139:1861–1872
286. Maretzky T, Reiss K, Ludwig A et al (2005) ADAM10 mediates E-cadherin shedding and regulates epithelial cell-cell adhesion, migration, and beta-catenin translocation. Proc Natl Acad Sci U S A 102:9182–9187. doi:10.1073/pnas.0500918102
287. Noë V, Fingleton B, Jacobs K et al (2001) Release of an invasion promoter E-cadherin fragment by matrilysin and stromelysin-1. J Cell Sci 114:111–118
288. Ryniers F, Stove C, Goethals M et al (2002) Plasmin produces an E-cadherin fragment that stimulates cancer cell invasion. Biol Chem 383:159–165. doi:10.1515/BC.2002.016
289. Symowicz J, Adley BP, Gleason KJ et al (2007) Engagement of collagen-binding integrins promotes matrix metalloproteinase-9-dependent E-cadherin ectodomain shedding in ovarian carcinoma cells. Cancer Res 67:2030–2039. doi:10.1158/0008-5472.CAN-06-2808
290. Damsky CH, Richa J, Solter D et al (1983) Identification and purification of a cell surface glycoprotein mediating intercellular adhesion in embryonic and adult tissue. Cell 34:455–466
291. Wheelock MJ, Buck CA, Bechtol KB, Damsky CH (1987) Soluble 80-kd fragment of cell-CAM 120/80 disrupts cell-cell adhesion. J Cell Biochem 34:187–202. doi:10.1002/jcb.240340305
292. Lee KH, Choi EY, Hyun MS et al (2007) Association of extracellular cleavage of E-cadherin mediated by MMP-7 with HGF-induced in vitro invasion in human stomach cancer cells. Eur Surg Res 39:208–215. doi:10.1159/000101452
293. Najy AJ, Day KC, Day ML (2008) The ectodomain shedding of E-cadherin by ADAM15 supports ErbB receptor activation. J Biol Chem 283:18393–18401. doi:10.1074/jbc.M801329200
294. Whittard JD, Craig SE, Mould AP et al (2002) E-cadherin is a ligand for integrin alpha-2beta1. Matrix Biol 21:525–532
295. Gocheva V, Zeng W, Ke D et al (2006) Distinct roles for cysteine cathepsin genes in multistage tumorigenesis. Genes Dev 20:543–556. doi:10.1101/gad.1407406
296. Van Aken E, De Wever O, Correia da Rocha AS, Mareel M (2001) Defective E-cadherin/catenin complexes in human cancer. Virchows Arch 439:725–751
297. Covington MD, Burghardt RC, Parrish AR (2006) Ischemia-induced cleavage of cadherins in NRK cells requires MT1-MMP (MMP-14). Am J Physiol Renal Physiol 290:F43–F51. doi:10.1152/ajprenal.00179.2005
298. Monea S, Jordan BA, Srivastava S et al (2006) Membrane localization of membrane type 5 matrix metalloproteinase by AMPA receptor binding protein and cleavage of cadherins. J Neurosci 26:2300–2312. doi:10.1523/JNEUROSCI.3521-05.2006
299. Reiss K, Maretzky T, Ludwig A et al (2005) ADAM10 cleavage of N-cadherin and regulation of cell-cell adhesion and beta-catenin nuclear signalling. EMBO J 24:742–752. doi:10.1038/sj.emboj.7600548

300. Uemura K, Kihara T, Kuzuya A et al (2006) Characterization of sequential N-cadherin cleavage by ADAM10 and PS1. Neurosci Lett 402:278–283. doi:10.1016/j.neulet.2006.04.018
301. Marambaud P, Wen PH, Dutt A et al (2003) A CBP binding transcriptional repressor produced by the PS1/epsilon-cleavage of N-cadherin is inhibited by PS1 FAD mutations. Cell 114:635–645
302. Cifuentes-Diaz C, Nicolet M, Goudou D et al (1994) N-cadherin expression in developing, adult and denervated chicken neuromuscular system: accumulations at both the neuromuscular junction and the node of Ranvier. Development 120:1–11
303. Utton MA, Eickholt B, Howell FV et al (2001) Soluble N-cadherin stimulates fibroblast growth factor receptor dependent neurite outgrowth and N-cadherin and the fibroblast growth factor receptor co-cluster in cells. J Neurochem 76:1421–1430
304. Chetty R, Serra S (2008) Nuclear E-cadherin immunoexpression: from biology to potential applications in diagnostic pathology. Adv Anat Pathol 15:234–240. doi:10.1097/PAP.0b013e31817bf566
305. Natalwala A, Spychal R, Tselepis C (2008) Epithelial-mesenchymal transition mediated tumourigenesis in the gastrointestinal tract. World J Gastroenterol 14:3792–3797
306. Salahshor S, Naidoo R, Serra S et al (2008) Frequent accumulation of nuclear E-cadherin and alterations in the Wnt signaling pathway in esophageal squamous cell carcinomas. Mod Pathol 21:271–281. doi:10.1038/modpathol.3800990
307. Serra S, Salahshor S, Fagih M et al (2007) Nuclear expression of E-cadherin in solid pseudopapillary tumors of the pancreas. JOP 8:296–303
308. Céspedes MV, Larriba MJ, Pavón MA et al (2010) Site-dependent E-cadherin cleavage and nuclear translocation in a metastatic colorectal cancer model. Am J Pathol 177:2067–2079. doi:10.2353/ajpath.2010.100079
309. Orsulic S, Huber O, Aberle H et al (1999) E-cadherin binding prevents beta-catenin nuclear localization and beta-catenin/LEF-1-mediated transactivation. J Cell Sci 112(Pt 8):1237–1245
310. Sadot E, Simcha I, Shtutman M et al (1998) Inhibition of beta-catenin-mediated transactivation by cadherin derivatives. Proc Natl Acad Sci U S A 95:15339–15344
311. Maetzel D, Denzel S, Mack B et al (2009) Nuclear signalling by tumour-associated antigen EpCAM. Nat Cell Biol 11:162–171. doi:10.1038/ncb1824
312. Munz M, Baeuerle PA, Gires O (2009) The emerging role of EpCAM in cancer and stem cell signaling. Cancer Res 69:5627–5629. doi:10.1158/0008-5472.CAN-09-0654
313. Denzel S, Maetzel D, Mack B et al (2009) Initial activation of EpCAM cleavage via cell-to-cell contact. BMC Cancer 9:402. doi:10.1186/1471-2407-9-402
314. Rashid MG, Sanda MG, Vallorosi CJ et al (2001) Posttranslational truncation and inactivation of human E-cadherin distinguishes prostate cancer from matched normal prostate. Cancer Res 61:489–492
315. Rios-Doria J, Day KC, Kuefer R et al (2003) The role of calpain in the proteolytic cleavage of E-cadherin in prostate and mammary epithelial cells. J Biol Chem 278:1372–1379. doi:10.1074/jbc.M208772200
316. Steinhusen U, Weiske J, Badock V et al (2001) Cleavage and shedding of E-cadherin after induction of apoptosis. J Biol Chem 276:4972–4980. doi:10.1074/jbc.M006102200
317. Dorudi S, Sheffield JP, Poulsom R et al (1993) E-cadherin expression in colorectal cancer. An immunocytochemical and in situ hybridization study. Am J Pathol 142:981–986
318. Perl AK, Wilgenbus P, Dahl U et al (1998) A causal role for E-cadherin in the transition from adenoma to carcinoma. Nature 392:190–193. doi:10.1038/32433
319. Shitashige M, Satow R, Honda K et al (2008) Regulation of Wnt signaling by the nuclear pore complex. Gastroenterology 134:1961–7191, 1971.e1–4. doi:10.1053/j.gastro.2008.03.010
320. Elzagheid A, Algars A, Bendardaf R et al (2006) E-cadherin expression pattern in primary colorectal carcinomas and their metastases reflects disease outcome. World J Gastroenterol 12:4304–4309
321. Lombaerts M, van Wezel T, Philippo K et al (2006) E-cadherin transcriptional downregulation by promoter methylation but not mutation is related to epithelial-to-mesenchymal transition in breast cancer cell lines. Br J Cancer 94:661–671. doi:10.1038/sj.bjc.6602996

322. Wang Y, Shang Y (2013) Epigenetic control of epithelial-to-mesenchymal transition and cancer metastasis. Exp Cell Res 319:160–169. doi:10.1016/j.yexcr.2012.07.019
323. Berx G, van Roy F (2009) Involvement of members of the cadherin superfamily in cancer. Cold Spring Harb Perspect Biol 1:a003129. doi:10.1101/cshperspect.a003129
324. Strathdee G (2002) Epigenetic versus genetic alterations in the inactivation of E-cadherin. Semin Cancer Biol 12:373–379
325. Darwanto A, Kitazawa R, Maeda S, Kitazawa S (2003) MeCP2 and promoter methylation cooperatively regulate E-cadherin gene expression in colorectal carcinoma. Cancer Sci 94:442–447
326. Hendrich B, Tweedie S (2003) The methyl-CpG binding domain and the evolving role of DNA methylation in animals. Trends Genet 19:269–277. doi:10.1016/S0168-9525(03)00080-5
327. Koizume S, Tachibana K, Sekiya T et al (2002) Heterogeneity in the modification and involvement of chromatin components of the CpG island of the silenced human CDH1 gene in cancer cells. Nucleic Acids Res 30:4770–4780
328. Brabletz T, Hlubek F, Spaderna S et al (2005) Invasion and metastasis in colorectal cancer: epithelial-mesenchymal transition, mesenchymal-epithelial transition, stem cells and beta-catenin. Cells Tissues Organs 179:56–65. doi:10.1159/000084509
329. Bender CM, Pao MM, Jones PA (1998) Inhibition of DNA methylation by 5-aza-2'-deoxycytidine suppresses the growth of human tumor cell lines. Cancer Res 58:95–101
330. Jüttermann R, Li E, Jaenisch R (1994) Toxicity of 5-aza-2'-deoxycytidine to mammalian cells is mediated primarily by covalent trapping of DNA methyltransferase rather than DNA demethylation. Proc Natl Acad Sci U S A 91:11797–11801
331. Kurkjian C, Kummar S, Murgo AJ (2008) DNA methylation: its role in cancer development and therapy. Curr Probl Cancer 32:187–235. doi:10.1016/j.currproblcancer.2008.08.002
332. Graff JR, Herman JG, Lapidus RG et al (1995) E-cadherin expression is silenced by DNA hypermethylation in human breast and prostate carcinomas. Cancer Res 55:5195–5199
333. Yoshiura K, Kanai Y, Ochiai A et al (1995) Silencing of the E-cadherin invasion-suppressor gene by CpG methylation in human carcinomas. Proc Natl Acad Sci U S A 92:7416–7419
334. Thiery JP, Acloque H, Huang RYJ, Nieto MA (2009) Epithelial-mesenchymal transitions in development and disease. Cell 139:871–890. doi:10.1016/j.cell.2009.11.007
335. Tiwari N, Gheldof A, Tatari M, Christofori G (2012) EMT as the ultimate survival mechanism of cancer cells. Semin Cancer Biol 22:194–207. doi:10.1016/j.semcancer.2012.02.013
336. Nieto MA (2011) The ins and outs of the epithelial to mesenchymal transition in health and disease. Annu Rev Cell Dev Biol 27:347–376. doi:10.1146/annurev-cellbio-092910-154036
337. Peinado H, Olmeda D, Cano A (2007) Snail, Zeb and bHLH factors in tumour progression: an alliance against the epithelial phenotype? Nat Rev Cancer 7:415–428. doi:10.1038/nrc2131
338. Peña C, García JM, Larriba MJ et al (2009) SNAI1 expression in colon cancer related with CDH1 and VDR downregulation in normal adjacent tissue. Oncogene 28:4375–4385. doi:10.1038/onc.2009.285
339. Viré E, Brenner C, Deplus R et al (2006) The Polycomb group protein EZH2 directly controls DNA methylation. Nature 439:871–874. doi:10.1038/nature04431
340. Cano A, Pérez-Moreno MA, Rodrigo I et al (2000) The transcription factor snail controls epithelial-mesenchymal transitions by repressing E-cadherin expression. Nat Cell Biol 2:76–83. doi:10.1038/35000025
341. Herranz N, Pasini D, Díaz VM et al (2008) Polycomb complex 2 is required for E-cadherin repression by the Snail1 transcription factor. Mol Cell Biol 28:4772–4781. doi:10.1128/MCB.00323-08
342. Bolós V, Peinado H, Pérez-Moreno MA et al (2003) The transcription factor Slug represses E-cadherin expression and induces epithelial to mesenchymal transitions: a comparison with Snail and E47 repressors. J Cell Sci 116:499–511
343. Hajra KM, Chen DY-S, Fearon ER (2002) The SLUG zinc-finger protein represses E-cadherin in breast cancer. Cancer Res 62:1613–1618

344. Hemavathy K, Guru SC, Harris J et al (2000) Human Slug is a repressor that localizes to sites of active transcription. Mol Cell Biol 20:5087–5095
345. Shi Y, Sawada J, Sui G et al (2003) Coordinated histone modifications mediated by a CtBP co-repressor complex. Nature 422:735–738. doi:10.1038/nature01550
346. Van Grunsven LA, Michiels C, Van de Putte T et al (2003) Interaction between Smad-interacting protein-1 and the corepressor C-terminal binding protein is dispensable for transcriptional repression of E-cadherin. J Biol Chem 278:26135–26145. doi:10.1074/jbc.M300597200
347. Grooteclaes ML, Frisch SM (2000) Evidence for a function of CtBP in epithelial gene regulation and anoikis. Oncogene 19(33): 3823–3828. http://doi.org/10.1038/sj.onc.1203721
348. Peña C, García JM, García V et al (2006) The expression levels of the transcriptional regulators p300 and CtBP modulate the correlations between SNAIL, ZEB1, E-cadherin and vitamin D receptor in human colon carcinomas. Int J Cancer 119:2098–2104. doi:10.1002/ijc.22083
349. Fu J, Qin L, He T et al (2011) The TWIST/Mi2/NuRD protein complex and its essential role in cancer metastasis. Cell Res 21:275–289. doi:10.1038/cr.2010.118
350. Ching-Wen L, Shih-Han K, Pan-Chyr Y (2014) The MiRNAs and epithelial-mesenchymal transition in cancers. Curr Pharm Des 20(33):5309–5318
351. Díaz-López A, Moreno-Bueno G, Cano A (2014) Role of microRNA in epithelial to mesenchymal transition and metastasis and clinical perspectives. Cancer Manag Res 6:205–216. doi:10.2147/CMAR.S38156
352. Lu J, Getz G, Miska EA et al (2005) MicroRNA expression profiles classify human cancers. Nature 435:834–838. doi:10.1038/nature03702
353. Ma L, Young J, Prabhala H et al (2010) miR-9, a MYC/MYCN-activated microRNA, regulates E-cadherin and cancer metastasis. Nat Cell Biol 12:247–256. doi:10.1038/ncb2024
354. Bracken CP, Gregory PA, Kolesnikoff N et al (2008) A double-negative feedback loop between ZEB1-SIP1 and the microRNA-200 family regulates epithelial-mesenchymal transition. Cancer Res 68:7846–7854. doi:10.1158/0008-5472.CAN-08-1942
355. Burk U, Schubert J, Wellner U et al (2008) A reciprocal repression between ZEB1 and members of the miR-200 family promotes EMT and invasion in cancer cells. EMBO Rep 9:582–589. doi:10.1038/embor.2008.74
356. Kong W, Yang H, He L et al (2008) MicroRNA-155 is regulated by the transforming growth factor beta/Smad pathway and contributes to epithelial cell plasticity by targeting RhoA. Mol Cell Biol 28:6773–6784. doi:10.1128/MCB.00941-08
357. Kim NH, Kim HS, Li X-Y et al (2011) A p53/miRNA-34 axis regulates Snail1-dependent cancer cell epithelial-mesenchymal transition. J Cell Biol 195:417–433. doi:10.1083/jcb.201103097
358. Kim T, Veronese A, Pichiorri F et al (2011) p53 regulates epithelial-mesenchymal transition through microRNAs targeting ZEB1 and ZEB2. J Exp Med 208:875–883. doi:10.1084/jem.20110235
359. Liu X, Wang C, Chen Z et al (2011) MicroRNA-138 suppresses epithelial-mesenchymal transition in squamous cell carcinoma cell lines. Biochem J 440:23–31. doi:10.1042/BJ20111006
360. Davalos V, Moutinho C, Villanueva A et al (2012) Dynamic epigenetic regulation of the microRNA-200 family mediates epithelial and mesenchymal transitions in human tumorigenesis. Oncogene 31:2062–2074. doi:10.1038/onc.2011.383
361. Díaz-Martín J, Díaz-López A, Moreno-Bueno G et al (2014) A core microRNA signature associated with inducers of the epithelial-to-mesenchymal transition. J Pathol 232:319–329. doi:10.1002/path.4289
362. Wee EJH, Peters K, Nair SS et al (2012) Mapping the regulatory sequences controlling 93 breast cancer-associated miRNA genes leads to the identification of two functional promoters of the Hsa-mir-200b cluster, methylation of which is associated with metastasis or hormone receptor status in adva. Oncogene 31:4182–4195. doi:10.1038/onc.2011.584
363. Larue L, Ohsugi M, Hirchenhain J, Kemler R (1994) E-cadherin null mutant embryos fail to form a trophectoderm epithelium. Proc Natl Acad Sci U S A 91:8263–8267

364. Brasch J, Harrison OJ, Honig B, Shapiro L (2012) Thinking outside the cell: how cadherins drive adhesion. Trends Cell Biol 22:299–310. doi:10.1016/j.tcb.2012.03.004
365. Takeichi M (1995) Morphogenetic roles of classic cadherins. Curr Opin Cell Biol 7:619–627
366. Pieters T, van Roy F (2014) Role of cell-cell adhesion complexes in embryonic stem cell biology. J Cell Sci 127:2603–2613. doi:10.1242/jcs.146720
367. Jaenisch R, Young R (2008) Stem cells, the molecular circuitry of pluripotency and nuclear reprogramming. Cell 132:567–582. doi:10.1016/j.cell.2008.01.015
368. Li D, Zhou J, Wang L et al (2010) Integrated biochemical and mechanical signals regulate multifaceted human embryonic stem cell functions. J Cell Biol 191:631–644. doi:10.1083/jcb.201006094
369. Bao S, Tang F, Li X et al (2009) Epigenetic reversion of post-implantation epiblast to pluripotent embryonic stem cells. Nature 461:1292–1295. doi:10.1038/nature08534
370. Takahashi K, Yamanaka S (2006) Induction of pluripotent stem cells from mouse embryonic and adult fibroblast cultures by defined factors. Cell 126:663–676. doi:10.1016/j.cell.2006.07.024
371. Chen T, Yuan D, Wei B et al (2010) E-cadherin-mediated cell-cell contact is critical for induced pluripotent stem cell generation. Stem Cells 28:1315–1325. doi:10.1002/stem.456
372. Chen J, Liu J, Yang J et al (2011) BMPs functionally replace Klf4 and support efficient reprogramming of mouse fibroblasts by Oct4 alone. Cell Res 21:205–212. doi:10.1038/cr.2010.172
373. De Craene B, Berx G (2013) Regulatory networks defining EMT during cancer initiation and progression. Nat Rev Cancer 13:97–110. doi:10.1038/nrc3447
374. Koopmansch B, Berx G, Foidart J-M et al (2013) Interplay between KLF4 and ZEB2/SIP1 in the regulation of E-cadherin expression. Biochem Biophys Res Commun 431:652–657. doi:10.1016/j.bbrc.2013.01.070
375. Yori JL, Johnson E, Zhou G et al (2010) Kruppel-like factor 4 inhibits epithelial-to-mesenchymal transition through regulation of E-cadherin gene expression. J Biol Chem 285:16854–16863. doi:10.1074/jbc.M110.114546
376. Soncin F, Mohamet L, Eckardt D et al (2009) Abrogation of E-cadherin-mediated cell-cell contact in mouse embryonic stem cells results in reversible LIF-independent self-renewal. Stem Cells 27:2069–2080. doi:10.1002/stem.134
377. Martello G, Sugimoto T, Diamanti E et al (2012) Esrrb is a pivotal target of the Gsk3/Tcf3 axis regulating embryonic stem cell self-renewal. Cell Stem Cell 11:491–504. doi:10.1016/j.stem.2012.06.008
378. Wray J, Hartmann C (2012) WNTing embryonic stem cells. Trends Cell Biol 22:159–168. doi:10.1016/j.tcb.2011.11.004
379. Herlyn D, Herlyn M, Steplewski Z, Koprowski H (1979) Monoclonal antibodies in cell-mediated cytotoxicity against human melanoma and colorectal carcinoma. Eur J Immunol 9:657–659. doi:10.1002/eji.1830090817
380. Trzpis M, McLaughlin PM, van Goor H et al (2008) Expression of EpCAM is up-regulated during regeneration of renal epithelia. J Pathol 216:201–208. doi:10.1002/path.2396
381. Balzar M, Bakker HA, Briaire-de-Bruijn IH et al (1998) Cytoplasmic tail regulates the intercellular adhesion function of the epithelial cell adhesion molecule. Mol Cell Biol 18:4833–4843
382. Maaser K, Borlak J (2008) A genome-wide expression analysis identifies a network of EpCAM-induced cell cycle regulators. Br J Cancer 99:1635–1643. doi:10.1038/sj.bjc.6604725
383. Münz M, Kieu C, Mack B et al (2004) The carcinoma-associated antigen EpCAM upregulates c-myc and induces cell proliferation. Oncogene 23:5748–5758. doi:10.1038/sj.onc.1207610
384. Hanna J, Saha K, Pando B et al (2009) Direct cell reprogramming is a stochastic process amenable to acceleration. Nature 462:595–601. doi:10.1038/nature08592
385. Hanna JH, Saha K, Jaenisch R (2010) Pluripotency and cellular reprogramming: facts, hypotheses, unresolved issues. Cell 143:508–525. doi:10.1016/j.cell.2010.10.008
386. Lu T-Y, Lu R-M, Liao M-Y et al (2010) Epithelial cell adhesion molecule regulation is associated with the maintenance of the undifferentiated phenotype of human embryonic stem cells. J Biol Chem 285:8719–8732. doi:10.1074/jbc.M109.077081

387. Reya T, Clevers H (2005) Wnt signalling in stem cells and cancer. Nature 434:843–850. doi:10.1038/nature03319
388. Yamashita T, Budhu A, Forgues M, Wang XW (2007) Activation of hepatic stem cell marker EpCAM by Wnt-beta-catenin signaling in hepatocellular carcinoma. Cancer Res 67:10831–10839. doi:10.1158/0008-5472.CAN-07-0908
389. Kroepil F, Dulian A, Vallböhmer D et al (2013) High EpCAM expression is linked to proliferation and lauren classification in gastric cancer. BMC Res Notes 6:253. doi:10.1186/1756-0500-6-253
390. Nakagawa S, Takeichi M (1998) Neural crest emigration from the neural tube depends on regulated cadherin expression. Development 125:2963–2971
391. Suyama K, Shapiro I, Guttman M, Hazan RB (2002) A signaling pathway leading to metastasis is controlled by N-cadherin and the FGF receptor. Cancer Cell 2:301–314
392. Tran NL, Adams DG, Vaillancourt RR, Heimark RL (2002) Signal transduction from N-cadherin increases Bcl-2. Regulation of the phosphatidylinositol 3-kinase/Akt pathway by homophilic adhesion and actin cytoskeletal organization. J Biol Chem 277:32905–32914. doi:10.1074/jbc.M200300200
393. Gout S, Marie C, Lainé M et al (2004) Early enterocytic differentiation of HT-29 cells: biochemical changes and strength increases of adherens junctions. Exp Cell Res 299:498–510. doi:10.1016/j.yexcr.2004.06.008
394. Schreider C, Peignon G, Thenet S et al (2002) Integrin-mediated functional polarization of Caco-2 cells through E-cadherin--actin complexes. J Cell Sci 115:543–552
395. Ekblom M, Falk M, Salmivirta K et al (1998) Laminin isoforms and epithelial development. Ann N Y Acad Sci 857:194–211
396. Timpl R (1996) Macromolecular organization of basement membranes. Curr Opin Cell Biol 8:618–624
397. Turck N, Lefebvre O, Gross I et al (2006) Effect of laminin-1 on intestinal cell differentiation involves inhibition of nuclear nucleolin. J Cell Physiol 206:545–555. doi:10.1002/jcp.20501
398. Houde M, Laprise P, Jean D et al (2001) Intestinal epithelial cell differentiation involves activation of p38 mitogen-activated protein kinase that regulates the homeobox transcription factor CDX2. J Biol Chem 276:21885–21894. doi:10.1074/jbc.M100236200
399. Shih DQ, Targan SR, McGovern D (2008) Recent advances in IBD pathogenesis: genetics and immunobiology. Curr Gastroenterol Rep 10:568–575
400. Koch S, Nusrat A (2012) The life and death of epithelia during inflammation: lessons learned from the gut. Annu Rev Pathol 7:35–60. doi:10.1146/annurev-pathol-011811-120905
401. Scharl M, Rogler G (2012) Inflammatory bowel disease pathogenesis: what is new? Curr Opin Gastroenterol 28(4):301–309. http://doi.org/10.1097/MOG.0b013e328353e61e
402. Fries W, Mazzon E, Squarzoni S et al (1999) Experimental colitis increases small intestine permeability in the rat. Lab Invest 79:49–57
403. Marin ML, Geller SA, Greenstein AJ et al (1983) Ultrastructural pathology of Crohn's disease: correlated transmission electron microscopy, scanning electron microscopy, and freeze fracture studies. Am J Gastroenterol 78:355–364
404. Bouma G, Strober W (2003) The immunological and genetic basis of inflammatory bowel disease. Nat Rev Immunol 3:521–533. doi:10.1038/nri1132
405. Gassler N, Rohr C, Schneider A et al (2001) Inflammatory bowel disease is associated with changes of enterocytic junctions. Am J Physiol Gastrointest Liver Physiol 281:G216–G228
406. Hermiston ML, Gordon JI (1995) Inflammatory bowel disease and adenomas in mice expressing a dominant negative N-cadherin. Science 270:1203–1207
407. Zbar AP, Simopoulos C, Karayiannakis AJ (2004) Cadherins: an integral role in inflammatory bowel disease and mucosal restitution. J Gastroenterol 39:413–421. doi:10.1007/s00535-004-1335-8
408. Jankowski JA, Bedford FK, Boulton RA et al (1998) Alterations in classical cadherins associated with progression in ulcerative and Crohn's colitis. Lab Invest 78:1155–1167
409. Radford-Smith G, Jewell DP (1996) Cytokines and inflammatory bowel disease. Baillieres Clin Gastroenterol 10:151–164

410. Issenman RM, Jenkins RT, Radoja C (1993) Intestinal permeability compared in pediatric and adult patients with inflammatory bowel disease. Clin Invest Med 16:187–196
411. Jenkins RT, Ramage JK, Jones DB et al (1988) Small bowel and colonic permeability to 51Cr-EDTA in patients with active inflammatory bowel disease. Clin Invest Med 11:151–155
412. Miki K, Moore DJ, Butler RN et al (1998) The sugar permeability test reflects disease activity in children and adolescents with inflammatory bowel disease. J Pediatr 133:750–754
413. Munkholm P, Langholz E, Hollander D et al (1994) Intestinal permeability in patients with Crohn's disease and ulcerative colitis and their first degree relatives. Gut 35:68–72
414. Wang F, Graham WV, Wang Y et al (2005) Interferon-gamma and tumor necrosis factor-alpha synergize to induce intestinal epithelial barrier dysfunction by up-regulating myosin light chain kinase expression. Am J Pathol 166:409–419
415. Bruewer M, Utech M, Ivanov AI et al (2005) Interferon-gamma induces internalization of epithelial tight junction proteins via a macropinocytosis-like process. FASEB J 19:923–933. doi:10.1096/fj.04-3260com
416. McCormick DA, Horton LW, Mee AS (1990) Mucin depletion in inflammatory bowel disease. J Clin Pathol 43:143–146
417. Morson BC (1972) Rectal biopsy in inflammatory bowel disease. N Engl J Med 287:1337–1339. doi:10.1056/NEJM197212282872607
418. Renes IB, Verburg M, Van Nispen DJPM et al (2002) Epithelial proliferation, cell death, and gene expression in experimental colitis: alterations in carbonic anhydrase I, mucin MUC2, and trefoil factor 3 expression. Int J Colorectal Dis 17:317–326. doi:10.1007/s00384-002-0409-4
419. Van der Sluis M, De Koning BAE, De Bruijn ACJM et al (2006) Muc2-deficient mice spontaneously develop colitis, indicating that MUC2 is critical for colonic protection. Gastroenterology 131:117–129. doi:10.1053/j.gastro.2006.04.020
420. Mastroianni JR, Ouellette AJ (2009) Alpha-defensins in enteric innate immunity: functional Paneth cell alpha-defensins in mouse colonic lumen. J Biol Chem 284:27848–27856. doi:10.1074/jbc.M109.050773
421. Wehkamp J, Schmid M, Fellermann K, Stange EF (2005) Defensin deficiency, intestinal microbes, and the clinical phenotypes of Crohn's disease. J Leukoc Biol 77:460–465. doi:10.1189/jlb.0904543
422. Sandow MJ, Whitehead R (1979) The Paneth cell. Gut 20:420–431
423. Wada R (2009) Proposal of a new hypothesis on the development of colorectal epithelial neoplasia: nonspecific inflammation--colorectal Paneth cell metaplasia--colorectal epithelial neoplasia. Digestion 79(Suppl 1):9–12. doi:10.1159/000167860
424. Feagins LA, Souza RF, Spechler SJ (2009) Carcinogenesis in IBD: potential targets for the prevention of colorectal cancer. Nat Rev Gastroenterol Hepatol 6:297–305. doi:10.1038/nrgastro.2009.44
425. Jankowski JA, Bedford FK, Kim YS (1997) Changes in gene structure and regulation of E-cadherin during epithelial development, differentiation, and disease. Prog Nucleic Acid Res Mol Biol 57:187–215
426. Kaler P, Godasi BN, Augenlicht L, Klampfer L (2009) The NF-kappaB/AKT-dependent Induction of Wnt Signaling in Colon Cancer Cells by Macrophages and IL-1beta. Cancer Microenviron. doi:10.1007/s12307-009-0030-y
427. Lee G, Goretsky T, Managlia E et al (2010) Phosphoinositide 3-kinase signaling mediates beta-catenin activation in intestinal epithelial stem and progenitor cells in colitis. Gastroenterology 139:869–881, 881.e1–9. doi:10.1053/j.gastro.2010.05.037
428. Nava P, Koch S, Laukoetter MG et al (2010) Interferon-gamma regulates intestinal epithelial homeostasis through converging beta-catenin signaling pathways. Immunity 32:392–402. doi:10.1016/j.immuni.2010.03.001
429. He XC, Yin T, Grindley JC et al (2007) PTEN-deficient intestinal stem cells initiate intestinal polyposis. Nat Genet 39:189–198. doi:10.1038/ng1928
430. Early DS, Fontana L, Davidson NO (2008) Translational approaches to addressing complex genetic pathways in colorectal cancer. Transl Res 151:10–16. doi:10.1016/j.trsl.2007.09.002

431. Lee S, Bang S, Song K, Lee I (2006) Differential expression in normal-adenoma-carcinoma sequence suggests complex molecular carcinogenesis in colon. Oncol Rep 16:747–754
432. Fodde R (2002) The APC gene in colorectal cancer. Eur J Cancer 38:867–871
433. Scott RJ, van der Luijt R, Spycher M et al (1995) Novel germline APC gene mutation in a large familial adenomatous polyposis kindred displaying variable phenotypes. Gut 36:731–736
434. Sparks AB, Morin PJ, Vogelstein B, Kinzler KW (1998) Mutational analysis of the APC/beta-catenin/Tcf pathway in colorectal cancer. Cancer Res 58:1130–1134
435. Lee H, Flaherty P, Ji HP (2013) Systematic genomic identification of colorectal cancer genes delineating advanced from early clinical stage and metastasis. BMC Med Genomics 6:54. doi:10.1186/1755-8794-6-54
436. Powell SM, Zilz N, Beazer-Barclay Y et al (1992) APC mutations occur early during colorectal tumorigenesis. Nature 359:235–237. doi:10.1038/359235a0
437. Schneikert J, Behrens J (2007) The canonical Wnt signalling pathway and its APC partner in colon cancer development. Gut 56:417–425. doi:10.1136/gut.2006.093310
438. Morin PJ, Sparks AB, Korinek V et al (1997) Activation of beta-catenin-Tcf signaling in colon cancer by mutations in beta-catenin or APC. Science 275:1787–1790
439. Behrens J (2005) The role of the Wnt signalling pathway in colorectal tumorigenesis. Biochem Soc Trans 33:672–675. doi:10.1042/BST0330672
440. Machado JC, Soares P, Carneiro F et al (1999) E-cadherin gene mutations provide a genetic basis for the phenotypic divergence of mixed gastric carcinomas. Lab Invest 79:459–465
441. Schuhmacher C, Becker I, Oswald S et al (1999) Loss of immunohistochemical E-cadherin expression in colon cancer is not due to structural gene alterations. Virchows Arch 434:489–495
442. Hazan RB, Phillips GR, Qiao RF et al (2000) Exogenous expression of N-cadherin in breast cancer cells induces cell migration, invasion, and metastasis. J Cell Biol 148:779–790
443. Cheung LWT, Leung PCK, Wong AST (2010) Cadherin switching and activation of p120 catenin signaling are mediators of gonadotropin-releasing hormone to promote tumor cell migration and invasion in ovarian cancer. Oncogene 29:2427–2440. doi:10.1038/onc.2009.523
444. Batistatou A, Charalabopoulos AK, Scopa CD et al (2006) Expression patterns of dysadherin and E-cadherin in lymph node metastases of colorectal carcinoma. Virchows Arch 448:763–767. doi:10.1007/s00428-006-0183-8
445. Ikeguchi M, Makino M, Kaibara N (2001) Clinical significance of E-cadherin-catenin complex expression in metastatic foci of colorectal carcinoma. J Surg Oncol 77:201–207
446. Pieters T, van Hengel J, van Roy F (2012) Functions of p120ctn in development and disease. Front Biosci (Landmark Ed) 17:760–783
447. Thoreson MA, Reynolds AB (2002) Altered expression of the catenin p120 in human cancer: implications for tumor progression. Differentiation 70:583–589. doi:10.1046/j.1432-0436.2002.700911.x
448. Bellovin DI, Bates RC, Muzikansky A et al (2005) Altered localization of p120 catenin during epithelial to mesenchymal transition of colon carcinoma is prognostic for aggressive disease. Cancer Res 65:10938–10945. doi:10.1158/0008-5472.CAN-05-1947
449. Shibata T, Kokubu A, Sekine S et al (2004) Cytoplasmic p120ctn regulates the invasive phenotypes of E-cadherin-deficient breast cancer. Am J Pathol 164:2269–2278. doi:10.1016/S0002-9440(10)63783-2
450. Strumane K, Bonnomet A, Stove C et al (2006) E-cadherin regulates human Nanos1, which interacts with p120ctn and induces tumor cell migration and invasion. Cancer Res 66:10007–10015. doi:10.1158/0008-5472.CAN-05-3096
451. Bonnomet A, Polette M, Strumane K et al (2008) The E-cadherin-repressed hNanos1 gene induces tumor cell invasion by upregulating MT1-MMP expression. Oncogene 27:3692–3699. doi:10.1038/sj.onc.1211035
452. Efstathiou JA, Liu D, Wheeler JM et al (1999) Mutated epithelial cadherin is associated with increased tumorigenicity and loss of adhesion and of responsiveness to the motogenic trefoil factor 2 in colon carcinoma cells. Proc Natl Acad Sci U S A 96:2316–2321
453. Roy HK, Smyrk TC, Koetsier J et al (2005) The transcriptional repressor SNAIL is overexpressed in human colon cancer. Dig Dis Sci 50:42–46

454. Shioiri M, Shida T, Koda K et al (2006) Slug expression is an independent prognostic parameter for poor survival in colorectal carcinoma patients. Br J Cancer 94:1816–1822. doi:10.1038/sj.bjc.6603193
455. Wheeler JM, Kim HC, Efstathiou JA et al (2001) Hypermethylation of the promoter region of the E-cadherin gene (CDH1) in sporadic and ulcerative colitis associated colorectal cancer. Gut 48:367–371
456. Schackmann RCJ, Tenhagen M, van de Ven RAH, Derksen PWB (2013) p120-catenin in cancer – mechanisms, models and opportunities for intervention. J Cell Sci 126:3515–3525. doi:10.1242/jcs.134411
457. Ogasawara N, Tsukamoto T, Mizoshita T et al (2006) Mutations and nuclear accumulation of beta-catenin correlate with intestinal phenotypic expression in human gastric cancer. Histopathology 49:612–621. doi:10.1111/j.1365-2559.2006.02560.x
458. Karatzas G, Karayiannakis AJ, Syrigos KN et al (2000) Expression patterns of the E-cadherin-catenin cell-cell adhesion complex in gastric cancer. Hepatogastroenterology 47:1465–1469
459. Chung Y, Lam AKY, Luk JM et al (2007) Altered E-cadherin expression and p120 catenin localization in esophageal squamous cell carcinoma. Ann Surg Oncol 14:3260–3267. doi:10.1245/s10434-007-9511-8
460. Stairs DB, Bayne LJ, Rhoades B et al (2011) Deletion of p120-catenin results in a tumor microenvironment with inflammation and cancer that establishes it as a tumor suppressor gene. Cancer Cell 19:470–483. doi:10.1016/j.ccr.2011.02.007
461. Smalley-Freed WG, Efimov A, Short SP et al (2011) Adenoma formation following limited ablation of p120-catenin in the mouse intestine. PLoS One 6, e19880. doi:10.1371/journal.pone.0019880
462. Markowitz SD, Bertagnolli MM (2009) Molecular origins of cancer: Molecular basis of colorectal cancer. N Engl J Med 361:2449–2460. doi:10.1056/NEJMra0804588
463. Iwao K, Nakamori S, Kameyama M et al (1998) Activation of the beta-catenin gene by interstitial deletions involving exon 3 in primary colorectal carcinomas without adenomatous polyposis coli mutations. Cancer Res 58:1021–1026
464. Mirabelli-Primdahl L, Gryfe R, Kim H et al (1999) Beta-catenin mutations are specific for colorectal carcinomas with microsatellite instability but occur in endometrial carcinomas irrespective of mutator pathway. Cancer Res 59:3346–3351
465. Perchiniak EM, Groden J (2011) Mechanisms regulating microtubule binding, DNA replication, and apoptosis are controlled by the intestinal tumor suppressor APC. Curr Colorectal Cancer Rep 7:145–151. doi:10.1007/s11888-011-0088-z
466. Kourtidis A, Ngok SP, Anastasiadis PZ (2013) p120 catenin: an essential regulator of cadherin stability, adhesion-induced signaling, and cancer progression. Prog Mol Biol Transl Sci 116:409–432. doi:10.1016/B978-0-12-394311-8.00018-2
467. Zen K, Yasui K, Gen Y et al (2009) Defective expression of polarity protein PAR-3 gene (PARD3) in esophageal squamous cell carcinoma. Oncogene 28:2910–2918. doi:10.1038/onc.2009.148
468. Al-Saad S, Al-Shibli K, Donnem T et al (2008) The prognostic impact of NF-kappaB p105, vimentin, E-cadherin and Par6 expression in epithelial and stromal compartment in non-small-cell lung cancer. Br J Cancer 99:1476–1483. doi:10.1038/sj.bjc.6604713
469. Nolan ME, Aranda V, Lee S et al (2008) The polarity protein Par6 induces cell proliferation and is overexpressed in breast cancer. Cancer Res 68:8201–8209. doi:10.1158/0008-5472.CAN-07-6567
470. Viloria-Petit AM, David L, Jia JY et al (2009) A role for the TGFbeta-Par6 polarity pathway in breast cancer progression. Proc Natl Acad Sci U S A 106:14028–14033. doi:10.1073/pnas.0906796106
471. Giardiello FM, Brensinger JD, Tersmette AC et al (2000) Very high risk of cancer in familial Peutz-Jeghers syndrome. Gastroenterology 119:1447–1453
472. Hemminki A, Markie D, Tomlinson I et al (1998) A serine/threonine kinase gene defective in Peutz-Jeghers syndrome. Nature 391:184–187. doi:10.1038/34432

473. Sebbagh M, Santoni M-J, Hall B et al (2009) Regulation of LKB1/STRAD localization and function by E-cadherin. Curr Biol 19:37–42. doi:10.1016/j.cub.2008.11.033
474. Ten Klooster JP, Jansen M, Yuan J et al (2009) Mst4 and Ezrin induce brush borders downstream of the Lkb1/Strad/Mo25 polarization complex. Dev Cell 16:551–562. doi:10.1016/j.devcel.2009.01.016, S1534-5807(09)00044-6 [pii]
475. Baas AF, Kuipers J, van der Wel NN et al (2004) Complete polarization of single intestinal epithelial cells upon activation of LKB1 by STRAD. Cell 116:457–466, doi: S009286740400114X [pii]
476. Forster LF, Defres S, Goudie DR et al (2000) An investigation of the Peutz-Jeghers gene (LKB1) in sporadic breast and colon cancers. J Clin Pathol 53:791–793
477. Bardeesy N, Sinha M, Hezel AF et al (2002) Loss of the Lkb1 tumour suppressor provokes intestinal polyposis but resistance to transformation. Nature 419:162–167. doi:10.1038/nature01045
478. Shorning BY, Zabkiewicz J, McCarthy A et al (2009) Lkb1 deficiency alters goblet and paneth cell differentiation in the small intestine. PLoS One 4, e4264. doi:10.1371/journal.pone.0004264
479. Li R-X, Chen Z-H, Chen Z-K (2014) The role of EPH receptors in cancer-related epithelial-mesenchymal transition. Chin J Cancer 33:231–240. doi:10.5732/cjc.013.10108
480. Orsulic S, Kemler R (2000) Expression of Eph receptors and ephrins is differentially regulated by E-cadherin. J Cell Sci 113(Pt 1):1793–1802
481. Zantek ND, Azimi M, Fedor-Chaiken M et al (1999) E-cadherin regulates the function of the EphA2 receptor tyrosine kinase. Cell Growth Differ 10:629–638
482. Miura K, Nam J-M, Kojima C et al (2009) EphA2 engages Git1 to suppress Arf6 activity modulating epithelial cell-cell contacts. Mol Biol Cell 20:1949–1959. doi:10.1091/mbc.E08-06-0549
483. Hou F, Yuan W, Huang J et al (2012) Overexpression of EphA2 correlates with epithelial-mesenchymal transition-related proteins in gastric cancer and their prognostic importance for postoperative patients. Med Oncol 29:2691–2700. doi:10.1007/s12032-011-0127-2
484. Saito T, Masuda N, Miyazaki T et al (2004) Expression of EphA2 and E-cadherin in colorectal cancer: correlation with cancer metastasis. Oncol Rep 11:605–611
485. Yuan W, Chen Z, Wu S et al (2009) Expression of EphA2 and E-cadherin in gastric cancer: correlated with tumor progression and lymphogenous metastasis. Pathol Oncol Res 15:473–478. doi:10.1007/s12253-008-9132-y
486. Miao H, Li D-Q, Mukherjee A et al (2009) EphA2 mediates ligand-dependent inhibition and ligand-independent promotion of cell migration and invasion via a reciprocal regulatory loop with Akt. Cancer Cell 16:9–20. doi:10.1016/j.ccr.2009.04.009
487. Fang W, Bin IRC, Zhuang G et al (2008) Overexpression of EPHA2 receptor destabilizes adherens junctions via a RhoA-dependent mechanism. J Cell Sci 121:358–368. doi:10.1242/jcs.017145
488. Hiramoto-Yamaki N, Takeuchi S, Ueda S et al (2010) Ephexin4 and EphA2 mediate cell migration through a RhoG-dependent mechanism. J Cell Biol 190:461–477. doi:10.1083/jcb.201005141
489. Huang J, Xiao D, Li G et al (2014) EphA2 promotes epithelial-mesenchymal transition through the Wnt/β-catenin pathway in gastric cancer cells. Oncogene 33:2737–2747. doi:10.1038/onc.2013.238
490. Batlle E, Henderson JT, Beghtel H et al (2002) Beta-catenin and TCF mediate cell positioning in the intestinal epithelium by controlling the expression of EphB/ephrinB. Cell 111:251–263
491. Clevers H, Batlle E (2006) EphB/EphrinB receptors and Wnt signaling in colorectal cancer. Cancer Res 66:2–5. doi:10.1158/0008-5472.CAN-05-3849
492. Batlle E, Bacani J, Begthel H et al (2005) EphB receptor activity suppresses colorectal cancer progression. Nature 435:1126–1130. doi:10.1038/nature03626

493. Kumar SR, Scehnet JS, Ley EJ et al (2009) Preferential induction of EphB4 over EphB2 and its implication in colorectal cancer progression. Cancer Res 69:3736–3745. doi:10.1158/0008-5472.CAN-08-3232
494. Chiu S-T, Chang K-J, Ting C-H et al (2009) Over-expression of EphB3 enhances cell-cell contacts and suppresses tumor growth in HT-29 human colon cancer cells. Carcinogenesis 30:1475–1486. doi:10.1093/carcin/bgp133
495. Cortina C, Palomo-Ponce S, Iglesias M et al (2007) EphB-ephrin-B interactions suppress colorectal cancer progression by compartmentalizing tumor cells. Nat Genet 39:1376–1383. doi:10.1038/ng.2007.11
496. Berndorff D, Gessner R, Kreft B et al (1994) Liver-intestine cadherin: molecular cloning and characterization of a novel Ca(2+)-dependent cell adhesion molecule expressed in liver and intestine. J Cell Biol 125:1353–1369
497. Gessner R, Tauber R (2000) Intestinal cell adhesion molecules. Liver-intestine cadherin. Ann N Y Acad Sci 915:136–143
498. Freund JN, Domon-Dell C, Kedinger M, Duluc I (1998) The Cdx-1 and Cdx-2 homeobox genes in the intestine. Biochem Cell Biol 76:957–969
499. Angres B, Kim L, Jung R et al (2001) LI-cadherin gene expression during mouse intestinal development. Dev Dyn 221:182–193. doi:10.1002/dvdy.1146
500. Baumgartner W (2013) Possible roles of LI-Cadherin in the formation and maintenance of the intestinal epithelial barrier. Tissue Barrier 1, e23815. doi:10.4161/tisb.23815
501. Bartolmäs T, Hirschfeld-Ihlow C, Jonas S et al (2012) LI-cadherin cis-dimerizes in the plasma membrane Ca(2+) independently and forms highly dynamic trans-contacts. Cell Mol Life Sci 69:3851–3862. doi:10.1007/s00018-012-1053-y
502. Wendeler MW, Drenckhahn D, Gessner R, Baumgartner W (2007) Intestinal LI-cadherin acts as a Ca2+-dependent adhesion switch. J Mol Biol 370:220–230. doi:10.1016/j.jmb.2007.04.062
503. Baumgartner W, Wendeler MW, Weth A et al (2008) Heterotypic trans-interaction of LI- and E-cadherin and their localization in plasmalemmal microdomains. J Mol Biol 378:44–54. doi:10.1016/j.jmb.2008.02.023
504. Takamura M, Ichida T, Matsuda Y et al (2004) Reduced expression of liver-intestine cadherin is associated with progression and lymph node metastasis of human colorectal carcinoma. Cancer Lett 212:253–259. doi:10.1016/j.canlet.2004.03.016
505. Bartolomé RA, Barderas R, Torres S et al (2014) Cadherin-17 interacts with $\alpha 2\beta 1$ integrin to regulate cell proliferation and adhesion in colorectal cancer cells causing liver metastasis. Oncogene 33:1658–1669. doi:10.1038/onc.2013.117
506. Liu LX, Lee NP, Chan VW et al (2009) Targeting cadherin-17 inactivates Wnt signaling and inhibits tumor growth in liver carcinoma. Hepatology 50:1453–1463. doi:10.1002/hep.23143
507. Wang J, Yu J-C, Kang W-M et al (2012) The predictive effect of cadherin-17 on lymph node micrometastasis in pN0 gastric cancer. Ann Surg Oncol 19:1529–1534. doi:10.1245/s10434-011-2115-3
508. Mani SA, Guo W, Liao M-J et al (2008) The epithelial-mesenchymal transition generates cells with properties of stem cells. Cell 133:704–715. doi:10.1016/j.cell.2008.03.027
509. Regard JB, Cherman N, Palmer D et al (2011) Wnt/β-catenin signaling is differentially regulated by Gα proteins and contributes to fibrous dysplasia. Proc Natl Acad Sci U S A 108:20101–20106. doi:10.1073/pnas.1114656108
510. Thiery JP, Sleeman JP (2006) Complex networks orchestrate epithelial-mesenchymal transitions. Nat Rev Mol Cell Biol 7:131–142. doi:10.1038/nrm1835
511. Stemmer V, de Craene B, Berx G, Behrens J (2008) Snail promotes Wnt target gene expression and interacts with beta-catenin. Oncogene 27:5075–5080. doi:10.1038/onc.2008.140
512. Bates IR, Wiseman PW, Hanrahan JW (2006) Investigating membrane protein dynamics in living cells. Biochem Cell Biol 84:825–831. doi:10.1139/o06-189
513. Vasioukhin V, Fuchs E (2001) Actin dynamics and cell-cell adhesion in epithelia. Curr Opin Cell Biol 13:76–84

514. Vignjevic D, Schoumacher M, Gavert N et al (2007) Fascin, a novel target of beta-catenin-TCF signaling, is expressed at the invasive front of human colon cancer. Cancer Res 67:6844–6853. doi:10.1158/0008-5472.CAN-07-0929
515. Hugo H, Ackland ML, Blick T et al (2007) Epithelial--mesenchymal and mesenchymal--epithelial transitions in carcinoma progression. J Cell Physiol 213:374–383. doi:10.1002/jcp.21223
516. Nguyen TT-B, Ward JPT, Hirst SJ (2005) beta1-Integrins mediate enhancement of airway smooth muscle proliferation by collagen and fibronectin. Am J Respir Crit Care Med 171:217–223. doi:10.1164/rccm.200408-1046OC
517. Prall F (2007) Tumour budding in colorectal carcinoma. Histopathology 50:151–162. doi:10.1111/j.1365-2559.2006.02551.x
518. Brabletz T, Jung A, Reu S et al (2001) Variable beta-catenin expression in colorectal cancers indicates tumor progression driven by the tumor environment. Proc Natl Acad Sci U S A 98:10356–10361. doi:10.1073/pnas.171610498
519. Friedl P, Gilmour D (2009) Collective cell migration in morphogenesis, regeneration and cancer. Nat Rev Mol Cell Biol 10:445–457. doi:10.1038/nrm2720
520. Etienne-Manneville S (2014) Neighborly relations during collective migration. Curr Opin Cell Biol 30C:51–59. doi:10.1016/j.ceb.2014.06.004
521. Rørth P (2009) Collective cell migration. Annu Rev Cell Dev Biol 25:407–429. doi:10.1146/annurev.cellbio.042308.113231
522. Theveneau E, Mayor R (2013) Collective cell migration of epithelial and mesenchymal cells. Cell Mol Life Sci 70:3481–3492. doi:10.1007/s00018-012-1251-7
523. Kolega J (1981) The movement of cell clusters in vitro: morphology and directionality. J Cell Sci 49:15–32
524. Zaidel-Bar R, Itzkovitz S, Ma'ayan A et al (2007) Functional atlas of the integrin adhesome. Nat Cell Biol 9:858–867. doi:10.1038/ncb0807-858
525. Tarin D, Thompson EW, Newgreen DF (2005) The fallacy of epithelial mesenchymal transition in neoplasia. Cancer Res 65:5996–6000. doi:10.1158/0008-5472.CAN-05-0699, discussion 6000–1
526. Tsuji T, Ibaragi S, Hu G (2009) Epithelial-mesenchymal transition and cell cooperativity in metastasis. Cancer Res 69:7135–7139. doi:10.1158/0008-5472.CAN-09-1618
527. Friedl P, Locker J, Sahai E, Segall JE (2012) Classifying collective cancer cell invasion. Nat Cell Biol 14:777–783. doi:10.1038/ncb2548
528. Christiansen JJ, Rajasekaran AK (2006) Reassessing epithelial to mesenchymal transition as a prerequisite for carcinoma invasion and metastasis. Cancer Res 66:8319–8326. doi:10.1158/0008-5472.CAN-06-0410
529. Alexander S, Koehl GE, Hirschberg M et al (2008) Dynamic imaging of cancer growth and invasion: a modified skin-fold chamber model. Histochem Cell Biol 130:1147–1154. doi:10.1007/s00418-008-0529-1
530. Giampieri S, Manning C, Hooper S et al (2009) Localized and reversible TGFbeta signalling switches breast cancer cells from cohesive to single cell motility. Nat Cell Biol 11:1287–1296. doi:10.1038/ncb1973
531. Shintani Y, Fukumoto Y, Chaika N et al (2008) Collagen I-mediated up-regulation of N-cadherin requires cooperative signals from integrins and discoidin domain receptor 1. J Cell Biol 180:1277–1289. doi:10.1083/jcb.200708137
532. Aman A, Piotrowski T (2008) Wnt/beta-catenin and Fgf signaling control collective cell migration by restricting chemokine receptor expression. Dev Cell 15:749–761. doi:10.1016/j.devcel.2008.10.002
533. Daly AJ, McIlreavey L, Irwin CR (2008) Regulation of HGF and SDF-1 expression by oral fibroblasts--implications for invasion of oral cancer. Oral Oncol 44:646–651. doi:10.1016/j.oraloncology.2007.08.012
534. Lecaudey V, Gilmour D (2006) Organizing moving groups during morphogenesis. Curr Opin Cell Biol 18:102–107. doi:10.1016/j.ceb.2005.12.001

535. Orimo A, Gupta PB, Sgroi DC et al (2005) Stromal fibroblasts present in invasive human breast carcinomas promote tumor growth and angiogenesis through elevated SDF-1/CXCL12 secretion. Cell 121:335–348. doi:10.1016/j.cell.2005.02.034
536. Schmidt M, Paes K, De Mazière A et al (2007) EGFL7 regulates the collective migration of endothelial cells by restricting their spatial distribution. Development 134:2913–2923. doi:10.1242/dev.002576
537. Valentin G, Haas P, Gilmour D (2007) The chemokine SDF1a coordinates tissue migration through the spatially restricted activation of Cxcr7 and Cxcr4b. Curr Biol 17:1026–1031. doi:10.1016/j.cub.2007.05.020
538. Vitorino P, Meyer T (2008) Modular control of endothelial sheet migration. Genes Dev 22:3268–3281. doi:10.1101/gad.1725808
539. Haas P, Gilmour D (2006) Chemokine signaling mediates self-organizing tissue migration in the zebrafish lateral line. Dev Cell 10:673–680. doi:10.1016/j.devcel.2006.02.019
540. Hegerfeldt Y, Tusch M, Bröcker E-B, Friedl P (2002) Collective cell movement in primary melanoma explants: plasticity of cell-cell interaction, beta1-integrin function, and migration strategies. Cancer Res 62:2125–2130
541. Friedl P, Wolf K (2003) Tumour-cell invasion and migration: diversity and escape mechanisms. Nat Rev Cancer 3:362–374. doi:10.1038/nrc1075
542. Ilina O, Friedl P (2009) Mechanisms of collective cell migration at a glance. J Cell Sci 122:3203–3208. doi:10.1242/jcs.036525
543. Theveneau E, Mayor R (2012) Cadherins in collective cell migration of mesenchymal cells. Curr Opin Cell Biol 24:677–684. doi:10.1016/j.ceb.2012.08.002
544. Serrels A, Canel M, Brunton VG, Frame MC (2011) Src/FAK-mediated regulation of E-cadherin as a mechanism for controlling collective cell movement: insights from in vivo imaging. Cell Adh Migr 5:360–365
545. Canel M, Serrels A, Miller D et al (2010) Quantitative in vivo imaging of the effects of inhibiting integrin signaling via Src and FAK on cancer cell movement: effects on E-cadherin dynamics. Cancer Res 70:9413–9422. doi:10.1158/0008-5472.CAN-10-1454
546. Canel M, Serrels A, Frame MC, Brunton VG (2013) E-cadherin-integrin crosstalk in cancer invasion and metastasis. J Cell Sci 126:393–401. doi:10.1242/jcs.100115
547. Cai D, Chen S-C, Prasad M et al (2014) Mechanical feedback through E-cadherin promotes direction sensing during collective cell migration. Cell 157:1146–1159. doi:10.1016/j.cell.2014.03.045
548. Peglion F, Llense F, Etienne-Manneville S (2014) Adherens junction treadmilling during collective migration. Nat Cell Biol 16:639–651. doi:10.1038/ncb2985
549. Black PH (2002) Stress and the inflammatory response: a review of neurogenic inflammation. Brain Behav Immun 16:622–653
550. Gonzalez-Rey E, Chorny A, Delgado M (2007) Regulation of immune tolerance by anti-inflammatory neuropeptides. Nat Rev Immunol 7:52–63. doi:10.1038/nri1984
551. Gross KJ, Pothoulakis C (2007) Role of neuropeptides in inflammatory bowel disease. Inflamm Bowel Dis 13:918–932. doi:10.1002/ibd.20129
552. Lightman SL (2008) The neuroendocrinology of stress: a never ending story. J Neuroendocrinol 20:880–884. doi:10.1111/j.1365-2826.2008.01711.x
553. Santos J, Yates D, Guilarte M et al (2008) Stress neuropeptides evoke epithelial responses via mast cell activation in the rat colon. Psychoneuroendocrinology 33:1248–1256. doi:10.1016/j.psyneuen.2008.07.002
554. Neunlist M, Toumi F, Oreschkova T et al (2003) Human ENS regulates the intestinal epithelial barrier permeability and a tight junction-associated protein ZO-1 via VIPergic pathways. Am J Physiol Gastrointest Liver Physiol 285:G1028–G1036. doi:10.1152/ajpgi.00066.2003
555. Yuan P-Q, Wu SV, Wang L, Taché Y (2010) Corticotropin releasing factor in the rat colon: expression, localization and upregulation by endotoxin. Peptides 31:322–331. doi:10.1016/j.peptides.2009.11.012

556. Ducarouge B, Jacquier-Sarlin MR (2011) Stress neuromediators are key regulators of the intestinal barrier: link to inflammation and cancer. Trends Cell Mol Biol 6:59–88
557. Miller G, Chen E, Cole SW (2009) Health psychology: developing biologically plausible models linking the social world and physical health. Annu Rev Psychol 60:501–524. doi:10.1146/annurev.psych.60.110707.163551
558. Chang J, Adams MR, Clifton MS et al (2011) Urocortin 1 modulates immunosignaling in a rat model of colitis via corticotropin-releasing factor receptor 2. Am J Physiol Gastrointest Liver Physiol 300:G884–G894. doi:10.1152/ajpgi.00319.2010
559. Hisamatsu T, Inoue N, Yajima T et al (2007) Psychological aspects of inflammatory bowel disease. J Gastroenterol 42(Suppl 1):34–40. doi:10.1007/s00535-006-1937-4
560. Maunder RG, Levenstein S (2008) The role of stress in the development and clinical course of inflammatory bowel disease: epidemiological evidence. Curr Mol Med 8:247–252
561. Mawdsley JE, Rampton DS (2006) The role of psychological stress in inflammatory bowel disease. Neuroimmunomodulation 13:327–336. doi:10.1159/000104861
562. Boucher M-J, Laprise P, Rivard N (2005) Cyclic AMP-dependent protein kinase A negatively modulates adherens junction integrity and differentiation of intestinal epithelial cells. J Cell Physiol 202:178–190. doi:10.1002/jcp.20104
563. Yuan Z, McCauley R, Chen-Scarabelli C et al (2010) Activation of Src protein tyrosine kinase plays an essential role in urocortin-mediated cardioprotection. Mol Cell Endocrinol 325:1–7. doi:10.1016/j.mce.2010.04.013
564. Ducarouge B, Pelissier-Rota M, Lainé M et al (2013) CRF2 signaling is a novel regulator of cellular adhesion and migration in colorectal cancer cells. PLoS One 8, e79335. doi:10.1371/journal.pone.0079335
565. Castagliuolo I, Lamont JT, Qiu B et al (1996) Acute stress causes mucin release from rat colon: role of corticotropin releasing factor and mast cells. Am J Physiol 271:G884–G892
566. Pothoulakis C, Castagliuolo I, Leeman SE (1998) Neuroimmune mechanisms of intestinal responses to stress. Role of corticotropin-releasing factor and neurotensin. Ann N Y Acad Sci 840:635–648
567. Tanaka MN, Diaz BL, de Souza W, Morgado-Diaz JA (2008) Prostaglandin E2-EP1 and EP2 receptor signaling promotes apical junctional complex disassembly of Caco-2 human colorectal cancer cells. BMC Cell Biol 9:63. doi:10.1186/1471-2121-9-63
568. Larauche M, Kiank C, Tache Y (2009) Corticotropin releasing factor signaling in colon and ileum: regulation by stress and pathophysiological implications. J Physiol Pharmacol 60(Suppl 7):33–46
569. Kaprara A, Pazaitou-Panayiotou K, Kortsaris A, Chatzaki E (2010) The corticotropin releasing factor system in cancer: expression and pathophysiological implications. Cell Mol Life Sci 67:1293–1306. doi:10.1007/s00018-010-0265-2
570. Florio P, De Falco G, Leucci E et al (2006) Urocortin expression is downregulated in human endometrial carcinoma. J Endocrinol 190:99–105. doi:10.1677/joe.1.06726
571. Miceli F, Ranelletti FO, Martinelli E et al (2009) Expression and subcellular localization of CRH and its receptors in human endometrial cancer. Mol Cell Endocrinol 305:6–11. doi:10.1016/j.mce.2009.02.013
572. Van Aken J, Cuvelier CA, De Wever N et al (1993) Immunohistochemical analysis of E-cadherin expression in human colorectal tumours. Pathol Res Pract 189:975–978. doi:10.1016/S0344-0338(11)80667-9
573. Chen Y, Dubé CM, Rice CJ, Baram TZ (2008) Rapid loss of dendritic spines after stress involves derangement of spine dynamics by corticotropin-releasing hormone. J Neurosci 28:2903–2911. doi:10.1523/JNEUROSCI.0225-08.2008
574. Swinny JD, Valentino RJ (2006) Corticotropin-releasing factor promotes growth of brain norepinephrine neuronal processes through Rho GTPase regulators of the actin cytoskeleton in rat. Eur J Neurosci 24:2481–2490. doi:10.1111/j.1460-9568.2006.05129.x

575. Yang Y, Park H, Yang Y et al (2007) Enhancement of cell migration by corticotropin-releasing hormone through ERK1/2 pathway in murine melanoma cell line, B16F10. Exp Dermatol 16:22–27. doi:10.1111/j.1600-0625.2006.00511.x
576. Androulidaki A, Dermitzaki E, Venihaki M et al (2009) Corticotropin Releasing Factor promotes breast cancer cell motility and invasiveness. Mol Cancer 8:30. doi:10.1186/1476-4598-8-30
577. Bale TL, Vale WW (2004) CRF and CRF receptors: role in stress responsivity and other behaviors. Annu Rev Pharmacol Toxicol 44:525–557. doi:10.1146/annurev.pharmtox.44.101802.121410
578. Chatzaki E, Lambropoulou M, Constantinidis TC et al (2006) Corticotropin-releasing factor (CRF) receptor type 2 in the human stomach: protective biological role by inhibition of apoptosis. J Cell Physiol 209:905–911. doi:10.1002/jcp.20792
579. Jin L, Zhang Q, Guo R et al (2011) Different effects of corticotropin-releasing factor and urocortin 2 on apoptosis of prostate cancer cells in vitro. J Mol Endocrinol 47:219–227. doi:10.1530/JME-11-0048
580. Wang J, Li S (2007) Corticotropin-releasing factor family and its receptors: tumor therapeutic targets? Biochem Biophys Res Commun 362:785–788. doi:10.1016/j.bbrc.2007.08.014
581. Wang J, Xu Y, Xu Y et al (2008) Urocortin's inhibition of tumor growth and angiogenesis in hepatocellular carcinoma via corticotrophin-releasing factor receptor 2. Cancer Invest 26:359–368. doi:10.1080/07357900701788106
582. Arranz A, Venihaki M, Mol B et al (2010) The impact of stress on tumor growth: peripheral CRF mediates tumor-promoting effects of stress. Mol Cancer 9:261. doi:10.1186/1476-4598-9-261
583. Kaprara A, Pazaitou-Panayiotou K, Chemonidou MC et al (2010) Distinct distribution of corticotropin releasing factor receptors in human breast cancer. Neuropeptides 44:355–361. doi:10.1016/j.npep.2010.06.005
584. Baigent SM (2001) Peripheral corticotropin-releasing hormone and urocortin in the control of the immune response. Peptides 22:809–820
585. Saunders PR, Hanssen NP, Perdue MH (1997) Cholinergic nerves mediate stress-induced intestinal transport abnormalities in Wistar-Kyoto rats. Am J Physiol Gastrointest Liver Physiol 273:G486–G490
586. Kiliaan AJ, Saunders PR, Bijlsma PB et al (1998) Stress stimulates transepithelial macromolecular uptake in rat jejunum. Am J Physiol 275:G1037–G1044
587. Santos J, Saunders PR, Hanssen NP et al (1999) Corticotropin-releasing hormone mimics stress-induced colonic epithelial pathophysiology in the rat. Am J Physiol 277:G391–G399
588. Gareau MG, Jury J, Perdue MH (2007) Neonatal maternal separation of rat pups results in abnormal cholinergic regulation of epithelial permeability. Am J Physiol Gastrointest Liver Physiol 293:G198–G203. doi:10.1152/ajpgi.00392.2006
589. Demaude J, Levêque M, Chaumaz G et al (2009) Acute stress increases colonic paracellular permeability in mice through a mast cell-independent mechanism: involvement of pancreatic trypsin. Life Sci 84:847–852. doi:10.1016/j.lfs.2009.03.016
590. Brown J (1990) The muscarinic receptors. Humana, New York
591. Wessler I, Kirkpatrick CJ (2008) Acetylcholine beyond neurons: the non-neuronal cholinergic system in humans. Br J Pharmacol 154:1558–1571. doi:10.1038/bjp.2008.185
592. Schuller HM (2009) Is cancer triggered by altered signalling of nicotinic acetylcholine receptors? Nat Rev Cancer 9:195–205. doi:10.1038/nrc2590
593. Shah N, Khurana S, Cheng K, Raufman J-P (2009) Muscarinic receptors and ligands in cancer. Am J Physiol Cell Physiol 296:C221–C232. doi:10.1152/ajpcell.00514.2008
594. Von Rosenvinge EC, Raufman J-P (2011) Muscarinic receptor signaling in colon cancer. Cancers (Basel) 3:971–981. doi:10.3390/cancers3010971
595. Kopp R, Lambrecht G, Mutschler E et al (1989) Human HT-29 colon carcinoma cells contain muscarinic M3 receptors coupled to phosphoinositide metabolism. Eur J Pharmacol Mol Pharmacol 172:397–405. doi:10.1016/0922-4106(89)90021-7
596. O'Malley KE, Farrell CB, O'Boyle KM, Baird AW (1995) Cholinergic activation of Cl- secretion in rat colonic epithelia. Eur J Pharmacol 275:83–89

597. Raufman J-P, Samimi R, Shah N et al (2008) Genetic ablation of M3 muscarinic receptors attenuates murine colon epithelial cell proliferation and neoplasia. Cancer Res 68:3573–3578. doi:10.1158/0008-5472.CAN-07-6810
598. Frucht H, Jensen RT, Dexter D et al (1999) Human colon cancer cell proliferation mediated by the M3 muscarinic cholinergic receptor. Clin Cancer Res 5:2532–2539
599. Yang WL, Frucht H (2000) Cholinergic receptor up-regulates COX-2 expression and prostaglandin E(2) production in colon cancer cells. Carcinogenesis 21:1789–1793
600. Cheng K, Raufman J-P (2005) Bile acid-induced proliferation of a human colon cancer cell line is mediated by transactivation of epidermal growth factor receptors. Biochem Pharmacol 70:1035–1047. doi:10.1016/j.bcp.2005.07.023
601. Cheng K, Samimi R, Xie G et al (2008) Acetylcholine release by human colon cancer cells mediates autocrine stimulation of cell proliferation. Am J Physiol Gastrointest Liver Physiol 295:G591–G597. doi:10.1152/ajpgi.00055.2008
602. Cheng K, Xie G, Raufman J-P (2007) Matrix metalloproteinase-7-catalyzed release of HB-EGF mediates deoxycholyltaurine-induced proliferation of a human colon cancer cell line. Biochem Pharmacol 73:1001–1012. doi:10.1016/j.bcp.2006.11.028
603. Xie G, Cheng K, Shant J, Raufman J-P (2009) Acetylcholine-induced activation of M3 muscarinic receptors stimulates robust matrix metalloproteinase gene expression in human colon cancer cells. Am J Physiol Gastrointest Liver Physiol 296:G755–G763. doi:10.1152/ajpgi.90519.2008
604. Belo A, Cheng K, Chahdi A et al (2011) Muscarinic receptor agonists stimulate human colon cancer cell migration and invasion. Am J Physiol Gastrointest Liver Physiol 300:G749–G760. doi:10.1152/ajpgi.00306.2010
605. Calandrella SO, Barrett KE, Keely SJ (2005) Transactivation of the epidermal growth factor receptor mediates muscarinic stimulation of focal adhesion kinase in intestinal epithelial cells. J Cell Physiol 203:103–110. doi:10.1002/jcp.20190
606. Pelissier-Rota M, Lainé M, Ducarouge B, Bonaz B, Jacquier-Sarlin M (2013) Role of cholinergic receptors in colorectal cancer: potential therapeutic implications of vagus nerve stimulation? J Cancer Ther 2013(4):1116–1131. doi:10.4236/jct2013.46128
607. Cameron HL, Perdue MH (2007) Muscarinic acetylcholine receptor activation increases transcellular transport of macromolecules across mouse and human intestinal epithelium in vitro. Neurogastroenterol Motil 19:47–56. doi:10.1111/j.1365-2982.2006.00845.x
608. Boivin GP, Washington K, Yang K et al (2003) Pathology of mouse models of intestinal cancer: consensus report and recommendations. Gastroenterology 124:762–777. doi:10.1053/gast.2003.50094
609. Song P, Sekhon HS, Lu A et al (2007) M3 muscarinic receptor antagonists inhibit small cell lung carcinoma growth and mitogen-activated protein kinase phosphorylation induced by acetylcholine secretion. Cancer Res 67:3936–3944. doi:10.1158/0008-5472.CAN-06-2484
610. Serban DE (2014) Gastrointestinal cancers: influence of gut microbiota, probiotics and prebiotics. Cancer Lett 345:258–270. doi:10.1016/j.canlet.2013.08.013
611. Grimoud J, Durand H, de Souza S et al (2010) In vitro screening of probiotics and synbiotics according to anti-inflammatory and anti-proliferative effects. Int J Food Microbiol 144:42–50. doi:10.1016/j.ijfoodmicro.2010.09.007
612. Pothoulakis C (2009) Review article: anti-inflammatory mechanisms of action of Saccharomyces boulardii. Aliment Pharmacol Ther 30:826–833. doi:10.1111/j.1365-2036.2009.04102.x
613. Bergogne-Bérézin E (1995) Ecologic impact of antibiotherapy. Role of substitution microorganisms in the control of antibiotic-related diarrhea and colitis. Presse Med 24:145–148, 151–152, 155–156
614. Czerucka D, Piche T, Rampal P (2007) Review article: yeast as probiotics – Saccharomyces boulardii. Aliment Pharmacol Ther 26:767–778. doi:10.1111/j.1365-2036.2007.03442.x
615. Guslandi M (2005) Antibiotics for inflammatory bowel disease: do they work? Eur J Gastroenterol Hepatol 17:145–147

616. Canonici A, Siret C, Pellegrino E et al (2011) Saccharomyces boulardii improves intestinal cell restitution through activation of the α2β1 integrin collagen receptor. PLoS One 6, e18427. doi:10.1371/journal.pone.0018427
617. Dalmasso G, Cottrez F, Imbert V et al (2006) Saccharomyces boulardii inhibits inflammatory bowel disease by trapping T cells in mesenteric lymph nodes. Gastroenterology 131:1812–1825. doi:10.1053/j.gastro.2006.10.001
618. Lee SK, Kim YW, Chi S-G et al (2009) The effect of Saccharomyces boulardii on human colon cells and inflammation in rats with trinitrobenzene sulfonic acid-induced colitis. Dig Dis Sci 54:255–263. doi:10.1007/s10620-008-0357-0
619. Sougioultzis S, Simeonidis S, Bhaskar KR et al (2006) Saccharomyces boulardii produces a soluble anti-inflammatory factor that inhibits NF-kappaB-mediated IL-8 gene expression. Biochem Biophys Res Commun 343:69–76. doi:10.1016/j.bbrc.2006.02.080
620. Jahn HU, Ullrich R, Schneider T et al (1996) Immunological and trophical effects of Saccharomyces boulardii on the small intestine in healthy human volunteers. Digestion 57:95–104
621. Canonici A, Pellegrino E, Siret C et al (2012) Saccharomyces boulardii improves intestinal epithelial cell restitution by inhibiting αvβ5 integrin activation state. PLoS One 7, e45047. doi:10.1371/journal.pone.0045047
622. Hartnett L, Egan LJ (2012) Inflammation, DNA methylation and colitis-associated cancer. Carcinogenesis 33:723–731. doi:10.1093/carcin/bgs006
623. Terg R, Sambuelli A, Coronel E et al (2008) Prevalence of primary sclerosing cholangitis in patients with ulcerative colitis and the risk of developing malignancies. A large prospective study. Acta Gastroenterol Latinoam 38:26–33
624. Han YW, Ikegami A, Rajanna C et al (2005) Identification and characterization of a novel adhesin unique to oral fusobacteria. J Bacteriol 187:5330–5340. doi:10.1128/JB.187.15.5330-5340.2005
625. Xu M, Yamada M, Li M et al (2007) FadA from Fusobacterium nucleatum utilizes both secreted and nonsecreted forms for functional oligomerization for attachment and invasion of host cells. J Biol Chem 282:25000–25009. doi:10.1074/jbc.M611567200
626. Fardini Y, Wang X, Témoin S et al (2011) Fusobacterium nucleatum adhesin FadA binds vascular endothelial cadherin and alters endothelial integrity. Mol Microbiol 82:1468–1480. doi:10.1111/j.1365-2958.2011.07905.x
627. Rubinstein MR, Wang X, Liu W et al (2013) Fusobacterium nucleatum promotes colorectal carcinogenesis by modulating E-cadherin/β-catenin signaling via its FadA adhesin. Cell Host Microbe 14:195–206. doi:10.1016/j.chom.2013.07.012
628. Kurkjian C, Kummar S (2009) Advances in the treatment of metastatic colorectal cancer. Am J Ther 16:412–420. doi:10.1097/MJT.0b013e3181907ed9
629. Derycke L, De Wever O, Stove V et al (2006) Soluble N-cadherin in human biological fluids. Int J Cancer 119:2895–2900. doi:10.1002/ijc.22219
630. André T, Boni C, Mounedji-Boudiaf L et al (2004) Oxaliplatin, fluorouracil, and leucovorin as adjuvant treatment for colon cancer. N Engl J Med 350:2343–2351. doi:10.1056/NEJMoa032709
631. Kuebler JP, Wieand HS, O'Connell MJ et al (2007) Oxaliplatin combined with weekly bolus fluorouracil and leucovorin as surgical adjuvant chemotherapy for stage II and III colon cancer: results from NSABP C-07. J Clin Oncol 25:2198–2204. doi:10.1200/JCO.2006.08.2974
632. Brueckner B, Garcia Boy R, Siedlecki P et al (2005) Epigenetic reactivation of tumor suppressor genes by a novel small-molecule inhibitor of human DNA methyltransferases. Cancer Res 65:6305–6311. doi:10.1158/0008-5472.CAN-04-2957
633. Stresemann C, Brueckner B, Musch T et al (2006) Functional diversity of DNA methyltransferase inhibitors in human cancer cell lines. Cancer Res 66:2794–2800. doi:10.1158/0008-5472.CAN-05-2821
634. Subramaniam D, Thombre R, Dhar A, Anant S (2014) DNA methyltransferases: a novel target for prevention and therapy. Front Oncol 4:80. doi:10.3389/fonc.2014.00080

635. Ichii S, Horii A, Nakatsuru S et al (1992) Inactivation of both APC alleles in an early stage of colon adenomas in a patient with familial adenomatous polyposis (FAP). Hum Mol Genet 1:387–390. doi:10.1093/hmg/1.6.387
636. Laird PW, Jackson-Grusby L, Fazeli A et al (1995) Suppression of intestinal neoplasia by DNA hypomethylation. Cell 81:197–205. doi:10.1016/0092-8674(95)90329-1
637. Moser A, Pitot H, Dove W (1990) A dominant mutation that predisposes to multiple intestinal neoplasia in the mouse. Science (80-) 247:322–324. doi: 10.1126/science.2296722
638. Linhart HG, Lin H, Yamada Y et al (2007) Dnmt3b promotes tumorigenesis in vivo by gene-specific de novo methylation and transcriptional silencing. Genes Dev 21:3110–3122. doi:10.1101/gad.1594007
639. Eads CA, Nickel AE, Laird PW (2002) Complete genetic suppression of polyp formation and reduction of CpG-island hypermethylation in Apc(Min/+) Dnmt1-hypomorphic Mice. Cancer Res 62:1296–1299
640. Brock MV, Gou M, Akiyama Y et al (2003) Prognostic importance of promoter hypermethylation of multiple genes in esophageal adenocarcinoma. Clin Cancer Res 9:2912–2919
641. Corn PG, Smith BD, Ruckdeschel ES et al (2000) E-cadherin expression is silenced by 5' CpG island methylation in acute leukemia. Clin Cancer Res 6:4243–4248
642. Nam J-S, Ino Y, Kanai Y et al (2004) 5-aza-2'-deoxycytidine restores the E-cadherin system in E-cadherin-silenced cancer cells and reduces cancer metastasis. Clin Exp Metastasis 21:49–56
643. Fang MZ, Wang Y, Ai N et al (2003) Tea polyphenol (-)-epigallocatechin-3-gallate inhibits DNA methyltransferase and reactivates methylation-silenced genes in cancer cell lines. Cancer Res 63:7563–7570
644. Chuang JC, Yoo CB, Kwan JM et al (2005) Comparison of biological effects of non-nucleoside DNA methylation inhibitors versus 5-aza-2'-deoxycytidine. Mol Cancer Ther 4:1515–1520. doi:10.1158/1535-7163.MCT-05-0172
645. Huang Z-H, Li L-H, Yang F, Wang J-F (2007) Detection of aberrant methylation in fecal DNA as a molecular screening tool for colorectal cancer and precancerous lesions. World J Gastroenterol 13:950–954
646. Nakayama H, Hibi K, Takase T et al (2003) Molecular detection of p16 promoter methylation in the serum of recurrent colorectal cancer patients. Int J Cancer 105:491–493. doi:10.1002/ijc.11117
647. Zou H-Z, Yu B-M, Wang Z-W et al (2002) Detection of aberrant p16 methylation in the serum of colorectal cancer patients. Clin Cancer Res 8:188–191
648. Nakagawa H, Nuovo GJ, Zervos EE et al (2001) Age-related hypermethylation of the 5' region of MLH1 in normal colonic mucosa is associated with microsatellite-unstable colorectal cancer development. Cancer Res 61:6991–6995
649. Bird AP, Wolffe AP (1999) Methylation-induced repression--belts, braces, and chromatin. Cell 99:451–454
650. Tarmin L, Yin J, Harpaz N et al (1995) Adenomatous polyposis coli gene mutations in ulcerative colitis-associated dysplasias and cancers versus sporadic colon neoplasms. Cancer Res 55:2035–2038
651. Cameron EE, Bachman KE, Myöhänen S et al (1999) Synergy of demethylation and histone deacetylase inhibition in the re-expression of genes silenced in cancer. Nat Genet 21:103–107. doi:10.1038/5047
652. Jones PL, Veenstra GJ, Wade PA et al (1998) Methylated DNA and MeCP2 recruit histone deacetylase to repress transcription. Nat Genet 19:187–191. doi:10.1038/561
653. Yee KWL, Minden MDBJ (2007) Phase I trial of two sequence-specific schedules annual, of decitabine and vorinostat in patients with acute myeloid leukemia (AML). ASH Annu Meet Abstr 110:908
654. Glauben R, Batra A, Fedke I et al (2006) Histone hyperacetylation is associated with amelioration of experimental colitis in mice. J Immunol 176:5015–5022
655. Takai N, Desmond JC, Kumagai T et al (2004) Histone deacetylase inhibitors have a profound antigrowth activity in endometrial cancer cells. Clin Cancer Res 10:1141–1149

656. Glauben R, Batra A, Stroh T et al (2008) Histone deacetylases: novel targets for prevention of colitis-associated cancer in mice. Gut 57:613–622. doi:10.1136/gut.2007.134650
657. Hegazi RAF, Mady HH, Melhem MF et al (2003) Celecoxib and rofecoxib potentiate chronic colitis and premalignant changes in interleukin 10 knockout mice. Inflamm Bowel Dis 9:230–236
658. Tanaka T, Kohno H, Suzuki R et al (2003) A novel inflammation-related mouse colon carcinogenesis model induced by azoxymethane and dextran sodium sulfate. Cancer Sci 94:965–973
659. Furlan A, Monzani V, Reznikov LL et al (2011) Pharmacokinetics, safety and inducible cytokine responses during a phase 1 trial of the oral histone deacetylase inhibitor ITF2357 (givinostat). Mol Med 17:353–362. doi:10.2119/molmed.2011.00020
660. Bertrand P (2010) Inside HDAC with HDAC inhibitors. Eur J Med Chem 45:2095–2116. doi:10.1016/j.ejmech.2010.02.030
661. Cousens LS, Gallwitz D, Alberts BM (1979) Different accessibilities in chromatin to histone acetylase. J Biol Chem 254:1716–1723
662. Corfe BM (2012) Hypothesis: butyrate is not an HDAC inhibitor, but a product inhibitor of deacetylation. Mol Biosyst 8:1609–1612. doi:10.1039/c2mb25028d
663. Kondo K, Kohno N, Yokoyama A, Hiwada K (1998) Decreased MUC1 expression induces E-cadherin-mediated cell adhesion of breast cancer cell lines. Cancer Res 58:2014–2019
664. Barshishat M, Polak-Charcon S, Schwartz B (2000) Butyrate regulates E-cadherin transcription, isoform expression and intracellular position in colon cancer cells. Br J Cancer 82:195–203. doi:10.1054/bjoc.1999.0899
665. Masuda T, Saito H, Kaneko F et al (2000) Up-regulation of E-cadherin and I-catenin in human hepatocellular carcinoma cell lines by sodium butyrate and interferon-alpha. In Vitro Cell Dev Biol Anim 36:387–394
666. Mariadason JM, Barkla DH, Gibson PR (1997) Effect of short-chain fatty acids on paracellular permeability in Caco-2 intestinal epithelium model. Am J Physiol 272:G705–G712
667. Peng L, Li Z-R, Green RS et al (2009) Butyrate enhances the intestinal barrier by facilitating tight junction assembly via activation of AMP-activated protein kinase in Caco-2 cell monolayers. J Nutr 139:1619–1625. doi:10.3945/jn.109.104638
668. Wu S, Li RW, Li W, Li C-J (2012) Transcriptome characterization by RNA-seq unravels the mechanisms of butyrate-induced epigenomic regulation in bovine cells. PLoS One 7, e36940. doi:10.1371/journal.pone.0036940
669. Maslowski KM, Vieira AT, Ng A et al (2009) Regulation of inflammatory responses by gut microbiota and chemoattractant receptor GPR43. Nature 461:1282–1286. doi:10.1038/nature08530
670. Tao R, de Zoeten EF, Ozkaynak E et al (2007) Deacetylase inhibition promotes the generation and function of regulatory T cells. Nat Med 13:1299–1307. doi:10.1038/nm1652
671. Wong JMW, de Souza R, Kendall CWC et al (2006) Colonic health: fermentation and short chain fatty acids. J Clin Gastroenterol 40:235–243
672. Wilson AJ, Byun D-S, Nasser S et al (2008) HDAC4 promotes growth of colon cancer cells via repression of p21. Mol Biol Cell 19:4062–4075. doi:10.1091/mbc.E08-02-0139
673. Wilson AJ, Byun D-S, Popova N et al (2006) Histone deacetylase 3 (HDAC3) and other class I HDACs regulate colon cell maturation and p21 expression and are deregulated in human colon cancer. J Biol Chem 281:13548–13558. doi:10.1074/jbc.M510023200
674. Zhang B, Groffen J, Heisterkamp N (2007) Increased resistance to a farnesyltransferase inhibitor by N-cadherin expression in Bcr/Abl-P190 lymphoblastic leukemia cells. Leukemia 21:1189–1197. doi:10.1038/sj.leu.2404667
675. Williams E, Williams G, Gour BJ et al (2000) A novel family of cyclic peptide antagonists suggests that N-cadherin specificity is determined by amino acids that flank the HAV motif. J Biol Chem 275:4007–4012
676. Williams G, Williams E-J, Doherty P (2002) Dimeric versions of two short N-cadherin binding motifs (HAVDI and INPISG) function as N-cadherin agonists. J Biol Chem 277:4361–4367. doi:10.1074/jbc.M109185200
677. Mariotti A, Perotti A, Sessa C, Rüegg C (2007) N-cadherin as a therapeutic target in cancer. Expert Opin Investig Drugs 16:451–465. doi:10.1517/13543784.16.4.451

678. Visentin B, Vekich JA, Sibbald BJ et al (2006) Validation of an anti-sphingosine-1-phosphate antibody as a potential therapeutic in reducing growth, invasion, and angiogenesis in multiple tumor lineages. Cancer Cell 9:225–238. doi:10.1016/j.ccr.2006.02.023
679. Williams EJ, Williams G, Howell FV et al (2001) Identification of an N-cadherin motif that can interact with the fibroblast growth factor receptor and is required for axonal growth. J Biol Chem 276:43879–43886. doi:10.1074/jbc.M105876200
680. Kim JB, Islam S, Kim YJ et al (2000) N-Cadherin extracellular repeat 4 mediates epithelial to mesenchymal transition and increased motility. J Cell Biol 151:1193–1206
681. Cappuzzo F, Finocchiaro G, Rossi E et al (2008) EGFR FISH assay predicts for response to cetuximab in chemotherapy refractory colorectal cancer patients. Ann Oncol 19:717–723. doi:10.1093/annonc/mdm492
682. Frattini M, Saletti P, Romagnani E et al (2007) PTEN loss of expression predicts cetuximab efficacy in metastatic colorectal cancer patients. Br J Cancer 97:1139–1145. doi:10.1038/sj.bjc.6604009
683. Sartore-Bianchi A, Moroni M, Veronese S et al (2007) Epidermal growth factor receptor gene copy number and clinical outcome of metastatic colorectal cancer treated with panitumumab. J Clin Oncol 25:3238–3245. doi:10.1200/JCO.2007.11.5956
684. Chung KY, Shia J, Kemeny NE et al (2005) Cetuximab shows activity in colorectal cancer patients with tumors that do not express the epidermal growth factor receptor by immunohistochemistry. J Clin Oncol 23:1803–1810. doi:10.1200/JCO.2005.08.037
685. Cunningham D, Humblet Y, Siena S et al (2004) Cetuximab monotherapy and cetuximab plus irinotecan in irinotecan-refractory metastatic colorectal cancer. N Engl J Med 351:337–345. doi:10.1056/NEJMoa033025
686. Moroni M, Sartore-Bianchi A, Veronese S, Siena S (2008) EGFR FISH in colorectal cancer: what is the current reality? Lancet Oncol 9:402–403. doi:10.1016/S1470-2045(08)70109-8
687. Schrag D (2004) The price tag on progress--chemotherapy for colorectal cancer. N Engl J Med 351:317–319. doi:10.1056/NEJMp048143
688. Amado RG, Wolf M, Peeters M et al (2008) Wild-type KRAS is required for panitumumab efficacy in patients with metastatic colorectal cancer. J Clin Oncol 26:1626–1634. doi:10.1200/JCO.2007.14.7116
689. Benvenuti S, Sartore-Bianchi A, Di Nicolantonio F et al (2007) Oncogenic activation of the RAS/RAF signaling pathway impairs the response of metastatic colorectal cancers to anti-epidermal growth factor receptor antibody therapies. Cancer Res 67:2643–2648. doi:10.1158/0008-5472.CAN-06-4158
690. Downward J (2003) Targeting RAS signalling pathways in cancer therapy. Nat Rev Cancer 3:11–22. doi:10.1038/nrc969
691. Lièvre A, Bachet J-B, Le Corre D et al (2006) KRAS mutation status is predictive of response to cetuximab therapy in colorectal cancer. Cancer Res 66:3992–3995. doi:10.1158/0008-5472.CAN-06-0191
692. Schubbert S, Shannon K, Bollag G (2007) Hyperactive Ras in developmental disorders and cancer. Nat Rev Cancer 7:295–308. doi:10.1038/nrc2109
693. Baselga J, Rosen N (2008) Determinants of RASistance to anti-epidermal growth factor receptor agents. J Clin Oncol 26:1582–1584. doi:10.1200/JCO.2007.15.3700
694. Aligayer H, Boyd DD, Heiss MM et al (2002) Activation of Src kinase in primary colorectal carcinoma: an indicator of poor clinical prognosis. Cancer 94:344–351. doi:10.1002/cncr.10221
695. Cartwright CA, Kamps MP, Meisler AI et al (1989) pp60c-src activation in human colon carcinoma. J Clin Invest 83:2025–2033. doi:10.1172/JCI114113
696. Cartwright CA, Meisler AI, Eckhart W (1990) Activation of the pp60c-src protein kinase is an early event in colonic carcinogenesis. Proc Natl Acad Sci U S A 87:558–562
697. Talamonti MS, Roh MS, Curley SA, Gallick GE (1993) Increase in activity and level of pp60c-src in progressive stages of human colorectal cancer. J Clin Invest 91:53–60. doi:10.1172/JCI116200
698. Pories S, Jaros K, Steele G et al (1992) Oncogene-mediated transformation of fetal rat colon in vitro. Oncogene 7:885–893

699. Irby RB, Mao W, Coppola D et al (1999) Activating SRC mutation in a subset of advanced human colon cancers. Nat Genet 21:187–190. doi:10.1038/5971
700. Irby RB, Yeatman TJ (2000) Role of Src expression and activation in human cancer. Oncogene 19:5636–5642. doi:10.1038/sj.onc.1203912
701. Griffiths GJ, Koh MY, Brunton VG et al (2004) Expression of kinase-defective mutants of c-Src in human metastatic colon cancer cells decreases Bcl-xL and increases oxaliplatin- and Fas-induced apoptosis. J Biol Chem 279:46113–46121. doi:10.1074/jbc.M408550200
702. Kopetz S, Lesslie DP, Dallas NA et al (2009) Synergistic activity of the SRC family kinase inhibitor dasatinib and oxaliplatin in colon carcinoma cells is mediated by oxidative stress. Cancer Res 69:3842–3849. doi:10.1158/0008-5472.CAN-08-2246
703. Chen J, Elfiky A, Han M et al (2014) The role of Src in colon cancer and its therapeutic implications. Clin Colorectal Cancer 13:5–13. doi:10.1016/j.clcc.2013.10.003
704. Golas JM, Lucas J, Etienne C et al (2005) SKI-606, a Src/Abl inhibitor with in vivo activity in colon tumor xenograft models. Cancer Res 65:5358–5364. doi:10.1158/0008-5472.CAN-04-2484
705. Chen J (2008) Is Src the key to understanding metastasis and developing new treatments for colon cancer? Nat Clin Pract Gastroenterol Hepatol 5:306–307. doi:10.1038/ncpgasthep1141
706. Golubovskaya VM, Gross S, Kaur AS et al (2003) Simultaneous inhibition of focal adhesion kinase and SRC enhances detachment and apoptosis in colon cancer cell lines. Mol Cancer Res 1:755–764
707. Jia J, Starodub A, Cushman I et al (2013) Dual inhibition of αV integrins and Src kinase activity as a combination therapy strategy for colorectal cancer. Anticancer Drugs 24:237–250. doi:10.1097/CAD.0b013e32835d29fd
708. Leu T-H, Su SL, Chuang Y-C, Maa M-C (2003) Direct inhibitory effect of curcumin on Src and focal adhesion kinase activity. Biochem Pharmacol 66:2323–2331
709. Abram CL, Lowell CA (2008) The diverse functions of Src family kinases in macrophages. Front Biosci 13:4426–4450
710. Dougan M, Dranoff G (2009) Immune therapy for cancer. Annu Rev Immunol 27:83–117. doi:10.1146/annurev.immunol.021908.132544
711. De Wever O, Mareel M (2003) Role of tissue stroma in cancer cell invasion. J Pathol 200:429–447. doi:10.1002/path.1398
712. De Wever O, Nguyen Q-D, Van Hoorde L et al (2004) Tenascin-C and SF/HGF produced by myofibroblasts in vitro provide convergent pro-invasive signals to human colon cancer cells through RhoA and Rac. FASEB J 18:1016–1018. doi:10.1096/fj.03-1110fje
713. Muraoka RS, Dumont N, Ritter CA et al (2002) Blockade of TGF-beta inhibits mammary tumor cell viability, migration, and metastases. J Clin Invest 109:1551–1559. doi:10.1172/JCI15234
714. Won J, Kim H, Park EJ et al (1999) Tumorigenicity of mouse thymoma is suppressed by soluble type II transforming growth factor beta receptor therapy. Cancer Res 59:1273–1277
715. Yang Y-A, Dukhanina O, Tang B et al (2002) Lifetime exposure to a soluble TGF-beta antagonist protects mice against metastasis without adverse side effects. J Clin Invest 109:1607–1615. doi:10.1172/JCI15333
716. Hjelmeland MD, Hjelmeland AB, Sathornsumetee S et al (2004) SB-431542, a small molecule transforming growth factor-beta-receptor antagonist, inhibits human glioma cell line proliferation and motility. Mol Cancer Ther 3:737–745
717. Inman GJ, Nicolás FJ, Callahan JF et al (2002) SB-431542 is a potent and specific inhibitor of transforming growth factor-beta superfamily type I activin receptor-like kinase (ALK) receptors ALK4, ALK5, and ALK7. Mol Pharmacol 62:65–74
718. Laping NJ, Grygielko E, Mathur A et al (2002) Inhibition of transforming growth factor (TGF)-beta1-induced extracellular matrix with a novel inhibitor of the TGF-beta type I receptor kinase activity: SB-431542. Mol Pharmacol 62:58–64
719. Matsuyama S, Iwadate M, Kondo M et al (2003) SB-431542 and Gleevec inhibit transforming growth factor-beta-induced proliferation of human osteosarcoma cells. Cancer Res 63:7791–7798

720. Giannakis M, Stappenbeck TS, Mills JC et al (2006) Molecular properties of adult mouse gastric and intestinal epithelial progenitors in their niches. J Biol Chem 281:11292–11300. doi:10.1074/jbc.M512118200
721. O'Brien CA, Pollett A, Gallinger S, Dick JE (2007) A human colon cancer cell capable of initiating tumour growth in immunodeficient mice. Nature 445:106–110. doi:10.1038/nature05372
722. Schepers A, Clevers H (2012) Wnt signaling, stem cells, and cancer of the gastrointestinal tract. Cold Spring Harb Perspect Biol 4:a007989. doi:10.1101/cshperspect.a007989
723. Todaro M, Francipane MG, Medema JP, Stassi G (2010) Colon cancer stem cells: promise of targeted therapy. Gastroenterology 138:2151–2162. doi:10.1053/j.gastro.2009.12.063
724. Dalerba P, Dylla SJ, Park I-K et al (2007) Phenotypic characterization of human colorectal cancer stem cells. Proc Natl Acad Sci U S A 104:10158–10163. doi:10.1073/pnas.0703478104
725. Visvader JE, Lindeman GJ (2008) Cancer stem cells in solid tumours: accumulating evidence and unresolved questions. Nat Rev Cancer 8:755–768. doi:10.1038/nrc2499
726. Hirsch D, Barker N, McNeil N et al (2014) LGR5 positivity defines stem-like cells in colorectal cancer. Carcinogenesis 35:849–858. doi:10.1093/carcin/bgt377
727. May R, Riehl TE, Hunt C et al (2008) Identification of a novel putative gastrointestinal stem cell and adenoma stem cell marker, doublecortin and CaM kinase-like-1, following radiation injury and in adenomatous polyposis coli/multiple intestinal neoplasia mice. Stem Cells 26:630–637. doi:10.1634/stemcells.2007-0621
728. Nakanishi Y, Seno H, Fukuoka A et al (2013) Dclk1 distinguishes between tumor and normal stem cells in the intestine. Nat Genet 45:98–103. doi:10.1038/ng.2481
729. Deeb KK, Trump DL, Johnson CS (2007) Vitamin D signalling pathways in cancer: potential for anticancer therapeutics. Nat Rev Cancer 7:684–700. doi:10.1038/nrc2196
730. Thorne J, Campbell MJ (2008) The vitamin D receptor in cancer. Proc Nutr Soc 67:115–127. doi:10.1017/S0029665108006964
731. Abe E, Miyaura C, Sakagami H et al (1981) Differentiation of mouse myeloid leukemia cells induced by 1 alpha,25-dihydroxyvitamin D3. Proc Natl Acad Sci 78:4990–4994. doi:10.1073/pnas.78.8.4990
732. Colston K, Colston MJ, Feldman D (1981) 1,25-dihydroxyvitamin D3 and malignant melanoma: the presence of receptors and inhibition of cell growth in culture. Endocrinology 108:1083–1086. doi:10.1210/endo-108-3-1083
733. GARLAND CF, GARLAND FC (1980) Do Sunlight and Vitamin D Reduce the Likelihood of Colon Cancer? Int J Epidemiol 9:227–231. doi:10.1093/ije/9.3.227
734. Freedman DM, Looker AC, Chang S-C, Graubard BI (2007) Prospective study of serum vitamin D and cancer mortality in the United States. J Natl Cancer Inst 99:1594–1602. doi:10.1093/jnci/djm204
735. Gandini S, Boniol M, Haukka J et al (2011) Meta-analysis of observational studies of serum 25-hydroxyvitamin D levels and colorectal, breast and prostate cancer and colorectal adenoma. Int J Cancer 128:1414–1424. doi:10.1002/ijc.25439
736. Giovannucci E (2013) Epidemiology of vitamin D and colorectal cancer. Anticancer Agents Med Chem 13:11–19
737. Jenab M, Bueno-de-Mesquita HB, Ferrari P et al (2010) Association between pre-diagnostic circulating vitamin D concentration and risk of colorectal cancer in European populations:a nested case–control study. BMJ 340:b5500. doi:10.1136/bmj.b5500
738. Platz EA, Hankinson SE, Hollis BW et al (2000) Plasma 1,25-dihydroxy- and 25-hydroxyvitamin D and adenomatous polyps of the distal colorectum. Cancer Epidemiol Biomarkers Prev 9:1059–1065
739. Wu K, Feskanich D, Fuchs CS et al (2007) A nested case control study of plasma 25-hydroxyvitamin D concentrations and risk of colorectal cancer. J Natl Cancer Inst 99:1120–1129. doi:10.1093/jnci/djm038
740. Feldman D, Krishnan AV, Swami S et al (2014) The role of vitamin D in reducing cancer risk and progression. Nat Rev Cancer 14:342–357. doi:10.1038/nrc3691
741. Pereira F, Larriba MJ, Muñoz A (2012) Vitamin D and colon cancer. Endocr Relat Cancer 19:R51–R71. doi:10.1530/ERC-11-0388

742. Brehier A, Thomasset M (1988) Human colon cell line HT-29: characterisation of 1,25-dihydroxyvitamin D3 receptor and induction of differentiation by the hormone. J Steroid Biochem 29:265–270. doi:10.1016/0022-4731(88)90275-0
743. Díaz GD, Paraskeva C, Thomas MG et al (2000) Apoptosis is induced by the active metabolite of vitamin D3 and its analogue EB1089 in colorectal adenoma and carcinoma cells: possible implications for prevention and therapy. Cancer Res 60:2304–2312
744. Giuliano AR, Franceschi RT, Wood RJ (1991) Characterization of the vitamin D receptor from the Caco-2 human colon carcinoma cell line: effect of cellular differentiation. Arch Biochem Biophys 285:261–269. doi:10.1016/0003-9861(91)90358-P
745. Halline AG, Davidson NO, Skarosi SF et al (1994) Effects of 1,25-dihydroxyvitamin D3 on proliferation and differentiation of Caco-2 cells. Endocrinology 134:1710–1717. doi:10.1210/endo.134.4.8137734
746. Fujita H, Sugimoto K, Inatomi S et al (2008) Tight junction proteins claudin-2 and -12 are critical for vitamin D-dependent Ca2+ absorption between enterocytes. Mol Biol Cell 19:1912–1921. doi:10.1091/mbc.E07-09-0973
747. Palmer HG (2001) Vitamin D3 promotes the differentiation of colon carcinoma cells by the induction of E-cadherin and the inhibition of beta-catenin signaling. J Cell Biol 154:369–388. doi:10.1083/jcb.200102028
748. Pálmer HG, Sánchez-Carbayo M, Ordóñez-Morán P et al (2003) Genetic signatures of differentiation induced by 1alpha,25-dihydroxyvitamin D3 in human colon cancer cells. Cancer Res 63:7799–7806
749. Álvarez-Díaz S, Larriba MJ, López-Otín C, Muñoz A (2010) Vitamin D: proteases, protease inhibitors and cancer. Cell Cycle 9:32–37. doi:10.4161/cc.9.1.10266
750. Egan JB, Thompson PA, Vitanov MV et al (2010) Vitamin D receptor ligands, adenomatous polyposis coli, and the vitamin D receptor FokI polymorphism collectively modulate beta-catenin activity in colon cancer cells. Mol Carcinog 49:337–352. doi:10.1002/mc.20603
751. Shah S, Islam MN, Dakshanamurthy S et al (2006) The molecular basis of vitamin D receptor and beta-catenin crossregulation. Mol Cell 21:799–809. doi:10.1016/j.molcel.2006.01.037
752. Aguilera O, Peña C, García JM et al (2007) The Wnt antagonist DICKKOPF-1 gene is induced by 1alpha,25-dihydroxyvitamin D3 associated to the differentiation of human colon cancer cells. Carcinogenesis 28:1877–1884. doi:10.1093/carcin/bgm094
753. Ordóñez-Morán P, Alvarez-Díaz S, Valle N et al (2010) The effects of 1,25-dihydroxyvitamin D3 on colon cancer cells depend on RhoA-ROCK-p38MAPK-MSK signaling. J Steroid Biochem Mol Biol 121:355–361. doi:10.1016/j.jsbmb.2010.02.031
754. Ordóñez-Morán P, Larriba MJ, Pálmer HG et al (2008) RhoA-ROCK and p38MAPK-MSK1 mediate vitamin D effects on gene expression, phenotype, and Wnt pathway in colon cancer cells. J Cell Biol 183:697–710. doi:10.1083/jcb.200803020
755. Cross HS, Bareis P, Hofer H et al (2001) 25-Hydroxyvitamin D3-1α-hydroxylase and vitamin D receptor gene expression in human colonic mucosa is elevated during early cancerogenesis. Steroids 66:287–292. doi:10.1016/S0039-128X(00)00153-7
756. González-Sancho JM, Larriba MJ, Ordóñez-Morán P et al (2006) Effects of 1alpha,25-dihydroxyvitamin D3 in human colon cancer cells. Anticancer Res 26:2669–2681
757. Modica S, Gofflot F, Murzilli S et al (2010) The intestinal nuclear receptor signature with epithelial localization patterns and expression modulation in tumors. Gastroenterology 138:636–648, 648.e1–12. doi: 10.1053/j.gastro.2009.09.060
758. Shabahang M, Buras RR, Davoodi F et al (1993) 1,25-Dihydroxyvitamin D3 receptor as a marker of human colon carcinoma cell line differentiation and growth inhibition. Cancer Res 53:3712–3718
759. Sheinin Y, Kaserer K, Wrba F et al (2000) In situ mRNA hybridization analysis and immunolocalization of the vitamin D receptor in normal and carcinomatous human colonic mucosa: relation to epidermal growth factor receptor expression. Virchows Arch 437:501–507. doi:10.1007/s004280000275

760. Larriba MJ, Valle N, Pálmer HG et al (2007) The inhibition of Wnt/beta-catenin signalling by 1alpha,25-dihydroxyvitamin D3 is abrogated by Snail1 in human colon cancer cells. Endocr Relat Cancer 14:141–151. doi:10.1677/ERC-06-0028
761. Pálmer HG, Larriba MJ, García JM et al (2004) The transcription factor SNAIL represses vitamin D receptor expression and responsiveness in human colon cancer. Nat Med 10:917–919. doi:10.1038/nm1095
762. Larriba MJ, Martín-Villar E, García JM et al (2009) Snail2 cooperates with Snail1 in the repression of vitamin D receptor in colon cancer. Carcinogenesis 30:1459–1468. doi:10.1093/carcin/bgp140
763. Pan Y-Z, Gao W, Yu A-M (2009) MicroRNAs regulate CYP3A4 expression via direct and indirect targeting. Drug Metab Dispos 37:2112–2117. doi:10.1124/dmd.109.027680
764. Baffa R, Fassan M, Volinia S et al (2009) MicroRNA expression profiling of human metastatic cancers identifies cancer gene targets. J Pathol 219:214–221. doi:10.1002/path.2586
765. Alvarez-Díaz S, Valle N, Ferrer-Mayorga G et al (2012) MicroRNA-22 is induced by vitamin D and contributes to its antiproliferative, antimigratory and gene regulatory effects in colon cancer cells. Hum Mol Genet 21:2157–2165. doi:10.1093/hmg/dds031
766. Abedin SA, Banwell CM, Colston KW et al (2006) Epigenetic corruption of VDR signalling in malignancy. Anticancer Res 26:2557–2566
767. Gaschott T, Stein J (2003) Short-chain fatty acids and colon cancer cells: the vitamin D receptor--butyrate connection. Recent Results Cancer Res 164:247–257
768. Gaschott T, Steinmeyer A, Steinhilber D, Stein J (2002) ZK 156718, a low calcemic, antiproliferative, and prodifferentiating vitamin D analog. Biochem Biophys Res Commun 290:504–509. doi:10.1006/bbrc.2001.6213
769. Kang W, Lee S, Jeon E et al (2011) Emerging role of vitamin D in colorectal cancer. World J Gastrointest Oncol 3:123–127. doi:10.4251/wjgo.v3.i8.123
770. Ordonez-Moran P, Larriba MJ, Pendas-Franco N et al (2005) Vitamin D and cancer: an update of in vitro and in vivo data. Front Biosci 10:2723–2749
771. Xu H, Posner GH, Stevenson M, Campbell FC (2010) Apc(MIN) modulation of vitamin D secosteroid growth control. Carcinogenesis 31:1434–1441. doi:10.1093/carcin/bgq098

# Chapter 5
# Cadherin-Based Cell-Cell Adhesions: Adhesion Structure, Signalling and Computational Modeling

**Sandeep Kumar and Shamik Sen**

**Abstract** Cadherins represent a large group (more than 100) of transmembrane cell surface glycoproteins that play central roles in tissue organization and morphogenesis by modulating cell-cell adhesions in a $Ca^{++}$ dependent manner. In this chapter, we review the various types of cadherins, interacting partners of cadherins, the structure and assembly of cadherin-mediated cell-cell adhesions, force-sensitive nature of cadherin bonds, and the role of cadherins in cancer. Finally, we discuss about some of the different computational and mathematical approaches used to explore different aspects of cell-cell adhesions, and their relationship to cellular processes relevant to cancer.

**Keywords** Extracellular matrix (ECM) • Cell-cell adhesion • Cell-ECM adhesion • Epithelial – mesenchymal transition (EMT) • Cellular automata (CA)

## 5.1 Cadherins: Sub-types and Interacting Partners

Cadherins constitute a large superfamily of proteins that can be grouped into classical and non-classical cadherins [1]. Classical cadherins were the first cadherins identified with roles in organisation of cell-cell junctions and include E-cadherin, N-cadherin, P-cadherin and VE cadherin. In contrast, desmosomal cadherins, T-cadherin, Fat cadherin and protocadherins are members of the non-classical subfamily [2]. The different types of cadherins are expressed in different tissues and perform different functions. E-cadherin plays very important role in defining epithelial tissue architecture and is generally found in cells of epithelial origin. N-cadherin, or neuronal cadherin, plays an important role in maintaining synaptic junctions [3], and is found in neurons, heart, skeletal muscle, and fibroblasts. N-cadherin mediated cell-cell communication plays very important role during embryonic development

S. Kumar • S. Sen (✉)
Department of Biosciences and Bioengineering, IIT Bombay, Mumbai, India
e-mail: shamiks@iitb.ac.in

 
M. Kandouz (ed.), *Intercellular Communication in Cancer*,
DOI 10.1007/978-94-017-7380-5_5

[4, 5]. P-cadherin plays very important role in enhancing cell migration and invasion in breast cancer cells and has been correlated with tumor aggressiveness [6], and is generally found at places like placenta, epidermis and breast epithelium. VE-cadherin, or vascular endothelial cadherin, is found in endothelial cells and plays an important role in vasculature permeability and vasculature remodeling, and promotes tumor cell proliferation and invasion by stimulating TGF-β signalling [7]. In contrast to E, N and VE cadherins, which all connect to the cytoskeleton in the cytoplasmic side, T-cadherin does not couple to the cytoskeleton. T-cadherin has been associated with cell migration and phenotypic changes [8]. It is mainly found in neuronal cells and muscle. Fat cadherin, possessing the longest extracellular domain among all cadherins, is found in epithelial tissues and the central nervous system. It regulates cell proliferation and planar cell polarity [9]. Lastly, protocadherins, the largest subfamily of cadherin superfamily, facilitates cell-cell communication required for tissue morphogenesis and formation of neuronal circuits during early vertebrate development. In addition, protocadherins have been observed to act like tumor suppressor in breast cancer [10, 11].

The classical cadherins comprise of five extracellular domains (EC1–EC5) followed by a single pass transmembrane domain. The extracellular domains are sensitive to $Ca^{2+}$ concentration with loss of $Ca^{2+}$ concentration rendering 'flexibility' to the E-cadherin extracellular domain, and increase in $Ca^{2+}$ making the E-cadherin extracellular domain more stiff [12]. Homophilic binding interactions between EC1 domains of individual cadherin molecules con-tribute to the specificity of their interaction, and contribute to several interesting phenomena. Firstly, in a population of cells expressing two different types of cadherin, cells expressing the same type of cadherin aggregate together thereby leading to sorting of the two cell populations [13]. Secondly, in a population of cells expressing different levels of the same type of cadherin, cells get segregated based on protein levels [14]. Heterotypic interactions, which are much weaker than homophilic interactions, have also been reported between cells expressing N-cadherin and R-cadherin [15], and cells expressing N-cadherin and E-cadherin [16]. These weak interactions may mediate the migration of cells of one type through other types of tissues.

The mechanism of cell-cell communication and mechanosensing/mechanocoupling provided by cadherins depends on the type of cadherin involved. There are several proteins which play important roles in making the cadherin junction functional, and are specific to the cadherin sub-type (refer to the review on the 'cadherin-adhesome' [17]). While E-cadherin mediated cell-cell communication uses catenins to connect the actin network of the cell to the cell-adhesion junction, desmosomal cadherins uses desmoplakin, plakoglobin and other proteins to connect the cell's intermediate filaments to the cell-adhesion junctions [18]. Figure 5.1 shows the assembly of E-cadherin-mediated cell-cell adhesions, which comprises of two principal components: cadherins connecting two cells, and the catenins connecting the cadherin in each cell to the actin cytoskeleton. Two of the catenins highly associated with E-cadherin mediated cell-cell communication are β-catenin and α-catenin, respectively. While β-catenin works as the bridge between α-catenin and E-cadherin, α-catenin links β-catenin with the actin cytoskeleton. Controlling

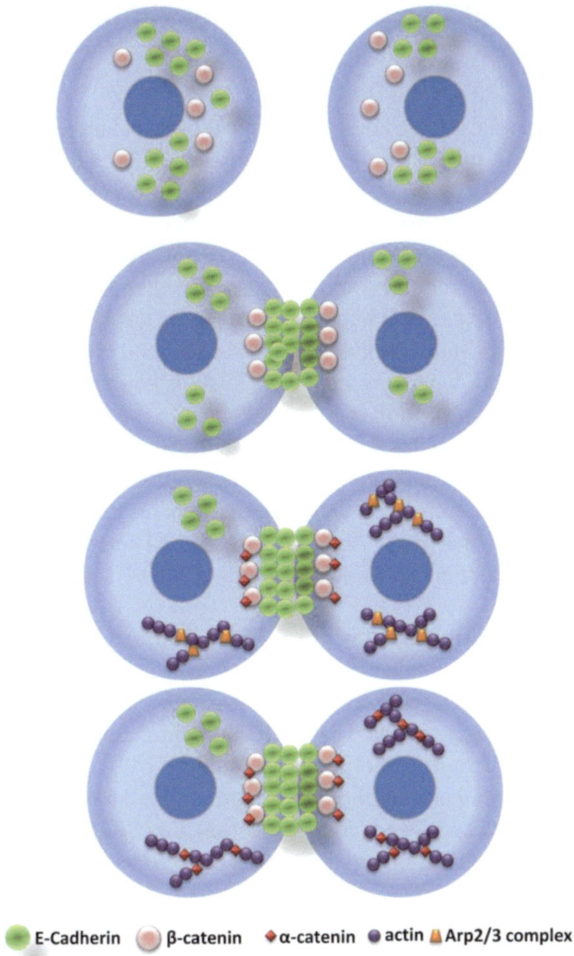


**Fig. 5.1 Formation of E-cadherin mediated cell-cell adhesion.** Formation of E-cadherin based cell-cell adhesion starts when two cells with enough adhesion potential (enough level of E-cadherin and b-catenin) come in contact. When two cells come in contact (i.e. contact surface area between cells is not zero) the E-cadherin in both the cells starts localizing to the contact site and form the E-cadherin based homophilic bond. After formation of homophilic bond the β-catenin starts localizing near intra-cellular domain of E-cadherin and gets attached to it to form E-cadherin-β-catenin (Eβ) complex. Formation of Eβ complex increases the a-catenin localization near Eβ. Increase in the concentration of α-catenin due to increased location perturb the actin network formation near Eβ based cell-cell junction by replacing Arp2/3 complex by α-catenin

the dynamics of β-catenin is one of the principal mechanisms through which several other signalling molecules control cadherin mediated cell-cell adhesions [19]. For example, phosphorylation of β-catenin via JNK causes disruption of cell-cell adhesions [20]. β-catenin also connects cadherin-junction kinetics with wnt signalling, one of the major cancer signalling pathways.

## 5.2 Structure and Dynamics of E-cadherin-Based Cell-Cell Adhesions

When E-cadherin-based adhesion is formed, it causes rapid increase in the β-catenin concentration at the site of intracellular domain of E-cadherin. Subsequently, E-cadherin-β-catenin (Eβ) bond is formed. The complex then causes localized increase in α-catenin levels. Upon reaching a critical concentration, α-catenin starts competing with the actin branching molecule Arp2/3 to get attached to the actin network. When α-catenin concentration increases further, it starts replacing Arp2/3 complex in the actin network. When α-catenin get attached to the actin network it stops the actin polymerization at that site and subsequently, causes perturbations in the actin dynamics and dependent processes such as cell migration. Apart from α and β-catenin, another protein which has been associated with E-cadherin based intercell communication is p120. Although the exact contribution of p120-catenin protein in E-cadherin based cell-cell adhesion in not much clear, it has been observed to play role in increasing the cadherin turnover to make stronger cell-cell adhesion [21–23]. Further, study done on MDCK cell line has demonstrated that snail induced epithelial to mesenchymal transition causes isoform pattern change in p120 [24, 25].

Studies on formation and dynamics of E-cadherin-based cell-cell adhesions have demonstrated the direct influence of the concentrations of E-cadherin and β-catenin in dictating the strength of cell-cell adhesions. One of the simplified models of E-cadherin-β-catenin (Eβ) mediated cell-cell communication has been presented by Chaplain and co-workers in a computational study [26]. In this study, E-cadherin was assumed to be in either one of the three forms: membrane anchored, cytoplasmic, and E-cadherin complexed with β-catenin, i.e., Eβ complex. The relative concentrations of these three fractions changes as the intensity of cell-cell communication (or adhesion) alters. Specifically, for realizing high cell-cell adhesion, the concentration of Eβ should be higher, with a corresponding decrease in the cytoplasmic fraction of E-cadherin. The membrane bound E-cadherin is a temporary state and works as intermediate form in transporting E-cadherin from cytoplasm to cell-cell adhesion site. The same study also demonstrated how this simple model can explain the generation of an 'EMT-wave' in an epithelial population. Additionally, phosphorylation of β-catenin by JNK [20] and Rac activation [27] both lead to loss of E-cadherin-mediated cell-cell adhesions.

## 5.3 Mechanobiology of Cadherin-Based Adhesions

It is well known that adherent cells actively sense the physical properties of their under-lying substrate through focal adhesions and reorganize their cytoskeletal apparatus accordingly [28–30]. This mechanoresponse has been documented in the enhanced cell spreading, motility and proliferation of a range of different cell types on substrates of increasing stiffness [31–33], stiffness-sensitive differentiation of mesenchymal stem cells [34], and in enhanced cancer invasion on substrates of higher stiffness [35, 36]. Interestingly, several reports have highlighted such mechanosensitive responses by cell-cell adhesions as well. On N-cadherin coated substrates of varying stiffness, N-cadherin-expressing myogenic cells were found to spread extensively, possess a well developed actin cytoskeleton and form robust cadherin-based adhesions [37]. In contrast, on softer substrates, cells were less spread and possessed a diffuse actin network. Furthermore, traction forces exerted by cells were directly correlated with the formation of cadherin adhesions, with larger tractions exerted on stiffer substrates. Interestingly, traction forces as well as cadherin adhesions decreased drastically upon inhibition of myosin II activity, indicative of mechanosensitive regulation of cadherin adhesions by the contractile actomyosin cytoskeleton. Similarly, in epithelial cells, E-cadherin has been shown to accumulate at zonula adherens in a myosin-dependent manner [38, 39], with size of junctions directly proportional to the tugging forces [40]. In addition, the cortical stiffness of cadherin adhesions also increases in a force-dependent manner [41]. Together, these results indicate that cadherin-based adhesions act as mechanosensors and adapt their strength to the stiffness of their microenvironment, with the actomyosin cytoskeleton playing an important role in the formation and stabilization of these adhesions. This is similar to the process of strengthening of cell-matrix adhesions [42].

While the mechanoresponsive behaviour of cadherin adhesions has been clearly demonstrated, it is not fully clear which is the mechanosensing element. As has been demonstrated in the enhanced vinculin binding of the focal adhesion protein talin upon stretch [43], it is possible that actomyosin forces may lead to conformational changes in the protein cadherin itself, thereby leading to subsequent signalling events. Alternatively, the mechanosensory element may be one of the numerous actin binding proteins which localize at cadherin junctions. Of these, α-catenin has been hypothesised to be the force-sensitive element as it binds to both F-actin and the cadherin complex [44, 45], and has been shown to recruit vinculin to cadherin junctions in a myosin-dependent manner [46]. Further, one of the centrally located epitopes of α-catenin was detected in the presence of myosin. One of the other important players is the protein vinculin, which is necessary for cortical stiffening when cadherin bonds are stimulated [47].

## 5.4 Cadherins in Cancer

Studies on the role of cadherins in cancer has demonstrated that multiple cadherins participate in influencing the behavior of tumor cells [48–50]. Since majority of cancers arise in epithelial tissues, therefore, expression of E-cadherin and/or its localization at cell-cell junctions is often altered in cancers, with cadherin dysfunction associated with cancer progression [51]. This has been established both in cell lines as well as in animal models [52]. Consistent with this, expression of a dominant-negative cadherin mutant in a mouse pancreatic β-cell tumor model accelerated tumor progression [53], with E-cadherin overexpression retarding this process [54]. In epithelial cells, repression of E-cadherin protein (at the gene level or at the protein level) causes disruption of cell-cell adhesion and triggers the process of epithelial to mesenchymal transition (EMT). Conversely, overexpression of E-cadherin has been suggested to be one of the way to control EMT and cancer progression [52, 55, 56]. Activation of Rac, JNK, WNT pathways have all been shown to cause disruption of E-cadherin-based cell-cell adhesions [57–59]. This blocking of intercellular communication triggers series of events which transforms cells with epithelial phenotype into cells with mesenchymal phenotype. These transformed cells have high migration potential, high proliferation rate, low death rate and high propensity to escape the tissue territory [60]. It has also been shown that the disruption of cell-cell adhesion in one cell of a population can lead to generate an 'EMT wave' in the entire population [26]. Disruption of cell-cell adhesion can also be induced by perturbing ECM properties. For example, increase in ligand density has been shown to negatively influence cell-cell adhesions in MCF-7 breast cancer cell line [61]. Collectively, these results illustrate an inverse relationship between E-cadherin expression and cancer progression.

Altered E-cadherin signalling influences cancer progression in multiple ways. Several studies have demonstrated the tight relationship between cell-cell adhesion of epithelial cells and their proliferation propensity. Disruption of E-cadherin-based cell-cell adhesions can induce cancerous growth by increasing cell proliferation. In pancreatic cancer cells, cell culture on collagen I-coated substrates induced disassembly of E-cadherin-based adhesions and led to nuclear localization of β-catenin. Nuclear accumulation of β-catenin subsequently in-creased cell proliferation by upregulating target genes cyclin D1 and c-myc [62]. Study by Liu et al. has implicated Rac1 as the intermediate molecule linking cell-cell adhesion and cell proliferation [63]. The same work has also shown that the strength of E-cadherin-based cell-cell adhesion was sensitive to population density, with cell-cell adhesion-dependent proliferation maximum at some intermediate population density. Work done by Perrais et al. offers a detailed account of how E-cadherin-based cell-cell communication is connected to the cell-proliferation circuitry [64]. In particular, the group has shown that E-cadherin ligation and binding of E-cadherin with β-catenin causes inhibition of cell-proliferation. Hermiston has also shown that, in adult vertebrate, increased expression of E-cadherin causes higher death rate and limits proliferation [65]. Together, these studies illustrate modulation of cell proliferation by E-cadherin.

Though E-cadherin downregulation is associated with cancer progression, it is important to note that E-cadherin suppression is not sufficient to induce tumorigenesis [66]. Instead, loss of E-cadherin is often associated with a corresponding gain of expression of N-cadherin, a process termed cadherin-switching. Switching of E-cadherin to N-cadherin is associated with EMT in cancer cells [67], which subsequently brings many phenotypical changes in cell favouring cancer progression [54, 68–70]. In a set of cadherin-switching experiments in mice in the intestinal epithelium, the number of differentiated cell types in the intestine drastically decreased postnatally together with the formation of highly proliferating polyps [71]. Similar experiments in alveolar cells of the mammary glands led to formation of mammary tumors [72]. In breast epithelial cells, overexpression of N-cadherin is capable of inducing scattering even in the presence of E-cadherin. While these results implicate N-cadherin as a prospective therapeutic target, contradictory results were observed in neuroblastoma cells where N-cadherin suppression correlated with metastatic dissemination [73]. Apart from E and N-cadherin, there are several other types of cadherins whose expressions change during cancer and influence cancer progression. For a detailed overview on the role of other types of cadherins in cancer, readers are urged to read a detailed review by van Roy [50].

While loss of E-cadherin triggers mesenchymal mode of migration, cancer cells may also move collectively, similar to collective cell migration observed in physiological processes like gastrulation, vessel formation, and wound healing [74]. In collective invasion by cancer cells, E-cadherin-based cell-cell adhesions are retained [75]. Collective invasion has been observed in squamous carcinomas, sarcomas and melanomas, as well as in carcinomas [76, 77]. Imaging of tumor cells in 3D cultures has revealed sprout formation by epithelial type tumor cells, structures which are also observed in clinical samples of advanced stage carcinomas. The structure of collectively invading cancer cells may range from strands of few cells in contact to broader masses, with combination of cell-cell adhesion, cell-ECM adhesion and proteolysis dictating the size and shape of the group. In 3D matrices, protease-dependent matrix degradation is key for ensuring collective migration [78–80], with matrix degradation effected by stromal fibroblasts as well [81]. The efficiency of collective migration is also dictated by the extent of coherence within the migrating population. Interestingly, studies on the E-cadherin stabilizing protein p120ctn in tumor cells have highlighted the contribution of matrix dimensionality as an important factor in influencing cell coherence [82].

## 5.5 Crosstalk Between Cell-Cell and Cell-Matrix Adhesions

Several reports have highlighted a crosstalk between cell-cell and cell-matrix adhesions [83]. Using the dual pipette assay, Dufour and co-workers demonstrated that integrin ligation stimulates formation of cell-cell adhesions in fibroblasts [84]. In direct contrast, formation of cell-matrix adhesions was found to inhibit formation of cell-cell adhesions in cancer cells when presented with a combination of cadherins and ECM proteins [85]. Interestingly, this suppression of cell-cell adhesion formation was found to be rigidity-dependent, with softer surfaces permitting formation of both types of adhesive structures. Further, deposition of ECM proteins on top of an epithelial monolayer led to a dramatic re-distribution of cell-cell adhesions from the top ECM-rich zone to the bottom ECM-free zone [86]. Several candidates link signals downstream of cell-matrix adhesions to regulate cell-cell adhesions. Of these, the signalling molecules Src and FAK, which localize at both the adhesive structures, deserve special mention [87]. Src and FAK can modulate the formation, dynamics and strength of cell-cell adhesions in one of three different ways: (i) by controlling E-cadherin expression through transcriptional control [88, 89]; (ii) by phosphorylating individual proteins that localize at cell-cell junctions [90, 91]; (iii) by regulating E-cadherin endocytosis and thereby its membrane localization [92, 93]. Signals downstream of Src and FAK impinge on various kinase signalling pathways (e.g., MAPK, MLCK and ROCK) which modulate the distribution of activated myosin within the cell [92]. Apart from Src and FAK, ILK and Rho GTPases are two of the other signalling molecules that modulate the crosstalk between integrin and cadherin-based adhesions [94, 95].

## 5.6 Computational Modeling of Cadherin-Based Adhesions

Mathematical modeling approaches have contributed significantly to our understanding of cancer [96–100]. Broadly speaking, the mathematical approaches applied to various biological problems fall in one of two categories: continuous and discrete models. This categorisation depends on the spatio-temporal scales of interest. Continuous models approach biological phenomena in terms of variation of fields, approximating populations of biological individuals as densities, which evolve subject to certain balance laws or diffusion equations. On the other hand, discrete models (e.g., cellular automata (CA)) approach the problem phenomenologically, focusing on population level, cell-level or sub-cellular level of abstraction preserving the behaviour of individual elements. To mitigate the deficiencies of these two approaches, hybrid models have evolved spanning a wide range of spatio-temporal scales with a sufficient level of accuracy, offering the advantages brought by the two approaches. There are two types of hybrid models. In the first type of hybrid models, discrete and continuous approaches are combined to describe different aspects of the system. This kind of approach has been used for modeling solid tumor growth and angiogenesis. In the second type of hybrid models, the same elements are described as objects in space in selected portions of the domain and as continuous variables in the others. Cellular Potts models (CPMs), represent one of the second type of models, where energy minimization drives evolution of the system [101].

Furthering our understanding of the role of cell-cell adhesion in cancer requires a seam-less integration of different types of experimental data with results from simulations and theoretical models. While the emergence of new tools, techniques and equipments, have made huge amounts of data generation possible, their analysis remains to be performed to gain in-sight and convert this data into knowledge. This is an area where usage of computer science approaches has benefited. For example, analysis of high throughput sequencing data has depended heavily on machine learning techniques for data interpretation. Computational modeling approaches have contributed significantly to our understanding of various aspects of cancer including intracellular signalling, role of MMPs in remodelling the stroma, heterogeneity in tumor cells due to stem cell differentiation dynamics and many more.

E-cadherin mediated cell-cell communication has multiple aspects spanning different time scales and length scales. Consequently, integrating the findings from different studies is something not so obvious and straightforward. In addition to providing a framework to integrate various disjoint studies to elucidate the mechanism of E-cadherin mediated cell-cell adhesion, mathematical models also have the potential to guide subsequent experimental research by helping in designing new 'potentially important' experiments. In this section, we briefly review two approaches that have been used for modeling different aspects of cell-cell adhesions, discuss the merits and demerits of these approaches. We also discuss about techniques used in mathematical models that have not directly addressed cell-cell adhesion, but can be adapted for modeling cell-cell communication.

CPMs have become a widely used technique for mathematical modeling of a wide range of biological phenomena, including avascular and vascular tumor growth [102, 103], wound healing [104], differential-adhesion driven cell rearrangement [105], and angiogenesis. One approach of modeling cell-cell adhesions in these processes is by accounting for cell-cell interactions through an adhesion energy term in the description for the total energy. For example, in an angiogenesis model presented by Amy et al. [106], cell-cell adhesion was modeled as a constant adhesion energy term. This amounts to all cells possessing equal cell-cell adhesion propensity. This may not always be true given the heterogeneity in the cancer cell population [26, 107]. Adhesivity of individual cells within a group of epithelial cells may also change as normal cells transform to cancer cells via processes like EMT. Additionally, cell-cell adhesion strength is dependent on the surface area of contact between two cells, and may hence differ depending on the extent of shape changes in the cells. At the molecular level, the strength of cell-cell adhesion is directly determined by the concentration of the E-cadherin-β-catenin complex. Taking these factors into consideration, Chaplain and co-workers have developed a hybrid model wherein the adhesion energy was dependent on the concentration of E-cadherin-β-catenin complex in the cell, which in turn, depended on the contact area between cells [26]. Using this model, the authors demonstrated how cell adhesiveness of individual cells was regulated by interactions between E-cadherin and β-catenin, and its implications in cell migration and invasion. Similar to using an energy description, a force description has also been used for studying dynamics of cell clusters, wherein individual cell movement was attributed to forces produced due to formation and breakage of bonds to neighbouring cells [107]. In this study, the movement of a cell within a cell cluster was dictated by the net force acting on the cell within a 'sensing radius', which was comparable to the length scale of the cell. Collectively, the studies above illustrate the simple approach of using an adhesive energy or force term for modeling cell-cell adhesions. These types of formulations can easily be modified to capture the dynamics of cell-cell adhesions in other contexts.

While the above approach is well suited for studying cell dynamics in various cellular processes, the effects of individual factors that regulate one or more aspects of cell-cell adhesion are ignored. Given the gamut of proteins and genes that influence cell signalling (discussed in previous sections), perturbation in these protein levels and gene expression may alter sub-cellular dynamics thereby influencing cell-cell adhesion. To probe and explore the influence of these proteins on cell-cell adhesion and to study the features of the protein-gene 'control-circuitry' responsible for controlling dynamics of cell-cell adhesion, kinetic modeling approaches are useful. Using such approach, for studying the influence of extracellular matrix (ECM) density on EMT, computational model of the protein-gene network regulating cell-cell adhesion through E-cadherin-β-catenin complex has been recently developed by our group [61]. In this model, first, the protein signalling network connecting cell- ECM interactions to cell-cell interactions was transformed into a system of ODEs using mass action kinetic based approach (Fig. 5.2a). This model was then used to study the influence of ligand density on cell-cell adhesion. Several molecular level signalling events were incorporated into our model including membrane-cytoplasmic shuttling of E-cadherin, activation of β-catening, formation of E-cadherin-β-catenin bond, and other post-translational modifications.

One of the major limitations of pure kinetic models is their inability to capture spatial dynamics. To overcome this limitation, if required, it is possible of using frameworks which provide option of modeling both kinetic and spatial aspects of the system. Discrete models (e.g., cellular automata, lattice based model, agent based models) are particularly suited for studying dynamics of cell population. In cellular automata (CA), the population is divided into automaton cells with each cell is evolved based on some pre-defined rule(s) [108]. CA has been used to model various aspects of cancer including tumor growth [109], role of ECM crosslinking on ECM degradation [110], cancer stem cell proliferation [111], tumor-stroma interactions in prostate cancer [112] and influence of fibroblasts in melanoma [113]. Using a similar cellular automaton model, we extended the systems level model to capture the crosstalk between cell-matrix and cell-cell adhesions in a cell population. In this model, each cell updated its state based on the ligand density and based on the state of its four orthogonal neighbours (Fig. 5.2b). The model was then used to predict the population dynamics for (i) a heterogeneous cell population sitting on a homogeneous substrate of given ligand density, (ii) a heterogeneous cell population sitting on a heterogeneous substrate with spatially varying ligand density. The model in-turn used the signalling network shown in Fig. 5.2a. Specifically, at every time-step (or iteration step), in addition to sensing its underlying ECM density and accordingly adjusting its adhesively (i.e., concentration of E-cadherin-β-catenin (Eβ) complex at the membrane) (Fig. 5.2a), each cell sensed the Eβ levels of neighbouring cells. The final adhesivity (i.e., Eβ levels) of the cell was set as the average of its ligand density-induced adhesivity and the adhesivity of neighbouring cells (Fig. 5.2c). The process was continued till the average population level Eβ reached a saturation (Fig. 5.2d).

This modeling framework allowed us to explore the tug of war between E-cadherin and ligand density in dictating the EMT potential of a single cell. While increase in ECM density led to reduction in Eβ levels, upregulation of E-cadherin successfully counterbalanced the effects of increase in ECM density. Though the results were qualitatively similar for both single cells as well as a cell population, the temporal dynamics were much gradual for a cell population. Doing these simulations for a heterogeneous cell population reminiscent of a primary tumor, allowed

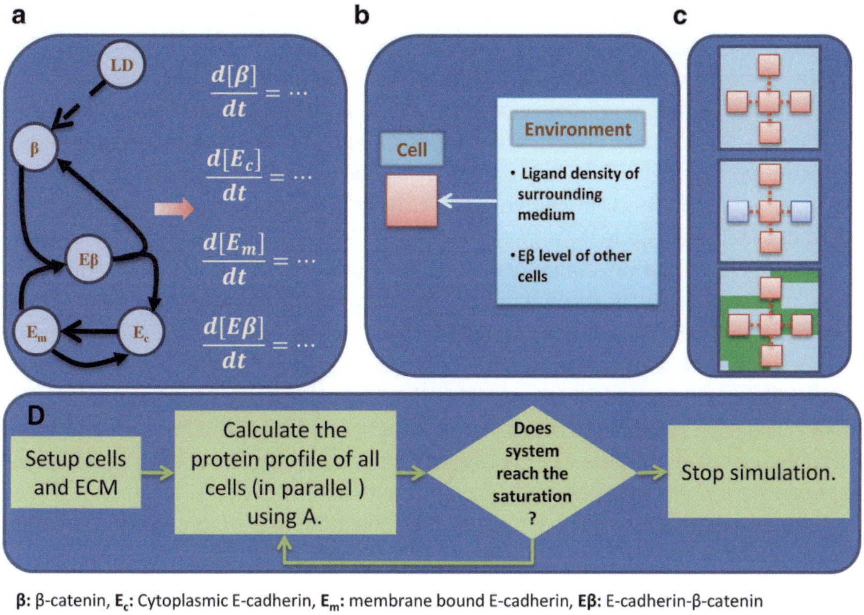


**Fig. 5.2 Combining mass action kinetic modeling and cellular automata modeling to study population dynamics of cell-cell adhesions for varying cell-matrix adhesion**. Figure shown the schematic of combining the mass action kinetic and cellular automata based discrete modelling techniques for capturing and cellular and population level spatio-temporal dynamics. (**a**). First the signalling network (inferred from the available data or curated) was transcoded into system of ordinary differential equations(ODEs) using mass action kinetic. Such a preliminary model can be use for exploring the temporal dynamics of different proteins involved in propagating the signal from cell-matrix adhesion to cell-cell adhesion site. (**b**). Cellular automata model was then developed in which each 'biological cell' was represented by a lattice cell. Each of these cells update the level of their proteins by sensing the environment (i.e. ligand density and protein level of surrounding cells) and then using some 'automaton rule' (e.g. average rule as used in [61]). (**c**). The cellular automata based population model was then used to probe the influence of cell-ECM adhesion on cell adhesion for (i) homogeneous cell population and homogeneous ECM, (ii) heterogeneous cell population and homogeneous ECM, and (iii) homogeneous cell population and heterogeneous ECM. (**d**). To evolve the population state (i.e. protein levels of all cells) each cell, in parallel, sense the ligand density of the surrounding and then calculate the concentration of cell proteins by simulating signalling network (by solving ODEs). This calculated value and the protein level of neighbouring cells (to incorporate neighbouring effect) were then used to calculate the final cell state and thereby system state

us to study the collective effects of ECM density and E-cadherin over-expression on cell scattering. Heterogeneity in the cell population was introduced by assigning different adhesivities (Eβ levels) to different cells within the population, with cells with very low adhesivities labeled as 'EMT cells'. Interestingly, our results suggested that EMT can be suppressed at low ligand densities even in the presence of a sizeable fraction of EMT cells. Further, the presence of EMT cells altered the sensitivity of E-cadherin increase in counterbalancing the effects of ECM density. Similar to the simulation results, a strong dependence of E-cadherin localization was observed experimentally in MCF-7 human breast cancer cell line, with a robust membrane presence at low ligand densities which decreased with increase in ECM density. Intriguingly, for the highly metastatic MDA-MB-231 breast cancer cells, which did not form any cell-cell adhesions, membrane localization of E-cadherin still exhibited a dependence on ECM density. Collectively, these results highlight the strong influence of ECM density on cell-cell adhesions and illustrates the utility of the modeling framework in studying sub-cellular and cellular aspects of cadherin-based cell-cell adhesions.

Although many of the components of E-cadherin-based cell-cell adhesions have been experimentally identified, how the individual components collectively contribute to modulate the spatiotemporal dynamics of cell-cell adhesions remains incompletely understood. For example, it has been shown that faster actin dynamics helps cell in developing clusters of cell-cell adhesions[114–116], no computational model has fully recapitulated the crosstalk between cell-cell adhesions and actin dynamics. Existing models [117, 118] may be suitably augmented for understanding kinetic events linking cell-cell adhesions with actin dynamics. Another aspect of cadherin bonds that has not been taken into consideration in any of the modeling studies is its complex force-dependent behaviour [119]. Given the homophilic and heterotypic interactions between different types of cadherins, it is also worth exploring how expression of different cadherins gives rise to heterogeneous sub-populations in the primary tumor, and how these sub-populations metastasize at distinct secondary locations [120–123].

## References

1. Hulpiao P, Redies CM (2009) Molecular evolution of the cadherin superfamily. Int J Biochem Cell Biol 41:349–369
2. Hirano S, Suzuki ST, Redies CM, Hirano S, Suzuki ST, Redies CM (2003) The cadherin superfamily in neural development: diversity, function and interaction with other molecules. Front Biosci 8:D306–D355
3. Tanaka H, Shan W, Phillips GR, Arndt K, Bozdagi O, Shapiro L, Huntley GW, Benson DL, Colman DR (2000) Molecular modification of N-cadherin in response to synaptic activity. Neuron 25(1):93–107
4. Hatta K, Takagi S, Fujisawa H, Takeichi M (1987) Spatial and temporal expression pat-tern of N-cadherin cell adhesion molecules correlated with morphogenetic processes of chicken embryos. Dev Biol 120(1):215–227
5. Takeichi M (1990) Cadherins: a molecular family important in selective cell-cell adhesion. Annu Rev Biochem 59:237–252
6. Falagas ME, Zarkadoulia E a, Ioannidou EN, George P, Christos C, Rafailidis PI (2007) P-cadherin expression in breast cancer: a review. Breast Cancer Res 9(5):214
7. Rezaei M, Katrin F, Ben W, Aleksandar K, Antje K-t (2012) Interplay between neural-cadherin and vascular endothelial-cadherin in breast cancer progression. Breast Cancer Res 14(6):R154
8. Rubina K, Talovskaya E, Cherenkov V, Ivanov D, Stambolsky D, Storozhevykh T (2005) LDL induces intracellular signalling and cell migration via atypical LDL-binding protein T-cadherin. Mol Cell Biochem 273(1–2):33–41

9. Takuji T, Masatoshi T (2005) New insights into fat cadherins. J Cell Sci 118(11):2347–2353
10. Yu JS, Koujak S, Nagase S, C-m L, Su T, Wang X, Keniry M, Memeo L (2008) PCDH8, the human homolog of PAPC, is a candidate tumor suppressor of breast cancer. Oncogene 27(34):4657–4665
11. Yu J, Cheng YY, Tao Q, Cheung KF, Lam CN, Geng H, Tian L-W, Wong YP, Tong JH, Ying J-M, Jin H, To KF, Chan F-c K, Sung JJ (2009) Methylation of protocadherin 10, a novel tumor suppressor, is associated with poor prognosis in patients with gastric cancer. Gastroenterology 136(2):640–651.e1
12. Courjean O, Guillaume C, Emilie P, Anne M, Sarah S, Noelle P, Jurgen¨ E, van Dorsselaer A, Hel´ene` F (2008) Modulation of E-cadherin monomer folding by cooperative binding of calcium ions. Biochemistry 47(8):2339–2349
13. Nose A, Tsuji K, Takeichi M (1990) Localization of specificity determining sites in cadherin cell adhesion molecules. Cell 61:147–155
14. Friedlander DR, Mge RM, Cunningham BA, Edelman GM (1989) Molecular evolution of the cadherin superfamily. Proc Natl Acad Sci U S A 86:7043–7047
15. Inuzuka H, Miyatani S, Takeichi M (1991) R-cadherin: a novel Ca(2+)-dependent cell-cell adhesion molecule expressed in the retina. Neuron 7(1):69–79
16. Katsamba P, Carroll K, Ahlsen G, Bahna F, Vendome J, Posy S, Rajebhosale M, Price S, Jessell TM, Ben-Shaul A, Shapiro L, Honig BH (2009) Linking molecular affinity and cellular specificity in cadherin-mediated adhesion. Proc Natl Acad Sci U S A 106(28):11594–11599
17. Zaidel-Bar R (2013) Cadherin adhesome at a glance. J Cell Sci 126:373–378
18. Garrod DR, Merritt AJ, Nie Z (2002) Desmosomal cadherins. Curr Opin Cell Biol 14:537–545
19. Ratheesh A, Yap AS (2012) A bigger picture: classical cadherins and the dynamic actin cytoskeleton. Nat Rev Mol Cell Biol 13:673–679
20. Lee M-h, Piyush K, Jun Q, Andreadis ST (2009) JNK phosphorylates β-catenin and regulates adherens junctions. FASEB J 09(0023):3874–3883
21. Reynolds AB, Agnes R-F (2004) Emerging roles for p120-catenin in cell adhesion and cancer. Oncogene 23(48):7947–7956
22. Ireton RC, Davis M a, van Hengel J, Mariner DJ, Kirk B, Thoreson M a, Anastasiadis PZ, Linsey M, Bundy LM, Linda S, Barbara G, van Roy F, Reynolds AB (2002) A novel role for p120 catenin in E-cadherin function. J Cell Biol 159(3):465–476
23. Davis M a, Ireton RC, Reynolds AB (2003) A core function for p120-catenin in cadherin turnover. J Cell Biol 163(3):525–534
24. Ohkubo T, Masayuki O (2004) The transcription factor Snail downregulates the tight junction components independently of E-cadherin downregulation. J Cell Sci 117(Pt 9):1675–1685
25. Bellovin DI, Bates RC, Alona M, Rimm DL, Mercurio AM (2005) Altered localization of p120-catenin during epithelial to mesenchymal transition of colon carcinoma is prognostic for aggressive disease. Cancer Res 65(23):10938–10945
26. Ramis-Conde I, Dirk D, Anderson AR, Chaplain MA (2008) Modeling the influence of the E-cadherin-β-catenin pathway in cancer cell invasion: a multi scale approach. Biophys J 95(1):155–165
27. Akhtar N, Hotchin NA (2001) RAC1 regulates adherens junctions through endocytosis of E-cadherin. Mol Biol Cell 12(April):847–862
28. Geiger B, Bershadsky AD, Pankov R, Yamada KM (2001) Transmembrane extracellular matrix-cytoskeleton crosstalk. Nat Rev Mol Cell Biol 2:793–805
29. Discher DE, Janmey P, Wang YL (2005) Tissue cells feel and respond to the stiffness of their substrate. Science 310:1139–1143
30. Geiger B, Spatz JP, Bershadsky AD (2009) Environmental sensing through focal adhesions. Nat Rev Mol Cell Biol 10:21–33
31. Pelham RJ, Yu-Li W (1997) Cell locomotion and focal adhesions are regulated by substrate flexibility. Proc Natl Acad Sci U S A 94(25):13661–13665

32. Peyton SR, Putnam AJ (2005) Extracellular matrix rigidity governs smooth muscle cell motility in a biphasic fashion. J Cell Physiol 204(1):198–209
33. Ulrich TA, de Juan-Pardo EM, Sanjay K (2009) The mechanical rigidity of the extracellular matrix regulates the structure, motility, and proliferation of glioma cells. Cancer Res 69(10):4167–4174
34. Engler AJ, Sen S, Sweeney HL, Discher DE (2006) Matrix elasticity directs stem cell lineage specification. Cell 126:677–689
35. Paszek MJ, Zahir N, Johnson KR, Lakins JN, Rozenberg GI, Gefen A, Reinhart-King CA, Margulies SS, Dembo M, Boettiger D, Hammer DA, Weaver VM (2005) Tensional homeostasis and the malignant phenotype. Cancer Cell 8:241–254
36. Levental KR, Yu H, Kass L, Lakins JN, Egeblad M, Erler JT, Fong SFT, Fong SFT, Csiszar E, Giaccia A, Weninger W, Yamaguchi M, Gasser DL, Weaver VM (2009) Matrix cross linking forces tumor progression by enhancing integrin signaling. Cell 139:891–906
37. Ladoux B, Anon E, Lambert M, Rabodzey A, Hersen P, Buguin A, Silberzan P, Mge RM (2010) Strength dependence of cadherin-mediated adhesions. Biophys J 98(4):534–542
38. Miyake Y, Inoue N, Nishimura K, Kinoshita N, Hosoya H, Yonemura S (2006) Actomyosin tension is required for correct recruitment of adherens junction components and zonula occludens formation. Exp Cell Res 312(9):1637–1650
39. Smutny M, Cox HL, Leerberg JM, Kovacs EM, Conti MA, Ferguson C, Hamilton NA, Parton RG, Adelstein RS, Yap AS (2010) Myosin II isoforms identify distinct functional modules that support integrity of the epithelial zonula adherens. Nat Cell Biol 12(7):696–702
40. Liu Z, Tan JL, Cohen DM, Yang MT, Sniadecki NJ, Ruiz SA, Nelson CM, Chen CS (2010) Mechanical tugging force regulates the size of cell-cell junctions. Proc Natl Acad Sci U S A 107:9944–9949
41. le Duc Q, Shi Q, Blonk I, Sonnenberg A, Wang N, Leckband D, de Rooij J (2010) Vinculin potentiates E-cadherin mechanosensing and is recruited to actin-anchored sites within adherens junctions in a myosin II-dependent manner. J Cell Biol 189(7):1107–1115
42. Balaban NQ, Schwarz US, Riveline D, Goichberg P, Tzur G, Sabanay I, Mahalu D, Safran S, Bershadsky A, Addadi L, Geiger B (2001) Force and focal adhesion assembly: a close relationship studied using elastic micropatterned substrates. Nat Cell Biol 3(5):466–472
43. del Rio A, Perez-Jimenez R, Liu R, Roca-Cusachs P, Fernandez JM, Sheetz MP (2009) Stretching single talin rod molecules activates vinculin binding. Science 323(5914):638–641
44. Watabe-Uchida M, Uchida N, Imamura Y, Nagafuchi A, Fujimoto K, Uemura T, Vermeulen S, van Roy F, Adamson ED, Takeichi M (1998) alpha-catenin-vinculin interaction functions to organize the apical junctional complex in epithelial cells. J Cell Biol 142(3):847–857
45. Huveneers HS, Oldenburg J, Spanjaard E, van der Krogt G, Grigoriev I, Akhmanova A, Rehmann H, de Rooij J (2012) Vinculin associates with endothelial VE-cadherin junctions to control force- dependent remodeling. J Cell Biol 196(5):641–652
46. Yonemura S, Wada Y, Watanabe T, Nagafuchi A, Shibata M (2010) alpha-catenin as a tension transducer that induces adherens junction development. Nat Cell Biol 12:533–542
47. Leerberg JM, Yap AS (2013) Vinculin, cadherin mechanotransduction and homeostasis of cell-cell junctions. Protoplasma 250(4):831
48. Jeanes A, Gottardi CJ, Yap AS (2008) Cadherins and cancer: how does cadherin dysfunction promote tumor progression? Oncogene 27:6920–6929
49. Berx G, van Roy F (2009) Involvement of members of the cadherin superfamily in cancer. Cold Spring Harb Perspect Biol 1(6):a003129
50. van Roy F (2014) Beyond e-cadherin: roles of other cadherin superfamily members in cancer. Nat Rev Cancer 14:121–134
51. Oikawa T, Atsuko N, Nobuyuki O (2013) Acquired expression of NFATc1 downregulates E-cadherin and promotes cancer cell invasion. Cancer Res 73:5100–5109
52. Vleminckx K, Vakaet L, Mareel M, Fiers W, van Roy F (1991) Genetic manipulation of E-cadherin expression by epithelial tumor cells reveals an invasion suppressor role. Cell 66(1):107–119

53. Perl A-K, Wilgenbus P, Dahl U, Semb H, Christofori G (1998) A causal role for E-cadherin in the transition from adenoma to carcinoma. Nature 392:190–193
54. Derksen PW, Liu X, Saridin F, van der Gulden H, Zevenhoven J, Evers B, van Beijnum JR, Griffioen AW, Vink J, Krimpenfort P, Peterse JL, Cardiff RD, Berns A, Jonkers J (2006) Somatic inactivation of e-cadherin and p53 in mice leads to metastatic lobular mammary carcinoma through induction of anoikis resistance and angiogenesis. Cancer Cell 10(5):437–449
55. Christofori G, Semb H (1999) The role of the cell-adhesion molecule E-cadherin as a tumour-suppressor gene. Trends Biochem Sci 24(2):73–76
56. Tripathi V, Popescu NC, Zimonjic DB (2014) DLC1 induces expression of E-cadherin in prostate cancer cells through Rho pathway and suppresses invasion. Oncogene 33(6):724–733
57. Ximei W, Xiaolin T, Kyu Sang J, Hilton MJ, Williams D a, Fanxin L (2008) Rac1 activation controls nuclear localization of β-catenin during canonical Wnt signaling. Cell 133(2):340–353
58. Crampton SP, Beibei W, Park EJ, Jai-Hyun K, Candice S, Waterman M-i L, Hughes CCW (2009) Integration of the β-catenin-dependent Wnt pathway with integrin signaling through the adaptor molecule Grb2. PLoS One 4(11):e7841
59. Heuberger J, Walter B (2010) Interplay of cadherin-mediated cell adhesion and canonical Wnt signaling. Cold Spring Harb Perspect Biol 2(2):a002915
60. Hanahan D, Weinberg R a (2011) Hallmarks of cancer: the next generation. Cell 144(5):646–674
61. Kumar S, Alakesh D, Shamik S (2014) Extracellular matrix density promotes EMT by weakening cell-cell adhesion. Mol Biosyst 10(4):838–850
62. Koenig A, Mueller C, Hasel C, Adler G, Menke A (2006) Collagen type i induces disruption of E-cadherin-mediated cell-cell contacts and promotes proliferation of pancreatic carcinoma cells. Cancer Res 66(9):4662–4671
63. Liu WF, Nelson CM, Pirone DM, Chen CS (2006) E-cadherin engagement stimulates proliferation via Rac1. J Cell Biol 173(3):431–441
64. Chen X, Mirna P-m, Gumbiner BM (2007) E-cadherin homophilic ligation inhibits cell growth and epidermal growth factor receptor signaling independently of other cell interactions. Mol Biol Cell 18(6):2013–2025
65. Hermiston ML, Wong MH, Gordon JI (1996) Forced expression of E-cadherin in the mouse intestinal epithelium slows cell migration and provides evidence for nonautonomous regulation of cell fate in a self-renewing system. Genes Dev 10(8):985–996
66. Tunggal JA, Helfrich I, Schmitz A, Schwarz H, Gnzel D, Fromm M, Kemler R, Krieg T, Niessen CM (2005) E-cadherin is essential for in vivo epidermal barrier function by regulating tight junctions. EMBO J 24(6):1146–1156
67. Tiwari N, Gheldof A, Tatari M, Christofori G (2012) EMT as the ultimate survival mechanism of cancer cells. Semin Cancer Biol 22(3):194–207
68. Eger A, Aigner K, Sonderegger S, Dampier B, Oehler S, Schreiber M, Berx G, Cano A, Beug H, Foisner R (2005) DeltaEF1 is a transcriptional repressor of E-cadherin and regulates epithelial plasticity in breast cancer cells. Oncogene 24(14):2375–2385
69. Gumbiner BM (2000) Regulation of cadherin adhesive activity. J Cell Biol 148(3):399–404
70. Hirohashi S (1998) Inactivation of the E-cadherin-mediated cell. Am J Pathol 153(2):333–339
71. Libusova L, Stemmler MP, Hierholzer A, Schwarz H, Kemler R (2010) N-cadherin can structurally substitute for e-cadherin during intestinal development but leads to polyp formation. Development 137:2297–2305
72. Kotb AM, Hierholzer A, Kemler R (2011) Replacement of E-cadherin by n-cadherin in the mammary gland leads to fibrocystic changes and tumor formation. Breast Cancer Res 13:R104

73. Lammens T, Swerts K, Derycke L, De Craemer A, De Brouwer S, De Preter K, Van R, Vandesompele J, Speleman F, Philipp J, Benoit Y, Beiske K, Bracke M, Laureys G (2012) N-cadherin in neuroblastoma disease: expression and clinical significance. PLoS One 7(2):e31206
74. Rorth P (2009) Collective cell migration. Ann Rev Cell Dev Biol 25:407–419
75. Friedl P, Locker J, Sahai E, Segall JE (2012) Classifying collective cancer cell invasion. Nat Cell Biol 14(8):777–783
76. Hegerfeldt Y, Tusch M, Brocker EB, Friedl P (2002) Collective cell movement in primary melanoma explants: plasticity of cell-cell interaction, beta1-integrin function, and migration strategies. Cancer Res 62:2125–2130
77. Friedl P, Gilmour D (2009) Collective cell migration in morphogenesis, regeneration and cancer. Nat Rev Mol Cell Biol 10:445–457
78. Nabeshima K, Inoue T, Shimao Y, Okada Y, Itoh Y, Seiki M, Koono M (2000) Front-cell-specific expression of membrane-type 1 matrix metalloproteinase and gelatinase a during cohort migration of colon carcinoma cells induced by hepatocyte growth factor/scatter factor. Cancer Res 60(13):3364–3369
79. Soulie P, Carrozzino F, Pepper MS, Strongin AY, Poupon MF, Montesano R (2005) Membrane-type–1 matrix metalloproteinase confers tumorigenicity on non malignant epithelial cells. Oncogene 24:1689–1697
80. Wolf K, Wu YI, Liu Y, Geiger J, Tam E, Overall C, Stack MS, Friedl P (2007) Multi-step pericellular proteolysis controls the transition from individual to collective cancer cell invasion. Nat Cell Biol 9(8):893–904
81. Gaggioli C, Hooper S, Hidalgo-Carcedo C, Grosse R, Marshall JF, Harrington K, Sahai E (2007) Fibroblast-led collective invasion of carcinoma cells with differing roles for RhoGTPases in leading and following cells. Nat Cell Biol 9(12):1392–1400
82. Macpherson IR, Hooper S, Serrels A, McGarry L, Ozanne BW, Harrington K, Frame MC, Sahai E, Brunton VG (2007) p120-catenin is required for the collective invasion of squamous cell carcinoma cells via a phosphorylation-independent mechanism. Oncogene 26(36):5214–5228
83. Burute M, Thery M (2012) Spatial segregation between cell-cell and cell-matrix adhesions. Curr Opin Cell Biol 24(5):628–636
84. Clara MR, Pincet F, Thiery JP, Dufour S (2010) Integrins stimulate E-cadherin-mediated intercellular adhesion by regulating Src-kinase activation and actomyosin contractility. J Cell Sci 123(5):712–722
85. Tsai J, Kam L (2009) Rigidity-dependent crosstalk between integrin and cadherin signaling. Biophys J 96:39–41
86. Ojakian GK, Radcliffe DR, Schwimmer R (2000) Integrin regulation of cell-cell adhesion during epithelial tubule formation. J Cell Sci 114:941–952
87. Mitra SK, Schlaepfer DD (2006) Integrin-regulated FAK-Src signaling in normal and cancer cells. Curr Opin Cell Biol 18:516–523
88. Menke A, Philippi C, Vogelmann R, Seidel B, Lutz MP, Adler G, Wedlich D (2001) Down-regulation of E-cadherin gene expression by collagen type I and type III in pancreatic cancer cell lines. Cancer Res 61:3508–3517
89. Li XY, Zhou X, Rowe RG, Hu Y, Schlaepfer DD, Ilic D, Dressler G, Park A, Guan L, Weiss SJ (2011) Snail1 controls epithelial–mesenchymal lineage commitment in focal adhesion kinase-null embryonic cells. J Cell Biol 195:729–738
90. Calautti E, Cabodi S, Stein PL, Hatzfeld M, Kedersha N, Paolo Dotto G (1998) Tyrosine phosphorylation and src family kinases control keratinocyte cell-cell adhesion. J Cell Biol 141:1449–1465
91. Playford MP, Vadali K, Cai X, Burridge K, Schaller MD (2008) Focal adhesion kinase regulates cell-cell contact formation in epithelial cells via modulation of Rho. Exp Cell Res 314:3187–3197

92. Avizienyte E, Fincham VJ, Brunton VG, Frame MC (2004) Src SH3/2 domain-mediated peripheral accumulation of Src and phospho-myosin is linked to deregulation of E-cadherin and the epithelial-mesenchymal transition. Mol Biol Cell 15:2794–2803
93. Palacios F, Tushir JS, Fujita Y, DSouza-Schorey C (2005) Lysosomal targeting of E-cadherin: a unique mechanism for the down-regulation of cell-cell adhesion during epithelial to mesenchymal transitions. Mol Biol Cell 25:389–402
94. Oloumi A, McPhee T, Dedhar S (2004) Regulation of E-cadherin expression and beta-catenin/Tcf transcriptional activity by the integrin-linked kinase. Biochim Biophys Acta 1691:1–15
95. Menke A, Giehl K (2012) Regulation of adherens junctions by Rho GTPases and p120-catenin. Arch Biochem Biophys 524:48–55
96. Quaranta V, Weaver AM, Cummings PT, Alexander ARA (2005) Mathematical modeling of cancer: the future of prognosis and treatment. Clin Chim Acta 357(2):173–179
97. Byrne HM (2010) Dissecting cancer through mathematics: from the cell to the animal model. Nat Rev Cancer 10:221–230
98. Gammon K (2012) Mathematical modelling: forecasting cancer. Nature 491(7425):66–67
99. Savage N (2012) Modelling: computing cancer. Nature 491(7425):S62–S63
100. Werner HMJ, Mills GB, Ram PT (2014) Cancer systems biology: a peek into the future of patient care? Nat Rev Clin Oncol 11(3):167–176
101. Glazier JA, Balter A, Poplawski NJ (2007) Magnetization to morphogenesis: a brief history of the Glazier-graner-Hogeweg model. Single Cell Based Models Biol Med:79–106
102. Turner S, Sherratt JA (2002) Intercellular adhesion and cancer invasion: a discrete simulation using the extended Potts model. J Theor Biol 216(1):85–100
103. Rubenstein BM, Kaufman LJ (2008) The role of extracellular matrix in glioma invasion: a cellular Potts model approach. Biophys J 95(12):5661–5680
104. Maini PK, Olsen L, Sherratt JA (2002) Mathematical models for cell-matrix interactions during dermal wound healing. Int J Bifurcations Chaos 12(9):2021–2029
105. Graner F, Glazier JA (1992) Simulation of biological cell sorting using a two dimensional extended Potts model. Phys Rev Lett 69(13):2013–2017
106. Bauer AL, Jackson TL, Yi J (2007) A cell-based model exhibiting branching and anastomosis during tumor-induced angiogenesis. Biophys J 92(9):3105–3121
107. Armstrong NJ, Painter KJ, Sherratt JA (2006) A continuum approach to modelling cell-cell adhesion. J Theor Biol 243(1):98–113
108. Moreira J, Deutsch A (2002) Cellular automaton models of tumor development: a critical review. Adv Comp Syst 05(02n03):247–267
109. Alarcon T, Byrne ´HM, Maini PK (2003) A cellular automaton model for tumour growth in inhomogeneous environment. J Theor Biol 225(2):257–274
110. Enderling H, Alexander NR, Clark ES, Branch KM, Lourdes E, Cornelia C, Jer´ome ˆ J, Nichole L, Zaman MH, Guelcher SA, Anderson AR, Weaver AM (2008) Dependence of invadopodia function on collagen fiber spacing and cross-linking: computational modeling and experimental evidence. Biophys J 95(5):2203–2218
111. Ninh AQ (2013) Two discrete stochastic cellular automata models of cancer stem cell proliferation. Int J Bios Bioch Bioinfo 3(5):5–8
112. Basanta D, Strand DW, Lukner RB, Franco OE, Cliffel DE, Ayala GE, Hayward SW, Anderson AR (2009) The role of transforming growth factor-beta-mediated tumor-stroma interactions in prostate cancer progression: an integrative approach. Cancer Res 69(17):7111–7120
113. Kim E, Rebecca V, Fedorenko IV, Messina JL, Mathew R, Maria-Engler SS, Basanta D, Smalley KS, Anderson AR (2013) Senescent fibroblasts in melanoma initiation and progression: an integrated theoretical, experimental, and clinical approach. Cancer Res 73:6874–6885
114. Vasioukhin V, Bauer C, Yin M, Fuchs E (2000) Directed actin polymerization is the driving force for epithelial cell-cell adhesion. Cell 100(2):209–219

115. Vasioukhin V, Fuchs E (2001) Actin dynamics and cell-cell adhesion in epithelia. Curr Opin Cell Biol 13(1):76–84
116. Zhang J, Betson M, Erasmus J, Zeikos K, Bailly M, Cramer LP, Braga Vania MM (2005) Actin at cell-cell junctions is composed of two dynamic and functional populations. J Cell Sci 118(Pt 23):5549–5562
117. Pantaloni D, Hills TL, Carlier M-F, Korn ED (1985) A model for actin polymerization and the kinetic effects of ATP hydrolysis. Proc Natl Acad Sci U S A 82(21):7207–7211
118. Cardelli L, Caron E, Phillips A (2008) A process model of actin polymerisation. Electron Notes Theory Comput Sci 1:127–144
119. Leckband D (2010) Design rules for biomolecular adhesion: lessons from force measurements. Annu Rev Chem Biomol Eng 1:365–389
120. Minn AJ, Kang Y, Serganova I, Gupta GP, Giri DD, Doubrovin M, Ponomarev V, Gerald L, Blasberg R, Massague´ J (2005) Distinct organ-specific metastatic potential of individual breast cancer cells and primary tumors. J Clin Invest 115(1):44–55
121. Gupta GP, Nguyen DX, Chiang AC, Bos PD, Kim JY, Nadal C, Gomis RR, Manova-Todorova K, Massague´ J (2007) Mediators of vascular remodelling co-opted for sequential steps in lung metastasis. Nature 446(7137):765–770
122. Bos PD, Zhang XH, Nadal C, Shu W, Gomis RR, Nguyen DX, Minn AJ, van de Vijver MJ, Gerald WL, Foekens JA, Massague´ J (2009) Genes that mediate breast cancer metastasis to the brain. Nature 459(7249):1005–1009
123. Nguyen DX, Bos PD, Massague´ J (2009) Metastasis: from dissemination to organ-specific colonization. Nat Rev Cancer 9(4):274–284

# Chapter 6
# E-Cadherin/Catenin Complex Modulations in Human Oral Cancer

**Ala-Eddin Al Moustafa and Amber Yasmeen**

**Abstract** Oral cancer is a common neoplasm in both male and female worldwide; the incidence and mortality of this cancer have increased over the past decades. Cancer metastasis is a major cause of morbidity and mortality in oral cancer patients, which is associated with the translocation of tumor cells from the initial neoplastic focus into neighboring host tissues. The E-cadherin/catenin complex plays an important role in maintaining epithelial integrity, thus disrupting this complex affect several signaling pathways including Wnt as well as the adhesive repertoire and consequently incites motility and invasion. Aberrant expression patterns of the E-cadherin/catenin complex are associated with a wide variety of human malignancies including oral cancer resulting from epithelial to mesenchymal transition (EMT). In this chapter we will overview the clinical description of oral metastatic cancer; furthermore, we will discuss the role of E-cadherin/catenin complex and its signaling pathways as well as EMT in the progression of this cancer.

**Keywords** Oral cancer • Cancer progression • E-cadherin/catenin complex and its signaling pathways • EMT

## 6.1 Introduction

Oral cancer is one of the most common non-communicable diseases worldwide with an estimated increase of 275,000 new cases each year [1]. This cancer is considered one of the most frequent types of malignancies in the head and neck (38 %) with an

A.-E. Al Moustafa (✉)
Medical Basic Science Department, College of Medicine & Biomedical Research Centre, Qatar University, Doha, Qatar

Oncology Department, Faculty of Medicine, McGill University, Montréal, QC, Canada

Syrian Research Cancer Centre of the Syrian Society against Cancer, Aleppo, Syria
e-mail: aalmoustafa@qu.edu.qa; ala-eddin.almoustafa@mcgill.ca

A. Yasmeen
Oncology Department, Faculty of Medicine, McGill University, Montréal, QC, Canada



M. Kandouz (ed.), *Intercellular Communication in Cancer*,
DOI 10.1007/978-94-017-7380-5_6

incidence of 75 % in male patients over the age of 60 years; about 95 % of these cancer cases are squamous cell carcinomas, which are of epithelial origin [2, 3]. "Oral carcinomas" is the term used for cancers that form in tissues of the oral cavity and the oropharynx [3, 4]. Oral squamous cell carcinomas are invasive lesions with the presence of perineural growth. It has a significant recurrence rate and frequently metastasizes to cervical lymph nodes [5]. Lymph node metastatic tumors occur in about 40 % of patients with oral cancer. Clinically, their manifestations are hidden in rates of 15–34 % [6, 7]. In general, tobacco, betel quid chewing, alcohol and virus infection including human papillomavirus (HPV) and Epstein-Barr virus (EBV) are regarded as major risk factors for oral cancer initiation and progression [4, 8, 9].

E-cadherins and catenins are important epithelial adhesion molecules. The E-cadherin protein is a transmembrane glycoprotein involved in calcium-dependent cellular adhesion. The cytoplasmic domain of E-cadherin can bind directly to either β- or γ-catenin, whereas α-catenin links E-cadherin to the actin cytoskeleton [10, 11]. Loss or reduction of E-cadherin-mediated adhesion is an important step in the development of invasion and metastasis in many types of carcinomas [11], including oral cancer [3, 12]. It has been reported that catenin and E-cadherin are under-expressed in oral carcinomas and nodal metastases [12–14]. On the other hand, the epithelial-mesenchymal transition (EMT) is believed to play a crucial role in cancer metastasis [15]; EMT refers to the loss of carcinoma epithelial phenotype and the acquisition of mesenchymal-associated features [16, 17]. As we mentioned above, in epithelial cells, the cytoplasmic tail of E-cadherin forms a dynamic complex with catenins which can affect several intracellular signal transduction pathways, including Wnt/β-catenin, PI3K/Akt, Rho GTPase, and NF-κB [16, 18]; these signaling pathways are considered as major mechanisms of the EMT initiation and therefore cancer progression. It is clear that oral cancer and especially its invasive form is a major health problem wherein the E-cadherin/catenin complex and EMT event play an important role in the development of this disease. In this chapter, we will overview the contributions of E-cadherin/catenin complex and EMT in the progression of human oral cancer.

## 6.2 E-Cadherin/Catenin Complex

### *6.2.1 Structure and Function*

Epithelial cells develop tight cell-cell adhesions as compared to other cell-types. They form distinct tissue layers through the formation of tissue boundaries and change tissue shapes causing cell rearrangements or the conversion between histological cell states and the long-range migration of cells. The stratified squamous epithelium that covers the oral cavity is especially well organized and form strong cell-cell adhesions mediated by cadherins [19]. Cadherins are a superfamily of calcium-dependent transmembrane proteins that are developmentally regulated and evolutionally conserved cell-cell adhesion molecules [11, 19]. Within the cadherin superfamily, epithelial-cadherin (E-cadherin) plays a decisive role in the development and maintenance of the epithelium at the adherence junction. A long extracellular and cytoplasmic domain characterizes the E-cadherin. Although the extracellular domain of E-cadherin establishes homophilic interactions between neighboring cells [20], its cytoplasmic tail associates with an array of intracellular proteins. These proteins link cell–cell adhesion to the actin–myosin network, vesicle transport, and cell polarity machinery. The best studied of these links is the binding of the cytoplasmic tail of E-cadherin to the Armadillo repeat protein β-catenin, which in turn binds α-catenin that interacts with actin and several actin-binding proteins [11, 21, 22]. Through the action of these intracellular binding partners, E-cadherin contacts modulate actin filament organization at the underlying cortex [23–25]. Signals generated at cell–cell junctions, such as those in response to changes in cell–cell contact, can also be transduced through the cytoplasmic tail of E-cadherin to the nucleus to alter gene expression [26, 27]. For example, β-catenin, a component of adherens junctions (AJs) and a transcriptional co-activator, has been implicated in the transduction of mechanical signals from junctions to the nucleus ([28]). Moreover, mechanical stimulation leads to β-catenin phosphorylation at the site of its interaction with E-cadherin and increased β-catenin nuclear localization, leading to the transcription of the oncogenes Myc as well as Twist1; these effects can be prevented by blocking β-catenin phosphorylation using Src kinase inhibitors [29, 30]. Additionally, earlier work has implicated E-cadherin, α-catenin, and vinculin in participating in a mechanosensory pathway that allows cells to modulate their actin cytoskeleton in response to applied force [22, 31]. These data suggest that the E-cadherin/catenin complex changes affect cell–cell contacts and mechanical stress and consequently cell invasion and metastasis through numerous signaling pathways.

### 6.2.2 *Signaling Pathways*

In addition to their role in adhesion, E-cadherin and its associated catenins are known modulators of several signaling pathways and molecules in normal epithelial and cancer cells, including Wnt, phosphoinositide 3-kinase (PI3K), Rho-family GTPases, and NF-κB [16, 18]. Each of these pathways has been described to influence several types of cells especially the epithelial phenotype (Fig. 6.1), suggesting that expression patterns of the E-cadherin/catenin complex in these cells could actively modulate their phenotype and consequently incite cell motility and invasion via the initiation of the EMT event [17].

The canonical Wnt/β-catenin signaling pathway is best known for its role during the developmental processes, nevertheless it can also regulate various functions in normal and cancer cells [16, 32]. In the absence of Wnt ligand binding to its receptor Frizzled and co-receptor LRP5 or LRP6, cytoplasmic β-catenin is phosphorylated by the kinases CK1 and GSK-3β, causing its ubiquitination and subsequent proteasomal degradation. However, once Wnt signaling is initiated, β-catenin phosphorylation is inhibited, leading to the accumulation of free β-catenin and its trans-

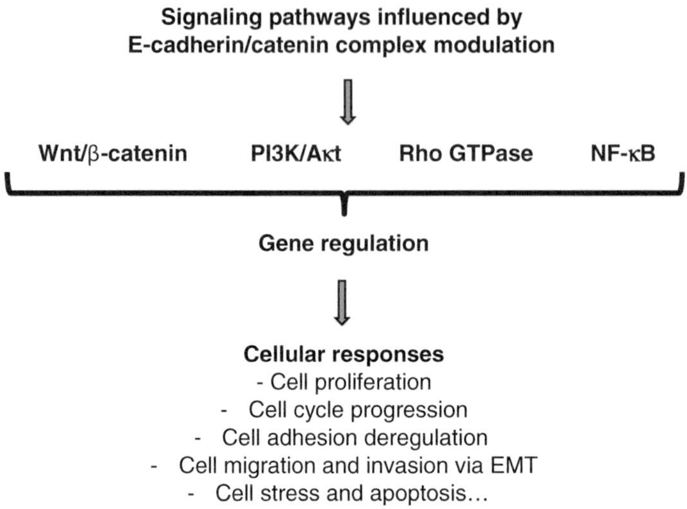


**Fig. 6.1** Intracellular signal transduction pathways associated with E-cadherin/catenin complex modulations. Wnt/β-catenin, PI3K/Akt, Rho GTPase, and NF-κB are major signaling pathways affected by E-cadherin and catenins. These signaling pathways ultimately alter the activity of multiple nuclear transcription factors, which can lead to several cellular responses including cell proliferation, cell cycle progression, cell adhesion, cell migration and invasion via EMT and cell stress and apoptosis

location to the nucleus. Formation of the active β-catenin/TCF/LEF transcription factor complex finally results in the expression of Wnt target genes and oncogenes in normal and cancer cells, respectively [15, 32].

It appears logical that E-cadherin expression patterns can suppress Wnt/β-catenin signaling by sequestering β-catenin at sites of cell-cell contact. Evidence suggests that the mere presence of the E-cadherin cytoplasmic domain, rather than E-cadherin adhesive properties, is required to inhibit Wnt/β-catenin–dependent gene expression [33]. Although direct binding of β-catenin to E-cadherin could be responsible for the sequestration effect, recent evidence suggests that E-cadherin also increases the interaction between caveolin-1 and β-catenin near the plasma membrane, resulting in reduced β-catenin/TCF/LEF-dependent transcription [34]. Moreover, E-cadherin–based adhesion increases the turnover of cytoplasmic β-catenin by promoting the activity of a β-catenin phosphodestruction complex localized near AJs [35]. However, outside AJs, unligated E-cadherin might actually be needed for initiating Wnt signaling, whereby the E-cadherin–p120-catenin complex controls the assembly of the Wnt signalosome [36]. Next, the activated CK1 kinase phosphorylates E-cadherin, thereby weakening the E-cadherin–β-catenin association and turning the E-cadherin–bound β-catenin pool into a signaling β-catenin pool. This could be particularly relevant for E-cadherin–expressing macrophages and epithelial cells. Indeed, Wnt/β-catenin signaling is generally considered to induce an anti-inflammatory phenotype in macrophages [37] as well as EMT event in epithelial cells, [15, 17].

Conversely, several lines of evidence indicate that the engagement of E-cadherin results in the activation of PI3K and Akt in keratinocytes and carcinoma cells [38–40]. Although the exact molecular mechanisms involved are not entirely clear; however, current data suggest that nascent cell-cell contact formation via E-cadherin results in the activation of the c-Src kinase [41], leading to phosphorylation and subsequent recruitment of the PI3K p85 subunit to AJs [38, 40]. However, the link between E-cadherin expression and PI3K activity is not always unequivocal and might be context-dependent. Indeed, E-cadherin has been shown to up-regulate PTEN expression via β-catenin–mediated Egr1 regulation causing β-catenin recruitment to cell-cell junctions and the suppression of PI3K/Akt signaling and growth arrest in ovarian cancer and mammary epithelial cells [42, 43]. We believe that the same activity may occur in human oral epithelial and cancer cells; however, E-cadherin/catenin complex modulations by PI3K and Akt signaling pathways in human oral epithelial cells remain to be investigated.

On the other hand, considerable evidence from multiple groups has demonstrated a recruitment and activation of Rac1 at sites of E-cadherin contact, but some controversy exists on the mechanism leading to Rac1 activation. Although Rac1 activation has been suggested to be a downstream event of c-Src and PI3K activity in some investigations [44, 45], other data suggest a PI3K-independent but rather p120-catenin–dependent mechanism [46]. Irrespective of the mechanism involved, Rac1 appears to be instrumental in stabilizing AJs. p120-catenin itself displays intricate Rho GTPase-modifying properties. Free p120-catenin is able to inhibit RhoA, but this effect is lost on p120-catenin association with E-cadherin [47], possibly explaining the known RhoA activation by nascent E-cadherin cell-cell contacts. E-cadherin loss and release of p120-catenin activates Rac1-MAPK signaling pathway promoting transformed cell growth [48]. Of note, some studies show that E-cadherin negatively regulates cell proliferation and migration by inhibiting RhoA and Cdc42, again illustrating the relevance of E-cadherin outside of cell signaling context [49, 50]. Interestingly, in human squamous carcinoma cells, migration and adhesion are tightly regulated by Rho-family GTPases wherein P-cadherin is involved in this process [51].

It is important to mention that the E-cadherin/catenin complex appears to possess the ability to down-modulate NF-κB activity. In distinct cellular systems, it was shown that a forced overexpression of E-cadherin reduces NF-κB activation, whereas loss of E-cadherin results in an increased activity of this transcription factor [50, 52]. Diversity might exist at the mechanistic level, with a possible requirement for p38 MAPK60 or RhoA-activated protein kinase D156 as intermediates between E-cadherin and NF-κB; Nevertheless, NF-κB suppression might result from an association with the E-cadherin/catenin complex [52]. In this respect, p120-catenin might serve as a docking site, considering the massive NF-κB activation, inflammation, and neoplastic formation *in vitro* and *in vivo* [53, 54].

Overall, these studies indicate that the E-cadherin/catenin complex signaling pathways play an important role in the progression of numerous human carcinomas including oral cancer through their initiation of the EMT.

## 6.3 Epithelial-Mesenchymal Transition and Cancer Progression

Cancer metastasis is a complex process by which cancer cells spread from a primary tumor to other organs and tissues, forming viable secondary deposits of cancer. In most cases, invading cancer cells lose their polarity and intercellular adhesions and acquire a more motile phenotype (Fig. 6.2) as a part of an epithelial-to-mesenchymal switch [55, 56]. While EMT was identified in the context of embryogenesis almost three decades ago, its underlying molecular mechanisms have only begun to be understood quite recently with the discovery of its role in tumor invasiveness [57].

A hallmark of EMT is the functional loss of the adherence junction protein E-cadherin. Although transcriptional control of E-cadherin during EMT appears to be prevalent, recent studies point to the existence of intricate relationships between epigenetic, transcriptional and translational mechanisms [58]. In general, E-cadherin expression is inhibited by a number of transcription factors, but only a small set of these factors is known to regulate it directly. The main groups of transcription factors that bind to the E-cadherin promoter and directly repress its transcription, which will be referred to hereafter as EMT-activating transcription factors (EMT-ATFs), are the ZEB (ZEB1, ZEB2) and Snail (Snail1, Snail2, Snail3) families of zinc finger proteins and the Twist family of bHLH factors (Twist1, Twist2) [55, 56]. While E12/E47 and Tbx3 also bind to the E-cadherin promoter [59, 60], the ability of Goosecoid and HMGA2 to repress E-cadherin expression and induce an EMT

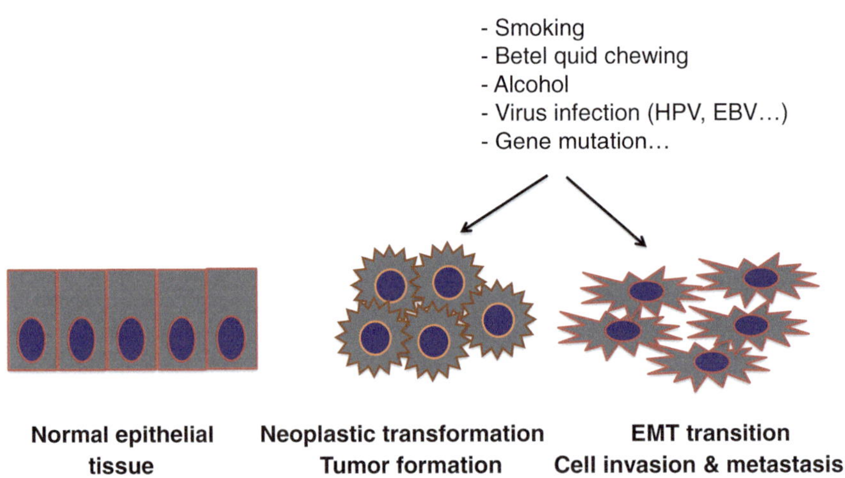


**Fig. 6.2** Cell transformation & Epithelial-mesenchymal transition (EMT) initiation. Cell transformation and consequently tumor formation, including oral cancer, is a multipronged process that requires the cooperation of several gene alterations due to several factors including smoking, alcohol consumption as well as virus and bacterial infection; these factors unfortunately can also enhance cancer progression and metastasis through the initiation of EMT which has been recently demonstrated by several investigations

seems to be mediated by other EMT-ATFs [61, 62]. Most of these transcription factors were originally identified as regulators of embryogenesis and cell differentiation and only later recognized for their role in cancer progression. Still, the upstream regulatory signals and downstream targets of EMT-ATFs in cancer largely concur with those during embryogenesis [57, 58].

E-cadherin mediates homotypic intercellular adhesion and interacts with intracellular proteins to establish and coordinate the morphology, polarity and function of epithelial cells [63, 64]. The down-regulation of E-cadherin is inherent to EMT, but EMT also entails the down-regulation of other epithelial specification genes, such as components of tight and gap junctions, desmosomes, cytokeratins, etc. [55, 56]. In parallel, there is an induction of mesenchymal markers (e.g., N-cadherin, cadherin-11), reorganization of the cytoskeleton (e.g., switch from cytokeratins to vimentin), and the synthesis of extracellular matrix components and metalloproteases [15, 56].

Many signals unleashing an EMT during embryogenesis are also active in cancer. Thus, stimuli such as TGFβ, FGF, EGF, IGF, HGF, PDGF, estrogens, Wnt, Shh, inflammatory cytokines or hypoxia as well as oncogenes like RasV12, ErbB2 or mutant p53 may be involved in EMT during cancer progression [15, 57]. Triggering and maintenance of the mesenchymal state requires cooperation between several of these pathways through autocrine signaling loops [65]. These growth factors, inflammatory, hypoxic and oncogenic signals, along with an increasing number of microRNAs, converge in the induction of EMT-ATFs either at the transcriptional level or by increasing their protein or mRNA stability. E-cadherin expression and function could be down-regulated by loss-of-function mutations, but modulation of EMT during embryogenesis and cancer progression mostly involves the participation of EMT-ATFs providing a high degree of functional redundancy and plasticity [57, 58]. E-cadherin was also found to be silenced by promoter hypermethylation in a number of carcinomas. Some of the signals triggering EMT activate or repress the expression of non-coding microRNAs, which in turn modulate tumor invasion and metastasis by regulating EMT-ATFs transcripts, targeting E-cadherin itself, or altering the expression of small GTPases or extracellular matrix receptors [66].

Down-regulation of E-cadherin is a critical initial step in EMT, not only because of the disruption of adherens junctions but also because loss of E-cadherin reinforces the EMT process by inducing the expression of Twist1 and ZEB1 in a feed-forward loop [67]. EMT-ATFs do not simply repress E-cadherin but are able to orchestrate the entire EMT program, inhibiting and activating a wide array of epithelial and mesenchymal genes, respectively [68–71].

Expression of EMT markers by primary human tumors correlates with enhanced invasiveness and poor clinical prognosis. Cancer cells at the tumor invasive edge that have transitioned through an EMT secrete cytokines and proteases that promote angiogenesis, remodel the peritumoral extracellular matrix, and activate non-neoplastic stromal cells. In turn, stromal cells release factors that reinforce the EMT in cancer cells and foster survival, growth, and invasiveness of the tumor, thus creating a reciprocal influence between the tumor and its microenvironment [72, 73]. EMT is required in order to shift cancer cells from a collective type of invasion, where cells retain E-cadherin and some intercellular adhesions, to an individual mesenchymal type of invasion, a change that depends on TGFβ signals at the tumor–host interface [74, 75]. While both models of invasion allow cancer cells to reach the lymphatic circulation, optimal hematogenous dissemination only occurs in EMT-mediated individual invasion [72, 75].

In addition to the mesenchymal switch, the molecular reprogramming encompassed by the EMT also endows cancer cells with stem-like characteristics [65, 76]. Many of the signals controlling normal stem cell homeostasis are inducers of EMT and seem to contribute to the generation and maintenance of cancer stem cells (CSCs). Overall, CSCs are thought to be important in the genesis of primary tumors and metastasis and could also be at the root of tumoral chemo-resistance and recurrence [77, 78].

As in embryogenesis, the EMT taking place during cancer progression is a reversible process. At the site of micro-metastasis, epigenetic changes and the absence of EMT-inducing signals lead to the re-expression of particular sets of microRNAs and the down-regulation of EMT-ATFs, allowing cancer cells to regain the epithelial characteristics of the primary tumor, through a mesenchymal-to-epithelial transition (MET), allowing them to grow to form a secondary tumor [79, 80]. Recent evidence indicates that EMT and metastatic dissemination of cancer cells may in fact start very early in tumor progression [58, 81]. Collectively, these data suggest that the EMT event is a major key regulator of cell invasion and metastasis in human carcinomas including oral malignancy.

## 6.4 E-Cadherin/Catenin Complex and Oral Cancers

As mentioned above, perturbations in orchestrated modulation of cell adhesion cause defects in tissue architecture that play critical roles in cancer development [55, 57]. The dissociation of the E-cadherin/catenin complex from the cell membrane is important in malignant progression. Down-regulation of membranous E-cadherin and β-catenin, and cytoplasmic/nuclear accumulation of β-catenin, have been previously reported in several cancers and hold promise as prognostic markers [60]. Several studies have shown that the E-cadherin/catenin complex expression is down regulated in human oral cancer [12–14]; however, there is a very limited number of investigations regarding the signaling pathway alterations of this complex in human oral cancer.

Although, it is well known that gene deregulation and interaction is the main key in cell alteration, transformation and consequently tumor formation of human carcinomas including oral. In parallel, the development of oral cancer is a multistep process involving interactions between several external factors such as tobacco-associated intra-oral carcinogens, areca nut, betel quid and alcohol consumption, as well as viral infections [4, 8, 9]. Oral carcinomas are often preceded by clinically evident lesions (often leukoplakia) with 5–10 times greater risk of multiple cancers in patients with oral carcinomas preceded by leukoplakia, which often progress to cancer if untreated [82, 83].

**Table 6.1** Clinical characteristics of human oral cancer (HOC) patients studied [84]

| Case no. | Cancer site | Histopathological type | Sex | Age |
|---|---|---|---|---|
| HOC-002 | Left side of floor of mouth and gingiva | Well differentiated SCC | M | 76 |
| HOC-008 | Left side of soft palate with tonsil and uvula | Poorly differentiated SCC | M | 42 |
| HOC-011 | Left side of tongue | Moderately to Poorly differentiated SCC | M | 55 |
| HOC-021 | Left mandible | Well differentiated SCC | M | 63 |

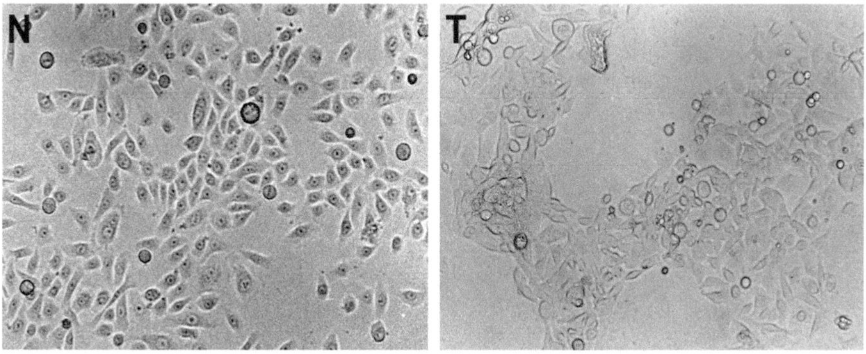


**Fig. 6.3** Primary normal and cancer cells of HOC-021 samples. Normal (N) and cancer (T) cells displayed an epithelial and epithelial-like morphology, respectively [84]

In order to identify genes involved in human oral carcinogenesis, we compared the gene expression profile in matched primary human normal oral epithelial cells and oral cancer cells from the same patients. As a first step, we isolated eight paired primary oral cell lines from four patients, and in each case, we obtained four oral cancer cells with corresponding normal epithelial cells (Table 6.1). More than 95 % of the primary normal cells displayed an epithelial morphology, in contrast to the primary cancer cells that displayed an epithelial-like morphology (Fig. 6.3). We used the cDNA array technique to identify genes that are differentially expressed in these cancer cell lines in comparison to the matched normal, non-cancerous, epithelial cell lines from the same patients. We identified 213 genes out of 12,530, which are regulated differently between matched normal and cancer cells. Ninety-one genes, with intensity ratios between 2.5 and 200, were up-regulated, while 122 genes were down-regulated with ratios between −200 and −2.5. Thus, 0.7 % of the 12 530 genes were over expressed and 1 % were under expressed in the cancer cells. It is therefore notable that the expression levels of 98.3 % of all genes studied in this experiment did not differ between normal and malignant squamous oral cells taken from the same patient. The altered genes included those associated with signal transduction, cell structure, cell cycle, transcription, apoptosis, metabolism, and cell-cell adhesion [84]. In general, most over-expressed genes in cancer cells encode growth factors and cell structure, while under-expressed genes are involved in cell-cell adhesion and motility, apoptosis and metabolism. Among these groups of genes, we subsequently focused on selected genes involved in cell-cell interaction and cell motility, because of the broad implication of these classes of genes in carcinogenesis and tumor invasion [85]. We investigated the expression of nine selected genes from the cell-cell adhesion and motility group by Western and/or RT-PCR. These genes included claudin-7, connexin 31.1, Wnt-5a, N-cadherin, E-cadherin, α-catenin, β-catenin, γ-catenin and fibronectin. As expected from the array analysis, Wnt-5a, N-cadherin and fibronectin were found to be up-regulated; whereas claudin-7, E-cadherin, α-catenin, β-catenin, γ-catenin, and connexin 31.1 were down-regulated in cancer versus normal cells at the protein and/or RNA levels [84].

Numerous studies demonstrated that E-cadherin/catenins complex is down-regulated in human cancer and metastasis including oral cancer [11, 13, 85]. Our Microarray data confirmed that the E-cadherin/catenin complex genes are down-regulated in human oral cancer cells in comparison with normal oral epithelial cells obtained from the same patients. Moreover, we reported for the first time that claudin-7 and connexin 31.1 are down–regulated in human oral cancer cells versus normal oral epithelial cells [84]. In conclusion, our array data provide a large body of information regarding gene expression profiles associated with human oral carcinogenesis, and also represent a source of potential targets for oral cancer prevention and/or therapeutics.

On the other hand, the matched normal/cancer cell lines provided us with a model to study human oral carcinogenesis. This is relevant, because we have developed a novel model derived from normal human oral epithelial (NOE) cells to study the mechanisms of oral cancer. Using this model, we established that E6/E7 of high-risk HPV type 16 cooperate with the ErbB-2 receptor to induce cellular transformation and morphological conversion from epithelial to epithelial-like "mesenchymal" of human NOE cells. In contrast, neither the expression of E6/E7 nor over-expression of ErbB-2 alone was able to induce cell transformation and morphological alteration [86]. We also demonstrated that E6/E7/ErbB-2 cooperation down-regulates E-cadherin and catenins expression and induces de-localization of β-catenin from the undercoat membrane to the nucleus in NOE cells. Meanwhile, we reported that cyclin D1 and c-myc are targets of E6/E7/ErbB-2 cooperation via the conversion of β-catenin's role from a cell-cell adhesion molecule to a transcriptional regulator [86]. Earlier studies reported that these genes are targets of the nuclear β-catenin, via its interaction with the Tcf/Lef family of transcription factors (124, 125). We revealed that D-type cyclins (D1, D2 and D3) are essential for cell transformation induced by E6/E7/ErbB-2 cooperation in human NOE and mouse normal embryonic fibroblast (NEF) cells [87, 88]. Recently, we found that the cooperation effect of E6/E7 with ErbB-2 occurs via β-catenin phosphorylation by pp60 (c-Src) and STAT3 activation [89], which consequently can provoke cell invasion and motility. Therefore, we were able to show that Src inhibitor (SKI-606) can restore the expression of β-catenin on the undercoat membrane to act as cell-cell adhesion and consequently block cell motility induced by E6/E7/ErbB-2 cooperation [10, 89].

In conclusion these studies suggest that E6/E7 of HPV type 16 cooperate with ErbB-2 in oral carcinogenesis, at least in part, via the conversion of β-catenin from a cell adhesion to a nuclear function to act as a potential transcriptional regulator. This conversion leads to the up-regulation of cyclin D1, c-myc and other onco-proteins necessary for the alteration of the E-cadherin/catenin complex and cell transformation and motility of NOE cells.

Earlier studies revealed that approximately 35 % of human oral cancer cases are positive for high-risk HPVs [4, 90, 91]. In general, it is well known that high-risk HPV onco-proteins, E5, E6 and E7, provoke cellular alteration and lead to HPV-induced cell transformation, in cooperation with other Oncogenes, and consequently tumor initiation and progression [86, 92, 93]. Previous investigations demonstrated that the E5 onco-protein could play an important role in cell alteration through its interaction with EGF-R1 signaling pathways (MAP kinas and PI3K/Akt) and pro-apoptotic proteins [94, 95], which could enhance tumor progression through EMT. On the other hand, E6 and E7 are assumed to work together in HPVs-infected cells [96]. Both E6 and E7 have functions that deregulate cell cycle, apoptosis and cell adhesion, through their interaction with p53 and pRb and other members of the pocket protein family [92, 97], which could also initiate the EMT and cancer progression.

In order to investigate the incidence of high-risk HPVs in human oral cancer in the Syrian population, we examined the presence of these viruses and their onco protein E6/E7 in a cohort of 80 oral cancer tissue samples from Syria by immuno-histochemistry and Tissue Microarray methodologies. Our data revealed that 43 % of these cancers are positives for high-risk HPVs, which represent invasive oral carcinomas [98].

On the other hand, it has been reported that EBV is present in more than 95 % of nasopharyngeal carcinomas worldwide [99–101]. EBV-infected cells express the latency III program of gene products, including six EBV nuclear antigens (EBNA1, -2, -3A, -3B, -3C and -LP) as well as three latent membrane proteins (LMP1, -2A, and -2B), and multiple non-coding RNAs (EBERs and miRNAs) [102–104].

The differential expression pattern of these latent genes defines the distinct latency programs of cancers associated with EBV [105, 106]. For instance, type II latency is characterized by a more restricted latent gene expression pattern (EBNA1, LMP1 and -2), and is associated with Hodgkin's lymphoma and nasopharyngeal as well as other carcinomas [100, 101, 107, 108]. Thus, LMP1 is considered the major EBV-encoded oncogenic protein, as it induces multitude effects promoting cell growth, protecting cells from apoptosis, enhancing cell motility, promoting angio-genesis and it is frequently expressed in EBV-associated human oral carcinomas [109, 110]. Recent studies have reported that the presence of EBV is associated with human invasive carcinomas of nasopharyngeal [111–113]. Moreover, earlier investigations demonstrated that LMP1 and EBNA1 onco-proteins of EBV enhance cancer progression and metastasis of nasopharyngeal malignancy through the initiation of the EMT event (Fig. 6.2) via EGF-R1, Akt and/or Wnt/beta-catenin signalling pathways [114–116].

Based on previous studies on the role of HPV and EBV in human oral cancer and in order to explore the cooperation effect between high-risk HPV and EBV, we are investigating the effect of E6/E7 of HPV type 16, which is the most frequent high-risk HPV type worldwide, with the LMP1 of EBV in NOE cells. Our preliminary data show that E6/E7 cooperate with LMP1 to induce cell proliferation and deregulate cell cycle progression as well as E-cadherin/catenin complex.

Finally, it is important to highlight the role of other human oral cancer risk factors such tobacco and alcohol consumption in the EMT initiation and consequently E-cadherin/catenin complex deregulation. Recent studies have pointed out that smoking could enhance the progression of several human carcinomas including oral through the down regulation of E-cadherin and the initiation of EMT [117–119]. Meanwhile, it is well known that alcohol consumption is associated with the risk of numerous progressive human cancers; thus, it has been revealed that alcohol can initiate EMT and consequently deregulate the E-cadherin/catenin complex by EGF-R/Snail mediated pathway as well as other growth factors such TGF-β [120, 121].

Collectively, these data suggest that the E-cadherin/catenin complex deregulation plays an important role in human oral carcinogenesis and metastasis via the initiation EMT, which could be affected by several oncogenes and/or onco-virus infections such as high-risk HPVs and EBV, in addition to other major risk factors of oral cancer such as tobacco and alcohol consuming.

## 6.5 Conclusion and Perspectives

This chapter presented substantial evidence that the E-cadherin/catenin complex is down regulated in human oral cancer, thereby enhancing the progression of this cancer through the initiation of the EMT event. However, we believe that further studies are required to elucidate the cadherin/catenin complex signaling pathways and their association with the EMT in human oral carcinogenesis and metastasis. In parallel, developing new *in vitro* and *in vivo* models such as cell lines and animal models are necessary to identify the exact role of each element in this complex in the EMT and the initiation and progression of human oral carcinomas which can lead to generate new targets to manage this cancer and other human carcinomas.

Alternatively and with regards to oral malignancy as well as other human carcinomas prevention, we assume that the elimination of a number of known risk factors especially tobacco, betel quid chewing, alcohol and onco-virus infections such as high-risk HPVs and EBV could diminish the development of these malignancies and diminish their metastases [98, 122, 123]. Therefore, we firmly believe that prevention of high-risk HPVs and EBV infection by using the presently available and/or upcoming vaccines could greatly reduce high-risk HPVs and EBV-associated cancers, including oral, and their progression to invasive form, which is responsible for the majority of cancer related deaths.

**Acknowledgments** We are grateful to Mrs. A. Kassab and Mr. B. Al Moustafa for their critical reading of the chapter. The research works from Dr. Al Moustafa's laboratory has been supported by the Canadian Institutes for Health Research, the Cancer Research Society Inc. of Canada, the National Colorectal Cancer Campaign and the Fonds de la Recherche en Santé du Québec (FRSQ-Réseau du Cancer).

## References

1. Parkin DM, Bray F, Ferlay J, Pisani P (2005) Global cancer statistics, 2002. CA Cancer J Clin 55(2):74–108
2. Bagan J, Sarrion G, Jimenez Y (2010) Oral cancer: clinical features. Oral Oncol 46(6):414–417
3. Noguti J, De Moura CF, De Jesus GP, Da Silva VH, Hossaka TA, Oshima CT, Ribeiro DA (2012) Metastasis from oral cancer: an overview. Cancer Genomics Proteomics 9(5):329–335
4. Al Moustafa AE, Chen D, Ghabreau L, Akil N (2009) Association between human papillomavirus and Epstein-Barr virus infections in human oral carcinogenesis. Med Hypotheses 73(2):184–186
5. Okura M, Aikawa T, Sawai NY, Iida S, Kogo M (2009) Decision analysis and treatment threshold in a management for the N0 neck of the oral cavity carcinoma. Oral Oncol 45(10):908–911
6. Fan S, Tang QL, Lin YJ, Chen WL, Li JS, Huang ZQ, Yang ZH, Wang YY, Zhang DM, Wang HJ, Dias-Ribeiro E, Cai Q, Wang L (2011) A review of clinical and histological parameters associated with contralateral neck metastases in oral squamous cell carcinoma. Int J Oral Sci 3(4):180–191
7. Lea J, Bachar G, Sawka AM, Lakra DC, Gilbert RW, Irish JC, Brown DH, Gullane PJ, Goldstein DP (2010) Metastases to level IIb in squamous cell carcinoma of the oral cavity: a systematic review and meta-analysis. Head Neck 32(2):184–190
8. Farris C, Petitte DM (2013) Head, neck, and oral cancer update. Home Healthc Nurse 31(6):322–328, quiz 328–330
9. Lambert R, Sauvaget C, de Camargo CM, Sankaranarayanan R (2011) Epidemiology of cancer from the oral cavity and oropharynx. Eur J Gastroenterol Hepatol 23(8):633–641
10. Al Moustafa AE, Kassab A, Darnel A, Yasmeen A (2008) High-risk HPV/ErbB-2 interaction on E-cadherin/catenin regulation in human carcinogenesis. Curr Pharm Des 14(22):2159–2172
11. Yasmeen A, Bismar TA, Al Moustafa AE (2006) ErbB receptors and epithelial-cadherin-catenin complex in human carcinomas. Future Oncol 2(6):765–781
12. Zaid KW (2014) Immunohistochemical assessment of E-cadherin and β-catenin in the histological differentiations of oral squamous cell carcinoma. Asian Pac J Cancer Prev 15(20):8847–8853
13. Lopes FF, da Costa Miguel MC, Pereira AL, da Cruz MC, de Almeida FR, Pinto LP, de Souza LB (2009) Changes in immunoexpression of E-cadherin and beta-catenin in oral squamous cell carcinoma with and without nodal metastasis. Ann Diagn Pathol 13(1):22–29
14. Sakamoto K, Imanishi Y, Tomita T, Shimoda M, Kameyama K, Shibata K, Sakai N, Ozawa H, Shigetomi S, Fujii R, Fujii M, Ogawa K (2012) Overexpression of SIP1 and downregulation of E-cadherin predict delayed neck metastasis in stage I/II oral tongue squamous cell carcinoma after partial glossectomy. Ann Surg Oncol 19(2):612–619
15. Al Moustafa AE, Yasmeen A, Ghabreau L, Akil N (2012) Does the Syrian population have to wait for the new generation of human papillomaviruses vaccine? Hum Vaccin Immunother 8(12):1867–1868

16. Smith A, Teknos TN, Pan Q (2013) Epithelial to mesenchymal transition in head and neck squamous cell carcinoma. Oral Oncol 49(4):287–292
17. Tsai JH, Yang J (2013) Epithelial-mesenchymal plasticity in carcinoma metastasis. Genes Dev 27(20):2192–2206
18. Van den Bossche J, Malissen B, Mantovani A, De Baetselier P, Van Ginderachter JA (2012) Regulation and function of the E-cadherin/catenin complex in cells of the monocyte-macrophage lineage and DCs. Blood 119(7):1623–1633
19. Thomas GJ, Speight PM (2001) Cell adhesion molecules and oral cancer. Crit Rev Oral Biol Med 12(6):479–498
20. Gumbiner B, Stevenson B, Grimaldi A (1988) The role of the cell adhesion molecule uvomorulin in the formation and maintenance of the epithelial junctional complex. J Cell Biol 107(4):1575–1587
21. Bershadsky A (2004) Magic touch: how does cell-cell adhesion trigger actin assembly? Trends Cell Biol 14(11):589–593
22. Yonemura S, Wada Y, Watanabe T, Nagafuchi A, Shibata M (2010) alpha-catenin as a tension transducer that induces adherens junction development. Nat Cell Biol 12(6):533–542
23. Baum B, Perrimon N (2001) Spatial control of the actin cytoskeleton in drosophila epithelial cells. Nat Cell Biol 3(10):883–890
24. Drees F, Pokutta S, Yamada S, Nelson WJ, Weis WI (2005) Alpha-catenin is a molecular switch that binds E-cadherin-beta-catenin and regulates actin-filament assembly. Cell 123(5):903–915
25. Perez-Moreno M, Jamora C, Fuchs E (2003) Sticky business: orchestrating cellular signals at adherens junctions. Cell 112(4):535–548
26. Balda MS, Matter K (2009) Tight junctions and the regulation of gene expression. Biochim Biophys Acta 1788(4):761–767
27. Bhatt T, Rizvi A, Batta SP, Kataria S, Jamora C (2013) Signaling and mechanical roles of E-cadherin. Cell Commun Adhes 20(6):189–199
28. Fagotto F (2013) Looking beyond the Wnt pathway for the deep nature of β-catenin. EMBO Rep 14(5):422–433
29. Whitehead J, Vignjevic D, Fütterer C, Beaurepaire E, Robine S, Farge E (2008) Mechanical factors activate beta-catenin-dependent oncogene expression in APC mouse colon. HFSP J 2(5):286–294
30. Yasmeen A, Alachkar A, Dekhil H, Gambacorti-Passerini C, Al Moustafa AE (2010) Locking Src/Abl tyrosine kinase activities regulate cell differentiation and invasion of human cervical cancer cells expressing E6/E7 oncoproteins of high-risk HPV. J Oncol 2010:1–10
31. le Duc Q, Shi Q, Blonk I, Sonnenberg A, Wang N, Leckband D, de Rooij J (2010) Vinculin potentiates E-cadherin mechanosensing and is recruited to actin-anchored sites within adherens junctions in a myosin II-dependent manner. J Cell Biol 189(7):1107–1115
32. Staal FJ, Luis TC, Tiemessen MM (2008) WNT signalling in the immune system: WNT is spreading its wings. Nat Rev Immunol 8(8):581–593
33. Gottardi CJ, Wong E, Gumbiner BM (2001) E-cadherin suppresses cellular transformation by inhibiting beta-catenin signaling in an adhesion-independent manner. J Cell Biol 153(5):1049–1059
34. Torres VA, Tapia JC, Rodriguez DA, Lladser A, Arredondo C, Leyton L, Quest AF (2007) E-cadherin is required for caveolin-1-mediated down-regulation of the inhibitor of apoptosis protein survivin via reduced beta-catenin-Tcf/Lef-dependent transcription. Mol Cell Biol 27(21):7703–7717
35. Maher MT, Flozak AS, Stocker AM, Chenn A, Gottardi CJ (2009) Activity of the beta-catenin phosphodestruction complex at cell-cell contacts is enhanced by cadherin-based adhesion. J Cell Biol 186(2):219–228
36. Casagolda D, Del Valle-Pérez B, Valls G, Lugilde E, Vinyoles M, Casado-Vela J, Solanas G, Batlle E, Reynolds AB, Casal JI, de Herreros AG, Duñach M (2010) A p120-catenin-CK1epsilon complex regulates Wnt signaling. J Cell Sci 123(Pt 15):2621–2631

37. Scheel C, Eaton EN, Li SH, Chaffer CL, Reinhardt F, Kah KJ, Bell G, Guo W, Rubin J, Richardson AL, Weinberg RA (2011) Paracrine and autocrine signals induce and maintain mesenchymal and stem cell states in the breast. Cell 145(6):926–940
38. De Santis G, Miotti S, Mazzi M, Canevari S, Tomassetti A (2009) E-cadherin directly contributes to PI3K/AKT activation by engaging the PI3K-p85 regulatory subunit to adherens junctions of ovarian carcinoma cells. Oncogene 28(9):1206–1217
39. Laprise P, Langlois MJ, Boucher MJ, Jobin C, Rivard N (2004) Down-regulation of MEK/ERK signaling by E-cadherin-dependent PI3K/Akt pathway in differentiating intestinal epithelial cells. J Cell Physiol 199(1):32–39
40. Xie Z, Bikle DD (2007) The recruitment of phosphatidylinositol 3-kinase to the E-cadherin-catenin complex at the plasma membrane is required for calcium-induced phospholipase C-gamma1 activation and human keratinocyte differentiation. J Biol Chem 282(12):8695–8703
41. McLachlan RW, Kraemer A, Helwani FM, Kovacs EM, Yap AS (2007) E-cadherin adhesion activates c-src signaling at cell-cell contacts. Mol Biol Cell 18(8):3214–3223
42. Fournier MV, Fata JE, Martin KJ, Yaswen P, Bissell MJ (2009) Interaction of E-cadherin and PTEN regulates morphogenesis and growth arrest in human mammary epithelial cells. Cancer Res 69(10):4545–4552
43. Lau MT, Klausen C, Leung PC (2011) E-cadherin inhibits tumor cell growth by suppressing PI3K/Akt signaling via β-catenin-Egr1-mediated PTEN expression. Oncogene 30(24):2753–2766
44. Fukuyama T, Ogita H, Kawakatsu T, Inagaki M, Takai Y (2006) Activation of Rac by cadherin through the c-Src-Rap1-phosphatidylinositol 3-kinase-Vav2 pathway. Oncogene 25(1):8–19
45. Nakagawa M, Fukata M, Yamaga M, Itoh N, Kaibuchi K (2001) Recruitment and activation of rac1 by the formation of E-cadherin-mediated cell-cell adhesion sites. J Cell Sci 114(Pt 10):1829–1838
46. Goodwin M, Kovacs EM, Thoreson MA, Reynolds AB, Yap AS (2003) Minimal mutation of the cytoplasmic tail inhibits the ability of E-cadherin to activate Rac but not phosphatidylinositol 3-kinase: direct evidence of a role for cadherin-activated Rac signaling in adhesion and contact formation. J Biol Chem 278(23):20533–20539
47. Anastasiadis PZ, Moon SY, Thoreson MA, Mariner DJ, Crawford HC, Zheng Y, Reynolds AB (2000) Inhibition of RhoA by p120 catenin. Nat Cell Biol 2(9):637–644
48. Soto E, Yanagisawa M, Marlow LA, Copland JA, Perez EA, Anastasiadis PZ (2008) p120 catenin induces opposing effects on tumor cell growth depending on E-cadherin expression. J Cell Biol 183(4):737–749
49. Asnaghi L, Vass WC, Quadri R, Day PM, Qian X, Braverman R, Papageorge AG, Lowy DR (2010) E-cadherin negatively regulates neoplastic growth in non-small cell lung cancer: role of Rho GTPases. Oncogene 29(19):2760–2771
50. Cowell CF, Yan IK, Eiseler T, Leightner AC, Döppler H, Storz P (2009) Loss of cell-cell contacts induces NF-kappaB via RhoA-mediated activation of protein kinase D1. J Cell Biochem 106(4):714–728
51. Johnson E, Theisen CS, Johnson KR, Wheelock MJ (2004) R-cadherin influences cell motility via Rho family GTPases. J Biol Chem 279(30):31041–31049
52. Solanas G, Porta-de-la-Riva M, Agustí C, Casagolda D, Sánchez-Aguilera F, Larriba MJ, Pons F, Peiró S, Escrivà M, Muñoz A, Duñach M, de Herreros AG, Baulida J (2008) E-cadherin controls beta-catenin and NF-kappaB transcriptional activity in mesenchymal gene expression. J Cell Sci 121(Pt 13):2224–2234
53. Perez-Moreno M, Davis MA, Wong E, Pasolli HA, Reynolds AB, Fuchs E (2006) p120-catenin mediates inflammatory responses in the skin. Cell 124(3):631–644
54. Stairs DB, Bayne LJ, Rhoades B, Vega ME, Waldron TJ, Kalabis J, Klein-Szanto A, Lee JS, Katz JP, Diehl JA, Reynolds AB, Vonderheide RH, Rustgi AK (2011) Deletion of p120-catenin results in a tumor microenvironment with inflammation and cancer that establishes it as a tumor suppressor gene. Cancer Cell 19(4):470–483

55. Kalluri R, Weinberg RA (2009) The basics of epithelial-mesenchymal transition. J Clin Invest 119(6):1420–1428
56. Nieto MA, Cano A (2012) The epithelial-mesenchymal transition under control: global programs to regulate epithelial plasticity. Semin Cancer Biol 22(5–6):361–368
57. Thiery JP, Acloque H, Huang RY, Nieto MA (2009) Epithelial-mesenchymal transitions in development and disease. Cell 139(5):871–890
58. Sánchez-Tilló E, Liu Y, de Barrios O, Siles L, Fanlo L, Cuatrecasas M, Darling DS, Dean DC, Castells A, Postigo A (2012) EMT-activating transcription factors in cancer: beyond EMT and tumor invasiveness. Cell Mol Life Sci 69(20):3429–3456
59. Perez-Moreno MA, Locascio A, Rodrigo I, Dhondt G, Portillo F, Nieto MA, Cano A (2001) A new role for E12/E47 in the repression of E-cadherin expression and epithelial-mesenchymal transitions. J Biol Chem 276(29):27424–27431
60. Rodriguez M, Aladowicz E, Lanfrancone L, Goding CR (2008) Tbx3 represses E-cadherin expression and enhances melanoma invasiveness. Cancer Res 68(19):7872–7881
61. Hartwell KA, Muir B, Reinhardt F, Carpenter AE, Sgroi DC, Weinberg RA (2006) The Spemann organizer gene, Goosecoid, promotes tumor metastasis. Proc Natl Acad Sci U S A 103(50):18969–18974
62. Thuault S, Valcourt U, Petersen M, Manfioletti G, Heldin CH, Moustakas A (2006) Transforming growth factor-beta employs HMGA2 to elicit epithelial-mesenchymal transition. J Cell Biol 174(2):175–183
63. Baum B, Georgiou M (2011) Dynamics of adherens junctions in epithelial establishment, maintenance, and remodeling. J Cell Biol 192(6):907–917
64. Capaldo CT, Macara IG (2007) Depletion of E-cadherin disrupts establishment but not maintenance of cell junctions in Madin-Darby canine kidney epithelial cells. Mol Biol Cell 18(1):189–200
65. Schaale K, Neumann J, Schneider D, Ehlers S, Reiling N (2011) Wnt signaling in macrophages: augmenting and inhibiting mycobacteria-induced inflammatory responses. Eur J Cell Biol 90(6–7):553–559
66. Iorio MV, Croce CM (2012) MicroRNA dysregulation in cancer: diagnostics, monitoring and therapeutics. A comprehensive review. EMBO Mol Med 4(3):143–159
67. Onder TT, Gupta PB, Mani SA, Yang J, Lander ES, Weinberg RA (2008) Loss of E-cadherin promotes metastasis via multiple downstream transcriptional pathways. Cancer Res 68(10):3645–3654
68. Batlle E, Sancho E, Francí C, Domínguez D, Monfar M, Baulida J, García De Herreros A (2000) The transcription factor snail is a repressor of E-cadherin gene expression in epithelial tumour cells. Nat Cell Biol 2(2):84–89
69. Moreno-Bueno G, Cubillo E, Sarrió D, Peinado H, Rodríguez-Pinilla SM, Villa S, Bolós V, Jordá M, Fabra A, Portillo F, Palacios J, Cano A (2006) Genetic profiling of epithelial cells expressing E-cadherin repressors reveals a distinct role for Snail, Slug, and E47 factors in epithelial-mesenchymal transition. Cancer Res 66(19):9543–9556
70. Vandewalle C, Comijn J, De Craene B, Vermassen P, Bruyneel E, Andersen H, Tulchinsky E, Van Roy F, Berx G (2005) SIP1/ZEB2 induces eMT by repressing genes of different epithelial cell-cell junctions. Nucleic Acids Res 33(20):6566–6578
71. Yang J, Mani SA, Donaher JL, Ramaswamy S, Itzykson RA, Come C, Savagner P, Gitelman I, Richardson A, Weinberg RA (2004) Twist, a master regulator of morphogenesis, plays an essential role in tumor metastasis. Cell 117(7):927–939
72. Friedl P, Alexander S (2011) Cancer invasion and the microenvironment: plasticity and reciprocity. Cell 147(5):992–1009
73. Hanahan D, Coussens LM (2012) Accessories to the crime: functions of cells recruited to the tumor microenvironment. Cancer Cell 21(3):309–322
74. Bierie B, Moses HL (2006) Tumour microenvironment: tGFbeta: the molecular Jekyll and Hyde of cancer. Nat Rev Cancer 6(7):506–520

75. Giampieri S, Manning C, Hooper S, Jones L, Hill CS, Sahai E (2009) Localized and reversible TGFbeta signalling switches breast cancer cells from cohesive to single cell motility. Nat Cell Biol 11(11):1287–1296
76. Mani SA, Guo W, Liao MJ, Eaton EN, Ayyanan A, Zhou AY, Brooks M, Reinhard F, Zhang CC, Shipitsin M, Campbell LL, Polyak K, Brisken C, Yang J, Weinberg RA (2008) The epithelial-mesenchymal transition generates cells with properties of stem cells. Cell 133(4):704–715
77. Brabletz T, Jung A, Spaderna S, Hlubek F, Kirchner T (2005) Opinion: migrating cancer stem cells – an integrated concept of malignant tumour progression. Nat Rev Cancer 5(9):744–749
78. Magee JA, Piskounova E, Morrison SJ (2012) Cancer stem cells: impact, heterogeneity, and uncertainty. Cancer Cell 21(3):283–296
79. Brabletz T (2012) To differentiate or not–routes towards metastasis. Nat Rev Cancer 12(6):425–436
80. Brabletz T, Jung A, Reu S, Porzner M, Hlubek F, Kunz-Schughart LA, Knuechel R, Kirchner T (2001) Variable beta-catenin expression in colorectal cancers indicates tumor progression driven by the tumor environment. Proc Natl Acad Sci U S A 98(18):10356–10361
81. Rhim AD, Mirek ET, Aiello NM, Maitra A, Bailey JM, McAllister F, Reichert M, Beatty GL, Rustgi AK, Vonderheide RH, Leach SD, Stanger BZ (2012) EMT and dissemination precede pancreatic tumor formation. Cell 148(1–2):349–361
82. Amagasa T, Yamashiro M, Uzawa N (2011) Oral premalignant lesions: from a clinical perspective. Int J Clin Oncol 16(1):5–14
83. Kumar A, Cascarini L, McCaul JA, Kerawala CJ, Coombes D, Godden D, Brennan PA (2013) How should we manage oral leukoplakia? Br J Oral Maxillofac Surg 51(5):377–383
84. Al Moustafa AE, Alaoui-Jamali MA, Batist G, Hernandez-Perez M, Serruya C, Alpert L, Black MJ, Sladek R, Foulkes WD (2002) Identification of genes associated with head and neck carcinogenesis by cDNA microarray comparison between matched primary normal epithelial and squamous carcinoma cells. Oncogene 21(17):2634–2640
85. Al Moustafa AE, Yen L, Benlimame N, Alaoui-Jamali MA (2002) Regulation of E-cadherin/catenin complex patterns by epidermal growth factor receptor modulation in human lung cancer cells. Lung Cancer 37(1):49–56
86. Al Moustafa AE, Foulkes WD, Benlimame N, Wong A, Yen L, Bergeron J, Batist G, Alpert L, Alaoui-Jamali MA (2004) E6/E7 proteins of HPV type 16 and ErbB-2 cooperate to induce neoplastic transformation of primary normal oral epithelial cells. Oncogene 23(2):350–358
87. Al Moustafa AE, Foulkes WD, Wong A, Jallal H, Batist G, Yu Q, Herlyn M, Sicinski P, Alaoui-Jamali MA (2004) Cyclin D1 is essential for neoplastic transformation induced by both E6/E7 and E6/E7/ErbB-2 cooperation in normal cells. Oncogene 23(30):5252–5256
88. Yasmeen A, Bismar TA, Dekhil H, Ricciardi R, Kassab A, Gambacorti-Passerini C, Al Moustafa AE (2007) ErbB-2 receptor cooperates with E6/E7 oncoproteins of HPV type 16 in breast tumorigenesis. Cell Cycle 6(23):2939–2943
89. Yasmeen A, Hosein AN, Yu Q, Al Moustafa AE (2007) Critical role for D-type cyclins in cellular transformation induced by E6/E7 of human papillomavirus type 16 and E6/E7/ErbB-2 cooperation. Cancer Sci 98(7):973–977
90. Ragin CC, Modugno F, Gollin SM (2007) The epidemiology and risk factors of head and neck cancer: a focus on human papillomavirus. J Dent Res 86(2):104–114
91. Termine N, Panzarella V, Falaschini S, Russo A, Matranga D, Lo Muzio L, Campisi G (2008) HPV in oral squamous cell carcinoma vs head and neck squamous cell carcinoma biopsies: a meta-analysis (1988–2007). Ann Oncol 19(10):1681–1690
92. Doorbar J (2013) Latent papillomavirus infections and their regulation. Curr Opin Virol 3(4):416–421
93. Moody CA, Laimins LA (2010) Human papillomavirus oncoproteins: pathways to transformation. Nat Rev Cancer 10(8):550–560

94. Kim SH, Juhnn YS, Kang S, Park SW, Sung MW, Bang YJ, Song YS (2006) Human papillomavirus 16 E5 up-regulates the expression of vascular endothelial growth factor through the activation of epidermal growth factor receptor, MEK/ ERK1,2 and PI3K/Akt. Cell Mol Life Sci 63(7–8):930–938
95. Suprynowicz FA, Disbrow GL, Krawczyk E, Simic V, Lantzky K, Schlegel R (2008) HPV-16 E5 oncoprotein upregulates lipid raft components caveolin-1 and ganglioside GM1 at the plasma membrane of cervical cells. Oncogene 27(8):1071–1078
96. Stacey SN, Jordan D, Williamson AJ, Brown M, Coote JH, Arrand JR (2000) Leaky scanning is the predominant mechanism for translation of human papillomavirus type 16 E7 oncoprotein from E6/E7 bicistronic mRNA. J Virol 74(16):7284–7297
97. Ghittoni R, Accardi R, Hasan U, Gheit T, Sylla B, Tommasino M (2010) The biological properties of E6 and E7 oncoproteins from human papillomaviruses. Virus Genes 40(1):1–13
98. Al Moustafa AE, Al-Awadhi R, Missaoui N, Adam I, Durusoy R, Ghabreau L, Akil N, Ahmed HG, Yasmeen A, Alsbeih G (2014) Human papillomaviruses-related cancers: presence and prevention strategies in the middle east and north African regions. Hum Vaccin Immunother 10(7):1812–1821
99. Ng RH, Ngan R, Wei WI, Gullane PJ, Phillips J (2014) Trans-oral brush biopsies and quantitative PCR for EBV DNA detection and screening of nasopharyngeal carcinoma. Otolaryngol Head Neck Surg 150(4):602–609
100. Slots J, Saygun I, Sabeti M, Kubar A (2006) Epstein-Barr virus in oral diseases. J Periodontal Res 41(4):235–244
101. Whitney BM, Chan AT, Rickinson AB, Lee SP, Lin CK, Johnson PJ (2002) Frequency of Epstein-Barr virus-specific cytotoxic T lymphocytes in the blood of Southern Chinese blood donors and nasopharyngeal carcinoma patients. J Med Virol 67(3):359–363
102. Kulwichit W, Edwards RH, Davenport EM, Baskar JF, Godfrey V, Raab-Traub N (1998) Expression of the Epstein-Barr virus latent membrane protein 1 induces B cell lymphoma in transgenic mice. Proc Natl Acad Sci U S A 95(20):11963–11968
103. Murata T, Tsurumi T (2013) Epigenetic modification of the Epstein-Barr virus BZLF1 promoter regulates viral reactivation from latency. Front Genet 4:53
104. Young LS, Rickinson AB (2004) Epstein-Barr virus: 40 years on. Nat Rev Cancer 4(10):757–768
105. Münz C, Moormann A (2008) Immune escape by Epstein-Barr virus associated malignancies. Semin Cancer Biol 18(6):381–387
106. Thompson MP, Kurzrock R (2004) Epstein-Barr virus and cancer. Clin Cancer Res 10(3):803–821
107. Amarante MK, Watanabe MA (2009) The possible involvement of virus in breast cancer. J Cancer Res Clin Oncol 135(3):329–337
108. Michelow P, Wright C, Pantanowitz L (2012) A review of the cytomorphology of Epstein-Barr virus-associated malignancies. Acta Cytol 56(1):1–14
109. Dawson CW, Port RJ, Young LS (2012) The role of the EBV-encoded latent membrane proteins LMP1 and LMP2 in the pathogenesis of nasopharyngeal carcinoma (NPC). Semin Cancer Biol 22(2):144–153
110. Morris MA, Dawson CW, Young LS (2009) Role of the Epstein-Barr virus-encoded latent membrane protein-1, LMP1, in the pathogenesis of nasopharyngeal carcinoma. Future Oncol 5(6):811–825
111. Endo K, Kondo S, Shackleford J, Horikawa T, Kitagawa N, Yoshizaki T, Furukawa M, Zen Y, Pagano JS (2009) Phosphorylated ezrin is associated with EBV latent membrane protein 1 in nasopharyngeal carcinoma and induces cell migration. Oncogene 28(14):1725–1735
112. Mazouni C, Fina F, Romain S, Ouafik L, Bonnier P, Brandone JM, Martin PM (2011) Epstein-Barr virus as a marker of biological aggressiveness in breast cancer. Br J Cancer 104(2):332–337

113. Pegtel DM, Subramanian A, Sheen TS, Tsai CH, Golub TR, Thorley-Lawson DA (2005) Epstein-Barr-virus-encoded LMP2A induces primary epithelial cell migration and invasion: possible role in nasopharyngeal carcinoma metastasis. J Virol 79(24):15430–15442
114. Kong QL, Hu LJ, Cao JY, Huang YJ, Xu LH, Liang Y, Xiong D, Guan S, Guo BH, Mai HQ, Chen QY, Zhang X, Li MZ, Shao JY, Qian CN, Xia YF, Song LB, Zeng YX, Zeng MS (2010) Epstein-Barr virus-encoded LMP2A induces an epithelial-mesenchymal transition and increases the number of side population stem-like cancer cells in nasopharyngeal carcinoma. PLoS Pathog 6(6), e1000940
115. QingLing Z, LiNa Y, Li L, Shuang W, YuFang Y, Yi D, Divakaran J, Xin L, YanQing D (2011) LMP1 antagonizes WNT/β-catenin signalling through inhibition of WTX and promotes nasopharyngeal dysplasia but not tumourigenesis in LMP1(B95-8) transgenic mice. J Pathol 223(5):574–583
116. Raab-Traub N (2012) Novel mechanisms of EBV-induced oncogenesis. Curr Opin Virol 2(4):453–458
117. Milara J, Peiró T, Serrano A, Cortijo J (2013) Epithelial to mesenchymal transition is increased in patients with COPD and induced by cigarette smoke. Thorax 68(5):410–420
118. Nagathihalli NS, Massion PP, Gonzalez AL, Lu P, Datta PK (2012) Smoking induces epithelial-to-mesenchymal transition in non-small cell lung cancer through HDAC-mediated downregulation of E-cadherin. Mol Cancer Ther 11(11):2362–2372
119. Yu CC, Chang YC (2013) Enhancement of cancer stem-like and epithelial-mesenchymal transdifferentiation property in oral epithelial cells with long-term nicotine exposure: reversal by targeting SNAIL. Toxicol Appl Pharmacol 266(3):459–469
120. Forsyth CB, Tang Y, Shaikh M, Zhang L, Keshavarzian A (2011) Role of snail activation in alcohol-induced iNOS-mediated disruption of intestinal epithelial cell permeability. Alcohol Clin Exp Res 35(9):1635–1643
121. Giannelli G, Villa E, Lahn M (2014) Transforming growth factor-β as a therapeutic target in hepatocellular carcinoma. Cancer Res 74(7):1890–1894
122. Al Moustafa AE, Achkhar A, Yasmeen A (2012) EGF-receptor signaling and epithelial-mesenchymal transition in human carcinomas. Front Biosci (Schol Ed) 4:671–684
123. Al Moustafa AE, Ghabreau L, Akil N, Rastam S, Alachkar A, Yasmeen A (2014) High-risk HPVs and human carcinomas in the Syrian population. Front Oncol 4:68
124. Shtutman M, Zhurinsky J, Simcha I, Albanese C, D'Amico M, Pestell R, Ben-Ze'ev A (1999) The cyclin D1 gene is a target of the beta-catenin/LEF-1 pathway. Proc Natl Acad Sci U S A 96(10):5522–5527
125. Lin SY, Xia W, Wang JC, Kwong KY, Spohn B, Wen Y, Pestell RG, Hung MC (2000) Beta-catenin, a novel prognostic marker for breast cancer: its roles in cyclin D1 expression and cancer progression. Proc Natl Acad Sci U S A 97(8):4262–4266

# Chapter 7
# Tumor-Derived Exosomes in Metastasis Signaling and Implications

**Yuan Hang Cao, Krikor Bijian, Maisa I. Alkailani, and Moulay A. Alaoui-Jamali**

**Abstract** Genome-wide studies for decoding molecular mechanisms of tumor cell-host interactions associated with cancer progression have revealed a determinant role for exosome-mediated intercellular communications in metastasis signaling. Exosomes are specialized plasma membrane-bound nanovesicles secreted under physiological conditions by various cells to regulate tissue homeostasis and immune response. In cancer, overproduction of exosomes and/or changes in their contents are common features of many cancer cell types and can bear unique molecular signatures reminiscent to the cancer phenotype. Exosomes mediate paracrine signaling required for cross-talks between tumor cells and the host stromal and infiltrating inflammatory and immune cells within the tumor tissue microenvironment as well as distant metastatic niches. Exosomes promote cancer invasiveness at multiple levels of the metastatic cascades favoring a permissive environment for cell survival and growth. This review focuses specifically on the implication of cancer-associated exosomes for cell signaling that promotes metastasis development. For detailed biochemical and molecular studies addressing exosome biology, we refer the reader to selected seminal studies in the field.

**Keywords** Cancer • Metastasis • Exosomes • Therapeutics

## Abbreviation

| | |
|---|---|
| Alix | ALG-2-interacting protein X |
| c-Met | Hepatocyte growth factor receptor |
| EGFR | Epidermal growth factor receptor |
| FasL | Fas ligand |
| HER-2 | Human epidermal growth factor receptor 2 |

Y.H. Cao • K. Bijian • M.I. Alkailani • M.A. Alaoui-Jamali (✉)
Faculty of Medicine, Departments of Medicine and Oncology, Segal Cancer Center and Lady Davis Institute for Medical Research of the Jewish General Hospital, McGill University, Montreal, Canada
e-mail: moulay.alaoui-jamali@mcgill.ca



M. Kandouz (ed.), *Intercellular Communication in Cancer*,
DOI 10.1007/978-94-017-7380-5_7

| | |
|---|---|
| HIF-1α | Hypoxia-inducible factor 1-alpha |
| hnRNPA2B1 | Heterogeneous nuclear ribonucleo protein A2B1 |
| ICAM1 | Intercellular adhesion molecule 1 |
| LAMP | Lysosome-associated membrane protein |
| LFA1 | Lymphocyte function-associated antigen 1 |
| Mdh1 | Malate dehydrogenase 1 |
| MDR1 | Multidrug resistance protein 1 |
| MFGE8 | Milk fat globule-EGF factor 8 protein |
| MMP | Matrix metallopeptidase |
| MVB | Multivesicular bodies |
| PD-1 | Programmed cell death protein 1 |
| Pgm1 | Phosphoglucomutase |
| PK | Pyruvate kinases |
| TGF-β | Transforming growth factor beta |
| TGF-βR | Transforming growth factor beta receptor |
| TNFα | Tumor necrosis factor-alpha |
| Tsg101 | Tumor susceptibility gene 101 protein |
| VEGF | Vascular endothelial growth factor |
| VEGR | Vascular endothelial growth factor receptor |

## 7.1 Exosome Biogenesis and Composition

Exosomes are small plasma membrane nanovesicles of 30–100 nm in diameter secreted by virtually every cell type and can be detected in most body fluids e.g. blood, breast milk and cerebrospinal fluid, urine and saliva. They are distinct from other cell membrane-shed vesicles such as microvesicles and apoptotic bodies and can be easily distinguished/isolated by electron microscopy, size distribution via sucrose gradient ultracentrifugation and nanoparticle tracking analysis (Fig. 7.1).

Exosomes are formed in the late endocytic cell compartment by inward/reverse "budding" of the lining endosomal membrane leading to the formation of MVB, initiated by the endosomal sorting complex for transport (ESCRT) [1, 2] (Fig. 7.1). Ubiquitinated proteins transported by the Trans-Golgi Network (TGN), or by inducible vesicular trafficking, are recruited to the surface of endosomes by ESCRT0 to form a complex with ESCRTI and ESCRTII and then trigger inward budding, allowing RNA and proteins to be delivered inside the vesicle. The vesicle is then cleaved through the recruitment of ESCRTIII to become an intraluminal vesicle (ILV) inside the MVB. Proteins in the ESCRT complex like Alix101 and TSG101 can be incorporated into ILV during this process. MVBs can also be formed through alternative mechanisms such as creamide-induced vesicle formation. The conversion of sphingolipids into ceramide by sphingomyelinases on endosome membranes triggers inward budding and leads to the formation of intraluminal vesicles [3]. Once formed, MVBs can fuse with the plasma membrane to release their intraluminal vesicles in

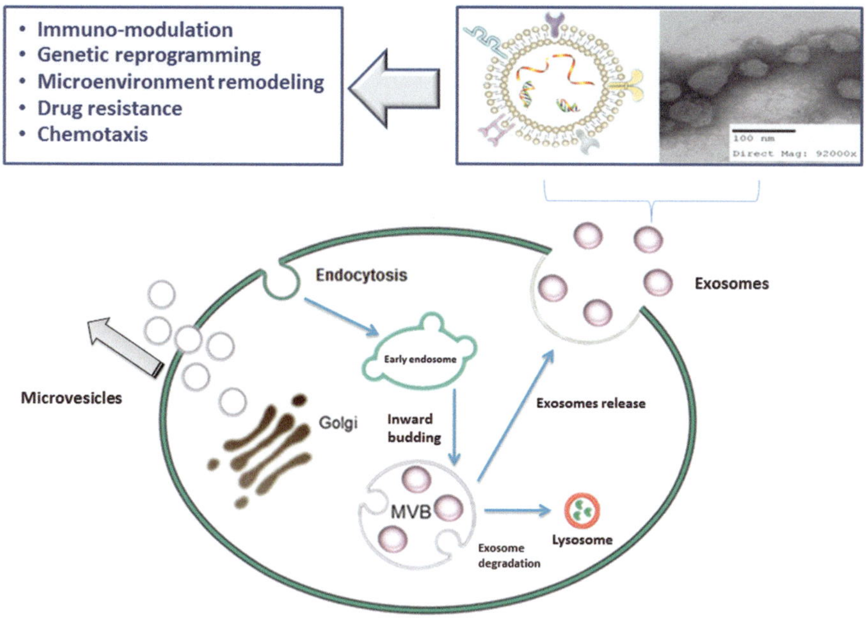


**Fig. 7.1 Exosomesbiogenesis.** These small extracellular vesicles are able to transport proteins and genetic materials (RNA, DNA and miRNA) between cells. They are formed through inward budding of early endosome, within MVB (multivesicular bodies). Depending on the composition of MVBs, they can undergo degradation by fusing with lysosomes or released into the extracellular space

the form of exosomes into the extracellular space, or alternatively fuse with lysosomes for degradation [4, 5].

Despite their small size, along with the modest thickness of their phospholipid bilayer (≈5 nm), over 4000 proteins have been identified in exosomes (www.exoCarta) and the protein composition can vary greatly between cell types and pathological conditions (Fig. 7.2) [6]. Prominient exosome proteins include several tetraspannins, heat shock proteins, MVB biogenesis proteins (Alix, and TSG101) and members of the Rab GTPase family involved in MVB trafficking and release [7–9]. Many of these proteins such as tetraspannins CD63 and CD81 and heat shock protein 70 are used to confirm exosome identity [10, 11].

In addition to proteins, exosomes mediate the transfer of RNA molecules from host to recipient cells. This transfer of genetic RNA material was initially demonstrated in mast cell-derived exosomes where microarray analysis identified a total of 1300 genes [12]. Moreover, the presence of small-RNAs and micro-RNAs (miRNA) have been identified in exosomal RNA extracts. These exosome-associated RNA molecules have been referred to as exosomal shuttle RNA (esRNA) [12], and have been identified in exosomes from a variety of normal and cancer cells, as well as biological fluids from cancer patients [13–17]. The transfer of esRNA occurs

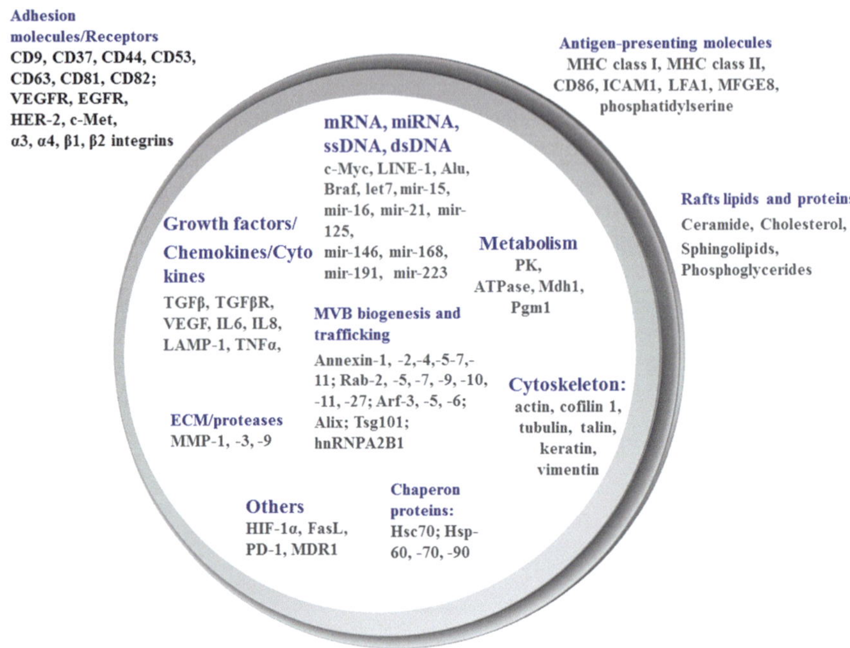


**Fig. 7.2 Composition ofexosomes.** The content of exosomes depends on the origin of the host cells from which they derive. In general, endosome specific proteins, lipids and proteins involved in MVB generation are enriched in exosomes. Many cell surface receptors and their ligands can also be found on exosomes, acting as signaling molecules in exosome mediated cell-cell interactions. Functional messenger RNA, microRNA, ssDNA and dsDNA can also be found in exosomes

through fusion of exosomes with the cell membrane of recipient cells and subsequent release of the exosomal content into the cytosol, which has been very well documented showing the transfer of luciferin from the exosomal lumen to recipient dendritic cells [18]. At present, over 764 miRNAs, 2300 mRNAs and some rare ribosomal RNAs have been identified in exosomes; most are listed on ExoCarta, the online database for esRNA [19]. This broad content of genetic material in exosomes highlights their potential impact on the regulation of gene expression in recipient cells.

## 7.2 Exosomes Implication in Metastasis Signaling

During the metastatic process, cancer cell-host interaction is conceptually referred to as the "seed and soil" relationship, where cancer cells represent the "seed" and their receptive microenvironment the "soil" [20]. This interaction is a determining factor for tumor progression to metastasis. In this context, cancer cell-derived exosomes play a key role, both supportive and regulatory, in metastasis signaling [21]. First, circulating exosomes are capable of attaching to the surface of neighboring or distant recipient cells such as by binding to heparin sulfate proteoglycan (HSPG) receptors and then incorporate their content through endocytosis [15, 22, 23] to activate a broad range of mechanisms crucial to the metastatic cascade. These include remodeling of the tumor microenvironment, transition of epithelial to mesenchymal state or the reverse mesenchymal to epithelial state, homing and survival of disseminated circulating tumor cells in distant sites, reactivation of dormant cancer cells and induction of a pro-inflammatory response.

### 7.2.1 *Influence of Exosomes on Epithelial-Mesenchymal Transition (EMT)*

Mounting evidence support a role of exosomes in the activation of EMT [24–26], a process crucial for the selection of cancer cells with higher plasticity and metastatic activity [27–29]. For example, exosomes derived from nasopharyngeal carcinoma have been shown to regulate EMT features such as loss of E-cadherin and gain of N-cadherin in part via the transfer of hypoxia-inducible factor-1α (HIF1α) [30]. Incubation of TGFβ- positive exosomes has also been reported to trigger EMT in gastric cells [31]. This suggests that other inducers of EMT may be found in tumor-derived exosomes, including β-catenin and vimentin, which can further cooperate to induce EMT in recipient epithelial cancer cells [32, 33]. As such, Luga et al. [34] was able to demonstrate the ability of fibroblasts-derived exosomes to promote breast cancer cell invasiveness via activation of Wnt-planar cell polarity (PCP) signaling involving surface-bound Wnt11. Other exosome-associated factors such MMPs, VEGF and TGF-β have also been identified to facilitate EMT and metastasis via induction of extracellular matrix remolding, angiogenesis and stromal differentiation, respectively [35–38].

### 7.2.2 *Modulation of Immune and Inflammatory Response*

An important mechanism by which exosomes promote metastasis is through modulation of the immune and inflammatory responses. Tumor-derived exosomes have been implicated in the disruption of immune-surveillance through activation of immune suppressor cells such as myeloid-derived suppressor cells (MDSC) and inhibition of natural killer (NK) cells [39]. As such, Ochao et al. were able to demonstrate that exosomes released from renal cell carcinoma were found enriched in cytokine-like TGF-beta and prostaglandin E2, promoting MDSC differentiation [40]. Exosomal Hsp72 [41] and MyD88 [42] were also reported to be involved in exosomes-mediated MDSC activation and expansion. In addition to MDSC activation, inhibition of NK cells is crucial for cancer cells to survive under immune-surveillance. This was demonstrated by Liu et al., where administration of murine mammary carcinoma-derived exosomes to mice bearing tumors lead to NK cells inhibition by altering Jak3 signaling and cyclin D3 and IL-2 level [43]. Tumor-derived exosomes have also been reported to trigger apoptosis of activated T cells, thereby suppressing the immune response [44] and interfering directly with dendritic cell differentiation [45]. They have also been shown to transform $CD14^+$ monocyte into HLA-DR ($^{-/low}$) cells that lack MHC class II for antigen presentation and release immunosuppressive cytokines (IL-6, TGF-beta) to inhibit activated T cells [45]. Together these findings highlight the essential role of tumor-derived exosomes in evading the immune system, a key event in metastasis progression.

### 7.2.3 Modulation of Cell Signaling through the Transfer of Genetic Material

Direct transfer of circulating genetic material is an emerging mode of cell-to-cell communication. Circulating miRNAs released from cells can be detected in body fluids in the form of miRNA/protein complex that protects miRNA from degradation [46]. Compared to circulating miRNA, exosomes can carry more RNA molecules to distant tumor sites, providing enhanced stability and resistance to unfavourable conditions [47]. Moreover, instead of random uptake, exosomes can mediate directed, cell type specific RNA delivery via cell surface receptor recognition [23]. In addition, sorting of miRNA into exosomes is regulated by sumoylated heterogeneous nuclear ribonucleoprotein A2B1 (hnRNPA2B1), through motif recognition [48]. Several studies support the crucial role of exosomes in regulating gene expression at distant sites through the transfer of genetic material from host cells. For instance, Skog et al. demonstrated that transfer of EGFRv3 mRNA by exosomes to recipient cells led to the production of functional EGFRv3 protein and metastasis signals [15]. Moreover, exosomes derived from Epstein-Barr virus transformed B lymphocytes [49] and lentiviral vector short hairpin RNA transformed epithelial cells [50], were shown to deliver miRNA to recipient cells and cause significant decreases in gene expression and protein activity through RNA interference. Therefore, due to their significant contribution to the regulation of gene expression, cancer derived exosomal miRNAs were heavily investigated for their influence on metastasis. In this context, Zhang et al. demonstrated that the direct transfer of exosomes carried miR-150 from cancer to human mammary epithelial cell, which led to the inhibition of c-Myb expression and caused an increased cell migration [51]. Moreover, exosomes secreted from IL4-activated macrophages promoted increased invasiveness of SKBR3 and MDA-MB-231 breast cancer cells, which was mediated by the exosome delivered miR-223, known to target *Mef2c* for degradation and to activate nuclear β-catenin levels [52]. Surprisingly, exosomal miRNA can also regulate the immune response by binding immune cell receptors triggering inflammatory reactions. This has clearly been demonstrated by Fabbri M et al. [53], where the exosome-mediated transfer of miR-21 and miR-29a bound and activated Toll-like receptors in mouse (*TLR7*) and human (*TLR8*) immune cells, leading to NF-kB activation, secretion of both TNFα and IL-6, and activation of CD69 promoting cancer progression [53].

## 7.3 Exosomes as a Source of Tumor Biomarkers and Therapeutic Targets

Taking into consideration that the content of circulating exosomes generally mirrors the molecular properties of the tumor of origin, exosomes represent an ideal source for discovery of diagnostic and therapeutic targets. Several studies have shown that protein markers detected in exosomes in blood and urine fluids from patients with melanoma, ovarian carcinoma, prostate, and kidney cancers have predictive values of aggressiveness [54–57]. An interesting example is the identification of HER2/neu tyrosine kinase receptor on exosomes derived from $HER2^+$ breast cancer [58, 59]. These $HER2^+$ exosomes are found to induce growth-stimulating signals to neighboring stromal cells at both primary and distant sites. In addition, Ciravolo et al. [60] demonstrated that the fact that exosomes display transmembrane receptors such as HER2 on their surface, this allows them to scavenge circulating anti-HER2 therapeutics such as Herceptin and hence reduce its bioavailability to cancer cells.

Tumor associated antigens carried by exosomes can also serve as potential biomarkers for prediction of cancer progression. In this context, Glod4, a protein with a glyoxalase-domain, has been identified as a promising auto-antigen for colorectal cancer biomarkers based on immunoscoring analysis [61]. As well, exosome-associated microRNAs have a great potential to serve as biomarkers to predict cancer progression and recurrence (reviewed in [62]).

As for therapeutic targets, several studies have attempted to deplete circulating cancer-associated exosomes to interfere with metastasis signaling, such as using blood exosomes-binding lectins coupled to antibody [63]. In contrast, synthetic exosomes have also been tested to deliver therapeutics to brain tissue through the blood-cerebrospinal fluid barrier [64]. Through the addition of targeting ligands on the exosome surface, it is possible to refine exosomes to deliver therapeutic molecules to specific cell types or compartment [65, 66]. In this context, Alvarez-Erviti et al. were able to deliver BACE1 siRNA-containing exosomes to the brain by fusing the neuronal targeting ligand protein with the exosome membrane [65].

## 7.4 Summary and Perspectives

Exosomes are considered as an important component of cell-cell communication in cancer progression and metastasis. Their ability to transport tumor cell derived material is crucial for the establishment of a proper tumor microenvironment and progression toward metastasis. Tumor derived exosomes are also able to disrupt the immune system by altering immune cell behavior and promoting immune system escapes. Notably, exosomes are also produced in healthy dendritic cell, where they mediate an anti-tumor response by presenting tumor specific antigen to immune cells [47]. Exosomes secreted by dendritic cells maintain the capacity to interact with T lymphocytes through MHC II complexes located on the exosomal surfaces [36–38]. These MHC II-positive exosomes bind to MHC II negative dendritic cells stimulating T cells with their carried antigens. Antigen loaded exosomes can also interact with B cells and trigger an antibody response and activate T-cell proliferation [39]. Due to their antigen-presentation function, it is still debatable whether complete removal of exosomes from human body fluids is beneficial for the patients. Although the clinical implication of exosomes in the field of oncology remains to be established, exciting preclinical studies support tumor cell dependence on exosome-mediated cross-signaling, which under certain circumstances may create vulnerabilities that can be exploited to identify innovative therapeutics for advanced tumors.

Commercial exosome extraction kits (ExoQuick$^{tm}$, ExoSpin$^{tm}$, PureEXO$^{tm}$) have been developed to meet the growing research need for studying and exploring exosomes. These kits concentrate and precipitate exosomes through immune-capture by targeting several well known tetraspanins and heat shock proteins that are enriched in exosomes. However, the quality of exosome extracts varies between kits and often contain impurities generated from precipitation steps. Alternative extraction methods include immuno-bead capture and ultracentrifugation, both having their own limitations in term of sensitivity, yield and sample volume. Moreover, the expression level of commonly used exosome protein markers differs between cell type, body fluid and stress conditions. Due to their prodigious release into body fluids such as blood and saliva, and the highly heterogeneous nature of exosomes, standardized and reproducible methods for exosome capturing and enrichment are of considerable scientific and commercial interest.

## References

1. Raiborg C, Stenmark H (2009) The ESCRT machinery in endosomal sorting of ubiquitylated membrane proteins. Nature 458:445–452
2. Hurley JH, Hanson PI (2010) Membrane budding and scission by the ESCRT machinery: it's all in the neck. Nature Rev Mol Cell Biol 11:556–566
3. Trajkovic K, Hsu C, Chiantia S, Rajendran L, Wenzel D, Wieland F, Schwille P, Brügger B, Simons M (2008) Ceramide triggers budding of exosome vesicles into multivesicular endosomes. Science 319(5867):1244–1247. doi:10.1126/science.1153124
4. Denzer K, Kleijmeer MJ, Stoorvodel W, Heijnen HF, Geuze HJ (2000) Exosome: from internal vesicle of the multivesicular body to intercellular signaling device. J Cell Sci 113:3365–3374
5. Record M, Subra C, Silvente-Poirot S, Poirot M (2011) Exosomes as intercellular signalosomes and pharmacological effectors. Biochem Pharmacol 81:1171–1182
6. Simpson RJ, Jensen SS, Lim JW (2008) Proteomic profiling of exosomes: current perspectives. Proteomics 8(19):4083–4099. doi:10.1002/pmic.200800109
7. Ostrowski M, Carmo NB, Krumeich S, Fanget I, Raposo G, Savina A, Moita CF, Schauer K, Hume AN, Freitas RP, Goud B, Benaroch P, Hacohen N, Fukuda M, Desnos C, Seabra MC,

Darchen F, Amigorena S, Moita LF, Thery C (2010) Rab27a and Rab27b control different steps of the exosome secretion pathway. Nat Cell Biol 12(1):19–30. doi:10.1038/ncb2000
8. Savina A, Fader CM, Damiani MT, Colombo MI (2005) Rab11 promotes docking and fusion of multivesicular bodies in a calcium-dependent manner. Traffic 6:131–143
9. Lebrand C, Corti M, Goodson H, Cosson P, CavalliV MN, Faure J, Gruenberg J (2002) Late endosome motility depends on lipids via the small GTPase Rab7. EMBO J 21:1289–1300
10. Conde-Vancells J, Rodriguez-Suarez E, Embade N, Gil D, Matthiesen R, Valle M, Elortza F, Lu SC, Mato JM, Falcon-Perez JM (2008) Characterization and comprehensive proteome profiling of exosomes secreted by hepatocytes. J Proteome Res 7(12):5157–5166
11. Raposo G, Stoorvogel W (2013) Extracellular vesicles:exosomes, microvesicles, and friends. J Cell Biol 200:373–383
12. Valadi H et al (2007) Exosome-mediated transfer of mRNAs and microRNAs is a novel mechanism of genetic exchange between cells. Nat Cell Biol 9(6):654–659
13. Taylor DD, Gercel-Taylor C (2008) MicroRNA signatures of tumor-derived exosomes as diagnostic biomarkers of ovarian cancer. Gynecol Oncol 110(1):13–21
14. Lehmann BD et al (2008) Senescence-associated exosome release from human prostate cancer cells. Cancer Res 68(19):7864–7871
15. Skog J et al (2008) Glioblastoma microvesicles transport RNA and proteins that promote tumour growth and provide diagnostic biomarkers. Nat Cell Biol 10(12):1470–1476
16. Lässer C, Alikhani VS, Ekström K et al (2011) Human saliva, plasma and breast milk exosomes contain RNA: uptake by macrophages. J Transl Med 9:9
17. Kogure T, Lin WL, Yan IK et al (2011) Intercellular nanovesicle-mediated microRNA transfer:a mechanism of environmental modulation of hepatocellular cancer cell growth. Hepatology 54(4):1237–1248
18. Montecalvo A, Larregina AT, Shufesky WJ et al (2012) Mechanism of transfer of functional microRNAs between mouse dendritic cells via exosomes. Blood 119(3):756–766
19. Simpson RJ, Kalra H, Mathivanan S (2012) ExoCarta as a resource for exosomal research. J Extracell Vesicles. doi:10.3402/jev.v1i0.18374
20. Paget S (1989) The distribution of secondary growths in cancer of the breast. Cancer Metastasis Rev 8(2):98–101
21. Bobrie A, Théry C (2013) Unraveling the physiological functions of exosome secretion by tumors. Oncoimmunology 2(1):e22565
22. Koga K, Matsumoto K, Akiyoshi T, Kubo M, Yamanaka N, Tasaki A, Nakashima H, Nakamura M, Kuroki S, Tanaka M, Katano M (2005) Purification, characterization and biological significance of tumor-derived exosomes. Anticancer Res 25(6A):3703–3707
23. Christianson HC, Svensson KJ, van Kuppevelt TH, Li J-P, Belting M (2013) Cancer cell exosomes depend on cell-surface heparan sulfate proteoglycans for their internalization and functional activity. Proc Natl Acad Sci U S A 110(43):17380–17385
24. Thiery JP (2002) Epithelial-mesenchymal transitions in tumour progression. Nat Rev Cancer 2(6):442–454
25. Thiery JP, Acloque H, Huang RY, Nieto MA (2009) Epithelial-mesenchymal transitions in development and disease. Cell 139(5):871–890. doi:10.1016/j.cell.2009.11.007
26. Kalluri R1, Weinberg RA (2009) The basics of epithelial-mesenchymal transition. J Clin Invest 119(6):1420–1428. doi:10.1172/JCI39104
27. Mani SA, Guo W, Liao MJ, Eaton EN, Ayyanan A, Zhou AY, Brooks M, Reinhard F, Zhang CC, Shipitsin M, Campbell LL, Polyak K, Brisken C, Yang J, Weinberg RA (2008) The epithelial-mesenchymal transition generates cells with properties of stem cells. Cell 133(4):704–715. doi:10.1016/j.cell.2008.03.027
28. Chaffer CL1, Marjanovic ND, Lee T, Bell G, Kleer CG, Reinhardt F, D'Alessio AC, Young RA, Weinberg RA (2013) Poised chromatin at the ZEB1 promoter enables breast cancer cell plasticity and enhances tumorigenicity. Cell 154(1):61–74. doi:10.1016/j.cell.2013.06.005
29. Thiery JP, Lim CT (2013) Tumor dissemination: an EMT affair. Cancer Cell 23(3):272–273. doi:10.1016/j.ccr.2013.03.004

30. Aga M, Bentz GL, Raffa S, Torrisi MR, Kondo S, Wakisaka N, Yoshizaki T, Pagano JS, Shackelford J (2014) Exosomal HIF1α supports invasive potential of nasopharyngeal carcinoma-associated LMP1-positive exosomes. Oncogene. doi:10.1038/onc.2014.66
31. Gu J, Qian H, Shen L, Zhang X, Zhu W, Huang L, Yan Y, Mao F, Zhao C, Shi Y, Xu W (2012) Gastric cancer exosomes trigger differentiation of umbilical cord derived mesenchymal stem cells to carcinoma-associated fibroblasts through TGF-β/Smad pathway. PLoS One 7(12), e52465. doi:10.1371/journal.pone.0052465
32. Ramteke A, Ting H, Agarwal C, Mateen S, Somasagara R, Hussain A, Graner M, Frederick B, Agarwal R, Deep G (2013) Exosomes secreted under hypoxia enhance invasiveness and stemness of prostate cancer cells by targeting adherens junction molecules. Mol Carcinog. doi:10.1002/mc.22124
33. Jeppesen DK, Nawrocki A, Jensen SG, Thorsen K, Whitehead B, Howard KA, Dyrskjøt L, Ørntoft TF, Larsen MR, Ostenfeld MS (2014) Quantitative proteomics of fractionated membrane and lumen exosome proteins from isogenic metastatic and nonmetastatic bladder cancer cells reveal differential expression of EMT factors. Proteomics 14(6):699–712. doi:10.1002/pmic.201300452
34. Luga V, Zhang L, Viloria-Petit AM, Ogunjimi AA, Inanlou MR, Chiu E, Buchanan M, Hosein AN, Basik M, Wrana JL (2012) Exosomes mediate stromal mobilization of autocrine Wnt-PCP signaling in breast cancer cell migration. Cell 151(7):1542–1556. doi:10.1016/j.cell.2012.11.024
35. Mu W, Rana S, Zöller M (2013) Host matrix modulation by tumor exosomes promotes motility and invasiveness. Neoplasia 15(8):875–887
36. Grange C, Tapparo M, Collino F, Vitillo L, Damasco C, Deregibus MC, Tetta C, Bussolati B, Camussi G (2011) Microvesicles released from human renal cancer stem cells stimulate angiogenesis and formation of lung premetastatic niche. Cancer Res 71(15):5346–5356. doi:10.1158/0008-5472.CAN-11-0241
37. Hood JL, San RS, Wickline SA (2011) Exosomes released by melanoma cells prepare sentinel lymph nodes for tumor metastasis. Cancer Res 71(11):3792–3801
38. Webber J, Steadman R, Mason MD, Tabi Z, Clayton A (2010) Cancer exosomes trigger fibroblast to myofibroblast differentiation. Cancer Res 70(23):9621–9630. doi:10.1158/0008-5472.CAN-10-1722, Epub 2010 Nov 23
39. Robbins PD (2014) Morelli AE regulation of immune responses by extracellular vesicles. Nat Rev Immunol 14(3):195–208
40. Ochoa AC, Zea AH, Hernandez C, Rodriguez PC (2007) Arginase, prostaglandins, and myeloid-derived suppressor cells in renal cell carcinoma. Clin Cancer Res 13(2 Pt 2):721s–726s
41. Chalmin F, Ladoire S, Mignot G, Vincent J, Bruchard M, Remy-Martin JP, Boireau W, Rouleau A, Simon B, Lanneau D, De Thonel A, Multhoff G, Hamman A, Martin F, Chauffert B, Solary E, Zitvogel L, Garrido C, Ryffel B, Borg C, Apetoh L, Rébé C, Ghiringhelli F (2010) Membrane-associated Hsp72 from tumor-derived exosomes mediates STAT3-dependent immunosuppressive function of mouse and human myeloid-derived suppressor cells. J Clin Invest 120(2):457–471. doi:10.1172/JCI40483. Epub Jan 19
42. Hong EH, Chang SY, Lee BR, Kim YS, Lee JM, Kang CY, Kweon MN, Ko HJ (2013) Blockade of Myd88 signaling induces antitumor effects by skewing the immunosuppressive function of myeloid-derived suppressor cells. J Cancer Educ 132(12):2839–2848. doi:10.1002/ijc.27974
43. Liu C, Yu S, Zinn K, Wang J, Zhang L, Jia Y, Kappes JC, Barnes S, Kimberly RP, Grizzle WE, Zhang HG (2006) Murine mammary carcinoma exosomes promote tumor growth by suppression of NK cell function. J Immunol 176(3):1375–1385
44. Taylor DD, Gerçel-Taylor C, Lyons KS, Stanson J, Whiteside TL (2003) T-cell apoptosis and suppression of T-cell receptor/CD3-zeta by Fas ligand-containing membrane vesicles shed from ovarian tumors. Clin Cancer Res 9(14):5113–5119
45. Valenti R, Huber V, Filipazzi P, Pilla L, Sovena G, Villa A, Corbelli A, Fais S, Parmiani G, Rivoltini L (2006) Human tumor-released microvesicles promote the differentiation of myeloid

cells with transforming growth factor-beta-mediated suppressive activity on T lymphocytes. Cancer Res 66(18):9290–9298

46. Weber JA, Baxter DH, Zhang S, Huang DY, Huang KH, Lee MJ, Galas DJ, Wang K (2010) The microRNA spectrum in 12 body fluids. Clin Chem 56(11):1733–1741. doi:10.1373/clinchem.2010.147405
47. Keller S, Ridinger J, Rupp AK, Janssen JW, Altevogt P (2011) Body fluid derived exosomes as a novel template for clinical diagnostics. J Transl Med 9:86. doi:10.1186/1479-5876-9-86
48. Villarroya-Beltri C, Gutiérrez-Vázquez C, Sánchez-Cabo F, Pérez-Hernández D, Vázquez J, Martin-Cofreces N, Martinez-Herrera DJ, Pascual-Montano A, Mittelbrunn M, Sánchez-Madrid F (2013) Sumoylated hnRNPA2B1 controls the sorting of miRNAs into exosomes through binding to specific motifs. Nat Commun 4:2980. doi:10.1038/ncomms3980
49. Pegtel DM, Cosmopoulos K, Thorley-Lawson DA et al (2010) Functional delivery of viral miRNAs via exosomes. Proc Natl Acad Sci U S A 107(14):6328–6333
50. Pan Q, Ramakrishnaiah V, Henry S et al (2011) Hepatic cell-to-cell transmission of small silencing RNA can extend the therapeutic reach of RNA interference (RNAi). Gut 61(9):1330–1339. doi:10.1136/gutjnl-2011-300449
51. Zhang Y, Liu D, Chen X, Li J, Li L, Bian Z, Sun F, Lu J, Yin Y, Cai X, Sun Q, Wang K, Ba Y, Wang Q, Wang D, Yang J, Liu P, Xu T, Yan Q, Zhang J, Zen K, Zhang CY (2010) Secreted monocytic miR-150 enhances targeted endothelial cell migration. Mol Cell 39(1):133–144. doi:10.1016/j.molcel.2010.06.010
52. Yang M, Chen J, Su F et al (2011) Microvesicles secreted by macrophages shuttle invasion potentiating microRNAs into breast cancer cells. Mol Cancer 10:117
53. Fabbri M, Paone A, Calore F, Galli R, Croce CM (2013) A new role for microRNAs, as ligands of Toll-like receptors. RNA Biol 10(2):169–174
54. Peinado H, Alečković M, Lavotshkin S, Matei I, Costa-Silva B, Moreno-Bueno G, Hergueta-Redondo M, Williams C, García-Santos G et al (2012) Melanoma exosomes educate bone marrow progenitor cells toward a pro-metastatic phenotype through MET. Nat Med 8(6):883–891. doi:10.1038/nm.2753
55. Li J, Sherman-Baust CA, Tsai-Turton M, Bristow RE, Roden RB, Morin PJ (2009) Claudin-containing exosomes in the peripheral circulation of women with ovarian cancer. BMC Cancer 9:244. doi:10.1186/1471-2407-9-244
56. Lu Q, Zhang J, Allison R, Gay H, Yang WX, Bhowmick NA, Frelix G, Shappell S, Chen YH (2009) Identification of extracellular delta-catenin accumulation for prostate cancer detection. Prostate 69(4):411–418. doi:10.1002/pros.20902
57. Morrissey JJ, London AN, Luo J, Kharasch ED (2010) Urinary biomarkers for the early diagnosis of kidney cancer. Mayo Clin Proc 85(5):413–421. doi:10.4065/mcp.2009.0709
58. Qu JL1, Qu XJ, Zhao MF, Teng YE, Zhang Y, Hou KZ, Jiang YH, Yang XH, Liu YP (2009) Gastric cancer exosomes promote tumour cell proliferation through PI3K/Akt and MAPK/ERK activation. Dig Liver Dis 41(12):875–880. doi:10.1016/j.dld.2009.04.006
59. Klinke DJ 2nd, Kulkarni YM, Wu Y, Byrne-Hoffman C (2014) Inferring alterations in cell-to-cell communication in HER2+ breast cancer using secretome profiling of three cell models. Biotechnol Bioeng. doi:10.1002/bit.25238
60. Ciravolo V, Huber V, Ghedini GC, Venturelli E, Bianchi F, Campiglio M, Morelli D, Villa A, Della Mina P, Menard S, Filipazzi P, Rivoltini L, Tagliabue E, Pupa SM (2012) Potential role f HER2-overexpressing exosomes in countering trastuzumab-based therapy. J Cell Physiol 227(2):658–667. doi:10.1002/jcp.22773
61. Klein-Scory S, Kübler S, Diehl H, Eilert-Micus C, Reinacher-Schick A, Stühler K, Warscheid B, Meyer HE, Schmiegel W, Schwarte-Waldhoff I (2010) Immunoscreening of the extracellular proteome of colorectal cancer cells. BMC Cancer 10:70. doi:10.1186/1471-2407-10-70
62. Schwarzenbach H, Nishida N, Calin GA, Pantel K (2014) Clinical relevance of circulating cell-free microRNAs in cancer. Nat Rev Clin Oncol 11(3):145–156
63. Marleau AM, Chen CS, Joyce JA, Tullis RH (2012) Exosome removal as a therapeutic adjuvant in cancer. J Transl Med 10:134. doi:10.1186/1479-5876-10-134

64. Grapp M, Wrede A, Schweizer M, Hüwel S, Galla HJ, Snaidero N, Simons M, Bückers J, Low PS, Urlaub H, Gärtner J, Steinfeld R (2013) Choroid plexus transcytosis and exosome shuttling deliver folate into brain parenchyma. Nat Commun 4:2123. doi:10.1038/ncomms3123
65. Alvarez-Erviti L, Seow Y, Yin H, Betts C, Lakhal S, Wood MJ (2011) Delivery of siRNA to the mouse brain by systemic injection of targeted exosomes. Nat Biotechnol 29(4):341–345. doi:10.1038/nbt.1807
66. Delcayre A, Estelles A, Sperinde J, Roulon T, Paz P, Aguilar B, Villanueva J, Khine S, Le Pecq JB (2005) Exosome display technology: applications to the development of new diagnostics and therapeutics. Blood Cells Mol Dis 35(2):158–168

# Chapter 8
# Tunneling Nanotubes: Intercellular Conduits for Direct Cell-to-Cell Communication in Cancer

**Emil Lou and Subbaya Subramanian**

**Abstract** Tunneling nanotubes (TNTs) are actin-based cytoplasmic extensions that directly transfer cellular cargo between connected cells. Intercellular transport of signaling molecules and regulatory RNAs are an integral part of defining the tumor microenvironment and intratumoral heterogeneity. The TNT-based cellular mode of molecular signaling affects multiple cell functions including cellular reprograming, tumor evasion, and chemotherapy resistance. In this chapter we will discuss various concepts of intercellular communication in cancer via TNTs and highlight the potential role played by TNTs in cancer pathobiology.

**Keywords** Tunneling nanotubes • Intercellular nanotubes • Intercellular communication • Tumor-stroma interactions • Intercellular transfer • Long-range cellular communication • Tumor heterogeneity • Molecular networking • Horizontal transfer

## 8.1 Introduction

Tumor microenvironment, intratumoral heterogeneity, and tumor-stroma interactions play a major role in tumor evasion and chemotherapy resistance. There is growing evidence that subclone populations exist whose genetic composition varies significantly within a single tumor. These subclones may be transformed by receiving extracellular signals, which may arrive in the form of soluble signals such as cytokines or signals packaged in protective diffusible vessels such as exosomes or microvesicles. The efficient delivery of exogenous molecular oncogenic signals

E. Lou (✉)
Division of Hematology, Oncology and Transplantation, University of Minnesota,
Mayo Mail Code 480, 420 Delaware Street SE, Minneapolis, MN 55455, USA
e-mail: emil-lou@umn.edu

S. Subramanian
Department of Surgery, University of Minnesota, Minneapolis, MN 55455, USA



M. Kandouz (ed.), *Intercellular Communication in Cancer*,
DOI 10.1007/978-94-017-7380-5_8

leads to invasion and increasing potency of malignancies at the cellular level and rapid advancement and disease progression at the clinical level. Intercellular communication is a critical but underappreciated cellular mechanism of this process [1–3]. Detailed mechanisms of cell-to-cell communication in the complex, heterogeneous tumor microenvironment are not well established, but have become a growing field of interest over the past decade.

Cancer cell biology investigators are addressing significant gaps in knowledge of how cell-cell communication and 'molecular networking' take place in the malignant tumor environment [3, 4]. As cellular communication is vital to tumor formation, progression, and recurrence, it is important to comprehensively characterize various mechanisms of interactions at the cellular level. If these interactions are critical to coordinating players in the tumor matrix, then disrupting or '*cutting off*' lines of communication would represent an important and perhaps underdeveloped strategy for selective and more effective therapies.

Modes of intercellular communication are increasingly being investigated as potential targets for treatment in cancer. The most commonly studied avenues of cellular transfer between cancer cells include gap junctions and their connexin proteins, chemokines, cytokine messengers, and microvesicles or exosomes [5–11]. Understanding the molecular mechanisms of selective intercellular trafficking and their role in advancing tumor promotion and progression will be important to identifying new approaches to disrupting this critical communication for treatment. Confirmation and validation of this rationale using in vitro and *in vivo* preclinical cancer models would open a new avenue to treatment of invasive, difficult-to-treat hematologic and solid tumor malignancies.

Tunneling nanotubes (TNTs) are a novel alternative form of direct cell-to-cell communication and 'molecular networking' in the complex tumor microenvironment. TNTs were in fact first observed and described by Rustom et al. using cultures of PC12, a commonly used cell line derived from pheochromocytoma [12]. *TNTs are fine, long, filamentous actin-basedcytoplasmic extensionsthat connect distant or nearby cells not located in immediate proximity*, and serve as short and long-range conduits for efficient, direct transfer of cellular cargo between connected cells [12–15]. Structurally, TNTs are open-ended 'intercellular bridges' whose walls consist of a contiguous lipid bilayer that establishes a direct membranous 'pipeline' between connected cells. In a study of rat kidney cells, neural crest cells, and human endothelial vascular cells (HUVECs), some TNT connections were also found to interface directly at one end with connexin-43, a component of gap junctions embedded in the cellular plasma membrane [16]. In another study of TNTs in urothelial cells, adherens junctions were proposed as a site of anchoring for stabilizing TNTs as they extend outside the cell [17]. In some studies, TNTs may branch out in a '*fork-in-the-road*' pattern [13, 18], and it is relatively common for cells to be connected by multiple rather than single TNTs.

These long, thin cytoplasmic extensions form *de novo* and serve as conduits for intercellular shuttling of cargo such as proteins, Golgi vesicles, mitochondria, G-protein coupled receptors, and even prions [13, 19–21]. Characteristic morphologic features distinguishing TNTs from other actin-based structures include their small diameter (approximately 50–400 nm, but can be as wide as 800 nm), contiguous cell-to-cell cytoplasmic connections, and non-adherence to the substratum when cultivated in vitro [12, 13, 22]. It is also important to differentiate TNTs from other actin-based cellular protrusions; this differentiation is not always evident based purely on imaging, but this fact underscores the importance of careful microscopic examination of live cell culture for correct identification of TNTs. TNTs span the three-dimensional plane *in vitro* and are often out of the focal plane of adherent cells. Adjusting the fine focus of the microscope onto TNTs often delineates the TNT at the expense of the adherent cell in the background, a fact that actually clarifies that the structure is most likely a true TNT. Non-adherence is a key characteristic of TNTs and helps to distinguish them from more common adherent actin-based protrusions. Additionally, some TNTs (the ones tethered and attached between two anchored cells) can in fact 'oscillate' upon movement of the culture plate or dish, further demonstrating non-adherence [13]. TNTs are somewhat delicate and sensitive to stress factors including prolonged exposure to light as well as movement or shaking. This sensitivity provides some technical challenges to fixation of cells connected by TNTs for accurate evaluation. Our approach to overcome this issue is careful fixation using high-concentration paraformaldehyde (PFA) to ensure TNTs remain intact, per our previously reported protocol [13]. We then stain TNT-forming cells using immunofluorescence-tagged antibodies for microscopy evaluation.

Few studies have elucidated the mechanisms of TNT-mediated cellular transport, but there is growing interest in examining TNTs as a novel mode of intercellular communication among many cell types [15]. Interestingly, similarly long membrane-lined tubular structures called plasmodesmata have been extensively studied in plants. Plasmodesmata connect cytoplasts of adjacent plant cells across the cell wall and are conduits for similar molecular cargo, including RNA [23–27]. As we continue to investigate TNT biology, it is important to keep the lessons learned from the discovery of plasmodesmata in mind, as plasmodesmata and TNTs may share many aspects of cellular structure and function.

Effective cell-cell 'cross-talk' could be difficult to achieve in the setting of a heterogeneous tumor microenvironment solidified by an increase in interstitial fluid pressure. Another important consideration is that stromal, non-malignant cells comprise as much as 80–90 % of a given tumor's volume; in fact, higher proportion of stroma has been associated with worse prognosis in invasive solid tumor malignancies [28–35]; by logical extension, the malignant cells that compose 10–20 % of the tumor may not be in close enough proximity to exchange cellular information via gap junctions. The vast majority of studies examining function, mechanism, and structure of TNTs have studied non-cancer cells; thus, an important future topic of study will be to determine what properties of TNTs are specific to certain cell or organ types, including cancer, and what properties of TNTs are universal across the spectrum of cells. To date, examination of TNTs by other groups world-wide has been more focused on a wide variety of cell types including dendritic cells and monocytes [22, 36], mature macrophages [37, 38], T cells [14, 39, 40], B cells [41], neutrophils [42], neuronal cells [20], kidney cells [43], endothelial progenitor cells [44], mesothelial cells [45], cardiomyocytes [46], and mesenchymal stromal cells [47, 48]. Nearly all of the limited studies of TNTs in cancer have studied cultures of human or animal-derived cancer cell lines, including PC12 (rat pheochromocytoma/neuroendocrine cancer) [12], human prostate carcinoma [49], human lung carcinoma, [13, 18], chronic myelogenous leukemia [50] or other leukemia [51], melanoma [52], human urothelial carcinoma [53], glioblastoma [21], human laryngeal carcinoma [54], and human ovarian adenocarcinoma and breast cancer [55, 56]. Much of our own work on TNTs has used human malignant pleural mesothelioma (MPM) cells derived from primary specimens (pleural effusions, ascites fluids, and primary tumors) in addition to cell lines [13], intact tumor tissues from MPM and lung adenocarcinoma [13, 57], and ovarian adenocarcinoma and murine osteosarcoma [58]. There is certain to be diversity in the formation, organization, maintenance, and disassembly of these unique cellular extensions, and much of this information is open to discovery. For the purposes of this book chapter, which focuses specifically on intercellular communication in cancer, we discuss TNTs in the context of clinical and translational relevance to this disease process. The major themes discussed include the relevance of TNTs to cancer biology, the role of TNT-mediated transfer of cellular cargo including microRNAs in cancer, and the interactions of TNTs with exosomes. We also discuss the significance of tumor-stromal interactions via TNTs, potential biomarkers and mechanisms of TNT formation, and TNTs as potential targets for cancer-directed therapy.

## 8.2 TNTs as Conduits for Direct Transport of Cellular Cargo Between Malignant and Stromal Cells

TNTs are not unique to cancer, but there is a significant gap in knowledge about their role in cancer cell biology. Little is known to date regarding the effects on recipient cells after intercellular transfer via TNTs in cancer. Much of our group's focus has been on the evaluation of cell lines, primary tumor cells, and cells derived from malignant effusions harvested from human patients. We have demonstrated that TNTs form between cells from a variety of aggressive malignancies cultured *in vitro*, including cancers of the breast, ovary, cervix, prostate, pancreas, bone (osteosarcoma), and brain (glioblastoma), indicating that nanotubes are a potentially universal process across cancer types (Fig. 8.1). Recent evidence also implicates TNTs as being involved in several clinically relevant problems in cancer, including invasion and chemoresistance.

The most clinically relevant studies of TNT transfer have been performed in the setting of cardiomyocyte and stem cell cultures. Several groups have demonstrated successful calcium signal propagation and even stem cell rescue or reprogramming

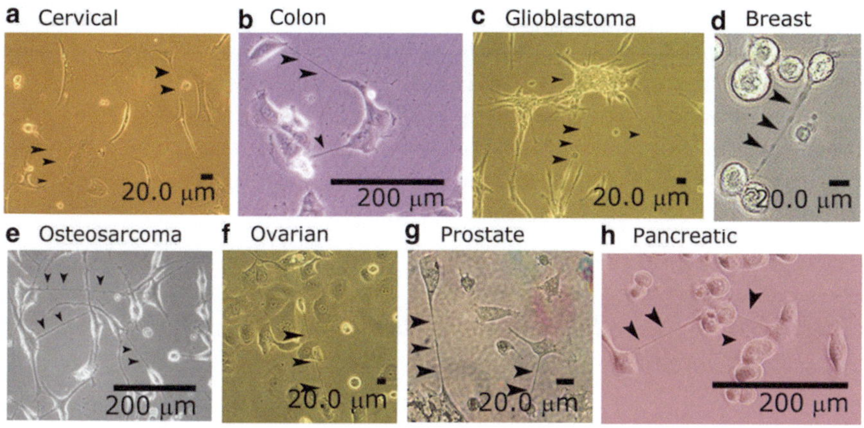


**Fig. 8.1 Composite of brightfield microscopic images oftunneling nanotubes(TNTs) forming and connecting cancer cells derived from a wide variety of cancers**. Established cancer cell lines and human primary malignant cells obtained from patients were cultured *in vitro* as described previously [13]. *Arrowheads* indicate location of TNTs. TNTs were seen between the following malignant cells: (**a**) cervical cancer cells derived from a pleural effusion specimen from a patient with advanced cervical adenocarcinoma (20× objective), (**b**) colon (DLD-1) (30×), (**c**) glioblastoma (U373MG) (20×), (**d**) breast cancer cells derived from an abdominal ascites specimen from a patient with metastatic breast carcinoma (20×), (**e**) osteosarcoma (K7M2) (20×), (**f**) ovarian (Ovcar3) (20×), (**g**) prostate cancer cells derived from a pleural effusion specimen from a patient with advanced prostate adenocarcinoma (40×), and (**h**) pancreatic (MIA PaCa-2) (30×)

of adult cardiomyocytes through TNT-mediated transfer of mitochondria or other signals [46, 59–63]. Transfer of mitochondria preferentially from stromal (endothelial) cells to malignant ovarian and breast cancer-derived cells similarly has been associated with chemotherapy resistance *in vitro* [56]. In addition, using time-lapse microscopy, we determined that TNTs can form de novo over 4–6 hours between mesothelioma cells in both close and long range – as many as five to six cell lengths apart (see movie S4, Lou et al. [13]). To visualize transfer of cellular cargo, we used fluorophores such as the lipophilic dyes DiI (red fluorescence) and DiO (green fluorescence) or mitochondria-specific stains (MitoTracker Red). Additionally, we also co-cultured mesothelioma cells transduced with red fluorescent protein (RFP) or green fluorescent protein (GFP)-expressing lentiviral vectors to examine for intercellular transport of proteins. We discovered that each of these cargo types – including lipophilic components from the cytosol, mitochondria, proteins, and even Golgi vesicles – could be transferred via TNTs in both unidirectional and bidirectional fashion [13]. Suppressing TNT formation through inhibition of the mTOR pathway negatively regulates such transfer. These studies, taken together, suggest that the intercellular transfer of mitochondria or other intercellular cargo may be an important mechanism of chemoresistance.

In addition to chemoresistance, TNTs may facilitate mediate malignant invasion. We have used modified wound-healing assays and demonstrated TNT formation along the leading invasive edge of mesothelioma cells *in vitro*. Time-lapse imaging revealed regular formation of TNTs by proliferating and migrating mesothelioma cells advancing to fill the gap [13]. This finding introduces the possibility that TNTs facilitate intercellular communication and the progression of malignancy at the leading edge of invasive tumors. Our ongoing work is further examining how direct and efficient communication takes place via TNTs connecting distant and proximal cells in the heterogeneous tumor matrix.

### *8.2.1 Direct MicroRNA Transfer Between Various Cell Types in Cancer*

We have also investigated the intercellular transfer of genetic materials, specifically microRNAs (miRNAs), which are small non-coding nucleic acids that have gained increased attention in the past decade for their posttranscriptional gene regulatory roles in cancer. miRNAs can be transferred between cells via lipid-based carriers including exosomes and microvesicles. We hypothesized that TNTs are also a significant conduit for miRNA transport and can serve as a more efficient mode of direct transfer of these regulatory RNA molecules. By tagging miRNAs with the fluorophore Alexa-488 and transfecting cells from several forms of cancer – including osteosarcoma and ovarian – we have succeeded in imaging transport of miRNAs within TNTs, and used time-lapse microscopy to demonstrate movement between connected cells [58]. Our finding opens new possibilities into identifying TNTs as a potential therapeutic target by disrupting TNT-mediated transfer of miRNAs, which stimulate increased cellular invasive capabilities and chemoresistance. To our knowledge, this was the first demonstration of miRNA transport via TNTs connecting malignant cells and also connecting malignant to stromal cells. Ongoing studies by our collaborative group are exploring effects of TNT-mediated transport of miRNA candidates associated with chemotherapy resistance in these cancers.

miRNAs can be found in various human body fluids. Recent studies also demonstrate that circulating miRNAs can be used as diagnostic and prognostic biomarkers for various cancer types [64]. Further, miRNAs are also implicated in cellular reprogramming. For instance, miR-320, a PTEN regulated miRNA, is active in stromal fibroblasts and plays a major role in reprogramming the tumor microenvironment and regulating tumor progression [65]. In the context of cellular reprogramming, the direct transfer of miRNAs between cells via TNTs will be more effective than passive diffusion-based transport mechanisms. Recently, we showed that oncogenic miR-19a (a miR-17-92 cluster member) could be directly transported between cancer cells and also between cancer and stromal cells via TNTs [58]. miR-19a was shown to be more oncogenic among the miR-17-92 cluster miRNAs [66]. Similarly, we also observed that miR-199a, with a potential role in chemoresistance, was also transported via nanotubes between ovarian normal and cancer cells [58]. These studies reiterate that direct transfer of miRNAs between cancer and stromal cells is an integral mechanism of tumor development and progression. The implications of direct miRNA transport between various cell types in cancer may also include the following:

1. Reprogramming of recipient cells towards malignant transformation that induces increased cell proliferation and metastasis.
2. Establishing a tumor microenvironment that is conducive for tumor development and progression. For example, a microenvironment in which T-cell activation and function is affected.
3. Specific cell-of-origin in tumor will be debatable as recipient cells can be efficiently reprogrammed to behave and function like donor cells.
4. Regulation of endogenous genes in the recipient cells by the miRNAs. In support of this idea, a recent study has shown that miR-223 can effectively regulate the endogenous ICAM-1 in the recipient cells [67].

Each miRNA is predicted to target and regulate hundreds of gene transcripts; in the context of intercellular miRNA transport, it is critical to establish the cell type within the tumor mass that expresses the miRNAs. For example, miR-143 and -145 are highly expressed in fibroblasts of normal colon tissue and are not expressed in normal colon epithelial cells. Hence, downregulation of miR-143/-145 in colon cancer (which are predominantly epithelial cells) may not be biologically relevant [68]. Since TNTs also have the potential to transport gene transcripts between cells, the functional viability of the transcripts in recipient cells in part depends on the miRNA composition of the recipient cell. There are numerous questions that need to be addressed in the transport of miRNAs and gene transcripts between cells.

1. Is there any selectivity in the transport of specific miRNAs between various cell types?
2. Are the miRNAs transported naked or via cellular vesicles or mitochondria? It can be noted that mitochondria are also reported to contain mature miRNAs [69].
3. Are the transported miRNA levels stochiometrically sufficient to regulate the endogenous transcript(s)?

## 8.3 Synergistic Communication in Cancer: Exosomes and TNTs Are Not Mutually Exclusive

We showed that exosomes and TNTs are not mutually exclusive forms of intercellular transfer of cargo in cancer, but rather they may work synergistically. We examined whether addition of exogenous exosomes induced a quantifiable increase in formation of TNTs. Mesothelioma cells co-cultured with exogenous tumor exosomes did in fact stimulate an increased rate of TNT formation; interestingly, addition of exosomes derived from benign mesothelial cells made no difference in TNT formation [57]. Electron microscopy revealed that exosomes were spatially located at the base of TNTs as well as in the extracellular environment. Based on initial studies characterizing the composition of TNTs, our subsequent studies also identified enrichment of lipid rafts – small intracytoplasmic cholesterol microdomains – in mesothelioma cells connected via nanotubes [57]. Lipid raft-associated proteins are present in the membranes of exosomes and have been implicated in formation of exosomes, packing of signaling cargo, and extrusion from a variety of cell types [70, 71]. Likewise, disruption of lipid rafts interferes with exosome activity and uptake by malignant cells [72]. As TNTs and exosomes are both in essence derived from the plasma membranes of cells – both cancerous or non-cancerous – the role of lipid rafts in mediating both exosomes and TNTs, and their mutual interactions, bears further study.

These findings implicated exosomes as potential chemotactic stimuli for TNT formation between malignant cells, and lipid rafts as a potential biomarker for TNTs. Other studies have affirmed the ability of cancer cell (melanoma)-derived exosomes to induce TNT formation between endothelial cells [52], supporting further the idea that cancer cells may communicate signals that induce TNT formation not only in other cancer cells, but also in other cells characteristic of the stroma-rich tumor microenvironment. An example of a process critical to invasion, progression, and survival of such cancers is angiogenesis, in which malignant and stromal cells foster the formation of new blood vessels which oxygenate and nourish tumors. As in the previously mentioned study, another group successfully demonstrated interaction of exosomes with TNTs between human umbilical endothelial cells (HUVECs). Using exosomes derived from a cell line of a hematologic malignancy, chronic myelogenous leukemia (CML), these exosomes appeared to enter and move within TNTs connecting remodeling HUVECs [50].

An important factor in advancement of our understanding of exosomes and TNTs is improved imaging technology, especially high-resolution microscopy that can identify previously unrecognized structures or sub-structures. Examples of such techniques include atomic force microscopy (AFM). Using AFM peak force imaging, abundant nanofilaments (10–20 nm in width) have been identified on exosomes derived from glioblastoma cells [73]. Whether such nanofilaments represent true TNTs as opposed to extensions used by the exosome for motility or structural support, such a finding raises the intriguing possibility that exosomes or microvesicles can themselves also form TNTs. Interestingly, the authors of this particular study demonstrated that exosomes derived from glioblastoma were taken up by recipient cells at a higher rate than normal control exosomes [73]; it is possible that nanofilaments were responsible for the higher rate of uptake by the cancer-derived exosomes.

As TNTs are not exclusive of other essential modes of communication, it will be essential to account and control for other forms of cell-cell transfer, including microvesicles and exosomes, when designing experiments. One approach our group has used to eliminate 'background noise' from other non-TNT forms of intercellular communication is the use of modified transwell migration/Boyden chambers and similar devices to culture cells that can form TNTs through cellulose or similar barriers, without allowing passage of exosomes. While modified cellulose or other membranes cannot completely exclude interchange of exosomes between the two populations, the degree of exosomal exchange is markedly decreased with this barrier, as compared to normal cell culture conditions, and thus dramatically decreases the effects of microvesicle/exosomal intercellular exchanges. After intercellular TNT-mediated exchange between barriers, separation of recipient cells can be achieved using FACS for further analysis [58].

## 8.4 Mechanisms of TNTs Formation

Ultrastructure and composition of TNTs have been examined most extensively in *in vitro* models of non-cancer cells, including neurons, and to a lesser extent in cancer [13]. It is generally accepted that all TNTs, regardless of cell of origin, are based on filamentous actin; reports on whether microtubules, composed of tubulin subunits, are present and play a role in TNT formation, extension, and disassembly have varied [18, 49]. Veranic et al. have provided an oft-cited designation of TNT subtypes to help distinguish the two common forms of TNTs described to date [74]. These categories include TNT subsets labeled as type I and type II nanotubes, with this categorization based on the context of formation of TNTs. Type I TNTs are essentially actin-based and initiated by a filopodial or similar cellular extension leading to formation of TNTs between distant cells; in their examination of urothelial cells, these researchers report TNTs of this type tend to be shorter and more dynamic than the alternate form, called type II, as the former include free-floating extensions in search of nearby cells for contact. Type II TNTs form and are maintained for a defined time as cells migrate via usual mechanisms of cell motility; reported characteristics include longer length, more stability, and presence of cytokeratin filaments, in addition to the fact that as they form the connected cells are already connected in some form [74]. In our own experience culturing malignant mesothelioma cells, we too have noted that type I nanotubes tend to be short and can quickly demonstrate transfer of cellular cargo via 'gondolas' which appear as dynamic hump-like bulges as cargo within TNTs is transferred from cell-to-cells (for representative time-lapse microscopic imaging of type I TNTs in mesothelioma, please see online video from the following reference [13]). As also seen in the study of urothelial cells, a type I TNT may be preceded by either filopodia (any cell form) or invadopodia (malignant cells), which appear to 'sense' the presence of nearby cells; in the above video of time-lapse imaging of mesothelioma cells, a thick, adherent invadopodium extends and then retracts, followed relatively quickly by formation of three new type I TNTs that attach to a nearby cluster of cells and transmit vesicular cargo through these TNTs in a unidirectional fashion [13]. This finding broaches the concept of efficient 'cell-seeking' behavior by cells prior to formation of TNTs, and the possibility of chemotactic gradients. Regardless, it appears that at least some components of this process may be specific to certain cell types based on origin (e.g. epithelial vs. mesothelial), organ (e.g. urothelium vs. lung), or especially disease process (benign vs. malignant). One important caveat that is of premium importance to cell biologists is to not confuse or dismiss the process of cytokinesis, in which the resulting intercellular connections between dividing cells can very easily be confused with type II TNTs. For this reason, time-lapse imaging is particularly important to distinguish cells that are in close proximity and forming type II TNTs, as compared with cells that undergo cell division via cytokinesis via contraction of an actin-myosin ring and undergo abscission (cytokinetic abscission reviewed by Guizetti et al. [75]).

Understanding of cellular environmental conditions that elicit TNTs *in vitro* has previously been limited. Forms of chemical TNT stimulation are diverse. PMA stimulates membrane bridge formation in NIH 3T3 fibroblasts [76]. Hydrogen peroxide creates oxidative stress, which stimulates TNT formation in astrocytes [77]. Serum depletion has also been used to elicit TNT formation [77]. Other stimulants of inflammation that have been implicated in TNT formation include tumor necrosis factor-alpha (TNF-α) [45]. The M-sec protein stimulates TNT formation in macrophages, and is linked to influx of calcium [37]. Galkina et al. used staurosporine, a general protein kinase inhibitor, to induce tubular extensions in neutrophils, and attributed this ability to a nitrous oxide-related mechanism [42]. However, the mechanism in that case may more likely be due to actin depolymerization, and the observation of cellular extensions may in fact be more similar to another microtubule-related phenomenon that has been termed 'microtentacles' [78–83]. Blebbistatin (a myosin II specific inhibitor S-(-)) has also been used to induce TNT formation and organelle transfer in normal rat kidney cells; in the same set of experiments, the myosin inhibitor 2,3-butanedione monoxime also induced TNTs, but inhibited transfer [43]. However, Wang and colleagues examined the intercellular transport along TNTs connecting lung carcinoma cells by tracking wheat germ agglutinin as a model drug carrier. Their study supported the role of myosin molecular motors in active actin-mediated transport of this carrier via TNTs [18]; this finding is consistent with the above reported finding of myosin inhibition interfering with TNT transfer activity but not with actual formation of TNTs. Interestingly, nanotube formation has also been stimulated using mechanical [84] or electric field [85–87] stimulation.

Our initial studies of mesothelioma cell culture identified TNTs hovering above the substratum and connecting mesothelioma cell lines cultured *in vitro* [13]. Early in the course of our study of TNTs, we set out to determine a method for reliably and reproducibly examining TNTs in vitro. We assessed several dozen combinations of media comprising various concentrations of sera (fetal calf serum), glucose, and also degrees of acidity, considering the fact that cancer cells are known to thrive and develop invasive phenotypes in acidic microenvironments [88]. We demonstrated that a low-serum, hyperglycemic medium, with or without acidification efficiently and reproducibly increased TNT formation, up to five-fold higher compared with normal passage medium [13]. Additionally, as we have proposed that TNTs are candidates for efficient intercellular communication at the advancing front of invasive malignancies [19], we have demonstrated that pre-culturing mesothelioma cells in low-serum, hyperglycemic medium facilitates microscopic visualization in modified scratch/wound-healing assays [13]. These TNTs demonstrated key distinguishing characteristics such as non-adherence, resistance to trypsinization, and spontaneous formation that was further stimulated under specific metabolic conditions. A hyperglycemic microenvironment was especially conducive to TNT formation, and this ability was enhanced in the presence of low serum. TNTs were most apparent in cultures of low or moderate cell density, as cultures that were more confluent impaired detection of nanotubes. These mesothelioma TNTs formed spontaneously and most noticeably during *in vitro* growth in semi-confluent cultures, but with higher prevalence in the acidified hyperglycemic and low-serum stressful microenvironment. TNTs formed within several hours of culturing cells once the cells began to adhere [13].

Recent studies have confirmed that hyperglycemia enhances invasive activity of malignant cells by inducing oxidative stress [89]. Genes responsible for glucose uptake and utilization, including enzymes responsible for aerobic glycolysis and glucose transporters, are notably overexpressed in a variety of malignant cell types, such as soft tissue sarcoma cells compared to normal pancreas cells and surrounding stromal cells [90], further supporting the physiologic relevance of studying cancer-specific, cell-cell interactions using hyperglycemic stimulation.

The fact that hyperglycemic medium stimulates cell-to-cell interactions and communication via TNTs is made even more significant by findings that high glucose levels lead to a reduction of intercellular exchange via gap junctions [91]. Gap junctions, which form connexin-based channels between adjacent cells in immediate proximity, are a well-studied entity in the context of intercellular communication. However, during invasive processes such as epithelial-to-mesenchymal transition (EMT), gap junctions are disrupted [92]. Enhancement of TNTs by hyperglycemia, a condition that downregulates gap junctions, is a notable finding in this context. The metabolic shift to aerobic glycolysis that is characteristic of tumors (i.e. the Warburg effect) results in decreased/acidic pericellular pH, which also increases invasiveness of malignant cells [93]. The approach of using hyperglycemic acidified medium, and determination that these factors help reproducibly stimulate nanotube formation, is our model for studying the function of TNTs *in vitro*.

## 8.5 Tumor-Stromal Interactions: Heterogeneity of TNTs

There is a greater understanding that stromal cells are active and dynamic components of tumor initiation, progression, and invasion. They are no longer considered to merely be passive structural components of the tumor infrastructure. This realization has paved the way for rational approaches to targeting the tumor-stroma complex as a whole. As much as 80–90 % of a malignant tumor is composed of the non-malignant and heterogeneous microenvironment including, but not limited to, inflammatory infiltrates, vascular endothelium, and stromal cells. Yet, there remains considerable uncertainty regarding how tumor-stroma exchange of cellular information takes place in this context.

The number of studies to date examining tumor-stromal cell interactions via TNTs is limited, but they do provide insight into the diversity of interactions and

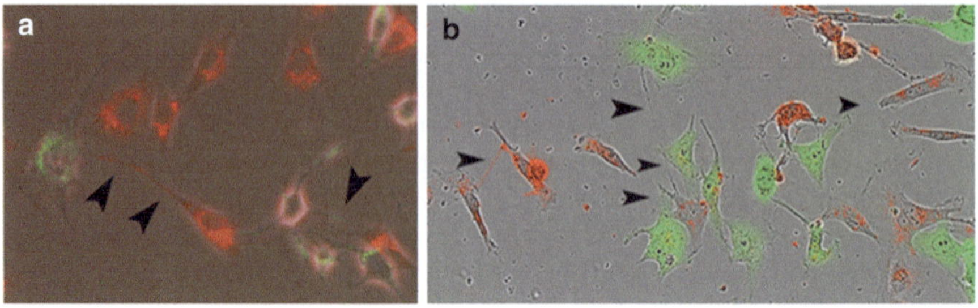


**Fig. 8.2 Tumor–stromal connections via TNTs**: TNT formation and consequent intercellular exchange of materials between osteosarcoma cells and osteoblasts. (**a**) TNT formation between osteosarcoma (K7M2 cell on *left*; *green* DiO) and a connected osteoblast cell (MC3T3, *red* DiI), with lipophilic *red*-stained cytosolic components in transit from the osteoblast to the osteosarcoma cell. Image taken after 24 hours of co-culture. (**b**) TNT formation between human osteosarcoma cells (MG63.2 cell line; *green* GFP), TNTs between human osteoblasts (*red*, DiI stain), and a TNT connection between osteosarcoma and osteoblast cells

ability of TNTs to form between these different cell types. In our studies, we have noted that cells derived from malignant pleural mesothelioma, when co-cultured with non-malignant mesothelial cells, form TNTs to each other, but do not participate in cross-talk with mesothelial cells; likewise, mesothelial cells form TNTs to each other in this context, but do not interact with malignant cells in co-culture [13]. This finding generates numerous questions regarding the presence of cell surface markers that participate in TNT formation and which differ between malignant and non-malignant cells. However, this finding is far from universal. We have demonstrated that osteosarcoma cells can in fact form TNTs with osteoblasts in culture (Fig. 8.2a) [58]. These findings provide evidence that, in some forms of cancer, TNTs can serve as conduits between malignant and stromal cells; exploring the results of this interaction will provide insight into a previously unrecognized form of intercellular communication taking place in the context of the heterogeneous tumor microenvironment. We have uncovered similar cross-talk between ovarian adenocarcinoma cells and ovarian epithelial cells in culture [58], indicating that whether tumor and stromal cells participate in TNT formation and communication may depend on site of origin, histology, and role of stroma in tumor progression.

## 8.6 Potential Biomarkers of TNTs

Potential biomarkers of TNTs are critical to the study of these delicate cellular structures. Our group's work has demonstrated that lipid rafts are enriched in mesothelioma cells forming TNTs [57]. The role of lipid rafts in cytoskeletal organization and assembly at the site of extrusion of exosomes and possibly TNTs makes lipid rafts a leading candidate for further study [94]. We also previously reported our examination of potential cell markers, including E-cadherin, which did not appear to have any association with TNTs [13]; however, downregulation of E-cadherin is an established marker of EMT, which we demonstrated in the same study to stimulate increased TNT formation [13]. Using malignant mesothelioma as a model system, our group has further investigated whether TNTs are associated with other well-established regulatory and structural components of cellular infrastructure. To identify other potential TNT markers, we examined the following regulatory and structural components of cellular infrastructure, particularly those involved in the invasive phenotype: CDC42 (a mediator of actin-based cytoplasmic extensions), merlin (a domain protein of the ezrin-radixin-moesin complex that is implicated in maintaining cell membrane stability, cell motility, and intercellular adhesions, and is encoded by the tumor suppressor gene NF2 [95]), P-selectin, microtubules (structural scaffolds involved in cell integrity and motility, assessed by analysis of acetyl-tubulin and beta-tubulin), AKT, and vimentin (a marker of EMT).

CDC42 is a key mediator of actin-based cytoplasmic extensions. CDC42 and its complementary factor RhoA have been implicated in propagation of apoptotic signals such as caspases via TNTs in autoreactive T lymphocytes [96]; in this study, inhibitors of RhoA led to reduced frequency of TNTs and consequent decreased intercellular exchange. Inhibition of CDC42 also reduced nanotube formation and reduced spread of cell death [96]. Our immunofluorescence (IF) staining for CDC42 demonstrated homogeneous distribution along the length of mesothelioma TNTs (Fig. 8.3a), indicating that this protein is likely integral to active actin filament organization implicated in formation of TNTs.

Surface markers including those responsible for cell adhesion (e.g. the cadherin family) are strong candidates to explain selective formation of TNTs between cells. Cadherins in particular are associated with both processes of carcinogenesis as well as local and distant cellular invasion of malignant cells [97]. E-cadherin contains

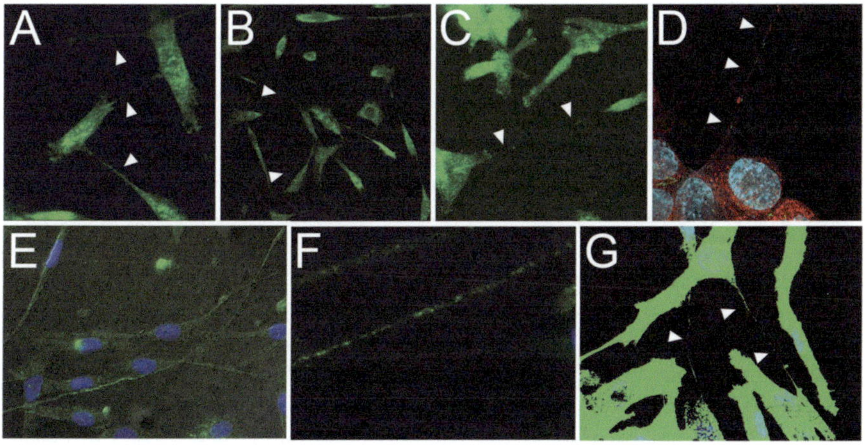


**Fig. 8.3 Further characterization of TNT composition and structure**. (**a**) CDC42, (**b**) NF2 (Merlin), (**c**) P-selectin, (**d**) acetyl-tubulin, (**e**) beta-tubulin, (**f**) higher magnification view of Beta-tubulin along the length of a TNT, (**g**) AKT

enhancer sequences that respond to the tumor promoter 12-O-tetradecanoylphorbol-13-acetate (TPA), an agent that recently has been shown to induce formation of TNT-like actin-based protrusions that connect spheroids formed from breast epithelial acini in a 3-dimensional model [98]. In this model system, the protrusions were dependent on activation of protein kinase C (PKC); notably, there was no disruption of E-cadherin cellular contact points [98]. A separate study determined that TPA-induced integrin activation reinforced coupling of the plasma membrane and cytoskeletal infrastructure; addition of TPA decreased formation of TNTs of CHO (Chinese hamster ovary cells) [99].

A specific isoform (PKCα) is associated with cellular communication and is overexpressed in mesothelioma compared with normal mesothelial pleural cells [100]. At the cellular level, PKCα regulates focal adhesions, and interacts directly with fascin, an actin-bundling protein that is focally expressed at the base of TNTs [13] and at the leading edge of migrating cells [101]. The cytoplasmic tail domain of E-cadherin also binds with intracytoplasmic beta-catenin, which we previously showed localizes within the cell body of mesothelioma cells forming TNTs [13]. While E-cadherin is essential to gap junction formation and maintenance [102, 103], their corresponding decrease in cell surface expression with EMT and invasive cells likely explains decreased IF expression of this marker in MSTO cells in our prior studies [13]. This finding is consistent with decreased expression of this marker with EMT, which we in turn have previously demonstrated leads to a higher rate of TNT formation [13]. We also postulate that this change provides further indication of the importance of TNTs for maintaining or restoring intercellular communication for separated cells following gap junction disruption. Furthermore, as it is known that the tumor promoter TPA (12-O-tetradecanoylphorbol-13-acetate) effectively blocks gap junction-mediated intercellular communication, it is reasonable to propose that its concurrent ability to indirectly stimulate formation of actin 'bridges' between spheroids, such as mammary acini, provides insight into the underlying mechanism of TNT formation, and also to the idea that TNTs can form structural tethers connecting acinus-type or similar aggregates in malignant effusions or ascites [104].

Neither NF2/merlin (Fig. 8.3b) nor P-selectin (Fig. 8.3c) was detected on immunofluorescence staining within or around TNTs. Merlin expression was not apparent within TNTs upon IF staining. However, as NF2 is inactivated in a majority of malignant mesotheliomas [105], it is unclear whether this lack of association of merlin with TNTs is specific to TNTs in mesothelioma or to TNTs in general. Interestingly, merlin provides a physical link to the actin cytoskeleton while mediating binding to cell surface glycoproteins such as CD44; while CD44 was upregulated in MSTO cells cultured in TNT-primed media, the negligible expression of merlin suggests that (at least in mesothelioma) this is not a key component of TNT formation. Mutated or phosphorylated forms of merlin induce an altered phenotype that includes elaboration of filopodial extensions [106]; thus, despite lack of merlin expression in mesothelioma TNTs in our study, merlin remains an interesting candidate for further evaluation in other cancer types. Further, lack or depletion of merlin increases signaling via the mTORC1 pathway, making such mesothelioma cells susceptible to inhibitors of this pathway [107]; this is especially important to take into account considering our group's prior discovery that treatment of MSTO cells with the mTOR inhibitor everolimus as well as metformin effectively suppressed TNT formation [13].

Microtubules are composed of α and β tubulin components that assemble as heterodimers to form structural scaffolds for cell integrity and cell motility. Their expression is increased in cancer, and microtubules participate in regulating cellular protrusions such as pseudopodia [108]. In one study examining human MPM tumors and patient outcomes, increased levels of βIII-tubulin correlated with worse progression-free survival [109]. This finding was confirmed in commonly used MPM cell lines, including MSTO-211H and H2052. Beta II tubulin was also overexpressed in MPM [110]. In this context, we stained TNT-forming MSTO-211H cells and visually confirmed the presence of acetylated tubulin (Fig. 8.3d) and beta tubulin (Fig. 8.3e, f) within TNTs. As TNTs are dynamic structures, it is important to take findings gleaned from the study of fixed cells or tissue in their proper context: that is, microtubules and actin filaments are dynamic components. Thus, the expression or quantity of either structural component will likely vary with formation and with disassembly of TNTs.

Phospho-Akt was minimal to negligible within TNTs, and only noted at all following overexposure of fluorescent signal of stained cells (Fig. 8.3g). Prior studies have suggested that activation of Akt plays a role in TNT formation in astrocytes [77], whereas other studies have demonstrated Akt-independent TNT induction in neuronal cell culture [111]. Regulation of mTOR in mesothelioma is thought to be independent of Akt activity [112], thus potentially explaining why IF staining did not localize Akt within the context of our study of TNTs.

In addition to microtubules and CDC42, vimentin is a potential marker of TNT formation. Vimentin is a widely acknowledged marker of EMT in malignant carcinomas, and is overexpressed (>2.3-fold higher) in mesothelioma compared with pleural mesothelium [113]. In thoracic malignancies especially, higher expression of vimentin is correlated with worse prognosis and rate of metastasis [114]. Further, demonstration of localization of vimentin at the invasive tumor-stromal interface in lung adenocarcinoma tumors supports this marker as a potential key player in TNT formation [114]. Indeed, our immunofluorescence studies revealed extensive and diffuse involvement of vimentin in TNTs that seems to support this view. Further studies should explore the role of vimentin in TNT formation during tumor invasion.

## 8.7 Disrupting TNT-Mediated Intercellular Communication as a Potential Target for Cancer-Directed Therapy

Identifying methods of suppressing or disrupting TNTs is a valuable approach for identifying key cellular components and molecular pathways of TNT formation and structural maintenance. Prior work by multiple groups has demonstrated that actin-depolymerizing agents such as cytochalasin B and D, latrunculin A, azide, colchicines, and tubulin inhibitors block TNT formation or traffic along TNTs [40, 76, 115, 116]. Actin-depolymerizing agents are the most common drugs used to assess effects of blocking TNT formation *in vitro*. Bukoreshtliev and colleagues used cytochalasin B to block TNT formation in PC12 cells [115]. Significantly, this drug appeared to have more of an effect on filopodia, and also prevented formation of new TNTs. However, the transport of vesicle-like structures still took place in the presence of cytochalasin B in established nanotubes, although overall traffic was decreased due to a decrease in the total number of nanotubes. This finding is relevant since both filopodia and TNTs are actin-dependent structures, but nanotubes contain less actin [115]. Cytochalasin D has been used to inhibit formation of membrane bridges in NIH 3T3 fibroblasts [76] and lymphocytes [40]. These results are consistent with our findings of lack of new TNT formation when using latrunculin A [13]. Other agents used to destabilize or prevent new formation of TNTs include azide, colchicines, and tubulin inhibitors such as nocodazole, which halt vesicular traffic along thin nanotubes in macrophages [116] and lymphocytes [40], although another group demonstrated that colchicine was not effective at altering membrane bridge morphology and neither was paclitaxel [76]. In T lymphocytes, toxin B of clostridium difficile and the CDC42-specific inhibitor secramine A blocked Fas-induced nanotube formation [96].

To identify potential metabolic pathways essential for TNT formation, our group identified several inhibitors of pathways that have been implicated in actin-based cell invasion. Metabolic pathways and their key mediators have been increasingly implicated in cancer etiology and propagation, and inhibitors of glycolysis and other such pathways have been proposed as potential cancer therapies. Metabolic inhibitors such as metformin have been suggested as a potential adjuvant treatment for solid tumors [117] to suppress glycolytic migration by inhibiting gluconeogenesis of glioma cells [118]. Metformin activates adenosine-monophosphate protein-activated protein kinase (AMPK) [119]. The mTOR pathway is also abnormally upregulated in numerous types of solid tumors and is also of great interest as a potential therapeutic target [119] due to the role of mTOR proteins in regulating cancer proliferation and invasion [120, 121]. They have also been implicated in actin organization, which furthers cancer cell invasion and metastasis. Recently, the mTOR and associated pathways have also been implicated in inducing TNT development in astrocytes under conditions of cellular stress [77]. Everolimus is an mTOR inhibitor in current use clinically for renal cell carcinoma [122]; it inhibits the mTORc1 protein. Migrastatin is a polyketide product initially derived from *Streptomyces*; synthetic analogs of migrastatin inhibit migration of cancer cells [123]. As noted already, migrastatin and its analogues target fascin and thereby block tumor progression [124]. In our work examining TNTs in mesothelioma in particular, we have assessed the effects of latrunculin A, metformin, everolimus, and migrastatin for this purpose. All four of these compounds significantly suppressed TNT formation by 72 h of incubation [13, 104].

## 8.8 Toward *In Vivo* Examination of TNTs in Cancer

The process of TNT formation in cancer cells, and the implication that TNTs promote invasion of malignant cells, has not been explored extensively. We demonstrated that TNTs can form between primary cancer cells derived from patients and cultivated *in vitro* [13]. Nearly all studies of TNTs have been performed *in vitro*. Thus, a common question that is broached is whether TNTs represent an artifact of *in vitro* culture, or whether they in fact are present *in vivo* in any clinically relevant disease setting (either in animal models or in humans). To date the only *in vivo* demonstration has been provided by an Australian group who have demonstrated formation of membranous nanotubes connecting dendritic cells in an animal model of inflammation of the cornea. Dendritic cells or myeloid cells expressing GFP or RFP interacted with each other or with macrophages following injection into mouse corneal stroma. Nanotube formation exchange was examined either *in vivo* or following corneal dissection and mounting for fluorescent microscopy examination. As a model of corneal injury, following induction of lipopolysaccharide (LPS)-induced inflammation, such exchange of fluorescent proteins did in fact take place, as the transfer of green and red fluorescent proteins to other cells via nanotubes produced cells that fluoresced in the yellow channel [125, 126]. There are obvious inherent challenges in evaluating the effects of such fine cellular extension *in vivo*. Thus, another avenue that has been successful used in the past few years is examination of *ex vivo* structures, including tumors resected from human patients and animal models of cancer.

To investigate TNTs as a physiologically relevant structure in human solid tumor malignancies, we successfully visualized TNTs in solid tumors resected from patients with mesothelioma and lung adenocarcinomas [13], providing the first evidence of the potential *in vivo* relevance of these cellular structures. Towards this end, we determined whether TNTs formed *in vivo* and could be detected in intact solid tumors from human patients. We microsectioned five intact samples of tumors obtained immediately after resection from patients with malignant pleural mesothelioma or poorly-differentiated lung adenocarcinoma. Three-dimensional reconstruction of these images revealed that TNTs were present in all five of the tumor specimens we had obtained from patients [13]. These nanotubes were noted to be long, thin, and suspended in the tumor matrix. Some TNTs were straight and others curved in a manner similar to the dendritic cell corneal model noted by Chinnery et al. [125]. Putative microvesicular cargo was visualized along the nanotubes as well, identical in appearance to what we had noted in TNTs *in vitro*. We and others have subsequently performed high-resolution microscopy and 3-dimensional reconstructions to confirm that nanotubes or nanotube-like structures are also present in resected tumors from animal models of cancer or from human patients with invasive malignancies such as an orthotopic model of murine osteosarcoma, human ovarian adenocarcinoma [56, 58], and laryngeal carcinoma [54].

## 8.9 Future Directions

The mechanisms of cell-to-cell communication in the complex heterogeneous tumor microenvironment are not well established. There is a significant gap in knowledge of how cell-cell communication and 'group-organization' takes place in the malignant tumor environment. The idea that interactions and transfer of molecular signals between malignant cells – and also between malignant and stromal cells – is crucial to invasion and progression of aggressive incurable malignancies is now recognized, but remains underdeveloped. Our current understanding of intercellular signaling via gap junctions, cytokines, exosomes, and microvesicles does not fully account for efficient communication between distant cells in the tumor matrix. TNTs are a novel candidate to elucidate how this process occurs.

The nomenclature for these structures has varied across publications from different research groups [19]. We propose that regardless of the various terms used – tunneling nanotubes, membrane nanotubes, intercellular bridges, etc. – they represent identical or at least similar structures that present an opportunity for an innovative and growing field of general cell biology, as well as cancer biology specifically. Examples of pertinent issues in cancer that can be addressed through investigation of TNTs include drug delivery of nano-sized cancer therapeutics [18] and development of cancer resistance to chemotherapy [19, 55, 56]. Developing 3-dimensional *in vitro* models to mimic aggressive solid malignant tumors, and also co-culture of malignant cells with a spectrum of stromal cells characteristic of the tumor microenvironment, will shed light on the mechanisms of TNT function. We anticipate that future studies in cancer will include examination of TNT formation between cancer cells co-cultured with stromal cells (e.g. fibroblasts) to more accurately reproduce tumor microenvironment conditions. It is estimated that stromal cells (e.g. cancer-associated fibroblasts, vascular endothelial cells, macrophages, etc.) comprise as much as 80–90 % of a malignant tumor's composition, further underscoring the need to identify modes of transport between malignant cells not located in immediate proximity [127, 128]. Elucidating TNTs and their role in the tumor microenvironment will explain how efficient communication and coordination takes place between distant and proximal cells in the tumor matrix.

In the context of tumor heterogeneity, mechanisms of communication represent an underexplored frontier that provides plausible explanations for how driving cellular factors, miRNAs, and other key signals promote tumor cell invasion in diseases such as colon, pancreatic, and ovarian cancers, and make these diseases extremely refractory to current standard treatments. As intra-tumor stroma proportion has been correlated with worse outcomes in several forms of cancer, future studies will include how TNTs fit into this cellular landscape. Approaches to the study of TNTs *in vitro* have, to date, varied widely across groups, and varies according to the types of cells being studied. As the field of study of TNTs in intercellular communication in cancer advances, methods will be adapted to fit specific cancer types of interest. Use of culture media designed to stimulate EMT also facilitate increased quantities of TNT formation [13]. To move this field forward, it is imperative to develop standard protocols for uniform approaches to the study of these unique, sensitive, and difficult-to-study cellular structures.

Investigating TNT function will also open the door to future exploration of clinically relevant issues such as tumor progression, tumor recurrence, and chemoresistance. For example, future studies will be designed to address (1) how cells at the leading edge of a three-dimensional invasive malignancy form TNTs to coordinate and synchronize unified tumor invasion, (2) how residual scattered malignant cells that remain following definitive surgical resection and chemotherapy treatment produce TNTs that may stimulate tumor recurrence, and (3) how malignant cells harness TNTs as conduits for propagation of resistance to chemotherapeutic agents, either through transfer of cellular chemoresistant factors, cargo that increases drug efflux (e.g. p-glycoprotein, a transporter protein encoded by the MDR1 gene [55] or through cell rescue from surrounding stromal cells). The recent discovery that TNTs can propagate antibiotic resistance of connected bacteria [129] provides further support for the idea that cancer cells may also be capable of acquiring genes that induce resistance to chemotherapy in this fashion.

In conclusion, the study of TNTs, a relatively new field in cancer cell biology, will produce important new knowledge about the tumor microenvironment that will be critical to advancing our understanding of how intercellular communication promotes progression, invasion, and recurrence of these malignancies.

**Acknowledgements** We thank our colleagues and research team for their helpful discussion and work on this topic, including Dr. Clifford Steer, Dr. Venugopal Thayanithy, and Snider Desir of the University of Minnesota, and Dr. Katia Manova-Todorova and Dr. Sho Fujisawa of Memorial Sloan-Kettering Cancer Center for helpful discussion and assistance with microscopy techniques. We also thank Michael Franklin, M.S., for critical review and assistance with editing this chapter. Due to space restrictions we could not cite many of the significant contributions made in this important and rapidly progressing field. This work was supported and funded by Minnesota Masonic Charities, Minnesota Medical Foundation (Grant number 4101-9225-12), Institutional Research Grant #118198-IRG-58-001-52-IRG94 and Research Grant RSG-13-381-01 from the American Cancer Society, the Deborah Powell Center for Women's Health at the University of Minnesota (Grant number PCWH-2013-002), the National Pancreas Foundation (Grant number 00035607), Karen Wyckoff Rein in Sarcoma Foundation (Grant number 1914), and the UMN Clinical and Translational Science Institute KL2 Scholar Award (to E.L.; Grant number 8UL1TR000114). Research reported in this publication was supported by the National Center for Advancing Translational Sciences of the National Institutes of Health Award Number UL1TR000114. The content is solely the responsibility of the authors and does not necessarily represent the official views of the National Institutes of Health.

## References

1. Kenny PA, Lee GY, Bissell MJ (2007) Targeting the tumor microenvironment. Front Biosci 12:3468–3474
2. Bissell MJ, Hines WC (2011) Why don't we get more cancer? A proposed role of the microenvironment in restraining cancer progression. Nat Med 17:320–329
3. Ruckert F, Grutzmann R, Pilarsky C (2012) Feedback within the inter-cellular communication and tumorigenesis in carcinomas. PLoS One 7, e36719

4. Axelrod R, Axelrod DE, Pienta KJ (2006) Evolution of cooperation among tumor cells. Proc Natl Acad Sci U S A 103:13474–13479
5. Hegmans JP, Bard MP, Hemmes A, Luider TM, Kleijmeer MJ et al (2004) Proteomic analysis of exosomes secreted by human mesothelioma cells. Am J Pathol 164:1807–1815
6. Bissell MJ, Radisky D (2001) Putting tumours in context. Nat Rev Cancer 1:46–54
7. Cottin S, Ghani K, de Campos-Lima PO, Caruso M (2010) Gemcitabine intercellular diffusion mediated by gap junctions: new implications for cancer therapy. Mol Cancer 9:141
8. Naus CC, Laird DW (2010) Implications and challenges of connexin connections to cancer. Nat Rev Cancer 10:435–441
9. Strassburg S, Hodson NW, Hill PI, Richardson SM, Hoyland JA (2012) Bi-directional exchange of membrane components occurs during co-culture of mesenchymal stem cells and nucleus pulposus cells. PLoS One 7, e33739
10. Bobrie A, Colombo M, Raposo G, Thery C (2011) Exosome secretion: molecular mechanisms and roles in immune responses. Traffic 12:1659–1668
11. Pap E, Pallinger E, Falus A (2011) The role of membrane vesicles in tumorigenesis. Crit Rev Oncol Hematol 79:213–223
12. Rustom A, Saffrich R, Markovic I, Walther P, Gerdes HH (2004) Nanotubular highways for intercellular organelle transport. Science 303:1007–1010
13. Lou E, Fujisawa S, Morozov A, Barlas A, Romin Y et al (2012) Tunneling nanotubes provide a unique conduit for intercellular transfer of cellular contents in human malignant pleural mesothelioma. PLoS One 7, e33093
14. Sowinski S, Jolly C, Berninghausen O, Purbhoo MA, Chauveau A et al (2008) Membrane nanotubes physically connect T cells over long distances presenting a novel route for HIV-1 transmission. Nat Cell Biol 10:211–219
15. Kwok R (2011) Cell biology: the new cell anatomy. Nature 480:26–28
16. Wang X, Veruki ML, Bukoreshtliev NV, Hartveit E, Gerdes HH (2010) Animal cells connected by nanotubes can be electrically coupled through interposed gap-junction channels. Proc Natl Acad Sci U S A 107:17194–17199
17. Lokar M, Iglic A, Veranic P (2010) Protruding membrane nanotubes: attachment of tubular protrusions to adjacent cells by several anchoring junctions. Protoplasma 246:81–87
18. Wang ZG, Liu SL, Tian ZQ, Zhang ZL, Tang HW et al (2012) Myosin-driven intercellular transportation of wheat germ agglutinin mediated by membrane nanotubes between human lung cancer cells. ACS Nano 6:10033–10041
19. Lou E, Fujisawa S, Barlas A, Romin Y, Manova-Todorova K et al (2012) Tunneling nanotubes: a new paradigm for studying intercellular communication and therapeutics in cancer. Commun Integr Biol 5:399–403
20. Gousset K, Schiff E, Langevin C, Marijanovic Z, Caputo A et al (2009) Prions hijack tunnelling nanotubes for intercellular spread. Nat Cell Biol 11:328–336
21. Guescini M, Leo G, Genedani S, Carone C, Pederzoli F et al (2012) Microvesicle and tunneling nanotube mediated intercellular transfer of g-protein coupled receptors in cell cultures. Exp Cell Res 318:603–613
22. Watkins SC, Salter RD (2005) Functional connectivity between immune cells mediated by tunneling nanotubules. Immunity 23:309–318
23. Brunkard JO, Runkel AM, Zambryski PC (2013) Plasmodesmata dynamics are coordinated by intracellular signaling pathways. Curr Opin Plant Biol 16:614–620
24. Hyun TK, Uddin MN, Rim Y, Kim JY (2011) Cell-to-cell trafficking of RNA and RNA silencing through plasmodesmata. Protoplasma 248:101–116
25. Nazim Uddin M, Kim JY (2013) Intercellular and systemic spread of RNA and RNAi in plants. Wiley Interdiscip Rev RNA 4:279–293
26. Zalepa-King L, Citovsky V (2013) A plasmodesmal glycosyltransferase-like protein. PLoS One 8, e58025
27. van Lent J, Storms M, van der Meer F, Wellink J, Goldbach R (1991) Tubular structures involved in movement of cowpea mosaic virus are also formed in infected cowpea protoplasts. J Gen Virol 72(Pt 11):2615–2623

28. Maitra A, Iacobuzio-Donahue C, Rahman A, Sohn TA, Argani P et al (2002) Immunohistochemical validation of a novel epithelial and a novel stromal marker of pancreatic ductal adenocarcinoma identified by global expression microarrays: sea urchin fascin homolog and heat shock protein 47. Am J Clin Pathol 118:52–59
29. Liang WS, Craig DW, Carpten J, Borad MJ, Demeure MJ et al (2012) Genome-wide characterization of pancreatic adenocarcinoma patients using next generation sequencing. PLoS One 7, e43192
30. Dekker TJ, van de Velde CJ, van Pelt GW, Kroep JR, Julien JP et al (2013) Prognostic significance of the tumor-stroma ratio: validation study in node-negative premenopausal breast cancer patients from the EORTC perioperative chemotherapy (POP) trial (10854). Breast Cancer Res Treat 139:371–379
31. de Kruijf EM, van Nes JG, van de Velde CJ, Putter H, Smit VT et al (2011) Tumor-stroma ratio in the primary tumor is a prognostic factor in early breast cancer patients, especially in triple-negative carcinoma patients. Breast Cancer Res Treat 125:687–696
32. Mesker WE, Liefers GJ, Junggeburt JM, van Pelt GW, Alberici P et al (2009) Presence of a high amount of stroma and downregulation of SMAD4 predict for worse survival for stage I-II colon cancer patients. Cell Oncol 31:169–178
33. Wu Y, Grabsch H, Ivanova T, Tan IB, Murray J et al (2013) Comprehensive genomic meta-analysis identifies intra-tumoural stroma as a predictor of survival in patients with gastric cancer. Gut 62:1100–1111
34. Song S, Nones K, Miller D, Harliwong I, Kassahn KS et al (2012) qpure: a tool to estimate tumor cellularity from genome-wide single-nucleotide polymorphism profiles. PLoS One 7:e45835
35. Huijbers A, Tollenaar RA, v Pelt GW, Zeestraten EC, Dutton S et al (2013) The proportion of tumor-stroma as a strong prognosticator for stage II and III colon cancer patients: validation in the VICTOR trial. Ann Oncol 24:179–185
36. Salter RD, Watkins SC (2006) Dynamic properties of antigen uptake and communication between dendritic cells. Immunol Res 36:211–220
37. Hase K, Kimura S, Takatsu H, Ohmae M, Kawano S et al (2009) M-Sec promotes membrane nanotube formation by interacting with Ral and the exocyst complex. Nat Cell Biol 11:1427–1432
38. Eugenin EA, Gaskill PJ, Berman JW (2009) Tunneling nanotubes (TNT) are induced by HIV-infection of macrophages: a potential mechanism for intercellular HIV trafficking. Cell Immunol 254:142–148
39. Sowinski S, Alakoskela JM, Jolly C, Davis DM (2011) Optimized methods for imaging membrane nanotubes between T cells and trafficking of HIV-1. Methods 53:27–33
40. Rudnicka D, Feldmann J, Porrot F, Wietgrefe S, Guadagnini S et al (2009) Simultaneous cell-to-cell transmission of human immunodeficiency virus to multiple targets through polysynapses. J Virol 83:6234–6246
41. Xu W, Santini PA, Sullivan JS, He B, Shan M et al (2009) HIV-1 evades virus-specific IgG2 and IgA responses by targeting systemic and intestinal B cells via long-range intercellular conduits. Nat Immunol 10:1008–1017
42. Galkina SI, Stadnichuk VI, Molotkovsky JG, Romanova JM, Sud'ina GF et al (2010) Microbial alkaloid staurosporine induces formation of nanometer-wide membrane tubular extensions (cytonemes, membrane tethers) in human neutrophils. Cell Adh Migr 4:32–38
43. Gurke S, Barroso JF, Hodneland E, Bukoreshtliev NV, Schlicker O et al (2008) Tunneling nanotube (TNT)-like structures facilitate a constitutive, actomyosin-dependent exchange of endocytic organelles between normal rat kidney cells. Exp Cell Res 314:3669–3683
44. Yasuda K, Park HC, Ratliff B, Addabbo F, Hatzopoulos AK et al (2010) Adriamycin nephropathy: a failure of endothelial progenitor cell-induced repair. Am J Pathol 176:1685–1695
45. Ranzinger J, Rustom A, Abel M, Leyh J, Kihm L et al (2011) Nanotube action between human mesothelial cells reveals novel aspects of inflammatory responses. PLoS One 6, e29537

46. Koyanagi M, Brandes RP, Haendeler J, Zeiher AM, Dimmeler S (2005) Cell-to-cell connection of endothelial progenitor cells with cardiac myocytes by nanotubes: a novel mechanism for cell fate changes? Circ Res 96:1039–1041
47. Plotnikov EY, Khryapenkova TG, Galkina SI, Sukhikh GT, Zorov DB (2010) Cytoplasm and organelle transfer between mesenchymal multipotent stromal cells and renal tubular cells in co-culture. Exp Cell Res 316:2447–2455
48. Cselenyak A, Pankotai E, Horvath EM, Kiss L, Lacza Z (2010) Mesenchymal stem cells rescue cardiomyoblasts from cell death in an in vitro ischemia model via direct cell-to-cell connections. BMC Cell Biol 11:29
49. Vidulescu C, Clejan S, O'Connor KC (2004) Vesicle traffic through intercellular bridges in DU 145 human prostate cancer cells. J Cell Mol Med 8:388–396
50. Mineo M, Garfield SH, Taverna S, Flugy A, De Leo G et al (2012) Exosomes released by K562 chronic myeloid leukemia cells promote angiogenesis in a Src-dependent fashion. Angiogenesis 15:33–45
51. Wu KF, Zheng GG, Ma XT, Song YH, Zhu XF (2011) Mechanism of leukemia relapse: novel insights on old problem. Zhongguo Shi Yan Xue Ye Xue Za Zhi 19:557–560
52. Hood JL, Pan H, Lanza GM, Wickline SA (2009) Paracrine induction of endothelium by tumor exosomes. Lab Invest 89:1317–1328
53. Lokar M, Kabaso D, Resnik N, Sepcic K, Kralj-Iglic V et al (2012) The role of cholesterol-sphingomyelin membrane nanodomains in the stability of intercellular membrane nanotubes. Int J Nanomedicine 7:1891–1902
54. Antanaviciute I, Rysevaite K, Liutkevicius V, Marandykina A, Rimkute L et al (2014) Long-distance communication between laryngeal carcinoma cells. PLoS One 9, e99196
55. Pasquier J, Galas L, Boulange-Lecomte C, Rioult D, Bultelle F et al (2012) Different modalities of intercellular membrane exchanges mediate cell-to-cell p-glycoprotein transfers in MCF-7 breast cancer cells. J Biol Chem 287:7374–7387
56. Pasquier J, Guerrouahen BS, Al Thawadi H, Ghiabi P, Maleki M et al (2013) Preferential transfer of mitochondria from endothelial to cancer cells through tunneling nanotubes modulates chemoresistance. J Transl Med 11:94
57. Thayanithy V, Babatunde V, Dickson EL, Wong P, Oh S et al (2014) Tumor exosomes induce tunneling nanotubes in lipid raft-enriched regions of human mesothelioma cells. Exp Cell Res 323:178–188
58. Thayanithy V, Dickson EL, Steer C, Subramanian S, Lou E (2014) Tumor-stromal cross talk: direct cell-to-cell transfer of oncogenic microRNAs via tunneling nanotubes. Transl Res 164:359–365
59. Acquistapace A, Bru T, Lesault PF, Figeac F, Coudert AE et al (2011) Human mesenchymal stem cells reprogram adult cardiomyocytes toward a progenitor-like state through partial cell fusion and mitochondria transfer. Stem Cells 29:812–824
60. Ma Z, Yang H, Liu H, Xu M, Runyan RB et al (2013) Mesenchymal stem cell-cardiomyocyte interactions under defined contact modes on laser-patterned biochips. PLoS One 8, e56554
61. Pankotai E, Cselenyak A, Ratosi O, Lorincz J, Kiss L et al (2012) The role of mitochondria in direct cell-to-cell connection dependent rescue of postischemic cardiomyoblasts. Mitochondrion 12:352–356
62. Figeac F, Lesault PF, Le Coz O, Damy T, Souktani R et al (2014) Nanotubular crosstalk with distressed cardiomyocytes stimulates the paracrine repair function of mesenchymal stem cells. Stem Cells 32:216–230
63. He K, Shi X, Zhang X, Dang S, Ma X et al (2011) Long-distance intercellular connectivity between cardiomyocytes and cardiofibroblasts mediated by membrane nanotubes. Cardiovasc Res 92:39–47
64. Steer CJ, Subramanian S (2012) Circulating microRNAs as biomarkers: a new frontier in diagnostics. Liver Transpl 18:265–269
65. Bronisz A, Godlewski J, Wallace JA, Merchant AS, Nowicki MO et al (2012) Reprogramming of the tumour microenvironment by stromal PTEN-regulated miR-320. Nat Cell Biol 14:159–167

66. Olive V, Bennett MJ, Walker JC, Ma C, Jiang I et al (2009) miR-19 is a key oncogenic component of mir-17-92. Genes Dev 23:2839–2849
67. Tabet F, Vickers KC, Cuesta Torres LF, Wiese CB, Shoucri BM et al (2014) HDL-transferred microRNA-223 regulates ICAM-1 expression in endothelial cells. Nat Commun 5:3292
68. Kent OA, McCall MN, Cornish TC, Halushka MK (2014) Lessons from miR-143/145: the importance of cell-type localization of miRNAs. Nucleic Acids Res 42:7528–7538
69. Kren BT, Wong PY, Sarver A, Zhang X, Zeng Y et al (2009) MicroRNAs identified in highly purified liver-derived mitochondria may play a role in apoptosis. RNA Biol 6:65–72
70. de Gassart A, Geminard C, Fevrier B, Raposo G, Vidal M (2003) Lipid raft-associated protein sorting in exosomes. Blood 102:4336–4344
71. Chen T, Guo J, Yang M, Zhu X, Cao X (2011) Chemokine-containing exosomes are released from heat-stressed tumor cells via lipid raft-dependent pathway and act as efficient tumor vaccine. J Immunol 186:2219–2228
72. Koumangoye RB, Sakwe AM, Goodwin JS, Patel T, Ochieng J (2011) Detachment of breast tumor cells induces rapid secretion of exosomes which subsequently mediate cellular adhesion and spreading. PLoS One 6, e24234
73. Sharma S, Das K, Woo J, Gimzewski JK (2014) Nanofilaments on glioblastoma exosomes revealed by peak force microscopy. J R Soc Interface 11:20131150
74. Veranic P, Lokar M, Schutz GJ, Weghuber J, Wieser S et al (2008) Different types of cell-to-cell connections mediated by nanotubular structures. Biophys J 95:4416–4425
75. Guizetti J, Gerlich DW (2010) Cytokinetic abscission in animal cells. Semin Cell Dev Biol 21:909–916
76. Jung S, Park J-Y, Joo J-H, Kim Y-M, Ha K-S (2011) Extracellular ultrathin fibers sensitive to intracellular reactive oxygen species: Formation of intercellular membrane bridges. Exp Cell Res 317:1763–1773
77. Wang Y, Cui J, Sun X, Zhang Y (2011) Tunneling-nanotube development in astrocytes depends on p53 activation. Cell Death Differ 18:732–742
78. Balzer EM, Whipple RA, Thompson K, Boggs AE, Slovic J et al (2010) c-Src differentially regulates the functions of microtentacles and invadopodia. Oncogene 29:6402–6408
79. Matrone MA, Whipple RA, Balzer EM, Martin SS (2010) Microtentacles tip the balance of cytoskeletal forces in circulating tumor cells. Cancer Res 70:7737–7741
80. Matrone MA, Whipple RA, Thompson K, Cho EH, Vitolo MI et al (2010) Metastatic breast tumors express increased tau, which promotes microtentacle formation and the reattachment of detached breast tumor cells. Oncogene 29:3217–3227
81. Whipple RA, Balzer EM, Cho EH, Matrone MA, Yoon JR et al (2008) Vimentin filaments support extension of tubulin-based microtentacles in detached breast tumor cells. Cancer Res 68:5678–5688
82. Whipple RA, Matrone MA, Cho EH, Balzer EM, Vitolo MI et al (2010) Epithelial-to-mesenchymal transition promotes tubulin detyrosination and microtentacles that enhance endothelial engagement. Cancer Res 70:8127–8137
83. Yoon JR, Whipple RA, Balzer EM, Cho EH, Matrone MA et al (2010) Local anesthetics inhibit kinesin motility and microtentacle protrusions in human epithelial and breast tumor cells. Breast Cancer Res Treat 129:691–701
84. Pontes B, Viana NB, Campanati L, Farina M, Neto VM et al (2008) Structure and elastic properties of tunneling nanotubes. Eur Biophys J 37:121–129
85. Castillo JA, Narciso DM, Hayes MA (2009) Bionanotubule formation from surface-attached liposomes using electric fields. Langmuir 25:391–396
86. Hayes MA, Pysher MD, Chen K (2007) Liposomes form nanotubules and long range networks in the presence of electric field. J Nanosci Nanotechnol 7:2283–2286
87. West J, Manz A, Dittrich PS (2008) Lipid nanotubule fabrication by microfluidic tweezing. Langmuir 24:6754–6758
88. Martinez-Zaguilan R, Seftor EA, Seftor RE, Chu YW, Gillies RJ et al (1996) Acidic pH enhances the invasive behavior of human melanoma cells. Clin Exp Metastasis 14:176–186

89. Li W, Ma Q, Li J, Guo K, Liu H et al (2011) Hyperglycemia enhances the invasive and migratory activity of pancreatic cancer cells via hydrogen peroxide. Oncol Rep 25:1279–1287
90. Chaika NV, Yu F, Purohit V, Mehla K, Lazenby AJ et al (2012) Differential expression of metabolic genes in tumor and stromal components of primary and metastatic loci in pancreatic adenocarcinoma. PLoS One 7, e32996
91. Manasson J, Tien T, Moore C, Kumar NM, Roy S (2013) High glucose-induced downregulation of connexin 30.2 promotes retinal vascular lesions: implications for diabetic retinopathy. Invest Ophthalmol Vis Sci 54:2361–2366
92. Lamouille S, Xu J, Derynck R (2014) Molecular mechanisms of epithelial-mesenchymal transition. Nat Rev Mol Cell Biol 15:178–196
93. Rozhin J, Sameni M, Ziegler G, Sloane BF (1994) Pericellular pH affects distribution and secretion of cathepsin B in malignant cells. Cancer Res 54:6517–6525
94. Delage E, Zurzolo C (2013) Exploring the role of lipids in intercellular conduits: breakthroughs in the pipeline. Front Plant Sci 4:504
95. Cooper J, Giancotti FG (2014) Molecular insights into NF2/Merlin tumor suppressor function. FEBS Lett 588:2743–2752
96. Arkwright PD, Luchetti F, Tour J, Roberts C, Ayub R et al (2010) Fas stimulation of T lymphocytes promotes rapid intercellular exchange of death signals via membrane nanotubes. Cell Res 20:72–88
97. Mesnil M, Yamasaki H (1993) Cell-cell communication and growth control of normal and cancer cells: evidence and hypothesis. Mol Carcinog 7:14–17
98. Klos KS, Warmka JK, Drachenberg DM, Chang L, Luxton GW et al (2014) Building bridges toward invasion: tumor promoter treatment induces a novel protein kinase C-dependent phenotype in MCF10A mammary cell acini. PLoS One 9, e90722
99. Tulla M, Helenius J, Jokinen J, Taubenberger A, Muller DJ et al (2008) TPA primes alpha2beta1 integrins for cell adhesion. FEBS Lett 582:3520–3524
100. Roe OD, Anderssen E, Helge E, Pettersen CH, Olsen KS et al (2009) Genome-wide profile of pleural mesothelioma versus parietal and visceral pleura: the emerging gene portrait of the mesothelioma phenotype. PLoS One 4, e6554
101. Anilkumar N, Parsons M, Monk R, Ng T, Adams JC (2003) Interaction of fascin and protein kinase Calpha: a novel intersection in cell adhesion and motility. EMBO J 22:5390–5402
102. Jongen WM, Fitzgerald DJ, Asamoto M, Piccoli C, Slaga TJ et al (1991) Regulation of connexin 43-mediated gap junctional intercellular communication by Ca2+ in mouse epidermal cells is controlled by E-cadherin. J Cell Biol 114:545–555
103. Mege RM, Matsuzaki F, Gallin WJ, Goldberg JI, Cunningham BA et al (1988) Construction of epithelioid sheets by transfection of mouse sarcoma cells with cDNAs for chicken cell adhesion molecules. Proc Natl Acad Sci U S A 85:7274–7278
104. Ady J, Desir S, Thayanithy V, Vogel RI, Moreira AL, Downey RJ et al (2014) Intercellular communication in malignant pleural mesothelioma: properties of tunneling nanotubes. Front Physiol 5:400
105. Cheng JQ, Lee WC, Klein MA, Cheng GZ, Jhanwar SC et al (1999) Frequent mutations of NF2 and allelic loss from chromosome band 22q12 in malignant mesothelioma: evidence for a two-hit mechanism of NF2 inactivation. Genes Chromosomes Cancer 24:238–242
106. Surace EI, Haipek CA, Gutmann DH (2004) Effect of merlin phosphorylation on neurofibromatosis 2 (NF2) gene function. Oncogene 23:580–587
107. Lopez-Lago MA, Okada T, Murillo MM, Socci N, Giancotti FG (2009) Loss of the tumor suppressor gene NF2, encoding merlin, constitutively activates integrin-dependent mTORC1 signaling. Mol Cell Biol 29:4235–4249
108. Tai CJ, Shen SC, Lee WR, Liao CF, Deng WP et al (2010) Increased cellular apoptosis susceptibility (CSE1L/CAS) protein expression promotes protrusion extension and enhances migration of MCF-7 breast cancer cells. Exp Cell Res 316:2969–2981
109. Ting S, Mairinger FD, Hager T, Welter S, Eberhardt WE et al (2013) ERCC1, MLH1, MSH2, MSH6, and betaIII-tubulin: resistance proteins associated with response and outcome to platinum-based chemotherapy in malignant pleural mesothelioma. Clin Lung Cancer 14:558 e553–567 e553

110. Oda E, Nakamura Y, Yamamoto M, Kojiro M (2005) Immunohistochemical distribution of tubulin beta II in human normal and neoplastic tissues. Kurume Med J 52:117–125
111. Gousset K, Marzo L, Commere PH, Zurzolo C (2013) Myo10 is a key regulator of TNT formation in neuronal cells. J Cell Sci 126:4424–4435
112. Wilson SM, Barbone D, Yang TM, Jablons DM, Bueno R et al (2008) mTOR mediates survival signals in malignant mesothelioma grown as tumor fragment spheroids. Am J Respir Cell Mol Biol 39:576–583
113. Singhal S, Wiewrodt R, Malden LD, Amin KM, Matzie K et al (2003) Gene expression profiling of malignant mesothelioma. Clin Cancer Res 9:3080–3097
114. Dauphin M, Barbe C, Lemaire S, Nawrocki-Raby B, Lagonotte E et al (2013) Vimentin expression predicts the occurrence of metastases in non small cell lung carcinomas. Lung Cancer 81:117–122
115. Bukoreshtliev NV, Wang X, Hodneland E, Gurke S, Barroso JF et al (2009) Selective block of tunneling nanotube (TNT) formation inhibits intercellular organelle transfer between PC12 cells. FEBS Lett 583:1481–1488
116. Onfelt B, Nedvetzki S, Benninger RK, Purbhoo MA, Sowinski S et al (2006) Structurally distinct membrane nanotubes between human macrophages support long-distance vesicular traffic or surfing of bacteria. J Immunol 177:8476–8483
117. Gonzalez-Angulo AM, Meric-Bernstam F (2010) Metformin: a therapeutic opportunity in breast cancer. Clin Cancer Res 16:1695–1700
118. Beckner ME, Gobbel GT, Abounader R, Burovic F, Agostino NR et al (2005) Glycolytic glioma cells with active glycogen synthase are sensitive to PTEN and inhibitors of PI3K and gluconeogenesis. Lab Invest 85:1457–1470
119. Tennant DA, Duran RV, Gottlieb E (2010) Targeting metabolic transformation for cancer therapy. Nat Rev Cancer 10:267–277
120. Kim EK, Yun SJ, Ha JM, Kim YW, Jin IH et al (2011) Selective activation of Akt1 by mammalian target of rapamycin complex 2 regulates cancer cell migration, invasion, and metastasis. Oncogene 30:2954–2963
121. Faoro L, Singleton PA, Cervantes GM, Lennon FE, Choong NW et al (2010) EphA2 mutation in lung squamous cell carcinoma promotes increased cell survival, cell invasion, focal adhesions, and mammalian target of rapamycin activation. J Biol Chem 285:18575–18585
122. Motzer RJ, Escudier B, Oudard S, Hutson TE, Porta C et al (2008) Efficacy of everolimus in advanced renal cell carcinoma: a double-blind, randomised, placebo-controlled phase III trial. Lancet 372:449–456
123. Oskarsson T, Nagorny P, Krauss IJ, Perez L, Mandal M et al (2010) Diverted total synthesis leads to the generation of promising cell-migration inhibitors for treatment of tumor metastasis: in vivo and mechanistic studies on the migrastatin core ether analog. J Am Chem Soc 132:3224–3228
124. Chen L, Yang S, Jakoncic J, Zhang JJ, Huang XY (2010) Migrastatin analogues target fascin to block tumour metastasis. Nature 464:1062–1066
125. Chinnery HR, Pearlman E, McMenamin PG (2008) Cutting edge: Membrane nanotubes in vivo: a feature of MHC class II+ cells in the mouse cornea. J Immunol 180:5779–5783
126. Seyed-Razavi Y, Hickey MJ, Kuffova L, McMenamin PG, Chinnery HR (2013) Membrane nanotubes in myeloid cells in the adult mouse cornea represent a novel mode of immune cell interaction. Immunol Cell Biol 91:89–95
127. Dvorak HF (1986) Tumors: wounds that do not heal. Similarities between tumor stroma generation and wound healing. N Engl J Med 315:1650–1659
128. Elenbaas B, Weinberg RA (2001) Heterotypic signaling between epithelial tumor cells and fibroblasts in carcinoma formation. Exp Cell Res 264:169–184
129. Ady J, Desir S, Thayanithy V, Vogel RI, Moreira AL, Downey RJ et al. (2014) Intercellular communication in malignant pleural mesothelioma: properties of tunneling nanotubes. Frontiers in Physiology 5:400

# Chapter 9
# Cell-Cell Fusion, Chemotaxis and Metastasis

**George E. Parris**

**Abstract** This review covers the concept of cell-cell fusion as an important feature of cancer progression and its consequences. The fusion of abnormal (mutated proliferating) cells with "task force" cells of the immune system (e.g., leukocytes and stem cells) that respond to tissue damage and stress has now been demonstrated. Evidence is being accumulated that these hybrid cells incorporate the motile and flexible characteristics of the "task force" cells with the lack of cell-cycle control of the proliferating cells. These characteristics may be the primary features that facilitate invasion and metastasis of cancers. The behavior of leukocytes and stem cells can, thus, be used to understand, predict and ultimately control the most dangerous features of cancer.

**Keywords** Cell-cell fusion • Chemotaxis • Metastasis • Inflammation • ADAM • TNFalpha • CXCL • CXCR • Leukocyte • Stem cell

## 9.1 Introduction

For the last 65 years, the mutations theory of cancer causation [1–4] has been the focus of risk assessment, prevention and treatment of cancer. Nonetheless, evidence is slowly accumulating that mutations are not sufficient to explain cancer [3, 5–7] and other mechanisms are being considered [8–12]. In particular, it is becoming evident that cell-cell fusion plays a significant role in progression (i.e., evolution) of cancers and may be the factor changing a benign neoplasm (caused by mutations) into a clinically significant cancer [13–30]. Evidence is even stronger that cell-cell fusion is common and perhaps predominant in the metastasis of tumors from the site of origin [19, 20, 28, 31–45]. The model that is emerging goes something like this:

**Step 1**. Mutations or other events (e.g., wounds and/or infections) create a disturbance in normal tissue characterized by proliferation. One of the responses to mutations at the cellular level is endoreduplication in which the cell tries to

G.E. Parris (✉)
Department of Chemistry, Montgomery College,
9601 Warfield Rd, Gaithersburg, MD 20882, USA
e-mail: antimony_121@Hotmail.com



M. Kandouz (ed.), *Intercellular Communication in Cancer*,
DOI 10.1007/978-94-017-7380-5_9

defend against loss of genetic information by doubling its genome (producing sable and resilient tetraploid cells [46, 47]) [30, 48–52]. The duplicate genome is available for homologous recombination, which increases the cell's tolerance to mutations [47, 53]. Various combinations of mutations and effects of xenobiotics can overcome telomere shortening and release the abnormal cells from internal control of the cell cycle. Nonetheless, the overall pattern of progression in the abnormal clones is loss of genetic information trending towards eventual senescence or extinction. However, along the way to extinction, the things that may be lost include the markers that identify the cell to the immune system and the abnormal cells may escape immune surveillance [54]. Cell-cell fusion (somatic sex) among the abnormal cells followed by further loss of chromosomes resulting in various metastable aneuploidy clones may facilitate local expansion of a cancer [15, 22, 30, 45, 55–57].

**Step 2**. Intercellular communications with the surrounding normal cells (stromal cells) cause various degrees of inflammation [30, 44, 45, 58–61]. Inflammation is accompanied by release of chemokines [62–104]. Chemokines attract mature leukocytes and progenitor cells (i.e., hematopoietic stem cells (HSCs) and mesenchymal stem cells (MSCs), a.k.a., multipotent stromal cells). The cells that converge on sites of inflammation represent a task force designed to stop blood loss, fight infection, clear damaged tissue, and repair damage. However, these cells bring with them tools that can be turned against the body: (i) they all have motile phenotypes, (ii) many of them are also fusogenic, (iii) some are immune-privileged, and (iv) some are progenitor cells that can differentiate.

**Step 3**. In the course of executing their various tasks, the motile and fusogenic stromal cells may produce new hybrids, which combine the lack of cell cycle control of the neoplasm with the motile phenotype of the cells from the task force [44, 45]. Suddenly, in essentially one step (not dozens of random mutations), a new motile cancer clone is formed, which may retain fusogenic and/or progenitor characteristics (i.e., cancer stem cells) [17, 19, 21, 24–26, 28–30, 40–42, 44, 45, 105–123].

**Step 4**. The motile cancer cells enter the blood or lymph responding to chemokines, which cause them to target specific organs and tissues, where they lodge and produce remote secondary tumors [19, 21, 36, 37, 39–44, 70, 86, 110, 124–131].

In this review, I will summarize the role played by cytokines in leading to cell-cell fusion and in metastasis.

## 9.2 Inflammation Attracts Motile "Task Force Cells"

Here I am introducing the term "Task Force Cells" to encompass those cells that circulate in the blood or lymph and are task to mitigate internal or external damage to the normal tissues. The major tasks of defense include (i) stopping blood flow in the event of a break in the skin or blood vessel; (ii) defeating invading cells, viruses, and toxins; (iii) clearing away damaged tissue/cells, (iv) regenerating viable tissues or patching with scar tissue. In terms of development, some of these cells are fully differentiated (platelets, leukocytes) and others are multi-potent (HSC or MSC). The stem cells can be generally classified as the HSC that back up the mature front line cells (i.e., providing replacements and reinforcements to leukocytes and clot-forming cells) and MSCs that will follow the initial response by forming replacement tissues of various types.

When blood vessels are broken, a complex response is initiated: platelets adhere to the endothelial walls and initiate blood clotting. They also release soluble P-selectin, which seems to prime leukocytes to initiate rolling [132, 133]. One possible mechanism of P-selectin activation would be to unmask the L-selectins on leukocytes. Concurrently, RNA is released from damaged or abnormal cells [134–137]. These signals, appear to initiate release of ADAM (A disintegrin and metalloproteinase, especially ADAM 10 and 17) [138–143] enzymes from local cells [138, 144–151]. The ADAMs cleave cell-surface CD44 (which tethers cells to the intercellular hyaluronan matrix [152–155]), cadherin (which tethers cells to one another [156–159]) and TNFalpha (which sends a apoptosis signal to nearby cells [160–163]).

Angiostatin (a 38 kDa protease fragment of plasminogen released by the action of matrix metalloproteinases (MMPs) and urokinase-type plasminogen activators (uPA)) [164, 165] and endostatin (a C-terminal, 20 kDa, zinc-binding protein cut from collagen XVIII by collagenase) [166, 167] were discovered to be naturally occurring inhibitors of angiogenesis in the 1990s. The mode of action of angiostatin and endostatin in reducing angiogenesis is still not known. One of their lesser known effects is to suppress migration of leukocytes, especially neutrophiles (a.k.a., polymorphonuclear leukocytes). Benelli et al. [168] showed that angiostatin reduced the migration of monocytes (which differentiate into macrophages) and blocked chemotaxis of neutrophiles. Similarly, in 1975 (22 years before the discovery of endostatin in 1997), Stecher [169] showed that degradation of collagen by collagenase produced a mixture of peptides that inhibited leukotactic activity.

ADAMs also cut the cell-surface proteins at bind CXCL16 to the surface of cells. Soluble CXCL16 shed from the surfaces of nearby cells quickly penetrates into the local blood vessels and triggers circulating leukocytes (expressing the receptor CXCR6) to roll to a stop at that point. The mechanism of leukocyte rolling to a stop on the endothelial cells that line blood vessels is complex and beyond the scope of this review. Nonetheless, I note that rolling involves activation of E-selectins on the endothelial cells, which cause adherence with complementary L-selectins on leukocytes [170–186]. As a rule, circulating stem cells (HSCs and MSCs) are not immediately snared by endothelial cells expressing E-selectins because they do not seem to naturally express L-selectins [185]. The association of stems cells to endothelial blood vessel walls appears to be dominated by interactions of integrins [187].

Tissue-resident mesenchymal stem cells (MSCs) near the site of stress are activated and release cytokines [188] especially CXCL12 (also and more descriptively known as stromal-derived factor-1, SDF-1). The resident MSCs and leukocytes also express a very strong receptor of CXCL12 known as CXCR7 [88, 189–193]. The presence of CXCR7 normally internalizes CXCL12, but under stress, there is a net release of CXCL12 into the blood. The gradient of CXCL12 facilitates the homing of (bone marrow-derived) stem cells expressing the CXCL12-receptor CXCR4 to the site of tissue damage/inflammation/stress.

## 9.3 Leukocytes and Stem Cells Fuse with Tumor Cells

Tumors apparently are initiated by an unfortunate combination of somatic mutations and/or persistent exposure to chemical agents (e.g., proteins produced by viral infection [194]) that provoke proliferation [195–200] while blocking apoptosis [194, 201, 202]. Although these neoplasms may grow (proliferate) rapidly, they progress (evolve) very slowly. Indeed, accumulation of mutations or induction of genomic instability in asexual clones appears to drive the neoplasm towards extinction as a result of progressive loss of genes that code proteins [23]. The process is known as Muller's ratchet and has been studied in viruses, prokaryotes, single-cell eukaryotes and individual chromosomes in sexually reproducing multicellular organisms (e.g., the Y chromosome) [203–209].

There are internal and external regulators of cell proliferation. Once internal and external control of the cell cycle has been overcome, a clone is free to evolve as a separate biological species. This important point was first made by Huxley in 1956 [210, 211] and was re-introduced by Duesberg et al. based on karyotype [56, 212, 213], but the most important point is that if a cancer clone is envisioned as a parasite species, then the normal laws of ecology and evolution will apply [18, 20]. As a rule, species that evolve faster have an ecological advantage; thus, we see that in established tumors, non-human (i.e., aneuploidy) karyotypes dominate over merely mutated human karyotypes. Gains and losses of whole chromosomes and genes represent a faster way to evolve than successive random point mutations [9, 10, 57, 214, 215]. Parris has pointed out that cell-cell fusion plays the role of sex in allowing recombination of genomes in the cancer species [18, 20, 29]. Sexual reproduction via cell-cell fusion is the necessary event needed to allow neoplasms to escape from the limitations of Muller's ratchet; and it provides a very rapid mode of acquiring complex phenotypes (e.g., motility, homing or drug-resistance) that would take numerous (unlikely) genetic and epigenetic modifications by any asexual mechanism of evolution. These phenotypes can often be compared to stem cells [28, 30, 44, 45].

Parris views sexual reproduction by cell-cell fusion to be the critical turning point in the evolution (progression) of insignificant benign neoplasia into rapidly progressing cancer. It is worth noting that the cell-cell fusion hypothesis and cell clone ecology model of cancer [20] provides a convenient explanation for the ideas of cancer stem cells (CSCs) [216]. The formation of a hybrid through cell-cell fusion is comparable to actions of a stem cell in producing a new clone. In particular, when one of the fusion partners happens to be a stem cell a pattern of behavior similar to differentiation into different tissue types may be expected of the hybrid [26, 28, 30, 41, 44, 114, 118, 125, 217–220].

Let us look at some of the experimental evidence for cell-cell fusion in vitro, in vivo and in actual cancers. There are a variety of ways that cell-cell fusion can be distinguished from endoreduplication, which also forms polyploidy cells with potentially unstable karyotypes, but which is unlikely to being new capabilities to the clone and appears to delay, but not stop Muller's ratchet. For the most part, cells have been marked with fluorescent proteins or in some cases male bone marrow donors have been introduced to female hosts. Let us consider some of the observations:

Stem cells and leukocytes have naturally fusogenic phenotypes [114, 218, 220–229]. Viral infections can also induce fusogenic behavior in somatic cells [18, 22, 28, 38, 117, 230–240]. There is extensive evidence that fusion of stem cells with tumor cells does, in fact, confer on the hybrid capabilities of the stem cell [30, 44, 105, 109, 118, 119, 218, 241–253]. This occurs even when cells of different species are fused [13, 14, 254–257], i.e., independent of karyotype or epigenetic state. Generally, the resulting polyploid genome is unstable and soon begins losing excess chromosomes via neosis (e.g., nuclear budding) [18, 55, 258–264]. The stem cell-somatic cell hybrids can undergo differentiation to multiple tissue types [114, 218, 220, 243, 244, 265, 266]. Leukocytes also fuse with tumor cells [36, 37, 267–270].

The consequences of fusion of stem cells, leukocytes, or virus-infected stromal cells with proliferating mutated somatic cells are dramatic. Endoreduplication and other processes that double the genome of an individual (normal) cell frequently result in stable polyploids that have the advantage of redundant genes, which may make the polyploids very resilient to genotoxic agents [46, 47]. On the other hand, cell-cell fusion is more likely to produce unstable genomes (in part because of lack of synchronization of the cell cycles of the fused cells, e.g., premature chromatin condensation [271, 272]) that give rise to a variety of aneuploidy clones. This may be the essential step in radical clone evolution (tumor progression). The mitotic catastrophe that gives rise to aneuploidy cells may either be neosis (i.e., nuclear budding to produce microcells, Raju cells [55, 258]) or chromothripsis with asymmetric mitosis [273–275]. The newly created aneuploid cells (i) may not be viable, (ii) may have unstable genomes that continue changing or (iii) may be relatively stable and persist for long periods of time. It is relevant to mention here that long-established cell lines (e.g., HeLa) actually undergo changes in karyotype (and likely chromosomal mutations) during their long "lives," which presumably converge on stable genomes [276–279]. Regardless, the aneuploid cells will be distinctly different from the cells of the host [9, 18, 56].

## 9.4 Cell-Cell Fusion and Evasion of the Immune System

Successful cancers must escape the immune system [280–282]. Although cancer cells can be viewed as new species, they have the unique advantage of having been derived from cells of the host. Thus, it should be noted that in the early stages of tumor development, the cells are very similar to normal human cells in terms of their cell-surface proteins and thus, they tend to escape the immune response. Moreover, hypoxia-induced expression of ADAM-10 may facilitate shedding of MHC class I-related chain molecules A and B and reduce the effectiveness of immune surveillance on tumor cells [283, 284]. As tumor cells evolve (i.e., become aneuploid), they may lose the MHC genes altogether [285].

Loss of MHC expression would be important for transmission of cancer among different individuals [286], but it is not critical for a parasite derived from the host. Normally, we would expect cells that have lost the MHC genes to still be under the control of natural killer (NK) cells of the immune system [287]. But, natural selection acting on randomly aneuploid cells circumvents this control in the following way: Normal cells express both the transforming growth factor-beta (TGF-beta) and its receptor (TGFBR2) [288] such that expression of TGF-beta automatically is received by the cell and suppresses proliferation. This is important because TGF-beta is a cytokine that suppresses NK cell maturation [289, 290]. In cells that are fusing and producing a variety of aneuploid daughters, a distinct genotype will be favored in which the cell loses its TGFBR2 gene [288, 291, 292] and acquires multiple copies of the TGF-beta gene [293–297]. High levels of TGF-beta gene are found in individual circulating tumor cells [298]. Indoleamine-2,3-dioxygenase (IDO) has also be found to regulate TGF-beta [101]. These cells, thus, proliferate rapidly and suppress the NK cells that would normally attack proliferating cells. This genotype and phenotype is readily assessable through cell-cell fusion, but would be difficult to achieve by mutations alone.

In the lymphatic system, antigen-presenting (mature) dendritic cells activate other immune cells. Some anti-cancer success has been achieved by artificially fusing cancer cells to mature dendritic cells [299–305]. Ironically, antigen-capturing (immature) dendritic cells, which sample surrounding cells looking for foreign antigen from parasites (viruses, bacteria and cancer cells) are fusion-prone and seem to produce dangerously aggressive hybrids [306]. It has been noted that fusion of immature dendritic cells with tumor cells produces hybrids that are more mobile that hybrids formed with mature dendritic cells [307]. Nonetheless, TGF-beta tends to immobilize dendritic cells and other motile immune cells in tumors and the stromal compartment and inhibits immune response in the tumor and lymph system [308, 309].

## 9.5 Passive Metastasis

Metastasis is the most dangerous characteristic that a neoplastic clone can acquire. While a clone is confined to a single locality, its damage is generally limited and treatable, but once it can move through the body its potential for harm is endless. Indeed, an important medical strategy might be directed at stopping metastasis instead of "curing" cancer.

Metastasis involves transportation of tumorigenic cells via the blood and lymph. Intercellular fluids normally flow from the arterial (high pressure) blood vessels into the tissues and drain from the tissues into the (low pressure) lymph system. In the lymph system, the intercellular fluids pass though the lymph nodes and enter the venous blood stream via the supraclavicular lymph ducts. In the immune response, fluid usually collects in the affected tissue (edema) because blood vessels are made more porous by TGF-beta-induced VEGF (formerly known as "vascular permeability factor" [260–265]), which appears to cause the pericytes (which seal the blood vessels) to retreat into the tissue [297, 310, 311]. In this situation, interstitial fluids may flow into the lymph system or directly back into the blood system.

In passive metastasis, tumor cells probably become dislodged as a result of the ongoing immune response to inflammation (described above) in which ADAM-10, -12 and -17 enzymes break the cadherin, CD44 and integrin connections to intercellular matrix and liberate cells [143, 153, 154, 160, 312–315]. Passive movement of these cells into the lymph and blood presumably occurs. Cells in the lymph first travel to the lymph node draining the tumor site and encounter lymphocytes that may have become activated by mature dendritic cells to tumor-derived antibodies. Here they may be destroyed or they may fuse with lymphocytes producing motile hybrids [316] (see below). Most single cells released into blood are soon destroyed [317] and the rest are apt to collect in the first capillaries encountered [318]; for cells in the body entering the venous circulation (including cells that passed through the lymph system), the first capillaries would usually be in the lungs (via pulmonary circulation). The liver is also designed to intercept cells and does receive metastases from the intestines (portal circulation). But, cells that become lodged in the lungs or liver (micrometastases) may not have the ability to penetrate into the tissue (extravasation) [317, 319–321]. Passively circulating tumor cells trapped in lung tissue, however, seem to acquire the ability to penetrate into remote tissue if accompanied by stromal cells or neutrophils [320, 322, 323]. It may be speculated that the effects of the stromal cells are the result of cell-cell fusion as discussed below.

## 9.6 Active Metastasis

Active metastasis requires that a tumor cell (i) be motile, (ii) possess homing capabilities and (iii) possess the ability to penetrate into the tissue of various organs (extravasation). These complex characteristics seem unlikely to arise *de novo* by accumulation of mutations in neoplastic cells. However, these characteristics are all possessed by fusogenic leukocytes and stem cells [324, 325]. Fusion among random tumor cells may occur as a result of viral infection or even the actions of ADAM9 and ADAM12 (a.k.a., meltrin-gamma and -alpha, respectively) [326–329]. These particular ADAMs have affinity for specific integrins [330–334] and are highly correlated with invasive tumors [313, 314, 335–337]. But, they do not seem to facilitate acquisition of the full complement of traits associated with the phenotype of active metastases. Thus, it has been proposed by Goldenberg [255–257], Pawelek and co-workers [19, 21, 39–42, 268, 269, 338], and others [20, 29, 38, 44, 125, 339–344] that fusion of tumor cells with leukocytes or (bone marrow-derived) stem cells (HSCs or MSCs) is the most likely (fastest) way for a tumor cell to acquire these traits. The viable hybrids (e.g., quasi-tetraploid) may also be resistant to genotoxic drugs, cytotoxic drugs or radiation [46, 120, 219, 345]. It has also been pointed out that slowly-proliferating tumor cells, which are likely to survive cytotoxic or genotoxic chemotherapy because they are not in the cell cycle during the period of maximum dose, might subsequently fuse with leukocytes or stem cells to produce a recurrence of neoplastic growth [18, 20, 24, 28, 219, 346].

The fate of motile cells is strongly affected by the gradients of chemokines in the vicinity of the primary tumor [347, 348]. The same gradient of CXCL12 and CXCL16 that brought the leukocytes and stem cells to the tumor tends to hold them and their fusion hybrids that express CXCR4 and/or CXCR6 there [88, 190, 349]. If the source of CXCL12/CXCL16 is suppressed, for example, by over expression of CXCR7, which facilitates the destruction of the chemokines [75, 87, 88, 92, 190, 192, 193, 350–355], motile cells including potentially metastatic tumor hybrids will begin to actively leave the tumor (primary or secondary). In particular, surgical removal of the primary tumor may inadvertently reduce the local concentration of chemokines and, thus, release motile tumor hybrids into the blood and lymph system.

It is interesting that this is exactly the same sort of argument that was made for angiostatin and endostatin [164, 167, 356], which are released from tumors and appears to suppress angiogenesis at remote site where micrometastases have become established. There are a number of anecdotal clinical observations and some systematic trials involving resection of primary tumors in humans. Most of these confirm the idea that removal of the primary tumor can be followed by growth of metastases [357–359]. There are also examples in mice [358] and cats [360]. The observations do not resolve the question of whether the individual metastases were founded before or after the primary tumor was removed, or both.

Successful metastasis requires tumorigenic cells that are both mobile and motile. While mobile (e.g., anchorage-independent) tumorigenic cells might passively enter the blood or lymph from a tumor and be randomly deposited in tissues by blood flow [35, 317, 320, 322, 361, 362], extravasation from the blood stream to a remote site requires (i) the active "skills" typical of motile cells (e.g., leukocytes or mesenchymal stem cells, MSCs) [363, 364], (ii) stimuli from the tissue that attract leukocytes [365, 366] and (iii) expression of glycoproteins that facilitate arrest of leukocytes such as E-selectin [367, 368]. Motile tumorigenic cells typically display chemokine receptors especially CXCR4 or CXCR6, which cause the cells to be attracted to sources of CXCL12 (a.k.a., stromal-derived factor-1, SDF-1) and CXCL16, respectively [76, 131, 369]. Homing tumorigenic cells might arise from appropriate mutations in differentiated cells or motile stem cells; or they might arise from fusion of MSCs or leukocytes with tumorigenic calls [40, 42]. The leukocyte fusion hypothesis of Pawelek et al. [40, 42, 43, 370] is particularly attractive because it provides a general "one-step" mechanism for conferring leukocyte-like properties onto any tumorigenic cell [324].

Regardless of their origin, motile cells expressing chemokine receptors follow chemokine gradients to tissues where the chemokines are released. The concentrations of chemokines in "background" tissues is suppressed by elimination mechanisms including expression of CXCR7, which scavenges CXC12 (SDF-1) and facilitates its destruction [371]. Bone marrow is predictably one of the strongest sources of secreted chemokines (especially CXCL12 and CXCL16) because it is the place where leukocytes and mesenchymal cells originate and are stored. Release of immune cells from bone marrow involves additional signaling via sphingosine-1-phosphate [372–374], which appears to over-ride the standing CXCL12/CXCL16 gradient. Thus, it is not surprising that bone marrow is one of the principal sites of metastasis [80, 86, 375–378] and metastasis there is not favored by blood flow patterns.

Remote sites of tissue damage, inflammation or immune response typically release chemotactic agents that also attract tumor hybrids of leukocytes and stem cells. These tissues, thus, are also likely to attract homing tumorigenic cells. For example, tumor cells home to injured tissue (sites of surgery or injection) during the healing process [365, 379–385].

## 9.7 Conclusions

Cell-cell fusion involves intercellular communication both in terms of sharing genetic information to produce hybrid cells with dangerous new abilities and in the sense that the hybrid cells frequently retain elements of the phenotype of the leukocyte or stem cell that fused with the cancer cell. Much of the behavior of cancer can be understood, predicted and potentially controlled by analogy with the behavior of leukocytes and stem cells.

For example, creating "traps" for migrating (motile) cancer cells by incorporating alpha- or beta-emitting isotopes on the inner surfaces of inert, hollow microspheres or tubes that are "baited" with slow release sources of CXC12 or similar chemokine that will attract motile cells might be an approach to improve brachytherapy.

## References


1. Strong LC (1949) A new theory of mutation and the origin of cancer. Yale J Biol Med 21(4):293–299
2. Berenblum I, Shubik P (1949) An experimental study of the initiating state of carcinogenesis, and a re-examination of the somatic cell mutation theory of cancer. Br J Cancer 3(1):109–118
3. Nordling CO (1955) Evidence regarding the multiple mutation theory of the cancer-inducing mechanism. Acta Genet Stat Med 5(2):93–104
4. Ames BN, Durston WE, Yamasaki E, Lee FD (1973) Carcinogens are mutagens: a simple test system combining liver homogenates for activation and bacteria for detection. Proc Natl Acad Sci U S A 70(8):2281–2285
5. Fardon JC (1953) A reconsideration of the somatic mutation theory of cancer in the light of some recent developments. Science 117(3043):441–445
6. Sonnenschein C, Soto AM (2000) Somatic mutation theory of carcinogenesis: why it should be dropped and replaced. Mol Carcinog 29(4):205–211
7. Soto AM, Sonnenschein C (2004) The somatic mutation theory of cancer: growing problems with the paradigm? Bioessays 26(10):1097–1107
8. Sell S (2004) Stem cell origin of cancer and differentiation therapy. Crit Rev Oncol Hematol 51(1):1–28
9. Duesberg P, Li R, Fabarius A, Hehlmann R (2005) The chromosomal basis of cancer. Cell Oncol 27(5–6):293–318
10. Nicholson JM, Duesberg P (2009) On the karyotypic origin and evolution of cancer cells. Cancer Genet Cytogenet 194(2):96–110
11. Heng HH, Stevens JB, Bremer SW, Ye KJ, Liu G, Ye CJ (2010) The evolutionary mechanism of cancer. J Cell Biochem 109(6):1072–1084
12. Huang S (2012) Tumor progression: chance and necessity in Darwinian and Lamarckian somatic (mutationless) evolution. Prog Biophys Mol Biol 110(1):69–86
13. Goldenberg DM, Bhan RD, Pavia RA (1971) In vivo human-hamster somatic cell fusion indicated by glucose 6-phosphate dehydrogenase and lactate dehydrogenase profiles. Cancer Res 31(8):1148–1152
14. Goldenberg DM, Pavia RA (1975) Oncogenesis by interspecific interaction of malignant murine and non-malignant hamster cells in vitro. Int J Cancer 15(2):282–300
15. Bremermann HJ (1979) Theory of spontaneous cell fusion. Sexuality in cell populations as an evolutionarily stable strategy. Applications to immunology and cancer. J Theor Biol 76(3):311–334
16. Busund LT, Killie MK, Bartnes K, Seljelid R (2002) Spontaneously formed tumorigenic hybrids of Meth A sarcoma and macrophages grow faster and are better vascularized than the parental tumor. Int J Cancer 100(4):407–413
17. Chakraborty AK, Pawelek JM (2003) GnT-V, macrophage and cancer metastasis: a common link. Clin Exp Metastasis 20(4):365–373
18. Parris GE (2005) Clinically significant cancer evolves from transient mutated and/or aneuploid neoplasia by cell fusion to form unstable syncytia that give rise to ecologically viable parasite species. Med Hypotheses 65(5):846–850
19. Pawelek JM (2005) Tumour-cell fusion as a source of myeloid traits in cancer. Lancet Oncol 6(12):988–993
20. Parris GE (2006) The cell clone ecology hypothesis and the cell fusion model of cancer progression and metastasis: history and experimental support. Med Hypotheses 66(1):76–83
21. Pawelek J, Chakraborty A, Lazova R, Yilmaz Y, Cooper D, Brash D, Handerson T (2006) Co-opting macrophage traits in cancer progression: a consequence of tumor cell fusion? Contrib Microbiol 13:138–155

22. Duelli DM, Padilla-Nash HM, Berman D, Murphy KM, Ried T, Lazebnik Y (2007) A virus causes cancer by inducing massive chromosomal instability through cell fusion. Curr Biol 17(5):431–437
23. Parris GE (2008) Cell-cell fusion is the rate-limiting step in causation and progression of clinically significant cancers. Cancer Genet Cytogenet 185(2):113
24. Dittmar T, Nagler C, Schwitalla S, Reith G, Niggemann B, Zanker KS (2009) Recurrence cancer stem cells--made by cell fusion? Med Hypotheses 73(4):542–547
25. Lu X, Kang Y (2009) Cell fusion as a hidden force in tumor progression. Cancer Res 69(22):8536–8539
26. Lu X, Kang Y (2011) Cell fusion hypothesis of the cancer stem cell. Adv Exp Med Biol 714:129–140
27. Zhang S, Mercado-Uribe I, Xing Z, Sun B, Kuang J, Liu J (2013) Generation of cancer stem-like cells through the formation of polyploid giant cancer cells. Oncogene 33:116–128
28. Dittmar T, Nagler C, Niggemann B, Zanker KS (2013) The dark side of stem cells. Triggering cancer progression by cell fusion. Curr Mol Med 13:735–750
29. Parris GE (2013) Historical perspective of cell-cell fusion in cancer initiation and progression. Crit Rev Oncog 18(1–2):1–18
30. Zhang S, Mercado-Uribe I, Xing Z, Sun B, Kuang J, Liu J (2014) Generation of cancer stem-like cells through the formation of polyploid giant cancer cells. Oncogene 33(1):116–128
31. Fukuchi K, Steiniger SC, Deryugina E et al (2010) Inhibition of tumor metastasis: functional immune modulation of the CUB domain containing protein 1. Mol Pharm 7(1):245–253
32. Larizza L, Schirrmacher V, Graf L, Pfluger E, Peres-Martinez M, Stohr M (1984) Suggestive evidence that the highly metastatic variant ESb of the T-cell lymphoma Eb is derived from spontaneous fusion with a host macrophage. Int J Cancer 34(5):699–707
33. Crowley CW, Cohen RL, Lucas BK, Liu G, Shuman MA, Levinson AD (1993) Prevention of metastasis by inhibition of the urokinase receptor. Proc Natl Acad Sci U S A 90(11):5021–5025
34. Zawadzki V, Perschl A, Rosel M, Hekele A, Zoller M (1998) Blockade of metastasis formation by CD44-receptor globulin. Int J Cancer 75(6):919–924
35. Goldberg SF, Harms JF, Quon K, Welch DR (1999) Metastasis-suppressed C8161 melanoma cells arrest in lung but fail to proliferate. Clin Exp Metastasis 17(7):601–607
36. Chakraborty AK, Sodi S, Rachkovsky M, Kolesnikova N, Platt JT, Bolognia JL, Pawelek JM (2000) A spontaneous murine melanoma lung metastasis comprised of host x tumor hybrids. Cancer Res 60(9):2512–2519
37. Chakraborty AK, Pawelek J, Ikeda Y, Miyoshi E, Kolesnikova N, Funasaka Y, Ichihashi M, Taniguchi N (2001) Fusion hybrids with macrophage and melanoma cells up-regulate N-acetylglucosaminyltransferase V, beta1-6 branching, and metastasis. Cell Growth Differ 12(12):623–630
38. Parris GE (2005) The role of viruses in cell fusion and its importance to evolution, invasion and metastasis of cancer clones. Med Hypotheses 64(5):1011–1014
39. Pawelek JM (2007) Viewing malignant melanoma cells as macrophage-tumor hybrids. Cell Adh Migr 1(1):2–6
40. Pawelek JM (2008) Cancer-cell fusion with migratory bone-marrow-derived cells as an explanation for metastasis: new therapeutic paradigms. Future Oncol 4(4):449–452
41. Pawelek JM, Chakraborty AK (2008) Fusion of tumour cells with bone marrow-derived cells: a unifying explanation for metastasis. Nat Rev Cancer 8(5):377–386
42. Pawelek JM, Chakraborty AK (2008) The cancer cell--leukocyte fusion theory of metastasis. Adv Cancer Res 101:397–444
43. Lazova R, Chakraborty A, Pawelek JM (2011) Leukocyte-cancer cell fusion: initiator of the warburg effect in malignancy? Adv Exp Med Biol 714:151–172
44. Xu MH, Gao X, Luo D, Zhou XD, Xiong W, Liu GX (2014) EMT and acquisition of stem cell-like properties are involved in spontaneous formation of tumorigenic hybrids between lung cancer and bone marrow-derived mesenchymal stem cells. PLoS One 9(2):e87893

45. Li H, Feng Z, Tsang TC et al (2014) Fusion of HepG2 cells with mesenchymal stem cells increases cancer associated and malignant properties: an in vivo metastasis model. Oncol Rep 32:539–547
46. Castedo M, Coquelle A, Vitale I, Vivet S, Mouhamad S, Viaud S, Zitvogel L, Kroemer G (2006) Selective resistance of tetraploid cancer cells against DNA damage-induced apoptosis. Ann N Y Acad Sci 1090:35–49
47. Dewhurst SM, McGranahan N, Burrell RA et al (2014) Tolerance of whole-genome doubling propagates chromosomal instability and accelerates cancer genome evolution. Cancer Discov 4(2):175–185
48. Gagos S, Iliopoulos D, Tseleni-Balafouta S, Agapitos M, Antachopoulos C, Kostakis A, Karayannakos P, Skalkeas G (1996) Cell senescence and a mechanism of clonal evolution leading to continuous cell proliferation, loss of heterozygosity, and tumor heterogeneity: studies on two immortal colon cancer cell lines. Cancer Genet Cytogenet 90(2):157–165
49. Illidge TM, Cragg MS, Fringes B, Olive P, Erenpreisa JA (2000) Polyploid giant cells provide a survival mechanism for p53 mutant cells after DNA damage. Cell Biol Int 24(9):621–633
50. Ivanov A, Cragg MS, Erenpreisa J, Emzinsh D, Lukman H, Illidge TM (2003) Endopolyploid cells produced after severe genotoxic damage have the potential to repair DNA double strand breaks. J Cell Sci 116(Pt 20):4095–4106
51. Schwerer MJ, Hemmer J, Kraft K, Maier H, Moller P, Barth TF (2003) Endoreduplication in conjunction with tumor progression in an aneuploid laryngeal squamous cell carcinoma. Virchows Arch 443(1):98–103
52. Puig PE, Guilly MN, Bouchot A et al (2008) Tumor cells can escape DNA-damaging cisplatin through DNA endoreduplication and reversible polyploidy. Cell Biol Int 32(9):1031–1043
53. Li X, Heyer WD (2008) Homologous recombination in DNA repair and DNA damage tolerance. Cell Res 18(1):99–113
54. Biragyn A, Longo DL (2012) Neoplastic “Black Ops”: cancer’s subversive tactics in overcoming host defenses. Semin Cancer Biol 22(1):50–59
55. Rajaraman R, Rajaraman MM, Rajaraman SR, Guernsey DL (2005) Neosis--a paradigm of self-renewal in cancer. Cell Biol Int 29(12):1084–1097
56. Duesberg P, Rasnick D (2000) Aneuploidy, the somatic mutation that makes cancer a species of its own. Cell Motil Cytoskeleton 47(2):81–107
57. Li R, Sonik A, Stindl R, Rasnick D, Duesberg P (2000) Aneuploidy vs. gene mutation hypothesis of cancer: recent study claims mutation but is found to support aneuploidy. Proc Natl Acad Sci U S A 97(7):3236–3241
58. Marx J (2004) Cancer research. Inflammation and cancer: the link grows stronger. Science 306(5698):966–968
59. Rosen SD (2004) Ligands for L-selectin: homing, inflammation, and beyond. Annu Rev Immunol 22:129–156
60. Davies PS, Powell AE, Swain JR, Wong MH (2009) Inflammation and proliferation act together to mediate intestinal cell fusion. PLoS One 4(8), e6530
61. Harkness T, Weaver BA, Alexander CM, Ogle BM (2013) Cell fusion in tumor development: accelerated genetic evolution. Crit Rev Oncog 18(1–2):19–42
62. Kioi M, Vogel H, Schultz G, Hoffman RM, Harsh GR, Brown JM (2010) Inhibition of vasculogenesis, but not angiogenesis, prevents the recurrence of glioblastoma after irradiation in mice. J Clin Invest 120(3):694–705
63. Smith MC, Luker KE, Garbow JR, Prior JL, Jackson E, Piwnica-Worms D, Luker GD (2004) CXCR4 regulates growth of both primary and metastatic breast cancer. Cancer Res 64(23):8604–8612
64. Kawada K, Sonoshita M, Sakashita H et al (2004) Pivotal role of CXCR3 in melanoma cell metastasis to lymph nodes. Cancer Res 64(11):4010–4017
65. Wang J, Sun Y, Song W, Nor JE, Wang CY, Taichman RS (2005) Diverse signaling pathways through the SDF-1/CXCR4 chemokine axis in prostate cancer cell lines leads to altered patterns of cytokine secretion and angiogenesis. Cell Signal 17(12):1578–1592

66. Kucia M, Ratajczak J, Ratajczak MZ (2005) Bone marrow as a source of circulating CXCR4+ tissue-committed stem cells. Biol Cell 97(2):133–146
67. Hensbergen PJ, Wijnands PG, Schreurs MW, Scheper RJ, Willemze R, Tensen CP (2005) The CXCR3 targeting chemokine CXCL11 has potent antitumor activity in vivo involving attraction of CD8+ T lymphocytes but not inhibition of angiogenesis. J Immunother 28(4):343–351
68. Liang Z, Yoon Y, Votaw J, Goodman MM, Williams L, Shim H (2005) Silencing of CXCR4 blocks breast cancer metastasis. Cancer Res 65(3):967–971
69. Kryczek I, Lange A, Mottram P et al (2005) CXCL12 and vascular endothelial growth factor synergistically induce neoangiogenesis in human ovarian cancers. Cancer Res 65(2):465–472
70. Chute JP (2006) Stem cell homing. Curr Opin Hematol 13(6):399–406
71. Vandercappellen J, Van Damme J, Struyf S (2008) The role of CXC chemokines and their receptors in cancer. Cancer Lett 267(2):226–244
72. Hu W, Zhen X, Xiong B, Wang B, Zhang W, Zhou W (2008) CXCR6 is expressed in human prostate cancer in vivo and is involved in the in vitro invasion of PC3 and LNCap cells. Cancer Sci 99(7):1362–1369
73. Lu Y, Wang J, Xu Y, Koch AE, Cai Z, Chen X, Galson DL, Taichman RS, Zhang J (2008) CXCL16 functions as a novel chemotactic factor for prostate cancer cells in vitro. Mol Cancer Res 6(4):546–554
74. Wang L, Wang Z, Yang B, Yang Q, Sun Y (2009) CXCR4 nuclear localization follows binding of its ligand SDF-1 and occurs in metastatic but not primary renal cell carcinoma. Oncol Rep 22(6):1333–1339
75. Maksym RB, Tarnowski M, Grymula K et al (2009) The role of stromal-derived factor-1--CXCR7 axis in development and cancer. Eur J Pharmacol 625(1–3):31–40
76. Lin S, Sun L, Hu J, Wan S, Zhao R, Yuan S, Zhang L (2009) Chemokine C-X-C motif receptor 6 contributes to cell migration during hypoxia. Cancer Lett 279(1):108–117
77. Fulton AM (2009) The chemokine receptors CXCR4 and CXCR3 in cancer. Curr Oncol Rep 11(2):125–131
78. Ishikawa T, Nakashiro K, Klosek SK, Goda H, Hara S, Uchida D, Hamakawa H (2009) Hypoxia enhances CXCR4 expression by activating HIF-1 in oral squamous cell carcinoma. Oncol Rep 21(3):707–712
79. Rettig MP, Ramirez P, Nervi B, DiPersio JF (2009) CXCR4 and mobilization of hematopoietic precursors. Methods Enzymol 460:57–90
80. Juarez JG, Thien M, Dela Pena A, Baraz R, Bradstock KF, Bendall LJ (2009) CXCR4 mediates the homing of B cell progenitor acute lymphoblastic leukaemia cells to the bone marrow via activation of p38MAPK. Br J Haematol 145(4):491–499
81. Mirisola V, Zuccarino A, Bachmeier BE, Sormani MP, Falter J, Nerlich A, Pfeffer U (2009) CXCL12/SDF1 expression by breast cancers is an independent prognostic marker of disease-free and overall survival. Eur J Cancer 45(14):2579–2587
82. Liekens S, Schols D, Hatse S (2010) CXCL12-CXCR4 axis in angiogenesis, metastasis and stem cell mobilization. Curr Pharm Des 16(35):3903–3920
83. Furusato B, Mohamed A, Uhlen M, Rhim JS (2010) CXCR4 and cancer. Pathol Int 60(7):497–505
84. Vaiselbuh SR (2010) How do leukemic stem cells find their niche? Pediatr Blood Cancer 55(2):218–219
85. Deng L, Chen N, Li Y, Zheng H, Lei Q (2010) CXCR6/CXCL16 functions as a regulator in metastasis and progression of cancer. Biochim Biophys Acta 1806(1):42–49
86. Hirbe AC, Morgan EA, Weilbaecher KN (2010) The CXCR4/SDF-1 chemokine axis: a potential therapeutic target for bone metastases? Curr Pharm Des 16(11):1284–1290
87. Hou KL, Hao MG, Bo JJ, Wang JH (2010) CXCR7 in tumorigenesis and progression. Chin J Cancer 29(4):456–459

88. Sun X, Cheng G, Hao M, Zheng J, Zhou X, Zhang J, Taichman RS, Pienta KJ, Wang J (2010) CXCL12 / CXCR4 / CXCR7 chemokine axis and cancer progression. Cancer Metastasis Rev 29(4):709–722
89. Singh AK, Arya RK, Trivedi AK, Sanyal S, Baral R, Dormond O, Briscoe DM, Datta D (2012) Chemokine receptor trio: CXCR3, CXCR4 and CXCR7 crosstalk via CXCL11 and CXCL12. Cytokine Growth Factor Rev 24:41–49
90. Oh YS, Kim HY, Song IC, Yun HJ, Jo DY, Kim S, Lee HJ (2012) Hypoxia induces CXCR4 expression and biological activity in gastric cancer cells through activation of hypoxia-inducible factor-1alpha. Oncol Rep 28(6):2239–2246
91. Shen B, Zheng MQ, Lu JW, Jiang Q, Wang TH, Huang XE (2013) CXCL12-CXCR4 promotes proliferation and invasion of pancreatic cancer cells. Asian Pac J Cancer Prev 14(9):5403–5408
92. Liao YX, Zhou CH, Zeng H, Zuo DQ, Wang ZY, Yin F, Hua YQ, Cai ZD (2013) The role of the CXCL12-CXCR4/CXCR7 axis in the progression and metastasis of bone sarcomas (Review). Int J Mol Med 32(6):1239–1246
93. Wang B, Wang W, Niu W et al (2014) SDF-1/CXCR4 axis promotes directional migration of colorectal cancer cells through upregulation of integrin alphavbeta6. Carcinogenesis 35(2):282–291
94. Li B, Xu W, Xu L, Jiang Z, Wen Z, Li K, Xiong S (2010) I-TAC is a dominant chemokine in controlling skin intragraft inflammation via recruiting CXCR3+ cells into the graft. Cell Immunol 260(2):83–91
95. Zhou WH, Hu WD, Wu ZQ, Zheng XM, Wang BC (2010) Role of CXCL16/CXCR6 axis in the metastasis of human prostate cancer. Zhonghua Yi Xue Za Zhi 90(14):947–951
96. Guo L, Cui ZM, Zhang J, Huang Y (2011) Chemokine axes CXCL12/CXCR4 and CXCL16/CXCR6 correlate with lymph node metastasis in epithelial ovarian carcinoma. Chin J Cancer 30(5):336–343
97. Isozaki T, Arbab AS, Haas CS, Amin MA, Arendt MD, Koch AE, Ruth JH (2013) Evidence that CXCL16 is a potent mediator of angiogenesis and is involved in endothelial progenitor cell chemotaxis : studies in mice with K/BxN serum-induced arthritis. Arthritis Rheum 65(7):1736–1746
98. Huang Y, Zhang J, Cui ZM, Zhao J, Zheng Y (2013) Expression of the CXCL12/CXCR4 and CXCL16/CXCR6 axes in cervical intraepithelial neoplasia and cervical cancer. Chin J Cancer 32(5):289–296
99. Moustakas A, Heldin P (2014) TGFbeta and matrix-regulated epithelial to mesenchymal transition. Biochim Biophys Acta 1840:2621–2634
100. Krstic J, Santibanez JF (2014) Transforming growth factor-beta and matrix metalloproteinases: functional interactions in tumor stroma-infiltrating myeloid cells. ScientificWorldJournal 2014:521754
101. Prendergast GC, Smith C, Thomas S, Mandik-Nayak L, Laury-Kleintop L, Metz R, Muller AJ (2014) Indoleamine 2,3-dioxygenase pathways of pathogenic inflammation and immune escape in cancer. Cancer Immunol Immunother 63:721–735
102. Watson D, Zhang GY, Hu M, Wang YM, Fletcher J, Sartor M, Alexander SI (2014) Transforming growth factor beta (TGFbeta) plays a crucial role in prolonging allograft survival in an allodepletion ("pruning") skin transplant model. Transpl Immunol 30:168–177
103. Matsumoto K, Ema M (2014) Roles of VEGF-A signalling in development, regeneration, and tumours. J Biochem 156:1–10
104. Sun X, Ingman WV (2014) Cytokine networks that mediate epithelial cell-macrophage crosstalk in the mammary gland: implications for development and cancer. J Mammary Gland Biol Neoplasia 19:191–201
105. Duran C, Talley PJ, Walsh J, Pigott C, Morton IE, Andrews PW (2001) Hybrids of pluripotent and nullipotent human embryonal carcinoma cells: partial retention of a pluripotent phenotype. Int J Cancer 93(3):324–332

106. Houghton J, Stoicov C, Nomura S et al (2004) Gastric cancer originating from bone marrow-derived cells. Science 306(5701):1568–1571
107. He X, Tsang TC, Pipes BL, Ablin RJ, Harris DT (2005) A stem cell fusion model of carcinogenesis. J Exp Ther Oncol 5(2):101–109
108. Chakraborty AK, Sousa Jde F, Chakraborty D, Funasaka Y, Bhattacharya M, Chatterjee A, Pawelek J (2006) GnT-V expression and metastatic phenotypes in macrophage-melanoma fusion hybrids is down-regulated by 5-Aza-dC: evidence for methylation sensitive, extragenic regulation of GnT-V transcription. Gene 374:166–173
109. Rizvi AZ, Swain JR, Davies PS, Bailey AS, Decker AD, Willenbring H, Grompe M, Fleming WH, Wong MH (2006) Bone marrow-derived cells fuse with normal and transformed intestinal stem cells. Proc Natl Acad Sci U S A 103(16):6321–6325
110. Wu XZ, Chen D, Xie GR (2007) Bone marrow-derived cells: roles in solid tumor. Minireview. Neoplasma 54(1):1–6
111. Korbling M, de Lima MJ, Thomas E, Khanna A, Najjar AM, Gu J, Gelovani JG, Broaddus R (2008) Fusion of circulating blood cells with solid-organ tissue cells in clinical stem cell transplants: a potential therapeutic model? Regen Med 3(2):157–164
112. Sinkovics JG (2009) Horizontal gene transfers and cell fusions in microbiology, immunology and oncology (review). Int J Oncol 35(3):441–465
113. Estrov Z (2009) Stem cells and somatic cells: reprogramming and plasticity. Clin Lymphoma Myeloma 9(Suppl 3):S319–S328
114. Xu D, Wang F, Gu H, Wang J, Guo Q, Zhang Y, Wang Z (2010) Hybrid cells differentiate to hepatic lineage cells and repair oxidative damage. Cell Mol Biol Lett 15(3):451–472
115. Howcroft TK, Zhang HG, Dhodapkar M, Mohla S (2011) Vesicle transfer and cell fusion: emerging concepts of cell-cell communication in the tumor microenvironment. Cancer Biol Ther 12(3):159–164
116. Ferrand J, Noel D, Lehours P, Prochazkova-Carlotti M, Chambonnier L, Menard A, Megraud F, Varon C (2011) Human bone marrow-derived stem cells acquire epithelial characteristics through fusion with gastrointestinal epithelial cells. PLoS One 6(5), e19569
117. Gao P, Zheng J (2011) Oncogenic virus-mediated cell fusion: new insights into initiation and progression of oncogenic viruses--related cancers. Cancer Lett 303(1):1–8
118. Ding J, Jin W, Chen C, Shao Z, Wu J (2012) Tumor associated macrophage x cancer cell hybrids may acquire cancer stem cell properties in breast cancer. PLoS One 7(7), e41942
119. Rappa G, Mercapide J, Lorico A (2012) Spontaneous formation of tumorigenic hybrids between breast cancer and multipotent stromal cells is a source of tumor heterogeneity. Am J Pathol 180(6):2504–2515
120. Berndt B, Zanker KS, Dittmar T (2013) Cell fusion is a potent inducer of aneuploidy and drug resistance in tumor cell/ normal cell hybrids. Crit Rev Oncog 18(1–2):97–113
121. Tan C, Dannull J, Nair SK, Ding E, Tyler DS, Pruitt SK, Lee WT (2013) Local secretion of IL-12 augments the therapeutic impact of dendritic cell-tumor cell fusion vaccination. J Surg Res 185:904–911
122. Barone M, Scavo MP, Maiorano E, Di Leo A, Francavilla A (2013) Bone marrow-derived stem cells and hepatocarcinogenesis in hepatitis B virus transgenic mice. Dig Liver Dis 46:243–250
123. Barry ER, Corry GN, Rasmussen TP (2010) Targeting DOT1L action and interactions in leukemia: the role of DOT1L in transformation and development. Expert Opin Ther Targets 14(4):405–418
124. Rachkovsky M, Pawelek J (1999) Acquired melanocyte stimulating hormone-inducible chemotaxis following macrophage fusion with Cloudman S91 melanoma cells. Cell Growth Differ 10(7):517–524
125. Ramakrishnan M, Mathur SR, Mukhopadhyay A (2013) Fusion-derived epithelial cancer cells express hematopoietic markers and contribute to stem cell and migratory phenotype in ovarian carcinoma. Cancer Res 73(17):5360–5370

126. Laubli H, Borsig L (2010) Selectins promote tumor metastasis. Semin Cancer Biol 20(3):169–177
127. Ali S, Lazennec G (2007) Chemokines: novel targets for breast cancer metastasis. Cancer Metastasis Rev 26(3–4):401–420
128. Bussard KM, Gay CV, Mastro AM (2008) The bone microenvironment in metastasis; what is special about bone? Cancer Metastasis Rev 27(1):41–55
129. Gassmann P, Haier J (2008) The tumor cell-host organ interface in the early onset of metastatic organ colonisation. Clin Exp Metastasis 25(2):171–181
130. Mousa SA, Petersen LJ (2009) Anti-cancer properties of low-molecular-weight heparin: preclinical evidence. Thromb Haemost 102(2):258–267
131. Zlotnik A, Burkhardt AM, Homey B (2011) Homeostatic chemokine receptors and organ-specific metastasis. Nat Rev Immunol 11(9):597–606
132. Wang HB, Wang JT, Zhang L et al (2007) P-selectin primes leukocyte integrin activation during inflammation. Nat Immunol 8(8):882–892
133. Chen M, Geng JG (2006) P-selectin mediates adhesion of leukocytes, platelets, and cancer cells in inflammation, thrombosis, and cancer growth and metastasis. Arch Immunol Ther Exp (Warsz) 54(2):75–84
134. Fischer S, Grantzow T, Pagel JI, Tschernatsch M, Sperandio M, Preissner KT, Deindl E (2012) Extracellular RNA promotes leukocyte recruitment in the vascular system by mobilising proinflammatory cytokines. Thromb Haemost 108(4):730–741
135. Fischer S, Gesierich S, Griemert B, Schanzer A, Acker T, Augustin HG, Olsson AK, Preissner KT (2013) Extracellular RNA liberates tumor necrosis factor-alpha to promote tumor cell trafficking and progression. Cancer Res 73(16):5080–5089
136. Wang J, Zhang KY, Liu SM, Sen S (2014) Tumor-associated circulating microRNAs as biomarkers of cancer. Molecules 19(2):1912–1938
137. Record M (2014) Intercellular communication by exosomes in placenta: a possible role in cell fusion? Placenta 35:297–302
138. Heydtmann M, Lalor PF, Eksteen JA, Hubscher SG, Briskin M, Adams DH (2005) CXC chemokine ligand 16 promotes integrin-mediated adhesion of liver-infiltrating lymphocytes to cholangiocytes and hepatocytes within the inflamed human liver. J Immunol 174(2):1055–1062
139. Hundhausen C, Schulte A, Schulz B et al (2007) Regulated shedding of transmembrane chemokines by the disintegrin and metalloproteinase 10 facilitates detachment of adherent leukocytes. J Immunol 178(12):8064–8072
140. Schramme A, Abdel-Bakky MS, Kampfer-Kolb N, Pfeilschifter J, Gutwein P (2008) The role of CXCL16 and its processing metalloproteinases ADAM10 and ADAM17 in the proliferation and migration of human mesangial cells. Biochem Biophys Res Commun 370(2):311–316
141. Crawford HC, Dempsey PJ, Brown G, Adam L, Moss ML (2009) ADAM10 as a therapeutic target for cancer and inflammation. Curr Pharm Des 15(20):2288–2299
142. Bret C, Hose D, Reme T et al (2011) Gene expression profile of ADAMs and ADAMTSs metalloproteinases in normal and malignant plasma cells and in the bone marrow environment. Exp Hematol 39(5):546–557.e548
143. Miller MA, Meyer AS, Beste MT et al (2013) ADAM-10 and -17 regulate endometriotic cell migration via concerted ligand and receptor shedding feedback on kinase signaling. Proc Natl Acad Sci U S A 110(22):E2074–E2083
144. Nakayama T, Hieshima K, Izawa D, Tatsumi Y, Kanamaru A, Yoshie O (2003) Cutting edge: profile of chemokine receptor expression on human plasma cells accounts for their efficient recruitment to target tissues. J Immunol 170(3):1136–1140
145. Ludwig A, Weber C (2007) Transmembrane chemokines: versatile 'special agents' in vascular inflammation. Thromb Haemost 97(5):694–703

146. Lehrke M, Millington SC, Lefterova M, Cumaranatunge RG, Szapary P, Wilensky R, Rader DJ, Lazar MA, Reilly MP (2007) CXCL16 is a marker of inflammation, atherosclerosis, and acute coronary syndromes in humans. J Am Coll Cardiol 49(4):442–449
147. Galkina E, Ley K (2007) Leukocyte influx in atherosclerosis. Curr Drug Targets 8(12):1239–1248
148. Hojo S, Koizumi K, Tsuneyama K et al (2007) High-level expression of chemokine CXCL16 by tumor cells correlates with a good prognosis and increased tumor-infiltrating lymphocytes in colorectal cancer. Cancer Res 67(10):4725–4731
149. Darash-Yahana M, Gillespie JW, Hewitt SM et al (2009) The chemokine CXCL16 and its receptor, CXCR6, as markers and promoters of inflammation-associated cancers. PLoS One 4(8), e6695
150. Borst O, Munzer P, Gatidis S et al (2012) The inflammatory chemokine CXC motif ligand 16 triggers platelet activation and adhesion via CXC motif receptor 6-dependent phosphatidylinositide 3-kinase/Akt signaling. Circ Res 111(10):1297–1307
151. Parsonage G, Machado LR, Hui JW et al (2012) CXCR6 and CCR5 localize T lymphocyte subsets in nasopharyngeal carcinoma. Am J Pathol 180(3):1215–1222
152. Goodison S, Urquidi V, Tarin D (1999) CD44 cell adhesion molecules. Mol Pathol 52(4):189–196
153. Takamune Y, Ikebe T, Nagano O, Nakayama H, Ota K, Obayashi T, Saya H, Shinohara M (2007) ADAM-17 associated with CD44 cleavage and metastasis in oral squamous cell carcinoma. Virchows Arch 450(2):169–177
154. Stamenkovic I, Yu Q (2009) Shedding light on proteolytic cleavage of CD44: the responsible sheddase and functional significance of shedding. J Invest Dermatol 129(6):1321–1324
155. Anderegg U, Eichenberg T, Parthaune T et al (2009) ADAM10 is the constitutive functional sheddase of CD44 in human melanoma cells. J Invest Dermatol 129(6):1471–1482
156. Deneberg S, Grovdal M, Karimi M et al (2010) Gene-specific and global methylation patterns predict outcome in patients with acute myeloid leukemia. Leukemia 24:932–941
157. Skubitz AP (2002) Adhesion molecules. Cancer Treat Res 107:305–329
158. Vestweber D (2007) Molecular mechanisms that control leukocyte extravasation through endothelial cell contacts. Ernst Schering Found Symp Proc 3:151–167
159. Solanas G, Cortina C, Sevillano M, Batlle E (2011) Cleavage of E-cadherin by ADAM10 mediates epithelial cell sorting downstream of EphB signalling. Nat Cell Biol 13(9):1100–1107
160. Dreymueller D, Pruessmeyer J, Groth E, Ludwig A (2012) The role of ADAM-mediated shedding in vascular biology. Eur J Cell Biol 91(6–7):472–485
161. Horiuchi K (2013) A brief history of tumor necrosis factor alpha--converting enzyme: an overview of ectodomain shedding. Keio J Med 62(1):29–36
162. Storci G, Sansone P, Mari S et al (2010) TNFalpha up-regulates SLUG via the NF-kappaB/HIF1alpha axis, which imparts breast cancer cells with a stem cell-like phenotype. J Cell Physiol 225:682–691
163. Christofferson DE, Li Y, Yuan J (2014) Control of life-or-death decisions by RIP1 kinase. Annu Rev Physiol 76:129–150
164. Folkman J (1995) Angiogenesis inhibitors generated by tumors. Mol Med 1(2):120–122
165. O'Reilly MS, Holmgren L, Shing Y et al (1994) Angiostatin: a circulating endothelial cell inhibitor that suppresses angiogenesis and tumor growth. Cold Spring Harb Symp Quant Biol 59:471–482
166. Beecken WD, Fernandez A, Joussen AM et al (2001) Effect of antiangiogenic therapy on slowly growing, poorly vascularized tumors in mice. J Natl Cancer Inst 93(5):382–387
167. O'Reilly MS, Boehm T, Shing Y et al (1997) Endostatin: an endogenous inhibitor of angiogenesis and tumor growth. Cell 88(2):277–285
168. Benelli R, Morini M, Carrozzino F, Ferrari N, Minghelli S, Santi L, Cassatella M, Noonan DM, Albini A (2002) Neutrophils as a key cellular target for angiostatin: implications for regulation of angiogenesis and inflammation. FASEB J 16(2):267–269

169. Stecher VJ (1975) The chemotaxis of selected cell types to connective tissue degradation products. Ann N Y Acad Sci 256:177–189
170. Abbassi O, Kishimoto TK, McIntire LV, Anderson DC, Smith CW (1993) E-selectin supports neutrophil rolling in vitro under conditions of flow. J Clin Invest 92(6):2719–2730
171. Dejana E, Breviario F, Caveda L (1994) Leukocyte-endothelial cell adhesive receptors. Clin Exp Rheumatol 12(Suppl 10):S25–S28
172. Sriramarao P, Norton CR, Borgstrom P, DiScipio RG, Wolitzky BA, Broide DH (1996) E-selectin preferentially supports neutrophil but not eosinophil rolling under conditions of flow in vitro and in vivo. J Immunol 157(10):4672–4680
173. Rainger GE, Wautier MP, Nash GB, Wautier JL (1996) Prolonged E-selectin induction by monocytes potentiates the adhesion of flowing neutrophils to cultured endothelial cells. Br J Haematol 92(1):192–199
174. Xie X, Raud J, Hedqvist P, Lindbom L (1997) In vivo rolling and endothelial selectin binding of mononuclear leukocytes is distinct from that of polymorphonuclear cells. Eur J Immunol 27(11):2935–2941
175. Tangemann K, Gunn MD, Giblin P, Rosen SD (1998) A high endothelial cell-derived chemokine induces rapid, efficient, and subset-selective arrest of rolling T lymphocytes on a reconstituted endothelial substrate. J Immunol 161(11):6330–6337
176. Kunkel EJ, Butcher EC (2002) Chemokines and the tissue-specific migration of lymphocytes. Immunity 16(1):1–4
177. Alon R, Feigelson S (2002) From rolling to arrest on blood vessels: leukocyte tap dancing on endothelial integrin ligands and chemokines at sub-second contacts. Semin Immunol 14(2):93–104
178. Renard M, Heutte F, Boutherin-Falson O, Finet M, Boisseau MR (2003) Induced changes of leukocyte slow rolling in an in flow pharmacological model of adhesion to endothelial cells. Biorheology 40(1–3):173–178
179. Ding Z, Issekutz TB, Downey GP, Waddell TK (2003) L-selectin stimulation enhances functional expression of surface CXCR4 in lymphocytes: implications for cellular activation during adhesion and migration. Blood 101(11):4245–4252
180. Biancone L, Cantaluppi V, Duo D, Deregibus MC, Torre C, Camussi G (2004) Role of L-selectin in the vascular homing of peripheral blood-derived endothelial progenitor cells. J Immunol 173(8):5268–5274
181. Stein JV, Nombela-Arrieta C (2005) Chemokine control of lymphocyte trafficking: a general overview. Immunology 116(1):1–12
182. Laudanna C, Alon R (2006) Right on the spot. Chemokine triggering of integrin-mediated arrest of rolling leukocytes. Thromb Haemost 95(1):5–11
183. Vestweber D (2007) Adhesion and signaling molecules controlling the transmigration of leukocytes through endothelium. Immunol Rev 218:178–196
184. Schaff U, Mattila PE, Simon SI, Walcheck B (2008) Neutrophil adhesion to E-selectin under shear promotes the redistribution and co-clustering of ADAM17 and its proteolytic substrate L-selectin. J Leukoc Biol 83(1):99–105
185. Sarkar D, Vemula PK, Teo GS, Spelke D, Karnik R, le Wee Y, Karp JM (2008) Chemical engineering of mesenchymal stem cells to induce a cell rolling response. Bioconjug Chem 19(11):2105–2109
186. Wiese G, Barthel SR, Dimitroff CJ (2009) Analysis of physiologic E-selectin-mediated leukocyte rolling on microvascular endothelium. J Vis Exp (24). doi:10.3791/1009
187. Yagi H, Soto-Gutierrez A, Parekkadan B, Kitagawa Y, Tompkins RG, Kobayashi N, Yarmush ML (2010) Mesenchymal stem cells: mechanisms of immunomodulation and homing. Cell Transplant 19(6):667–679
188. Brandau S, Jakob M, Hemeda H, Bruderek K, Janeschik S, Bootz F, Lang S (2010) Tissue-resident mesenchymal stem cells attract peripheral blood neutrophils and enhance their inflammatory activity in response to microbial challenge. J Leukoc Biol 88(5):1005–1015

189. Levoye A, Balabanian K, Baleux F, Bachelerie F, Lagane B (2009) CXCR7 heterodimerizes with CXCR4 and regulates CXCL12-mediated G protein signaling. Blood 113(24):6085–6093
190. Zheng K, Li HY, Su XL, Wang XY, Tian T, Li F, Ren GS (2010) Chemokine receptor CXCR7 regulates the invasion, angiogenesis and tumor growth of human hepatocellular carcinoma cells. J Exp Clin Cancer Res 29:31
191. Hoffmann F, Muller W, Schutz D, Penfold ME, Wong YH, Schulz S, Stumm R (2012) Rapid uptake and degradation of CXCL12 depend on CXCR7 carboxyl-terminal serine/threonine residues. J Biol Chem 287(34):28362–28377
192. Sanchez-Martin L, Sanchez-Mateos P, Cabanas C (2013) CXCR7 impact on CXCL12 biology and disease. Trends Mol Med 19(1):12–22
193. Berahovich RD, Zabel BA, Lewen S, Walters MJ, Ebsworth K, Wang Y, Jaen JC, Schall TJ (2014) Endothelial expression of CXCR7 and the regulation of systemic CXCL12 levels. Immunology 141(1):111–122
194. Levine AJ (2009) The common mechanisms of transformation by the small DNA tumor viruses: the inactivation of tumor suppressor gene products: p53. Virology 384(2):285–293
195. Lutz WK (1998) Dose–response relationships in chemical carcinogenesis: superposition of different mechanisms of action, resulting in linear-nonlinear curves, practical thresholds, J-shapes. Mutat Res 405(2):117–124
196. Cohen SM, Ohnishi T, Arnold LL, Le XC (2007) Arsenic-induced bladder cancer in an animal model. Toxicol Appl Pharmacol 222(3):258–263
197. Ames BN, Gold LS (1996) Re: E. Farber, Cell proliferation as a major risk factor for cancer: a concept of doubtful validity. Cancer Res., 55: 3759–3762, 1995. Cancer Res 56(18):4267–4269, author reply 4272–4264
198. Butterworth BE, Loury DJ, Smith-Oliver T, Cattley RC (1987) The potential role of chemically induced hyperplasia in the carcinogenic activity of the hypolipidemic carcinogens. Toxicol Ind Health 3(2):129–149
199. Ames BN (1989) Mutagenesis and carcinogenesis: endogenous and exogenous factors. Environ Mol Mutagen 14(Suppl 16):66–77
200. Ames BN, Gold LS (1990) Chemical carcinogenesis: too many rodent carcinogens. Proc Natl Acad Sci U S A 87(19):7772–7776
201. Marshall WL, Yim C, Gustafson E, Graf T, Sage DR, Hanify K, Williams L, Fingeroth J, Finberg RW (1999) Epstein-Barr virus encodes a novel homolog of the bcl-2 oncogene that inhibits apoptosis and associates with Bax and Bak. J Virol 73(6):5181–5185
202. Choi YB, Nicholas J (2008) Autocrine and paracrine promotion of cell survival and virus replication by human herpesvirus 8 chemokines. J Virol 82(13):6501–6513
203. Haigh J (1978) The accumulation of deleterious genes in a population – Muller's Ratchet. Theor Popul Biol 14(2):251–267
204. Chao L (1990) Fitness of RNA virus decreased by Muller's ratchet. Nature 348(6300):454–455
205. Andersson DI, Hughes D (1996) Muller's ratchet decreases fitness of a DNA-based microbe. Proc Natl Acad Sci U S A 93(2):906–907
206. Yuste E, Sanchez-Palomino S, Casado C, Domingo E, Lopez-Galindez C (1999) Drastic fitness loss in human immunodeficiency virus type 1 upon serial bottleneck events. J Virol 73(4):2745–2751
207. Charlesworth B, Charlesworth D (2000) The degeneration of Y chromosomes. Philos Trans R Soc Lond B Biol Sci 355(1403):1563–1572
208. Fontanari JF, Colato A, Howard RS (2003) Mutation accumulation in growing asexual lineages. Phys Rev Lett 91(21):218101
209. Takeuchi N, Kaneko K, Koonin EV (2014) Horizontal gene transfer can rescue prokaryotes from Muller's ratchet: benefit of DNA from dead cells and population subdivision. G3 (Bethesda) 4(2):325–339

210. Huxley J (1956) Cancer biology: comparative and genetic. Biol Rev Camb Philos Soc 31:474–514
211. Vincent MD (2010) The animal within: carcinogenesis and the clonal evolution of cancer cells are speciation events sensu stricto. Evolution 64(4):1173–1183
212. Duesberg P, Rausch C, Rasnick D, Hehlmann R (1998) Genetic instability of cancer cells is proportional to their degree of aneuploidy. Proc Natl Acad Sci U S A 95(23):13692–13697
213. Knauss S, Klein A (2012) From aneuploidy to cancer: the evolution of a new species? J Biosci 37(2):211–220
214. Lisanti MP, Martinez-Outschoorn UE, Chiavarina B et al (2010) Understanding the "lethal" drivers of tumor-stroma co-evolution: emerging role(s) for hypoxia, oxidative stress and autophagy/mitophagy in the tumor micro-environment. Cancer Biol Ther 10(6):537–542
215. Duesberg P, Li R, Fabarius A, Hehlmann R (2006) Aneuploidy and cancer: from correlation to causation. Contrib Microbiol 13:16–44
216. O'Connor ML, Xiang D, Shigdar S et al (2014) Cancer stem cells: a contentious hypothesis now moving forward. Cancer Lett 344(2):180–187
217. Colombo F, Baldan F, Mazzucchelli S et al (2011) Evidence of distinct tumour-propagating cell populations with different properties in primary human hepatocellular carcinoma. PLoS One 6(6), e21369
218. Lluis F, Cosma MP (2010) Cell-fusion-mediated somatic-cell reprogramming: a mechanism for tissue regeneration. J Cell Physiol 223(1):6–13
219. Nagler C, Zanker KS, Dittmar T (2011) Cell fusion, drug resistance and recurrence CSCs. Adv Exp Med Biol 714:173–182
220. Sanges D, Lluis F, Cosma MP (2011) Cell-fusion-mediated reprogramming: pluripotency or transdifferentiation? Implications for regenerative medicine. Adv Exp Med Biol 713:137–159
221. Chen KA, Laywell ED, Marshall G, Walton N, Zheng T, Steindler DA (2006) Fusion of neural stem cells in culture. Exp Neurol 198(1):129–135
222. Larsson LI, Bjerregaard B, Talts JF (2008) Cell fusions in mammals. Histochem Cell Biol 129(5):551–561
223. Li B, Bailey AS, Jiang S, Liu B, Goldman DC, Fleming WH (2010) Endothelial cells mediate the regeneration of hematopoietic stem cells. Stem Cell Res 4(1):17–24
224. Grompe M (2003) The role of bone marrow stem cells in liver regeneration. Semin Liver Dis 23(4):363–372
225. Koyanagi M, Brandes RP, Haendeler J, Zeiher AM, Dimmeler S (2005) Cell-to-cell connection of endothelial progenitor cells with cardiac myocytes by nanotubes: a novel mechanism for cell fate changes? Circ Res 96(10):1039–1041
226. Padron Velazquez JL (2006) Stem cell fusion as an ultimate line of defense against xenobiotics. Med Hypotheses 67(2):383–387
227. Chamberlain G, Fox J, Ashton B, Middleton J (2007) Concise review: mesenchymal stem cells: their phenotype, differentiation capacity, immunological features, and potential for homing. Stem Cells 25(11):2739–2749
228. Jansen KM, Pavlath GK (2008) Molecular control of mammalian myoblast fusion. Methods Mol Biol 475:115–133
229. Piquer-Gil M, Garcia-Verdugo JM, Zipancic I, Sanchez MJ, Alvarez-Dolado M (2009) Cell fusion contributes to pericyte formation after stroke. J Cereb Blood Flow Metab 29(3):480–485
230. Clancy EK, Barry C, Ciechonska M, Duncan R (2010) Different activities of the reovirus FAST proteins and influenza hemagglutinin in cell-cell fusion assays and in response to membrane curvature agents. Virology 397(1):119–129
231. Basanez G (2002) Membrane fusion: the process and its energy suppliers. Cell Mol Life Sci 59(9):1478–1490
232. Chen EH, Olson EN (2005) Unveiling the mechanisms of cell-cell fusion. Science 308(5720):369–373

233. Digel M, Sampaio KL, Jahn G, Sinzger C (2006) Evidence for direct transfer of cytoplasmic material from infected to uninfected cells during cell-associated spread of human cytomegalovirus. J Clin Virol 37(1):10–20
234. Podbilewicz B (2006) Cell fusion. WormBook 1–32
235. Sherer NM, Lehmann MJ, Jimenez-Soto LF, Horensavitz C, Pypaert M, Mothes W (2007) Retroviruses can establish filopodial bridges for efficient cell-to-cell transmission. Nat Cell Biol 9(3):310–315
236. Duelli D, Lazebnik Y (2007) Cell-to-cell fusion as a link between viruses and cancer. Nat Rev Cancer 7(12):968–976
237. Salsman J, Top D, Barry C, Duncan R (2008) A virus-encoded cell-cell fusion machine dependent on surrogate adhesins. PLoS Pathog 4(3), e1000016
238. Hu L, Plafker K, Vorozhko V, Zuna RE, Hanigan MH, Gorbsky GJ, Plafker SM, Angeletti PC, Ceresa BP (2009) Human papillomavirus 16 E5 induces bi-nucleated cell formation by cell-cell fusion. Virology 384(1):125–134
239. Richard JP, Leikina E, Langen R, Henne WM, Popova M, Balla T, McMahon HT, Kozlov M, Chernomordik L (2011) Intracellular curvature generating proteins in cell-to-cell fusion. Biochem J 440:185–193
240. Lawrence P, Perez BE, Drexler JF, Corman VM, Muller MA, Drosten C, Volchkov V (2014) Surface glycoproteins of the recently identified African Henipavirus promote viral entry and cell fusion in a range of Human, Simian and Bat cell lines. Virus Res 181:77–80
241. Tada M, Takahama Y, Abe K, Nakatsuji N, Tada T (2001) Nuclear reprogramming of somatic cells by in vitro hybridization with ES cells. Curr Biol 11(19):1553–1558
242. Ying QL, Nichols J, Evans EP, Smith AG (2002) Changing potency by spontaneous fusion. Nature 416(6880):545–548
243. Ambrosi DJ, Rasmussen TP (2005) Reprogramming mediated by stem cell fusion. J Cell Mol Med 9(2):320–330
244. Do JT, Han DW, Scholer HR (2006) Reprogramming somatic gene activity by fusion with pluripotent cells. Stem Cell Rev 2(4):257–264
245. Ambrosi DJ, Tanasijevic B, Kaur A, Obergfell C, O'Neill RJ, Krueger W, Rasmussen TP (2007) Genome-wide reprogramming in hybrids of somatic cells and embryonic stem cells. Stem Cells 25(5):1104–1113
246. Hanna J, Carey BW, Jaenisch R (2008) Reprogramming of somatic cell identity. Cold Spring Harb Symp Quant Biol 73:147–155
247. Matsumura H, Tada T (2008) Cell fusion-mediated nuclear reprogramming of somatic cells. Reprod Biomed Online 16(1):51–56
248. Sumer H, Nicholls C, Pinto AR, Indraharan D, Liu J, Lim ML, Liu JP, Verma PJ (2010) Chromosomal and telomeric reprogramming following ES-somatic cell fusion. Chromosoma 119(2):167–176
249. Sumer H, Jones KL, Liu J, Heffernan C, Tat PA, Upton KR, Verma PJ (2010) Reprogramming of somatic cells after fusion with induced pluripotent stem cells and nuclear transfer embryonic stem cells. Stem Cells Dev 19(2):239–246
250. Do JT, Scholer HR (2010) Cell fusion-induced reprogramming. Methods Mol Biol 636:179–190
251. Powell AE, Anderson EC, Davies PS, Silk AD, Pelz C, Impey S, Wong MH (2011) Fusion between Intestinal epithelial cells and macrophages in a cancer context results in nuclear reprogramming. Cancer Res 71(4):1497–1505
252. Li GC, Ye QH, Dong QZ, Ren N, Jia HL, Qin LX (2012) Mesenchymal stem cells seldomly fuse with hepatocellular carcinoma cells and are mainly distributed in the tumor stroma in mouse models. Oncol Rep 29:713–719
253. Wang Y, Fan H, Zhou B, Ju Z, Yu L, Guo L, Han J, Lu S (2012) Fusion of human umbilical cord mesenchymal stem cells with esophageal carcinoma cells inhibits the tumorigenicity of esophageal carcinoma cells. Int J Oncol 40(2):370–377

254. Chakraborty AK, de Freitas Sousa J, Espreafico EM, Pawelek JM (2001) Human monocyte x mouse melanoma fusion hybrids express human gene. Gene 275(1):103–106
255. Goldenberg DM, Pavia RA, Tsao MC (1974) In vivo hybridisation of human tumour and normal hamster cells. Nature 250(5468):649–651
256. Goldenberg DM, Zagzag D, Heselmeyer-Haddad KM, Berroa Garcia LY, Ried T, Loo M, Chang CH, Gold DV (2012) Horizontal transmission and retention of malignancy, as well as functional human genes, after spontaneous fusion of human glioblastoma and hamster host cells in vivo. Int J Cancer 131(1):49–58
257. Goldenberg DM, Gold DV, Loo M, Liu D, Chang CH, Jaffe ES (2013) Horizontal transmission of malignancy: in-vivo fusion of human lymphomas with hamster stroma produces tumors retaining human genes and lymphoid pathology. PLoS One 8(2), e55324
258. Sundaram M, Guernsey DL, Rajaraman MM, Rajaraman R (2004) Neosis: a novel type of cell division in cancer. Cancer Biol Ther 3(2):207–218
259. Rajaraman R, Guernsey DL, Rajaraman MM, Rajaraman SR (2006) Stem cells, senescence, neosis and self-renewal in cancer. Cancer Cell Int 6:25
260. Erenpreisa J, Cragg MS (2007) Cancer: a matter of life cycle? Cell Biol Int 31(12):1507–1510
261. Duncan AW, Dorrell C, Grompe M (2009) Stem cells and liver regeneration. Gastroenterology 137(2):466–481
262. Duncan AW, Hickey RD, Paulk NK, Culberson AJ, Olson SB, Finegold MJ, Grompe M (2009) Ploidy reductions in murine fusion-derived hepatocytes. PLoS Genet 5(2), e1000385
263. Duncan AW, Taylor MH, Hickey RD, Hanlon Newell AE, Lenzi ML, Olson SB, Finegold MJ, Grompe M (2010) The ploidy conveyor of mature hepatocytes as a source of genetic variation. Nature 467(7316):707–710
264. Duncan AW, Hanlon Newell AE, Smith L, Wilson EM, Olson SB, Thayer MJ, Strom SC, Grompe M (2012) Frequent aneuploidy among normal human hepatocytes. Gastroenterology 142(1):25–28
265. Do JT, Scholer HR (2006) Cell-cell fusion as a means to establish pluripotency. Ernst Schering Res Found Workshop 60:35–45
266. Guo J, Tecirlioglu RT, Nguyen L, Koh K, Jenkin G, Trounson A (2010) Reprogramming factors involved in hybrids and cybrids of human embryonic stem cells fused with hepatocytes. Cell Reprogram 12(5):529–541
267. Pawelek JM (1993) Melanoma as a macrophage/melanocyte hybrid and the symbiotic nature of eukaryotic cells. Melanoma Res 3(1):75–76
268. Sodi SA, Chakraborty AK, Platt JT et al (1998) Melanoma x macrophage fusion hybrids acquire increased melanogenesis and metastatic potential: altered N-glycosylation as an underlying mechanism. Pigment Cell Res 11(5):299–309
269. Rachkovsky M, Sodi S, Chakraborty A, Avissar Y, Bolognia J, McNiff JM, Platt J, Bermudes D, Pawelek J (1998) Melanoma x macrophage hybrids with enhanced metastatic potential. Clin Exp Metastasis 16(4):299–312
270. Pawelek JM (2000) Tumour cell hybridization and metastasis revisited. Melanoma Res 10(6):507–514
271. DiPaolo JA, Popescu NC (1976) Relationship of chromosome changes to neoplastic cell transformation. Am J Pathol 85(3):709–738
272. Hubner B, Strickfaden H, Muller S, Cremer M, Cremer T (2009) Chromosome shattering: a mitotic catastrophe due to chromosome condensation failure. Eur Biophys J 38(6):729–747
273. Holland AJ, Cleveland DW (2012) Chromoanagenesis and cancer: mechanisms and consequences of localized, complex chromosomal rearrangements. Nat Med 18(11):1630–1638
274. Storchova Z, Kuffer C (2008) The consequences of tetraploidy and aneuploidy. J Cell Sci 121(Pt 23):3859–3866
275. Lv L, Zhang T, Yi Q et al (2012) Tetraploid cells from cytokinesis failure induce aneuploidy and spontaneous transformation of mouse ovarian surface epithelial cells. Cell Cycle 11(15):2864–2875

276. Vogt M (1959) A study of the relationship between karyotype and phenotype in clones lines of strain hela. Genetics 44(6):1257–1270
277. Ghosh S, Ghosh I (1975) Variation of stemline karyotype in a HeLa cell line. Z Krebsforsch Klin Onkol Cancer Res Clin Oncol 84(2):129–133
278. Savelyeva L, Mamaeva S (1987) Heterogeneity and balance of chromosomes in human cell line M-HeLa-76: analysis of 100 karyotypes. Cancer Genet Cytogenet 28(2):311–325
279. Chen TR (1988) Re-evaluation of HeLa, HeLa S3, and HEp-2 karyotypes. Cytogenet Cell Genet 48(1):19–24
280. Cohen EP, Kim TS (1994) Neoplastic cells that express low levels of MHC class I determinants escape host immunity. Semin Cancer Biol 5(6):419–428
281. Algarra I, Garcia-Lora A, Cabrera T, Ruiz-Cabello F, Garrido F (2004) The selection of tumor variants with altered expression of classical and nonclassical MHC class I molecules: implications for tumor immune escape. Cancer Immunol Immunother 53(10):904–910
282. Prendergast GC (2008) Immune escape as a fundamental trait of cancer: focus on IDO. Oncogene 27(28):3889–3900
283. Chitadze G, Lettau M, Bhat J et al (2013) Shedding of endogenous MHC class I-related chain molecules A and B from different human tumor entities: heterogeneous involvement of the "a disintegrin and metalloproteases" 10 and 17. Int J Cancer 133(7):1557–1566
284. Barsoum IB, Hamilton TK, Li X, Cotechini T, Miles EA, Siemens DR, Graham CH (2011) Hypoxia induces escape from innate immunity in cancer cells via increased expression of ADAM10: role of nitric oxide. Cancer Res 71(24):7433–7441
285. Passlick B, Pantel K, Kubuschok B, Angstwurm M, Neher A, Thetter O, Schweiberer L, Izbicki JR (1996) Expression of MHC molecules and ICAM-1 on non-small cell lung carcinomas: association with early lymphatic spread of tumour cells. Eur J Cancer 32A(1):141–145
286. Belov K (2011) The role of the major histocompatibility complex in the spread of contagious cancers. Mamm Genome 22(1–2):83–90
287. Maghazachi AA (2010) Role of chemokines in the biology of natural killer cells. Curr Top Microbiol Immunol 341:37–58
288. Busch S, Acar A, Magnusson Y, Gregersson P, Ryden L, Landberg G (2013) TGF-beta receptor type-2 expression in cancer-associated fibroblasts regulates breast cancer cell growth and survival and is a prognostic marker in pre-menopausal breast cancer. Oncogene 34:27–38
289. Rajasekaran K, Chu H, Kumar P et al (2011) Transforming growth factor-beta-activated kinase 1 regulates natural killer cell-mediated cytotoxicity and cytokine production. J Biol Chem 286(36):31213–31224
290. Marcoe JP, Lim JR, Schaubert KL, Fodil-Cornu N, Matka M, McCubbrey AL, Farr AR, Vidal SM, Laouar Y (2012) TGF-beta is responsible for NK cell immaturity during ontogeny and increased susceptibility to infection during mouse infancy. Nat Immunol 13(9):843–850
291. Fang WB, Jokar I, Chytil A, Moses HL, Abel T, Cheng N (2011) Loss of one Tgfbr2 allele in fibroblasts promotes metastasis in MMTV: polyoma middle T transgenic and transplant mouse models of mammary tumor progression. Clin Exp Metastasis 28(4):351–366
292. Gold LI (1999) The role for transforming growth factor-beta (TGF-beta) in human cancer. Crit Rev Oncog 10(4):303–360
293. Pardali K, Moustakas A (2007) Actions of TGF-beta as tumor suppressor and pro-metastatic factor in human cancer. Biochim Biophys Acta 1775(1):21–62
294. Quintana-Bustamante O, Grueso E, Garcia-Escudero R, Arza E, Alvarez-Barrientos A, Fabregat I, Garcia-Bravo M, Meza NW, Segovia JC (2012) Cell fusion reprogramming leads to a specific hepatic expression pattern during mouse bone marrow derived hepatocyte formation in vivo. PLoS One 7(3), e33945
295. Zhang C, Zhang F, Tsan R, Fidler IJ (2009) Transforming growth factor-beta2 is a molecular determinant for site-specific melanoma metastasis in the brain. Cancer Res 69(3):828–835

296. Lu Y, Jiang F, Zheng X, Katakowski M, Buller B, To SS, Chopp M (2011) TGF-beta1 promotes motility and invasiveness of glioma cells through activation of ADAM17. Oncol Rep 25(5):1329–1335
297. Fang S, Pentinmikko N, Ilmonen M, Salven P (2012) Dual action of TGF-beta induces vascular growth in vivo through recruitment of angiogenic VEGF-producing hematopoietic effector cells. Angiogenesis 15(3):511–519
298. Powell AA, Talasaz AH, Zhang H et al (2012) Single cell profiling of circulating tumor cells: transcriptional heterogeneity and diversity from breast cancer cell lines. PLoS One 7(5), e33788
299. Vasir B, Borges V, Wu Z, Grosman D, Rosenblatt J, Irie M, Anderson K, Kufe D, Avigan D (2005) Fusion of dendritic cells with multiple myeloma cells results in maturation and enhanced antigen presentation. Br J Haematol 129(5):687–700
300. Zheng R, Shu S (2011) Immune response to cancer and its regulation in regional lymph nodes. J Surg Oncol 103(6):550–554
301. Yeheskely-Hayon D, Minai L, Golan L, Dann EJ, Yelin D (2013) Cell fusion: optically induced cell fusion using bispecific nanoparticles (small 22/2013). Small 9(22):3770
302. Mou Y, Xie H, Huang X, Han W, Ni Y, Su H, Wang Z, Hu Q (2013) Immunological suppression of head and neck carcinoma by dendritic cell tumor fusion vaccine. Oncol Lett 6(6):1799–1803
303. Mohamed YS, Dunnion D, Teobald I, Walewska R, Browning MJ (2012) In vitro evaluation of human hybrid cell lines generated by fusion of B-lymphoblastoid cells and ex vivo tumour cells as candidate vaccines for haematological malignancies. Vaccine 30(46):6578–6587
304. Cathelin D, Nicolas A, Bouchot A, Fraszczak J, Labbe J, Bonnotte B (2011) Dendritic cell-tumor cell hybrids and immunotherapy: what's next? Cytotherapy 13(7):774–785
305. Shu S, Cochran AJ, Huang RR, Morton DL, Maecker HT (2006) Immune responses in the draining lymph nodes against cancer: implications for immunotherapy. Cancer Metastasis Rev 25(2):233–242
306. Treilleux I, Blay JY, Bendriss-Vermare N et al (2004) Dendritic cell infiltration and prognosis of early stage breast cancer. Clin Cancer Res 10(22):7466–7474
307. Branham-O'Connor M, Li J, Kotturi HS, Yu X, Wagner TE, Wei Y (2010) Fusion induced reversal of dendritic cell maturation: an altered expression of inflammatory chemokine and chemokine receptors in dendritomas. Oncol Rep 23(2):545–550
308. Strobl H, Knapp W (1999) TGF-beta1 regulation of dendritic cells. Microbes Infect 1(15):1283–1290
309. Weber F, Byrne SN, Le S, Brown DA, Breit SN, Scolyer RA, Halliday GM (2005) Transforming growth factor-beta1 immobilises dendritic cells within skin tumours and facilitates tumour escape from the immune system. Cancer Immunol Immunother 54(9):898–906
310. Donovan D, Harmey JH, Toomey D, Osborne DH, Redmond HP, Bouchier-Hayes DJ (1997) TGF beta-1 regulation of VEGF production by breast cancer cells. Ann Surg Oncol 4(8):621–627
311. Shih SC, Ju M, Liu N, Mo JR, Ney JJ, Smith LE (2003) Transforming growth factor beta1 induction of vascular endothelial growth factor receptor 1: mechanism of pericyte-induced vascular survival in vivo. Proc Natl Acad Sci U S A 100(26):15859–15864
312. Maretzky T, Reiss K, Ludwig A, Buchholz J, Scholz F, Proksch E, de Strooper B, Hartmann D, Saftig P (2005) ADAM10 mediates E-cadherin shedding and regulates epithelial cell-cell adhesion, migration, and beta-catenin translocation. Proc Natl Acad Sci U S A 102(26):9182–9187
313. Huang J, Bridges LC, White JM (2005) Selective modulation of integrin-mediated cell migration by distinct ADAM family members. Mol Biol Cell 16(10):4982–4991
314. Mochizuki S, Okada Y (2007) ADAMs in cancer cell proliferation and progression. Cancer Sci 98(5):621–628

315. Pan Y, Han C, Wang C, Hu G, Luo C, Gan X, Zhang F, Lu Y, Ding X (2012) ADAM10 promotes pituitary adenoma cell migration by regulating cleavage of CD44 and L1. J Mol Endocrinol 49(1):21–33
316. Hirakawa S (2009) From tumor lymphangiogenesis to lymphvascular niche. Cancer Sci 100(6):983–989
317. Weiss L, Ward PM (1983) Cell detachment and metastasis. Cancer Metastasis Rev 2(2):111–127
318. Chambers AF, Groom AC, MacDonald IC (2002) Dissemination and growth of cancer cells in metastatic sites. Nat Rev Cancer 2(8):563–572
319. Hart IR, Talmadge JE, Fidler IJ (1981) Metastatic behavior of a murine reticulum cell sarcoma exhibiting organ-specific growth. Cancer Res 41(4):1281–1287
320. Duda DG, Duyverman AM, Kohno M, Snuderl M, Steller EJ, Fukumura D, Jain RK (2010) Malignant cells facilitate lung metastasis by bringing their own soil. Proc Natl Acad Sci U S A 107(50):21677–21682
321. Labelle M, Hynes RO (2012) The initial hours of metastasis: the importance of cooperative host-tumor cell interactions during hematogenous dissemination. Cancer Discov 2(12):1091–1099
322. Podsypanina K, Du YC, Jechlinger M, Beverly LJ, Hambardzumyan D, Varmus H (2008) Seeding and propagation of untransformed mouse mammary cells in the lung. Science 321(5897):1841–1844
323. Huh SJ, Liang S, Sharma A, Dong C, Robertson GP (2010) Transiently entrapped circulating tumor cells interact with neutrophils to facilitate lung metastasis development. Cancer Res 70(14):6071–6082
324. Lee N, Barthel SR, Schatton T (2014) Melanoma stem cells and metastasis: mimicking hematopoietic cell trafficking? Lab Invest 94(1):13–30
325. Adams DL, Martin SS, Alpaugh RK et al (2014) Circulating giant macrophages as a potential biomarker of solid tumors. Proc Natl Acad Sci U S A 111(9):3514–3519
326. Yagami-Hiromasa T, Sato T, Kurisaki T, Kamijo K, Nabeshima Y, Fujisawa-Sehara A (1995) A metalloprotease-disintegrin participating in myoblast fusion. Nature 377(6550):652–656
327. Huovila AP, Almeida EA, White JM (1996) ADAMs and cell fusion. Curr Opin Cell Biol 8(5):692–699
328. Harris HA, Murrills RJ, Komm BS (1997) Expression of meltrin-alpha mRNA is not restricted to fusagenic cells. J Cell Biochem 67(1):136–142
329. Abe E, Mocharla H, Yamate T, Taguchi Y, Manolagas SC (1999) Meltrin-alpha, a fusion protein involved in multinucleated giant cell and osteoclast formation. Calcif Tissue Int 64(6):508–515
330. Miller JB (1995) Developmental biology. Making one cell from two. Nature 377(6550):575–576
331. Nath D, Slocombe PM, Webster A, Stephens PE, Docherty AJ, Murphy G (2000) Meltrin gamma(ADAM-9) mediates cellular adhesion through alpha(6)beta(1)integrin, leading to a marked induction of fibroblast cell motility. J Cell Sci 113(Pt 12):2319–2328
332. Namba K, Nishio M, Mori K, Miyamoto N, Tsurudome M, Ito M, Kawano M, Uchida A, Ito Y (2001) Involvement of ADAM9 in multinucleated giant cell formation of blood monocytes. Cell Immunol 213(2):104–113
333. Thodeti CK, Frohlich C, Nielsen CK et al (2005) Hierarchy of ADAM12 binding to integrins in tumor cells. Exp Cell Res 309(2):438–450
334. Lafuste P, Sonnet C, Chazaud B, Dreyfus PA, Gherardi RK, Wewer UM, Authier FJ (2005) ADAM12 and alpha9beta1 integrin are instrumental in human myogenic cell differentiation. Mol Biol Cell 16(2):861–870
335. Micocci KC, Martin AC, Montenegro Cde F, Durante AC, Pouliot N, Cominetti MR, Selistre-de-Araujo HS (2013) ADAM9 silencing inhibits breast tumor cell invasion in vitro. Biochimie 95(7):1371–1378

336. Roy R, Rodig S, Bielenberg D, Zurakowski D, Moses MA (2011) ADAM12 transmembrane and secreted isoforms promote breast tumor growth: a distinct role for ADAM12-S protein in tumor metastasis. J Biol Chem 286(23):20758–20768
337. Rao VH, Kandel A, Lynch D, Pena Z, Marwaha N, Deng C, Watson P, Hansen LA (2012) A positive feedback loop between HER2 and ADAM12 in human head and neck cancer cells increases migration and invasion. Oncogene 31(23):2888–2898
338. Lazova R, Laberge GS, Duvall E et al (2013) A melanoma brain metastasis with a donor-patient hybrid genome following bone marrow transplantation: first evidence for fusion in human cancer. PLoS One 8(6), e66731
339. Dittmar T, Schwitalla S, Seidel J, Haverkampf S, Reith G, Meyer-Staeckling S, Brandt BH, Niggemann B, Zanker KS (2011) Characterization of hybrid cells derived from spontaneous fusion events between breast epithelial cells exhibiting stem-like characteristics and breast cancer cells. Clin Exp Metastasis 28(1):75–90
340. Dittmar T, Seidel J, Zaenker KS, Niggemann B (2006) Carcinogenesis driven by bone marrow-derived stem cells. Contrib Microbiol 13:156–169
341. Larizza L, Schirrmacher V, Pfluger E (1984) Acquisition of high metastatic capacity after in vitro fusion of a nonmetastatic tumor line with a bone marrow-derived macrophage. J Exp Med 160(5):1579–1584
342. Dezentje DA, Arking DE, Kortenhorst MS, West K, Chakravarti A, Kern SE (2009) Hybrids of aneuploid human cancer cells permit complementation of simple and complex cancer defects. Cancer Biol Ther 8(4):347–355
343. Man YG, Mason J, Harley R, Kim YH, Zhu K, Gardner WA (2011) Leukocyte-mediated cell dissemination and metastasis: findings from multiple types of human tumors. J Cell Biochem 112(4):1154–1167
344. Mi R, Pan C, Bian X, Song L, Tian W, Cao F, Yin J, Peng H, Ma J (2012) Fusion between tumor cells enhances melanoma metastatic potential. J Cancer Res Clin Oncol 138(10):1651–1658
345. Nagler C, Hardt C, Zanker KS, Dittmar T (2011) Co-cultivation of murine BMDCs with 67NR mouse mammary carcinoma cells give rise to highly drug resistant cells. Cancer Cell Int 11(1):21
346. Dittmar T, Zanker KS (2011) Introduction. Adv Exp Med Biol 714:1–3
347. Wendt MK, Cooper AN, Dwinell MB (2008) Epigenetic silencing of CXCL12 increases the metastatic potential of mammary carcinoma cells. Oncogene 27(10):1461–1471
348. Clatot F, Picquenot JM, Choussy O et al (2011) Intratumoural level of SDF-1 correlates with survival in head and neck squamous cell carcinoma. Oral Oncol 47:1062–1068
349. Xue TC, Han D, Chen RX, Zou JH, Wang Y, Tang ZY, Ye SL (2011) High expression of CXCR7 combined with Alpha fetoprotein in hepatocellular carcinoma correlates with extra-hepatic metastasis to lung after hepatectomy. Asian Pac J Cancer Prev 12(3):657–663
350. Boldajipour B, Mahabaleshwar H, Kardash E, Reichman-Fried M, Blaser H, Minina S, Wilson D, Xu Q, Raz E (2008) Control of chemokine-guided cell migration by ligand sequestration. Cell 132(3):463–473
351. Lee E, Han J, Kim K, Choi H, Cho EG, Lee TR (2012) CXCR7, not CXCR4, mediates SDF1-induced melanocyte migration. Pigment Cell Melanoma Res 26:58–66
352. Lee E, Han J, Kim K, Choi H, Cho EG, Lee TR (2013) CXCR7 mediates SDF1-induced melanocyte migration. Pigment Cell Melanoma Res 26(1):58–66
353. Mahabaleshwar H, Boldajipour B, Raz E (2008) Killing the messenger: the role of CXCR7 in regulating primordial germ cell migration. Cell Adh Migr 2(2):69–70
354. Moissoglu K, Majumdar R, Parent CA (2014) Cell migration: sinking in a gradient. Curr Biol 24(1):R23–R25
355. Zabel BA, Lewen S, Berahovich RD, Jaen JC, Schall TJ (2011) The novel chemokine receptor CXCR7 regulates trans-endothelial migration of cancer cells. Mol Cancer 10:73
356. Holmgren L, O'Reilly MS, Folkman J (1995) Dormancy of micrometastases: balanced proliferation and apoptosis in the presence of angiogenesis suppression. Nat Med 1(2):149–153

357. Peeters CF, de Waal RM, Wobbes T, Westphal JR, Ruers TJ (2006) Outgrowth of human liver metastases after resection of the primary colorectal tumor: a shift in the balance between apoptosis and proliferation. Int J Cancer 119(6):1249–1253
358. Schaefer C, Fuhrhop I, Schroeder M, Viezens L, Otten J, Fiedler W, Ruther W, Hansen-Algenstaedt N (2010) Microcirculation of secondary bone tumors in vivo: the impact of minor surgery at a distal site. J Orthop Res 28(11):1515–1521
359. Schaefer C, Schroeder M, Fuhrhop I, Viezens L, Otten J, Fiedler W, Ruther W, Hansen-Algenstaedt N (2011) Primary tumor dependent inhibition of tumor growth, angiogenesis, and perfusion of secondary breast cancer in bone. J Orthop Res 29(8):1251–1258
360. Phelps HA, Kuntz CA, Milner RJ, Powers BE, Bacon NJ (2011) Radical excision with five-centimeter margins for treatment of feline injection-site sarcomas: 91 cases (1998-2002). J Am Vet Med Assoc 239(1):97–106
361. Fornabaio DM, Alterman AL, Stackpole CW (1988) Metastatic dissemination of B16 melanoma: evidence that metastases can result from nonspecific trapping of disseminated tumor cells. Invasion Metastasis 8(1):1–16
362. Fogelquist S, Deutsch B, Groszek L, Valle EF, Stackpole CW (1991) Hemodynamic considerations in organ and tissue patterning of B16 melanoma systemic metastasis and colonization. Invasion Metastasis 11(5):261–272
363. Strell C, Entschladen F (2008) Extravasation of leukocytes in comparison to tumor cells. Cell Commun Signal 6:10
364. Jacobs PP, Sackstein R (2011) CD44 and HCELL: preventing hematogenous metastasis at step 1. FEBS Lett 585:3148–3158
365. Murphy P, Alexander P, Senior PV, Fleming J, Kirkham N, Taylor I (1988) Mechanisms of organ selective tumour growth by bloodborne cancer cells. Br J Cancer 57(1):19–31
366. Varani J (1982) Chemotaxis of metastatic tumor cells. Cancer Metastasis Rev 1(1):17–28
367. Uotani H, Yamashita I, Nagata T, Kishimoto H, Kashii Y, Tsukada K (2001) Induction of E-selectin after partial hepatectomy promotes metastases to liver in mice. J Surg Res 96(2):197–203
368. Sackstein R (2010) Directing stem cell trafficking via GPS. Methods Enzymol 479:93–105
369. Deng L, Chen N, Li Y, Zheng H, Lei Q (1806) CXCR6/CXCL16 functions as a regulator in metastasis and progression of cancer. Biochim Biophys Acta 1:42–49
370. Carman CV, Sage PT, Sciuto TE, de la Fuente MA, Geha RS, Ochs HD, Dvorak HF, Dvorak AM, Springer TA (2007) Transcellular diapedesis is initiated by invasive podosomes. Immunity 26(6):784–797
371. Naumann U, Cameroni E, Pruenster M, Mahabaleshwar H, Raz E, Zerwes HG, Rot A, Thelen M (2010) CXCR7 functions as a scavenger for CXCL12 and CXCL11. PLoS One 5(2):e9175
372. Ratajczak MZ, Lee H, Wysoczynski M, Wan W, Marlicz W, Laughlin MJ, Kucia M, Janowska-Wieczorek A, Ratajczak J (2010) Novel insight into stem cell mobilization-plasma sphingosine-1-phosphate is a major chemoattractant that directs the egress of hematopoietic stem progenitor cells from the bone marrow and its level in peripheral blood increases during mobilization due to activation of complement cascade/membrane attack complex. Leukemia 24(5):976–985
373. Li C, Kong Y, Wang H, Wang S, Yu H, Liu X, Yang L, Jiang X, Li L (2009) Homing of bone marrow mesenchymal stem cells mediated by sphingosine 1-phosphate contributes to liver fibrosis. J Hepatol 50(6):1174–1183
374. Ratajczak MZ, Kim CH, Abdel-Latif A, Schneider G, Kucia M, Morris AJ, Laughlin MJ, Ratajczak J (2011) A novel perspective on stem cell homing and mobilization: review on bioactive lipids as potent chemoattractants and cationic peptides as underappreciated modulators of responsiveness to SDF-1 gradients. Leukemia 26:63–72
375. Moll NM, Ransohoff RM (2010) CXCL12 and CXCR4 in bone marrow physiology. Expert Rev Hematol 3(3):315–322

376. Schulz C, von Andrian UH, Massberg S (2009) Hematopoietic stem and progenitor cells: their mobilization and homing to bone marrow and peripheral tissue. Immunol Res 44(1–3):160–168
377. Alix-Panabieres C, Riethdorf S, Pantel K (2008) Circulating tumor cells and bone marrow micrometastasis. Clin Cancer Res 14(16):5013–5021
378. Sacanna E, Ibrahim T, Gaudio M et al (2011) The role of CXCR4 in the prediction of bone metastases from breast cancer: a pilot study. Oncology 80(3–4):225–231
379. Aoki Y, Shimura H, Li H, Mizumoto K, Date K, Tanaka M (1999) A model of port-site metastases of gallbladder cancer: the influence of peritoneal injury and its repair on abdominal wall metastases. Surgery 125(5):553–559
380. Brundell S, Ellis T, Dodd T, Watson DI, Hewett PJ (2002) Hematogenous spread as a mechanism for the generation of abdominal wound metastases following laparoscopy. Surg Endosc 16(2):292–295
381. Lee JY, Murphy SM, Scanlon EF (1994) Effect of trauma on implantation of metastatic tumor in bone in mice. J Surg Oncol 56(3):178–184
382. Sampson WI (1976) Letter: Cancer at insulin injection site. JAMA 235(4):374
383. Skipper D, Jeffrey MJ, Cooper AJ, Taylor I, Alexander P (1988) Preferential growth of bloodborne cancer cells in colonic anastomoses. Br J Cancer 57(6):564–568
384. Worthy TS, Wynne EJ (1960) Metastatic carcinoma at the site of injection of penicillin. Br Med J 2(5207):1208–1209
385. Crowley JD, Still WJ (1960) Metastatic carcinoma at the site of injection of iron-dextran complex. Br Med J 1(5183):1411–1412

# Chapter 10
# Role of the Family of Ephs and Ephrins in Cell-Cell Communication in Cancer

**Mohamed Amessou and Mustapha Kandouz**

**Abstract** Modes of intercellular communication that require direct cell membrane-to-membrane interactions do not always involve the formation of junctional structures such as gap or tight junctions. An important example of such communication, involves the large family of proteins called Eph receptors (Erythropoietin-producing human hepatocellular carcinoma) and their ligands called Ephrins. In this review, we will introduce these proteins, their structural and functional features as well as their increasingly recognized role in cancer. Rather than providing a comprehensive report of all available data for each of the proteins and their characteristics, we will focus on introducing some key concepts of this protein family such as the fact that both receptors and ligands are membrane-embedded, thus eliciting bidirectional signaling, their dual and antagonistic functions in biological processes, and also their ability to engage in *cis* and *trans* interactions. We will also provide literature data to illustrate the genesis and transduction of signaling through Ephs and Ephrins. Finally, an important part of this chapter will be dedicated to integrating the functions of Ephs and Ephrins within the global system of cell-cell communication. These and other concepts will be addressed in view of their importance in understanding cancer progression and shedding light on the therapeutic pertinence of targeting these proteins. The authors apologize for overlooking important data available, because choices had to be made and focus was on illustrative examples rather than global survey of information.

**Keywords** Eph • Ephrin • Receptor • Ligand • Kinase • Bi-directional signaling • Gap junction • Tight junction • Adherens junction • Therapeutic targeting

---

M. Amessou
Department of Pathology, Wayne State University School of Medicine, Detroit, MI, USA

M. Kandouz (✉)
Department of Pathology, Wayne State University School of Medicine, Detroit, MI, USA

Karmanos Cancer Institute, Wayne State University, Detroit, MI, USA
e-mail: ag1764@wayne.edu


M. Kandouz (ed.), *Intercellular Communication in Cancer*,
DOI 10.1007/978-94-017-7380-5_10

## 10.1 Introduction

Although cell-cell junctions (i.e. Gap, tight, adherens junctions) are essential elements in intercellular communication, many proteins are involved in cell membrane-to-membrane communication, without necessarily depending on the formation of junctional structures per se. Among these, the Erythropoietin-producing human hepatocellular carcinoma (Eph) receptors and their ligands called Ephrins are of particular interest. Eph receptors constitute the largest family of receptor tyrosine kinases (RTKs) and include 14 Eph receptors associating with eight Ephrin ligands in mammals. Based on their sequence similarity and receptor-ligand binding affinities, Ephs and Ephrins are classified into two subfamilies (Fig. 10.1), A-type (EphA) and B-type (EphB) receptors, and A-type (EphrinA) and B-type (EphrinB) ligands. EphA receptors bind almost exclusively to EphrinAs and EphB receptors bind almost exclusively to EphrinBs [1] (Fig. 10.2). Nonetheless, though limited, there is crosstalk between A and B types. For example, the EphB2 receptor can bind EphrinB ligands as well as EphrinA5 [2]. There is also a structural difference between Ephrins; while EphrinAs are tethered to the cell surface by a glycosyl-phosphatidylinositol (GPI)-anchor, EphrinBs are bound to the membrane by a single transmembrane domain (Fig. 10.2). The extracellular domain of both A and B-type Eph receptors is made of a globular ligand-binding domain, a cysteine-rich region, and two fibronectin type III repeats, while the intracellular cytoplasmic domain includes a juxtamembrane region, a tyrosine kinase domain, a sterile $\alpha$ motif (SAM) protein–protein interaction domain, and a C-terminal PDZ-binding

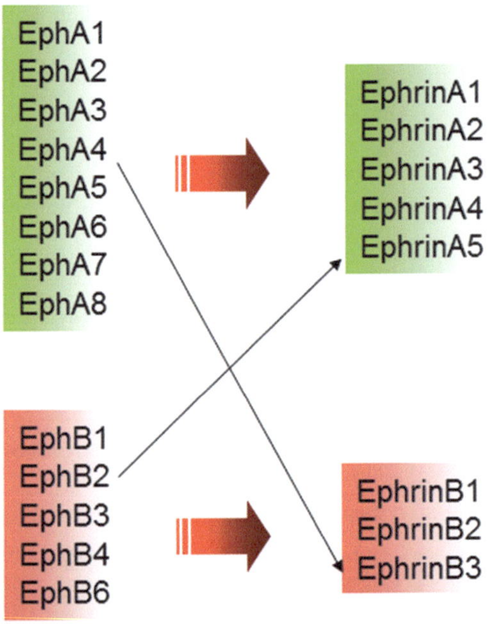


**Fig. 10.1** Eph-Ephrin interactions. A-type Eph receptors (A1-A8) bind A-type Ephrin ligands (A1-A5), while B-type Eph receptors (B1-5,6) bind B-type Ephrin ligands (B1-3). Exceptions to this family preferences include the binding of EphA4 to EphrinB3 in addition to EphrinAs and EphB2 binding to EphrinA5 in addition to EphrinBs

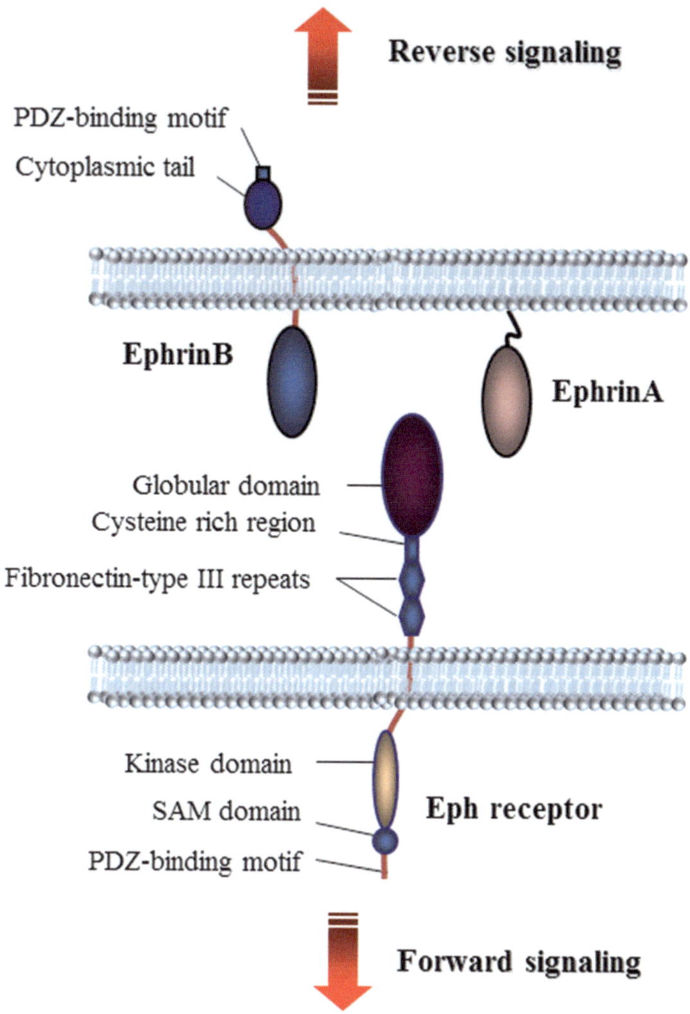


**Fig. 10.2** Domain structures and bidirectional signaling of Eph receptors and Ephrin ligands. EphrinAs are tethered to the cell surface by a glycosyl-phosphatidylinositol (GPI)-anchor, while EphrinBs are bound to the membrane by a single transmembrane domain. The A and B-type Eph receptors have an extracellular domain composed of a globular ligand-binding domain, a cysteine-rich region, and two fibronectin type III repeats, while the intracellular cytoplasmic domain includes a juxtamembrane region, a tyrosine kinase domain, a sterile α motif (SAM) protein–protein interaction domain, and a C-terminal PDZ-binding motif. Both ligands and receptors are membrane-anchored and, upon receptor-ligand interaction, bi-directional signaling events are initiated from both the receptor (forward signaling) and the ligand (reverse signaling)

motif (Fig. 10.2). A key characteristic of the Eph/Ephrin family is that both the receptors and the ligands are membrane-bound and, following receptor-ligand binding, interacting cells engage in a bi-directional signaling event initiated from both the receptor (forward signaling) and the ligand (reverse signaling) (Fig. 10.2), although the latter is relatively less understood than the former. This feature of signaling bi-directionality makes relativistic the distinction between "receptors" and "ligands", since based on functional considerations both Ephs and Ephrins could be viewed as receptors and ligands at the same time. However, for the sake of simplicity, we will refer to Ephs as receptors and Ephrins as ligands.

Ephs/Ephrins are involved in multiple biological processes in different organs and tissues, such as neuro-navigation, synaptic plasticity and response to nerve injury, regulation of angiogenesis, formation of spatial boundaries and tissue morphogenesis during embryogenesis, as well as the bone and skeletal development (for reviews see [3–20]). What could be considered as a core element in their function is their ability to regulate cell sorting and tissue distribution, by virtue of their role in cell-cell attraction and repulsion. Through this process, they can affect the intermingling between different cell populations (Fig. 10.3), an event of utmost importance in both normal physiology and disease onset and progression.

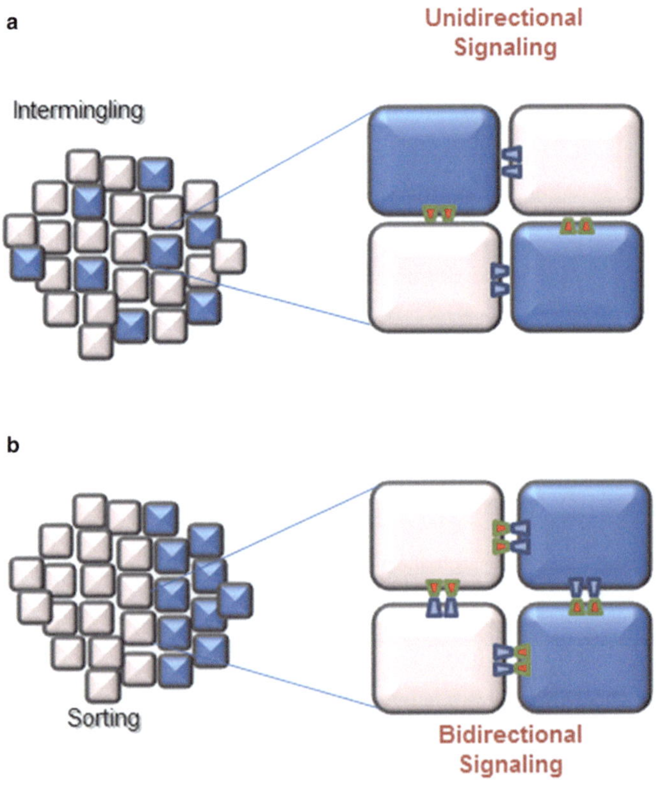


**Fig. 10.3** A simplified model of regulation of cell-cell communication by Eph/Ephrin interactions. (**a**) Unidirectional signaling: the absence of a pro-repulsion Eph/Ephrin signaling in a cell population, due to absence or inactivation of the receptor or ligand, allows it to intermingle with the population that is able to undergo such a signaling. (**b**) Bidirectional signaling: when both cell populations undergo signaling, the result is repulsion between the receptor- and ligand-expressing cells, thus sorting out the two distinct cell populations

## 10.2 Role of Eph Receptors and Ephrin Ligands in Cancer

The earliest hints of a possible role for Ephs/Ephrins in cancer came from reports of changes in their expression, examined at the protein or mRNA level in various cancers [21]. As illustrative examples, overexpression has been observed, such as of EphrinB2 in oesophageal squamous carcinomas [22], EphA2 in pancreatic cancer [23] and EphB2 in breast cancer [24]. Lower EphB6 expression is associated with an increased risk of developing metastases in early stage non-small cell lung cancer (NSCLC) [25]. Contrary to the expected overlap between their functions, the expression and function of different Ephs and Ephrins do not always evolve in the same direction. For example, high expression of EphA4 and low expression of EphB2 are correlated with colorectal cancer metastases to the liver [26].

A multitude of studies have subsequently shown the role of Eph/Ephrin signaling or lack thereof in cancer progression both *in vitro* and *in vivo*, and we will illustrate these functions with two examples only. Within the family, EphA2/EphrinA signaling is one of the most investigated pairs for its role in different processes such as cancer cell motility and invasiveness. Following EphrinA1 stimulation, EphA2 forms a complex with Src and FAK, resulting in FAK tyrosine phosphorylation, activation of a Rho-dependent actin and myosin contraction and cell repulsion [27]. Furthermore, EphrinA1 stimulation induces activation of the Abl tyrosine kinase, resulting in the phosphorylation of the CrkII adaptor and the disruption of its interaction with C3G, a Rap1 guanine nucleotide exchange factor. The result is a decrease in Rap1-GTP, cell rounding and loss of adhesion [28]. The second example is from the B-type receptors and ligands. EphB activation by EphrinB2 is involved in migration and invasion of glioma cells [29]. EphrinB2-mediated activation of EphB4 enhances migration of melanoma cells and increases RhoA activity. The use of a kinase-dead mutant to inhibit EphB4 signaling affects the organization of the actin cytoskeleton and inactivates RhoA, thus leading to inhibition of cell migration [30]. However, in breast cancer cells, EphB4/EphrinB2 signaling activates Abl family of tyrosine kinases and the Crk adapter protein, resulting in inhibition of cell viability, increase in proliferation, motility and invasion [31].

Ephs and Ephrins play a critical role in cardiovascular development and actually their role in the angiogenic remodeling of blood vessels was among the first aspects of their biology to be investigated [20, 32–35]. Among the different family members, the role of EphrinB2 is prominent. EphrinB2 levels are elevated in angiogenic endothelial cells [36, 37], where its activation induces migration of these cells [38]. EphrinB2 is important in the maintenance of vascular architecture via regulating cell migration and spatial organization [39]. Such findings have prompted the study of the role of Ephs/Ephrins in tumor-driven angiogenesis, in association with cancer metastatic progression. An indication of this role comes from data linking Eph/Ephrin expression and activation with angiogenesis and metastasis. For instance, expression of high levels of EphrinB2 correlates with higher tumor vascularization and metastasis in malignant melanoma [40]. In addition, a transgenic C57Bl/6/ASV-B murine model of hepatocellular carcinoma (HCC) development, showed increased levels of EphrinB2 in correlation with tumor progression, particularly in tumor-associated sinusoidal endothelial cells [41]. Similar conclusions were drawn from transgenic mice where wild type and dominant negative mutant EphrinB2 were expressed specifically in mammary epithelial cells, and which showed an effect on mammary gland vascularization; contrary to wild type EphrinB2 which induced superfluous but organized capillaries, the mutant EphrinB2 resulted in an irregular vasculature [42]. Moreover, using the same animal model crossed with the well-established NeuT transgenic animal model of mammary transformation, it was shown that EphrinB2 overexpression drives a metastatic phenotype [42]. Overall, there is a relatively abundant literature showing the role of Ephs and Ephrins in normal and tumor-associated angiogenesis and how it impacts cancer metastasis (Reviewed in [21]).

With the increasing interest in the role of cancer stem/progenitor cells, there are a growing number of studies assessing a potential function of Ephs/Ephrins within this research framework, in particular how it could be associated with the pre-metastatic evolution. The best known evidence comes from the intestinal epithelium and the role of EphB/EphrinB signaling in the positioning of cells within the stem cell niche [43]. High EphB2 expression actually contributes to a gene signature specific for adult intestinal stem cells (ISCs) that is associated with disease relapse in colorectal cancer patients [44]. Furthermore, EphB4 signaling, independently from EphrinB2 signaling has been shown to drive tumor metastasis in a model of NeuT mouse mammary tumors by impacting stem/progenitor cells prior to initiation of tumor formation [45]. On the other hand, EphrinB2 expression perturbs the regulation of the stem cell niche and cell differentiation [46].

## 10.3 Versatility and Key Features of Ephs and Ephrins

### *10.3.1 Bi-Directional Signaling in Homotypic and Heterotypic Cell-Cell Interactions*

A key particularity of Ephs and Ephrins is that, unlike growth factor receptors such as EGFR and PDGFR which are activated by free released soluble ligands, both the receptor and the ligand are membrane-bound and both undergo signaling into their respective cells; signaling that originates from the Eph receptor is called "forward signaling" and the one originating from the ligand is called "reverse signaling" (Fig. 10.2). This means that Ephs/Ephrins signaling has a functional impact on two cellular entities which could be either similar or different cell types, therefore likely to respond in different ways. With regard to cancer, Ephs and Ephrins are expressed in various cell types present within the tumor microenvironments, including endothelial cells of the angiogenic structure and different cellular components of the primary site as well as a multitude of cells in the metastatic niches. Furthermore, this function could be controlled through regulation of expression of different combinations of receptors and ligands in specific tissues under specific conditions, as well as either co-expression, mutual exclusion, or expression in complementary gradients within the same tissue. Keeping in mind that these proteins have first been identified as guidance molecules that are responsible for directing the movement of cells and cell sub-populations within their microenvironment, through processes of sorting and segregation, it would be legitimate to transpose this function as a framework for understanding their role during cancer progression and metastasis. Such a function would allow invasive and metastatic cells to bypass the boundaries of the primary tumor site and the endothelial system and would contribute to making permissible the intermingling with cells in the otherwise inhospitable metastatic niches.

To illustrate the pertinence of this model, let's take few examples of how Ephs/Ephrins regulate the process called contact inhibition of locomotion (CIL) [47], the process through which cells interrupt their unidirectional migration after colliding with other cells [48]. CIL is decreased in highly invasive cancer cells, thus allowing them to move through populations of stromal cells. On one hand, the homotypic interaction between EphA2 or EphA4 receptors and EphrinA ligands, both being embedded in prostate cancer cells, mediates CIL in these cells. On the other hand, the heterotypic interaction between tumor-embedded EphB3 and EphB4 and stromal cells-embedded EphrinB2 enhances the migratory potential of tumor cells. The net result is relief from CIL and increased ability of cancer cells to migrate and invade through the stroma, a process involved in metastasis [49]. Another example is that of metastatic prostate cancer cells PC3 co-cultured with fibroblasts; these cells fail to undergo CIL and to stop migrating thanks to the activation of EphB3 and EphB4 by EphrinB2, and the ensuing activation of Cdc42. Interestingly, homotypic contacts between PC3 cancer cells activate EphA-Rho-Rho kinase (ROCK) signaling, and result in CIL [50]. In another instance, a role in CIL has been shown for Nm23-H1 (Nonmetastatic protein 23), a tumor metastasis suppressor; it translocates to the sites of contact of U87MG glioblastoma cells with normal glial cells through association with α-catenin and N-cadherin. Interestingly, EphrinB1 has the ability to interfere with this mechanism, by inhibiting the association of Nm23-H1 with Tiam1 at the sites of contact between glioblastoma and glial cells, thus unblocking Rac1 activation, relieving cells from CIL and promoting their invasion [51].

During metastatic progression, when moving from angiogenic to metastatic sites such as the lung, brain, liver or bone marrow microenvironments, cancer cells are exposed to interactions with many cell types. These interactions could be impacted by differential expression and activation of Eph/Ephrins. Decreased levels of EphrinA1 are associated with preferential prostate cancer metastases to the bone [52]. EphB2 is activated in Waldenstrom's Macroglobulinemia (WM) cells, a rare type of non-Hodgkin lymphoma, while the EphrinB2 ligand is highly expressed on endothelial-cells and bone marrow stromal cells. It has been suggested that endothelial EphB2/EphrinB2 signaling might drive angiogenesis [53]. In another example, EphrinA4 is involved in regulating adhesion and transendothelial migration of Chronic lymphocytic leukemia (CLL) cells circulating between the blood and other tissues, while EphA2 is highly expressed in CD31 (+) endothelial cells of human lymph nodes. EphA2/EphrinA4 signaling reduces CLL cells' adhesion to ICAM-1 (intercellular adhesion molecule 1), VCAM-1 (vascular cell adhesion molecule 1), and other extracellular matrix molecules [54].

### 10.3.2 Cis *Versus* Trans *Interactions*

Another aspect in the biology of Ephs/Ephrins is the variety of their interaction types. In addition to the so-called "*trans*" interaction between an Eph and an Ephrin, with each of the partners being embedded in membranes of different neighboring cells, it has been shown that molecules coexpressed in the same cell can engage in lateral "*cis*" interactions and have a functional impact. For instance, in the nervous system, Eph receptor activation could be attenuated by *cis* interaction with Ephrins, thus playing a role in fine tuning Eph receptor activation during axon pathfinding and synaptic function [55–58]. In cancer cells, EphrinA3 inhibits the ability of EphA2 and EphA3 to bind Ephrins in *trans* and to become activated. EphA3 activation is attenuated by *cis* interaction with EphrinA3. It is suggested that *cis* and *trans* interactions do not show identical receptor-ligand selectivity. EphrinB2, which does not activate EphA3 in *trans*, can *cis*-inhibit both the EphB4 receptor and the A-type EphA3 receptor [19, 59]. Interestingly, a lung cancer mutation identified in the second fibronectin type III repeat of EphA3 was found to enhance its *cis* association with EphrinA3 [59]. It has therefore been hypothesized that *cis* interactions could play an opposite function to *trans* interactions, by attenuating the tumor suppressing functions of Eph receptors. The extent of the functional importance of *cis* interactions is not known and might be influenced by regulatory mechanisms such as expression levels and spatio-temporal regulation. For instance, not all neurons coexpressing Eph receptors and Ephrin ligands undergo *cis* interactions, and in fact in some neurons Ephs and Ephrins are found in distinct locations at microdomains of the plasma membrane [58, 60]. Whether such segregation occurs in cancer cells is not known. In general, the importance of *cis* interactions for the function of Ephs/Ephrins in cancer is yet to be uncovered. This question is also to be addressed in light of some technical limitations, most importantly the fact that immunohistochemical, immunoblotting or RT-PCR assessment of Ephs' and Ephrins' expression levels, do not distinguish between *cis* and *trans*-expressed proteins, thus rendering elusive to some extent the cancer prognostic value assigned to these proteins' expression levels.

### 10.3.3 *Duality of Eph/Ephrin Functions*

One of the least understood aspects of the Eph and Ephrin biology is their ability to show opposite functions in different contexts. This conclusion stems from the observation of apparently paradoxical effects of their expression and activation. For instance, in normal cells, although Eph/Ephrin interactions mainly result in repulsive and contact inhibitory functions, attractive and adhesive effects have also been reported [61–63]. In the developing visual system, EphrinBs behave as both attractants and repellents to retinal axons [64–66]. Using an *in vitro* axon outgrowth assay, depending on their concentration gradient, EphrinAs show both growth promotion and inhibition of retinal axons [67].

Also, there is increasing evidence that Ephs and Ephrins might play different roles at different stages of tumor progression, associated with various processes such as proliferation, survival, migration, differentiation and autophagy. In fact, in many cases, one protein is able to show opposite functions reminiscent of both a tumor promoter and a tumor suppressor protein [68, 69]. This behavior sets this family of RTKs apart from many other RTKs, which are traditionally viewed as oncogenes. This functional versatility has an important impact on the role of Ephs/Ephrins in cancer, as exemplified by EphA2/EphrinA signaling; while EphA2 overexpression has been linked to cancer progression [70–73], its activation can induce signaling indicative of tumor suppression such as inhibition of the Ras/ERK pathway, suppression of integrin function and induction of FAK (Focal Adhesion Kinase) dephosphorylation, inhibition of cell proliferation and migration, and induction of apoptosis [74–76]. Furthermore, *EphA2* KO (Knock out) mice were shown to have an increased susceptibility to skin carcinogenesis [77]. EphB4 was reported to have a positive effect on survival, invasion and migration, while it has also been suggested to be a tumor suppressor. When overexpressed in the prostate cancer cell line 22Rv1 and the non-transformed mammary epithelial cell line MCF-10A, EphB4 showed pro-oncogenic functions (*i.e.* anchorage-independent growth, migration and invasion) which were reversed by EphrinB2 [78]. The role of Ephs/Ephrins in the metastatic process can be dissociated from their role in migration, invasion or adhesion. While EphB4 has been reported to promote metastatic dissemination [79], metastasis but not migration of EphB4-expressing tumor cells or their adhesion to EphrinB2-expressing endothelial cells, requires the EphB4 forward signaling [80].

Another element of understanding of the dual functions of Eph receptors is the role of the Ephrin ligands. For example, ligand-dependent stimulation of EphA2 inhibits cell migration, while in absence of the ligand it promotes this function [81]. As seen in normal cells, there seems to be a role for ligand concentration gradients in cancer cells as well. Even the kinase-defective EphB6 exerts a biphasic effect in response to different concentrations of EphrinB2; at low ligand concentrations it promotes cell adhesion and migration, it however induces repulsion and inhibits migration at high ligand concentrations. It has been suggested that the observed differences could be explained by a differential tyrosine phosphorylation mediated by a Src family kinase [82].

### 10.3.4 Ligand-Independent Functions

Direct cell-cell contact is an essential feature in Ephs and Ephrins, and early studies have shown that only when these proteins are membrane-embedded or when they are artificially clustered, can they activate their binding partner [83–85]. In fact, soluble monomeric proteins can act as antagonists of activation [84]. Nevertheless, Ephs and Ephrins are able to undergo functions independently from each other. In fact, there are instances where Eph receptors are overexpressed in cancer cells but lack significant co-expression of cognate Ephrin ligands [31, 86–89]. There is also functional evidence for unidirectional signaling. This is illustrated for instance by the ability of EphA4 to induce cell sorting independently of interaction with Ephrins [90]. In prostate cancer cells, the dietary bioactive lipid Arachidonic Acid (AA) induces an increase in an Akt-dependent/ligand-independent activation of EphA2 [91]. Reduction in EphrinA expression, along with Akt-mediated phosphorylation of EphA2 on serine 897, is suggested to pave the way for ligand-independent pro-oncogenic functions of EphA2. However, EphrinA stimulation is able to restore EphA2's tumor suppressive signaling pathways (*i.e.* inhibition of PI3/Akt and Ras/ERK pathways). Within this framework, doxazosin, a small molecule agonist for EphA2 and EphA4, was developed to mimic ligand stimulation, thus inhibiting Akt and ERK kinase activities, triggering EphA2 receptor internalization, and suppressing haptotactic and chemotactic migration of prostate, breast, and glioma cancer cells [92]. EphrinA1 Ligand activation of EphA2 inhibits chemotactic migration of glioma and prostate cancer cells while, upon overexpression, EphA2 promotes migration in a ligand-independent manner that requires its phosphorylation on serine 897 by Akt. This has been put forward to explain the dual pro- and anti-oncogenic functions of EphA2 [81]. In MCF-10A mammary cells and 22Rv1 prostate cancer cells, EphB4 overexpression enhances anchorage-independent growth, migration and invasion and these effects are reversed by EphrinB2 [78]. This result supports the hypothesis that while ligand-independent EphB4 signaling is oncogenic, ligand-dependent signaling is tumor suppressive. EphrinB2 undergoes its effect partially by reducing EphB4 protein levels [78].

Eph receptors can also co-opt other signaling pathways, to undergo their own activation and signaling independently from Ephrins. In alveolar rhabdomyosarcoma, an aggressive myogenic childhood malignancy, a cross-talk of EphB4 with the Platelet-Derived Growth Factor Receptor PDGFRβ, was shown to facilitate PDGF ligand-dependent, Ephrin ligand-independent activation of EphB4, which results in activation of the Akt and Erk1/2 pathways and cell survival. This ligand-independent function contrasts with the EphrinB2-dependent EphB4 activation, which did not stimulate PDGFRβ, was unable to stimulate Akt or Erk, and induced apoptosis [81, 93]. In breast cancer cells, Ephexin4 functions as a guanine nucleotide exchange factor (GEF) for RhoG which recruits its effector ELMO2 and a Rac GEF Dock4 to form a complex with EphA2 at the tips of cortactin-rich protrusions, thus promoting a Rac-dependent, Ephrin ligand-independent cell migration and invasion toward epidermal growth factor [94].

### 10.3.5 Shedding and Release of Soluble Forms

Exceptionally, soluble forms of Ephrins have been reported which, upon cleavage by matrix metalloproteases, are released from the cell surface and are able to activate EphA receptors in a non-juxtacrine manner [95–98]. In normal cells, ADAM10 (A disintegrin and metalloproteinase 10) regulates the function of EphrinA2 in contact-mediated axon repulsion, by regulating its cleavage, which enables contact repulsion between Eph- and Ephrin-expressing neighboring cells [97]. ADAM10 constitutively binds to EphA3 and cleaves EphrinA5 expressed in *trans*-interacting cells [99]. It is suggested that, upon EphA3/EphrinA5 interaction, the interaction of ADAM10/EphrinA5 in *trans* is a mechanism for the selective recognition and cleavage of only EphA3-bound EphrinA5 [99]. EphrinA1 can be cleaved by the matrix metalloproteinase MMP-9 [98]. Similarly, EphB stimulates a sequential cleavage of EphrinB2 by a metalloproteinase and the PS1/γ-secretase system. This proteolysis generates an intracellular C-terminal peptide fragment EphrinB2/CTF2, which binds Src and inhibits its association with inhibitory kinase Csk, allowing autophosphorylation of Src at residue tyr418 [100]. Other cleaved members of the family include EphrinB3, cleaved by the serine protease Rhomboid-1 [101], EphrinB1 cleaved by the γ-secretase [102], EphB2 sequentially cleaved by the metalloproteinase/γ-secretase system [103–105], EphA4 [106, 107], and EphA2 [108]. As these studies were mostly performed in normal tissues, neither the extent nor the impact of Ephrin/Eph proteolytic cleavage and shedding are known in cancer tissues. EphrinA1 was found to be released from glioblastoma and breast adenocarcinoma cells as a soluble, monomeric protein that remains functional as a paracrine signal, since it is able to cause EphA2 internalization and downregulation, alteration of cell morphology and suppression of the Ras-MAPK signaling pathway [109]. The soluble form of EphrinA1 was found to have a growth-promoting activity in Hela and SKBR3 cells. Furthermore, the full-length and the soluble forms of EphrinA1 appear to have a different impact on the membrane localization of the EphA2 receptor; while the full length directs the receptor from the cell surface to sites of cell-cell contact, the soluble form favors cell surface localization away from cell-cell contacts [110]. EphA1 and EphrinA1 co-localize in pulmonary endothelial cells, and were shown to regulate vascular permeability and lung metastasis. ADAM12 binds EphA1 and Enhances EphrinA1 cleavage and shedding in response to TGF- β1 (transforming growth factor-β1) in lung primary tumors. Consequently, soluble EphrinA1 acts in an endocrine fashion, through release in the serum and interference with the endothelial EphA1/EphrinA1-mediated cell adhesion in the lungs, causing lung hyperpermeability and promoting tumor metastasis to the lungs, an effect that is significantly inhibited by depletion of soluble EphrinA1 using a neutralizing antibody [111].

## 10.4 Eph/Ephrin Signaling

### *10.4.1 Cell Surface Clustering and Signaling Platforms*

Eph receptors and Ephrin ligands are membrane-bound, and when not engaged in interactions, they are not part of compact membrane structures. Therefore, their assembly and interaction can only occur at sites of direct cell–cell contacts. A proposed model is that, as Eph receptors bind their Ephrin ligands, the mobility of the receptor-ligand complexes is reduced in comparison to the unbound receptor or ligand, thus resulting in the complex's accumulation at the cell-cell interface. With engagement of higher Ephrin concentrations, C-terminal-mediated receptor-receptor interactions lead to the formation of multimeric complexes [112]. Receptor signaling is then activated by transphosphorylation [113]. Upon their interaction, and unlike other RTKs that are activated by dimerization, Ephs/Ephrins constitute higher order clusters *in vivo*. The use of crystal structures of receptor ligand complexes has provided valuable information regarding aspects such as contact surfaces and binding affinity [114]. The crystallization of the complex formed between the EphA2 ectodomain and Ephrins A1 or A5 provides mechanistic information regarding this higher-order clustering structure [115]. However, relatively little is known about the function and components of this structure and how it works as a signaling platform. An indication of the role of the clusters in signaling is that when Eph receptors are stimulated with dimeric recombinant Ephrin–Fc fusion proteins, signaling is not efficiently activated, whereas preclustering of these soluble ligands is more effective, thanks to its effect in assembling larger Eph receptor clusters [85]. Nevertheless, it is interesting to note that Eph receptors do not always require ligand binding to undergo the formation of higher order structures. In fact, Eph receptor clustering and recruitment to the signal transduction cluster could occur in absence of ligand-induced conformational changes in the receptor extracellular domain [115, 116]. This could occur for instance when the receptor's expression levels are increased regardless of the presence of ligands. In summary, while Eph receptors can form clusters and undergo signaling independently from ligand binding, the latter's role, which also depends on membrane binding and clustering, seems to involve the ability to increase the concentration of cell surface receptor clusters. A model of nucleation/seeding has been suggested to explain Eph receptor clustering, whereby an even relatively low level of Ephrin binding could increase further Ephrin-independent Eph recruitment to the cluster [116, 117]. Another important aspect is that different Eph receptors seem to be involved in different types of clusters, a finding which could prove helpful in understanding Eph/Ephrin differential signaling and functional abilities [118]. In other respect, as evidenced by the use of different deletion mutants of EphA2, the receptor's localization at cell-cell contacts depends on its ectodomain and not on its intracellular domain [117]. A recent work making use of chemical dimerizers and fluorescence anisotropy to generate and visualize specific EphB2 cluster species in living cells, has revealed that while EphA2 has a propensity to form large signaling arrays, EphB2 forms small-sized yet functional clusters [119]. The study also showed that according to an ON–OFF mechanism model of EphB2 activation, the relative abundance of active multimers over inactive dimers determines the strength of EphB2 signaling and that its intracellular domains SAM and PDZ reduce clustering in the presence of extracellular Ephrins [119]. In other respect, and equally important is the finding that interaction with other Eph receptors in *cis* (*i.e.* on the same membrane) could also affect clustering [116]. Co-clustering of EphA and EphB receptors has been reported, and specific ligation of one receptor promotes recruitment and cross-activation of the other, as it is shown by EphA3 and EphB2 co-expression [120]. Whether there is different impact of heterotypic *versus* homotypic Eph clusters awaits further investigation. It is also not clear how the dynamic of clustering is regulated during different cellular processes, such as during cell-cell interaction, adhesion, invasion *etc*. For instance, how does an Eph receptor-expressing cell regulate its receptor clusters when moving into a gradient of cells expressing progressively increasing concentrations of Ephrin ligands?

What other non-Eph/Ephrin proteins are involved in Eph clusters is another important and open question. Ephs interact with other receptors, including RTKs such as EGFR [121], VEGFR [122], FGFR [123], CXCR4 [124], and the NMDA receptors [125, 126]. Interactions have also been reported with transmembrane metalloproteases such as ADAM10, ADAM13, and ADAM19, as well as matrix metalloproteinases MMP2/MMP9 [97, 105, 127–130].

As could be inferred from the belonging of Ephs/Ephrins to the RTK group, phosphorylation is a key feature in their roles in cancer. To cite few examples, EphB2 expression and phosphorylation are increased in glioma cells during migration and invasion [131]. One of its ligands, EphrinB2's expression and phosphorylation is also increased in human brain glioblastoma (GBM) tumors in comparison to their normal counterparts, and overexpression of this ligand in cell lines results in higher levels of tyrosine phosphorylation and stimulates cellular migration and invasion [29]. Tyrosine-phosphorylation of another EphB ligand, EphrinB1, is important for its pro-invasive function in scirrhous gastric cancer cells, via a Rac1 GTPase activation pathway [132]. Upon EphB1 interaction, EphrinB1 is tyrosine phosphorylated and transduces a pro-migratory signal in endothelial cells and induces an integrin $\alpha v\beta 3$ and $\alpha 5\beta 1$-mediated attachment [133]. From the A-type sub-family, EphA3, through its phosphotyrosine residue at position 602 (Y602), binds the adaptor protein Nck1, and this interaction could regulate cell migration and retraction [134].

Subsequently to receptor clustering and autophosphorylation, a signal transduction platform is activated. The repertoire of components of this platform and how they are coupled to the overall cell signaling network are far from being fully elucidated. However, many important aspects are currently known. The Src family associates with kinase-active Eph receptors largely in a kinase-dependent manner [135–141]. The role of PI3K/Akt and MAPK signaling pathways is also prominent. They mediate functions of Ephs and Ephrins in different processes such as cell death and autophagy [142–145]. The pro-migration and pro-invasion function of EphA2 involves phosphorylation by Akt [81]. EphrinA1 stimulates EphA2 tyrosine phosphorylation and degradation, and downstream mitogen-activated protein kinase (MAPK) activation in osteosarcoma cells, while the MAPK pathway in turn regulates EphrinA1 expression [146]. EphA2 forms a complex and cooperates with ErbB2 to promote tumor and metastatic progression, apparently via activation of Ras-MAPK signaling and RhoA GTPase [147]. EphrinA1 ligand activation of EphA2 inhibits ERK phosphorylation and decreases viability, via an EGFR/Ras/MAPK pathway [148, 149]. The activated fibroblast growth factor receptor (FGFR1) induces a ligand-independent phosphorylation of EphB2. By contrast, upon EphrinB1 ligand stimulation, FGFR1 signaling inhibits EphB2 phosphorylation through the MAPK pathway [150]. In addition to EphB2, another receptor EphA4 was shown to form a complex with FGFR1 in glioma cells, thus contributing to cell proliferation and migration through a mechanism involving the MAPK/Akt/Rac1/Cdc42 signaling pathways [123]. Interestingly, to further illustrate the importance of these pathways in Eph/Ephrin functions, even the kinase-deficient EphB6 also leads to MAPK activation in non-small cell lung cancer (NSCLC) cells [25]. By virtue of amino acid substitutions in its cytoplasmic "kinase domain", EphB6 is kinase-defective, [151, 152]. In spite of this deficiency, it has been shown that EphB6 is able to undergo tyrosine phosphorylation upon stimulation with the EphrinB1 ligand, thanks to transphosphorylation from the kinase-efficient EphB1, with whom it forms a stable hetero-complex [153]. EphB6 can also be tyrosine phosphorylated at its cytoplasmic domain by a Src family kinase [82].

### *10.4.2 Kinase-Independent Functions*

Studies suggest that Ephs/Ephrins may function in a kinase-independent manner. This is particularly true for the adhesive/attractive functions of Ephs/Ephrins. For instance, EphA7/EphrinA5 interaction results in cell repulsion. However, co-expression of a tyrosine kinase domain-truncated form of EphA7 suppresses tyrosine phosphorylation of the full-length receptor and induces adhesion rather than repulsion [63]. Also, EphA8 associates and localizes the p110gamma isoform of PI 3-kinase to the plasma membrane in a tyrosine kinase-independent fashion, thus eliciting signals required for $\alpha5\beta1$- or $\beta3$ integrins-mediated cell adhesion to fibronectin [154]. Based on such findings, it is hypothesized that tyrosine kinase-independent signaling could favor adhesion, while repulsion would be tyrosine kinase-dependent, although exceptions exist that relativize this trend, such as EphB1-mediated cell attachment to fibronectin that was found to be kinase-dependent [155]. In fact, kinase activity requirement is complex. For instance, on one hand kinase-deficiency or truncation of the cytoplasmic domain of EphA2, results in deficiency in EphrinA1-mediated cell rounding, retraction fiber formation, de-adhesion from the extracellular matrix, RhoA and Rac1 GTPase regulation, three-dimensional matrix invasion, and *in vivo* metastasis. On the other hand, EphA2 regulation of cell motility and invasion, or the formation of bone and visceral tumor colonies, is both EphA2 kinase-dependent and – independent [156]. While the kinase activity is necessary for EphB3-driven inhibition of integrin-mediated cell adhesion and induction of cell rounding, EphB3/EphrinB1 inhibitory role in HGF/SF directional cell migration involves a reduction in Rac1/Cdc42 activities and is kinase-independent [157].

## 10.5 Eph/Ephrins and Cell-Cell Junctions

### *10.5.1 Gap Junctions*

Although undergoing non-junctional interactions, Ephs and Ephrins are not isolated from the overall junction-based intercellular communication system. Gap junctional intercellular communications (GJICs) play essential roles in normal physiology as well as in the development of various diseases such as cancer [158]. The building blocks of gap junctions, the channels, are made of proteins known as Connexins. Although rudimentary, the information available so far regarding a connection between Ephs/Ephrins-mediated communication and GJIC or Connexins sheds an interesting light on the function of both cell-cell communication modes. In the zebrafish model, expression of Eph receptors and Ephrins regulates GJIC at the boundary between hindbrain cell populations, and while bidirectional signaling restricts the intermingling of adjacent cell populations, unidirectional signaling restricts GJIC [159]. In *Xenopus* embryos, both ectopic overexpression of EphrinB1 and blockade of Connexin32/Connexin34 gap junction channels using a dominant-negative mutant connexin, lead to the same result of loss of adhesion and dissociation of embryonic cells [160, 161]. Mutations in the X-linked EphrinB1 in humans cause the craniofrontonasal syndrome (CFNS), and it was shown that the mechanism involves an inhibition of GJIC at ectopic Ephrin boundaries and that upon Eph receptor engagement, EphrinB1 interacts with Connexin43 and regulates its distribution [162]. EphB4 was also shown to physically associate with Connexin43 and EphB activation inhibits GJIC between cardiomyocytes [163]. Nevertheless, the significance of a role for Ephs and Ephrins in GJIC in cancer is yet to uncover.

### *10.5.2 Tight Junctions*

Another important mode of cell-cell communication is the tight junction (TJ). TJs are structures that establish tight barriers between neighboring epithelial cells, thus limiting the flow of solutes across the epithelium. The link between these structures and Ephs/Ephrins is better established than with gap junctions. However, most data were obtained from non-cancer cells. The best known example is EphA2. Upon stimulation with recombinant EphrinA1-Fc, EphA2 phosphorylation was accompanied by disassembly of TJs in human brain microvascular endothelial cells (HBMECs). This effect is counteracted by either RNAi (RNA interference)-mediated silencing or expression of a kinase-inactive mutant of EphA2 [164]. This receptor associates with, tyrosine-phosphorylates and regulates the localization and function of the TJ protein Claudin4. Claudins are essential TJ proteins located on the apico-lateral region of epithelial cells and are involved in maintaining cell polarity. Subsequently the association of Claudin4 with ZO-1 (Zonula Occludens protein 1), another important TJ component, is decreased and Claudin4 is removed from sites of cell-cell contacts, thus increasing paracellular permeability [165]. Claudin4 inhibits EphA2 pro-oncogenic signaling by limiting the activity of ß-catenin and PI3K, thus preventing phosphorylation of EphA2 on S897 by AKT, a phosphorylation event involved in EphA2 functions in migration, invasion and metastasis [166]. In Non-Small Cells Lung Cancer (NSCLC) cells, activation of EphA2 signaling by EphrinA1 modulates Claudin2 gene expression via induction of Cdx-2, a tumor suppressor gene [167]. B-type Ephs and Ephrins are also connected to Claudins. *In vivo*, EphB2 and EphrinB2 co-localize with Claudins in epithelia at the point of morphogenetic cell-cell septation during development of the embryonic midline. The disruption of EphB2/EphrinB2 bidirectional signaling impairs cell-cell septation and closure events [168]. The function of EphrinB1 in stabilizing the morphology of adult cardiomyocytes (CM) is important for the cohesion of cardiac tissue architecture *in vivo*. In wild-type CMs, EphrinB1 interacts with a Claudin5/ZO-1 complex at the lateral membrane. However, in CM-specific EphrinB1 knockout mice, the complex disappears, and under pressure overload stress, these mice were found to exhibit striking tissue disorganization and to be prone to death [169]. *In vitro*, following the formation of cell-cell contacts, the cytoplasmic domain of EphrinB1 is phosphorylated on tyrosine residues via Claudin1 or Claudin4 with whom it creates a complex at their extracellular domains [170]. Evidence suggests that phosphorylation of EphrinB1 induced by Claudins is mediated by the Src family kinases [170]. In other respect, the Par polarity complex protein Par-6, a scaffolding protein, forms a complex with atypical protein kinase C (aPKC) and an active small GTPase Cdc42-GTP [171], to regulate TJ formation in the apical cell junctions [172]. EphrinB1 was shown to compete with Cdc42 for association with Par-6, and to disrupt the localization of TJ-associated proteins (ZO-1, Cingulin) at TJs, thus causing inactivation of the Par complex and loss of TJs [173, 174]. The EphrinB1/Par-6 interaction is disrupted by tyrosine phosphorylation of the intracellular domain of EphrinB1 [175].

Although these data were mainly obtained in normal physiology, taken together they strongly suggest a role for Ephs/Ephrins' interactions and activation in diseases such as cancers. Their deregulations could result in the dismantlement of tight junctions, loss in the definition of cell-cell boundaries, thus driving the detachment of tumor cells and invasion into other territories.

### *10.5.3 Adherens Junctions*

Adherens junctions (AJs) are epithelial cell-cell communication structures formed by epithelial cadherin (E-cadherin)–catenin complexes that bind to the cytoskeleton and regulatory and signaling partners, thus controlling cell-cell behavior [176, 177]. AJ complexes are constituted by three main protein families: transmembrane Cadherins, Armadillo proteins and cytoskeletal adaptors. Ephs and Ephrins coordinate with AJ proteins during various normal development processes. Differential Eph/Ephrin expression controls cadherin clustering at the vertebrate notochord-presomitic mesoderm boundary, preventing formation of adhesive bonds between cells of the two different types [178]. E-cadherin but not N-cadherin regulates the expression of several Ephs and Ephrins in embryonic stem (ES) cells. In epithelial cells, E-cadherin is required for EphA2 localization at cell-cell contacts and in the absence of functional E-cadherin, EphA2 localizes to the perinuclear region [179]. EphB2 and EphrinB are involved in the regenerative process within the peripheral nervous system, thanks to their cell sorting role; following nerve damage, EphB2/EphrinB signaling results in cell sorting of fibroblasts and Schwann cells, thus resulting in organized cell migration of Schwann cells that will guide axonal regrowth. These functions of EphB2/EphrinB signaling are mediated by the stemness factor Sox2 through N-cadherin relocalization to Schwann cell-cell contacts [180]. A coordination between the function of Ephs/Ephrins and N-cadherin has been shown during the spatiotemporal regulation of sympathetic ganglia formation [181]. Similarly, a role has been suggested for temporal and spatial patterns of adhesive N-CAM and N-cadherin and repulsive EphA4/EphrinB2 in dynamic morphological changes of somitogenesis, including intersomitic separation, boundary-shape evolution and sorting of misdifferentiated cells across compartment boundaries, during gastrulation [182]. EphA2/EphrinA5 interaction regulates the AJ complex by enhancing recruitment of β-catenin to N-cadherin thus ensuring an important role in lens development and maintenance [183]. Ectopic EphA4 activity in superficial blastula cells leads to loss of apical/basolateral polarity in early embryos of Xenopus laevis and it actually causes a preferential loss of AJs, in comparison to TJs [184]. A positive feedback loop has been proposed in which E-cadherin-based cell-cell contacts enhance EphA/EphrinA signaling. In turn, ligand activation and Tyr594 phosphorylation of EphA2, results in the latter's binding to Nck (noncatalytic region of tyrosine kinase) and Git1 (G protein-coupled receptor kinase-interacting protein 1), to suppress the activity of Arf6 (ADP-ribosylation factor 6), a protein important in the regulation of E-cadherin-based cell-cell adhesions, thus increasing cell compaction and apical-basal polarization of epithelial cells [185]. Another partner of Ephrins in TJ regulation is Pick1, an important AJ protein, and EphrinB1 expression is able to disrupt AJs in a manner that is blocked by the expression of Pick1 [186].

The role of Eph/Ephrin/AJ proteins' cooperation has also been addressed in certain pathological instances. Using knockout mice models, a role for EphA2/EphrinA5 in cataractogenesis has been reported. The mechanism involves a regulation of β-catenin signaling and E-cadherin by EphrinA5 to prevent lens anterior epithelial cells from undergoing the epithelial-mesenchymal transition (EMT) while EphA2 is involved in controlling the organization of lens fiber cells [187]. Also, using an EphA2 knockout mouse model, an EphA2/ Src/cortactin/F-actin signaling axis that affects E-cadherin junctions' basal-lateral distribution has been reported to control eye lens development and morphogenesis, thus providing a possible mechanism for cataracts [188]. Acute tobacco smoke exposure (TS) was shown to cause damage to bronchial airway epithelial cells (BAEpCs) by increasing EphA2 and EphrinA1 expression and decreasing E-cadherin expression, thus inducing cell-cell hyperpermeability, without affecting cell-matrix adhesions. SiRNA-mediated silencing of EphA2 expression attenuated TS-induced hyperpermeability in BAEpCs [189].

Reduced expression and/or mislocalization of E-cadherin, is associated with tumor progression and aggressive clinicopathological phenotypes in different cancer types. There is clear evidence that Ephs and Ephrins also contribute to this evolution. For instance, a significant relationship between the expression of EphA2 and E-cadherin, known to co-localize to sites of cell-cell contact, was observed in colorectal cancers and expression of both markers correlates closely with clinicopathological characteristics of cancer progression and metastasis [190]. Similarly, in gastric cancers, overexpression of EphA2 associates negatively with E-cadherin expression and positively with β-catenin and vimentin expression, and also with tumor progression clinicopathological parameters and poor survival outcome [191]. Furthermore, in these cancers, EphA2 induces upregulation of the EMT molecular markers N-cadherin and Snail, as well as the activation of Wnt/β-catenin signaling and expression of the latter's target genes TCF4, Cyclin-D1 and c-Myc [192]. EphA4 may also promote the motility and invasion of pancreatic cancer cells via the upregulation of Snail, as well as MMP2 upregulation and E-cadherin downregulation [193]. While in non-cancer epithelial cells, the expression and functioning of E-cadherin and EphA2 tyrosine-phosphorylation and localization to sites of cell-cell contacts appear to be inter-dependent, in breast cancer cells devoid of E-cadherin, EphA2 tyrosine-phosphorylation is decreased, and it is redistributed into membrane ruffles, an effect likely associated with a diminution of *trans* activation by Ephrin ligands. The normal pattern of EphA2 activation and localization is recovered by ectopic expression of E-cadherin [194]. Cells in the bottom of intestinal crypts near stem cell niches express EphB3. EphB3/EphrinB signaling drives changes consistent with a role in promoting mesenchymal-to-epithelial transition (MET). These include cytoskeleton re-organization, *i.e.* changing cell shape from spreading to cobble-like, patterning cortical actin cytoskeleton and polarizing E-cadherin and ZO-1. It also decreases mesenchymal markers (fibronectin and nuclear β-catenin), while increasing epithelial markers (ZO-1, E-cadherin and plakoglobin) and inactivating the EMT signaling pathway mediated by CrkL-Rac1. These functions in re-establishing epithelial cell-cell junctions have been suggested to contribute to EphB3-mediated tumor suppression [195]. In the intestinal epithelium, a triple interaction between EphB receptors, E-cadherin and the metalloproteinase ADAM10 at sites of adhesion with EphrinB1-expressing cells induces shedding of E-cadherin by ADAM10, which results in asymmetric localization of E-cadherin [130]. EphB2 and EphB3 genes are transcriptional targets of β-catenin and Tcf4 in both colorectal cancer (CRC) and normal intestinal cells. In the latter, EphB signaling controls the positioning of cell types along the crypt-villus axis, while in early stage CRC, it suppresses tumor progression. EphB receptors compartmentalize the expansion of CRC cells and restrict the spreading of EphB-expressing tumor cells into EphrinB1-positive territories *in vitro* and *in vivo*, through a mechanism dependent on E-cadherin-mediated adhesion. During CRC tumor progression, when

EphB expression is silenced, cells are enabled to escape the repulsive interactions with normal EphrinB1-expressing intestinal cells [196, 197]. The expression of E-cadherin and EphB3 is reduced in adenocarcinoma compared to dysplasia or healthy esophageal mucosa and simultaneous expression of both proteins is significantly inversely correlated to tumor stage. Furthermore, EphB3 expression results in redistribution of E-cadherin from the cytoplasm to the membrane. These data suggest a role for EphB3 in the dysplasia-adenocarcinoma transition, the infiltrative growth pattern and the development of lymph node metastases, via an effect on E-cadherin [198]. The role of EphA2 mammary tumorigenesis involves the destabilization of AJs in RhoA-dependent manner. Overexpression of a wild-type EphA2 in human mammary epithelial cells was shown to weaken E-cadherin-mediated cell-cell adhesion via RhoA GTPase activation. However, neither the total level of Cadherins nor the composition of the AJ complexes or the tyrosine phosphorylation of the cadherin complex components was affected. The EphA2-dependent Rho activation and destabilization of AJs is regulated via a signaling pathway that involves Src kinase, low molecular weight phosphotyrosine phosphatase (LMW-PTP) and p190 RhoGAP [199]. Stimulation of HT29 colon cancer cells with EphrinA1-Fc, results in tyrosine hyperphosphorylation of EphA2, E-cadherin, and β-catenin. This has been suggested to be involved in alleviating contact inhibition of HT29 cells [200]. VE-cadherin and EphA2 colocalize and interact in cell-cell adhesion junctions in melanoma cells, and VE-cadherin actually regulates the ability of EphA2 to interact with EphrinA1 and thus become phosphorylated. This concerted role of VE-cadherin and EphA2 is observed during a process found in highly aggressive melanoma cells, called vasculogenic mimicry (VM), which involves the formation of vasculogenic-like networks showing a gene expression profile associated with endothelial and epithelial cells. Knock down of VE-cadherin expression results in the reorganization of EphA2 on the cell surface, and its subsequent cytoplasmic accumulation and dephosphorylation [201]. Expression of EphA2, VE-cadherin, MMP-2 and MMP-9 proteins was shown to be associated with vasculogenic mimicry in pancreatic cancer. Furthermore, Ginsenoside Rg3, a trace tetracyclic triterpenoid saponin extracted from ginseng and which has anti-cancer activity, reduces the levels of vasculogenic mimicry, while down-regulating expression of EphA2, VE-cadherin, MMP2 and MMP9 [202].

Desmosomes are structurally close to AJs; they comprise three main protein families: cadherins, armadillo proteins and plakins, which are arranged in a similar manner to that of AJ complexes, and like AJs, desmosomes play an important role in cancer [203–205]. Keratinocytes' terminal differentiation involves EphrinA1-induced EphA2 signaling upon interaction with neighboring cells. Indeed, EphA2/EphrinA1 signaling restricts the lateral expansion of primary epidermal cell colonies without limiting their proliferation and promotes their compaction and stratification. This effect on differentiation is to a large extent due to the up-regulation of Desmoglein1, a desmosomal Cadherin that maintains the integrity and differentiated state of suprabasal keratinocytes in the epidermis [206].

## 10.6 Therapeutic Targeting of Ephs and Ephrins

The recent years have seen an increasing attempt to develop ways to target Ephs and Ephrins for cancer therapeutic purposes [207]. Although still in its infancy, the pertinence of such an approach is supported by experiments associating Ephs and Ephrins' expression downregulation or signaling inactivation with anti-tumor effects. Small-interfering RNA (siRNA)-mediated silencing of EphA2 expression inhibits the growth and migration and induces caspase-9-mediated apoptosis of malignant mesothelioma cells [208]. The function of Ephs and Ephrins in angiogenesis appears also to provide an interesting and promising window for intervention. For instance, using the RIP-Tag transgenic model of angiogenesis-dependent pancreatic islet cell carcinoma and the 4T1 model of metastatic mammary adenocarcinoma, it was reported that the blockade of EphA activation using EphA2-Fc or EphA3-Fc (soluble fusions between the extracellular domain of EphA2 and EphA3 respectively and the Fc region of IgG1 receptors), inhibits tumor angiogenesis by decreasing tumor vascular density, and inhibits tumor growth by decreasing cell proliferation, and increasing cell apoptosis, while also inhibiting migration of endothelial cells in response to tumor cells [209]. Similarly, a soluble monomeric EphB4 (sEphB4) reduces melanoma tumor growth and intratumoral microvessel density [210]. The extracellular domain of EphB4 was fused with human serum albumin (sEphB4-HSA) and used to block EphrinB2, thus leading to the inhibition of growth factors-induced migration and invasion of Kaposi sarcoma cells *in vitro*, and the reduction of blood vessel density, pericyte recruitment, vessel perfusion, and increase in hypoxia *in vivo* [211]. A soluble extracellular EphB4 fused to albumin (sEphB4-Alb) reduces vessel density and tumor growth [212]. A synthetic peptide consisting of a fusion of HIV-TAT and C-terminal amino acids 331–346 of EphrinB1 (PTD-EFNB1-C) once introduced inside cells, blocks EphrinB1-mediated signaling in scirrhous gastric cancer cells and peritoneal dissemination, through interfering with the ability of EphrinB1 to associate with an adaptor protein Dishevelled, and the blockade of RhoA activation [213]. Another approach is the use of antibodies. Anti-EphA2 high-affinity and highly specific monoclonal blocking antibodies, generated by differential screening of phage-Antibody libraries, were shown to inhibit tumor growth, angiogenesis and metastasis [214]. Another EphA2-agonistic monoclonal antibody, alone or in combination with paclitaxel, efficiently reduces microvascular density, proliferation, and VEGF levels, and increases endothelial cell apoptosis, thus inhibiting ovarian tumor growth *in vivo* [215]. Anti-EphB4 monoclonal antibodies, alone or in combination with the anti-VEGFA anti-angiogenic antibody bevacizumab, inhibit tumor angiogenesis, tumor growth and metastasis [216]. Last but not least, small molecule inhibitors have also been developed, including a kinase inhibitor of the EphB4 kinase, NVP-BHG712, which inhibits EphB4 kinase activity and inhibits VEGF driven vessel formation [217].

## 10.7 Conclusion

In spite of constituting the largest family of receptor tyrosine kinases, Ephs and Ephrins have significantly attracted cancer researchers' attention only in the last decade or so. Investigations have since largely justified this interest, as these receptors and ligands are involved in virtually all aspects of cancer biology, and regulate every homeostatic process, whether proliferation, apoptosis, differentiation, migration, invasion or autophagy, to name a few. Nevertheless, the extent of this role is yet to be uncovered. This task is rendered complex by the combination of a large number of family members, the possibility of different binding partners which is not necessarily moderated by potential redundancy, the existence of a bi-directional signaling and the role of heterotypic cell-cell contacts. The role of Ephs and Ephrins' interaction with other types of junctional and non-junctional intercellular communication is another important key question. With the advent of inhibitory approaches targeting these proteins, it will be necessary to address these and other pressing questions as well as to identify the determinants of these proteins' functional duality and versatility, before one can hope to take advantage of any therapeutic attempts.

## References

1. Gale NW, Holland SJ, Valenzuela DM, Flenniken A, Pan L, Ryan TE, Henkemeyer M, Strebhardt K, Hirai H, Wilkinson DG, Pawson T, Davis S, Yancopoulos GD (1996) Eph receptors and ligands comprise two major specificity subclasses and are reciprocally compartmentalized during embryogenesis. Neuron 17:9–19
2. Himanen JP, Chumley MJ, Lackmann M, Li C, Barton WA, Jeffrey PD, Vearing C, Geleick D, Feldheim DA, Boyd AW, Henkemeyer M, Nikolov DB (2004) Repelling class discrimination: ephrin-A5 binds to and activates EphB2 receptor signaling. Nat Neurosci 7:501–509
3. Du J, Fu C, Sretavan DW (2007) Eph/ephrin signaling as a potential therapeutic target after central nervous system injury. Curr Pharm Des 13:2507–2518
4. Frisen J, Holmberg J, Barbacid M (1999) Ephrins and their Eph receptors: multitalented directors of embryonic development. EMBO J 18:5159–5165
5. Holder N, Klein R (1999) Eph receptors and ephrins: effectors of morphogenesis. Development 126:2033–2044
6. O'Leary DD, Wilkinson DG (1999) Eph receptors and ephrins in neural development. Curr Opin Neurobiol 9:65–73
7. Yancopoulos GD, Klagsbrun M, Folkman J (1998) Vasculogenesis, angiogenesis, and growth factors: ephrins enter the fray at the border. Cell 93:661–664
8. Flenniken AM, Gale NW, Yancopoulos GD, Wilkinson DG (1996) Distinct and overlapping expression patterns of ligands for Eph-related receptor tyrosine kinases during mouse embryogenesis. Dev Biol 179:382–401
9. Friedman GC, O'Leary DD (1996) Eph receptor tyrosine kinases and their ligands in neural development. Curr Opin Neurobiol 6:127–133
10. Arvanitis D, Davy A (2008) Eph/ephrin signaling: networks. Genes Dev 22:416–429
11. Merlos-Suarez A, Batlle E (2008) Eph-ephrin signalling in adult tissues and cancer. Curr Opin Cell Biol 20:194–200
12. Pasquale EB (2008) Eph-ephrin bidirectional signaling in physiology and disease. Cell 133:38–52

13. Hruska M, Dalva MB (2012) Ephrin regulation of synapse formation, function and plasticity. Mol Cell Neurosci 50:35–44
14. Chen Y, Fu AK, Ip NY (2012) Eph receptors at synapses: implications in neurodegenerative diseases. Cell Signal 24:606–611
15. Matsuo K (2010) Eph and ephrin interactions in bone. Adv Exp Med Biol 658:95–103
16. Compagni A, Logan M, Klein R, Adams RH (2003) Control of skeletal patterning by ephrinB1-EphB interactions. Dev Cell 5:217–230
17. Davy A, Aubin J, Soriano P (2004) Ephrin-B1 forward and reverse signaling are required during mouse development. Genes Dev 18:572–583
18. Klein R (2009) Bidirectional modulation of synaptic functions by Eph/ephrin signaling. Nat Neurosci 12:15–20
19. Flanagan JG, Vanderhaeghen P (1998) The ephrins and Eph receptors in neural development. Annu Rev Neurosci 21:309–345
20. Wang HU, Chen ZF, Anderson DJ (1998) Molecular distinction and angiogenic interaction between embryonic arteries and veins revealed by ephrin-B2 and its receptor Eph-B4. Cell 93:741–753
21. Kandouz M (2012) The Eph/Ephrin family in cancer metastasis: communication at the service of invasion. Cancer Metastasis Rev 31:353–373
22. Tachibana M, Tonomoto Y, Hyakudomi R, Hyakudomi M, Hattori S, Ueda S, Kinugasa S, Yoshimura H (2007) Expression and prognostic significance of EFNB2 and EphB4 genes in patients with oesophageal squamous cell carcinoma. Dig Liver Dis 39:725–732
23. Mudali SV, Fu B, Lakkur SS, Luo M, Embuscado EE, Iacobuzio-Donahue CA (2006) Patterns of EphA2 protein expression in primary and metastatic pancreatic carcinoma and correlation with genetic status. Clin Exp Metastasis 23:357–365
24. Chukkapalli S, Amessou M, Dilly AK, Dekhil H, Zhao J, Liu Q, Bejna A, Thomas RD, Bandyopadhyay S, Bismar TA, Neill D, Azoulay L, Batist G, Kandouz M (2014) Role of the EphB2 receptor in autophagy, apoptosis and invasion in human breast cancer cells. Exp Cell Res 320:233–246
25. Yu J, Bulk E, Ji P, Hascher A, Koschmieder S, Berdel WE, Muller-Tidow C (2009) The kinase defective EPHB6 receptor tyrosine kinase activates MAP kinase signaling in lung adenocarcinoma. Int J Oncol 35:175–179
26. Oshima T, Akaike M, Yoshihara K, Shiozawa M, Yamamoto N, Sato T, Akihito N, Nagano Y, Fujii S, Kunisaki C, Wada N, Rino Y, Tanaka K, Masuda M, Imada T (2008) Overexpression of EphA4 gene and reduced expression of EphB2 gene correlates with liver metastasis in colorectal cancer. Int J Oncol 33:573–577
27. Parri M, Buricchi F, Giannoni E, Grimaldi G, Mello T, Raugei G, Ramponi G, Chiarugi P (2007) EphrinA1 activates a Src/focal adhesion kinase-mediated motility response leading to rho-dependent actino/myosin contractility. J Biol Chem 282:19619–19628
28. Huang X, Wu D, Jin H, Stupack D, Wang JY (2008) Induction of cell retraction by the combined actions of Abl-CrkII and Rho-ROCK1 signaling. J Cell Biol 183:711–723
29. Nakada M, Anderson EM, Demuth T, Nakada S, Reavie LB, Drake KL, Hoelzinger DB, Berens ME (2010) The phosphorylation of ephrin-B2 ligand promotes glioma cell migration and invasion. Int J Cancer 126:1155–1165
30. Yang NY, Pasquale EB, Owen LB, Ethell IM (2006) The EphB4 receptor-tyrosine kinase promotes the migration of melanoma cells through Rho-mediated actin cytoskeleton reorganization. J Biol Chem 281:32574–32586
31. Noren NK, Foos G, Hauser CA, Pasquale EB (2006) The EphB4 receptor suppresses breast cancer cell tumorigenicity through an Abl-Crk pathway. Nat Cell Biol 8:815–825
32. Zhang J, Hughes S (2006) Role of the ephrin and Eph receptor tyrosine kinase families in angiogenesis and development of the cardiovascular system. J Pathol 208:453–461
33. Aitsebaomo J, Portbury AL, Schisler JC, Patterson C (2008) Brothers and sisters: molecular insights into arterial-venous heterogeneity. Circ Res 103:929–939
34. Adams RH, Wilkinson GA, Weiss C, Diella F, Gale NW, Deutsch U, Risau W, Klein R (1999) Roles of ephrinB ligands and EphB receptors in cardiovascular development: demarcation of

arterial/venous domains, vascular morphogenesis, and sprouting angiogenesis. Genes Dev 13:295–306
35. Gerety SS, Wang HU, Chen ZF, Anderson DJ (1999) Symmetrical mutant phenotypes of the receptor EphB4 and its specific transmembrane ligand ephrin-B2 in cardiovascular development. Mol Cell 4:403–414
36. Gale NW, Baluk P, Pan L, Kwan M, Holash J, DeChiara TM, McDonald DM, Yancopoulos GD (2001) Ephrin-B2 selectively marks arterial vessels and neovascularization sites in the adult, with expression in both endothelial and smooth-muscle cells. Dev Biol 230:151–160
37. Shin D, Garcia-Cardena G, Hayashi S, Gerety S, Asahara T, Stavrakis G, Isner J, Folkman J, Gimbrone MA Jr, Anderson DJ (2001) Expression of ephrinB2 identifies a stable genetic difference between arterial and venous vascular smooth muscle as well as endothelial cells, and marks subsets of microvessels at sites of adult neovascularization. Dev Biol 230:139–150
38. Noren NK, Lu M, Freeman AL, Koolpe M, Pasquale EB (2004) Interplay between EphB4 on tumor cells and vascular ephrin-B2 regulates tumor growth. Proc Natl Acad Sci U S A 101:5583–5588
39. Foo SS, Turner CJ, Adams S, Compagni A, Aubyn D, Kogata N, Lindblom P, Shani M, Zicha D, Adams RH (2006) Ephrin-B2 controls cell motility and adhesion during blood-vessel-wall assembly. Cell 124:161–173
40. Vogt T, Stolz W, Welsh J, Jung B, Kerbel RS, Kobayashi H, Landthaler M, McClelland M (1998) Overexpression of Lerk-5/Eplg5 messenger RNA: a novel marker for increased tumorigenicity and metastatic potential in human malignant melanomas. Clin Cancer Res 4:791–797
41. Hainaud P, Contreres JO, Villemain A, Liu LX, Plouet J, Tobelem G, Dupuy E (2006) The role of the vascular endothelial growth factor-Delta-like 4 ligand/Notch4-ephrin B2 cascade in tumor vessel remodeling and endothelial cell functions. Cancer Res 66:8501–8510
42. Haldimann M, Custer D, Munarini N, Stirnimann C, Zurcher G, Rohrbach V, Djonov V, Ziemiecki A, Andres AC (2009) Deregulated ephrin-B2 expression in the mammary gland interferes with the development of both the glandular epithelium and vasculature and promotes metastasis formation. Int J Oncol 35:525–536
43. Holmberg J, Genander M, Halford MM, Anneren C, Sondell M, Chumley MJ, Silvany RE, Henkemeyer M, Frisen J (2006) EphB receptors coordinate migration and proliferation in the intestinal stem cell niche. Cell 125:1151–1163
44. Merlos-Suarez A, Barriga FM, Jung P, Iglesias M, Cespedes MV, Rossell D, Sevillano M, Hernando-Momblona X, da Silva-Diz V, Munoz P, Clevers H, Sancho E, Mangues R, Batlle E (2011) The intestinal stem cell signature identifies colorectal cancer stem cells and predicts disease relapse. Cell Stem Cell 8:511–524
45. Kaenel P, Schwab C, Mulchi K, Wotzkow C, Andres AC (2011) Preponderance of cells with stem cell characteristics in metastasising mouse mammary tumours induced by deregulated EphB4 and ephrin-B2 expression. Int J Oncol 38:151–160
46. Kaenel P, Antonijevic M, Richter S, Kuchler S, Sutter N, Wotzkow C, Strange R, Andres AC (2011) Deregulated ephrin-B2 signaling in mammary epithelial cells alters the stem cell compartment and interferes with the epithelial differentiation pathway. Int J Oncol 40(2):357–69
47. Batson J, Astin JW, Nobes CD (2013) Regulation of contact inhibition of locomotion by Eph-ephrin signalling. J Microsc 251:232–241
48. Mayor R, Carmona-Fontaine C (2010) Keeping in touch with contact inhibition of locomotion. Trends Cell Biol 20:319–328
49. Wang B (2011) Cancer cells exploit the Eph-ephrin system to promote invasion and metastasis: tales of unwitting partners. Sci Signal 4, e28
50. Astin JW, Batson J, Kadir S, Charlet J, Persad RA, Gillatt D, Oxley JD, Nobes CD (2010) Competition amongst Eph receptors regulates contact inhibition of locomotion and invasiveness in prostate cancer cells. Nat Cell Biol 12:1194–1204
51. Tanaka M, Kuriyama S, Aiba N (2012) Nm23-H1 regulates contact inhibition of locomotion, which is affected by ephrin-B1. J Cell Sci 125:4343–4353

52. Morrissey C, True LD, Roudier MP, Coleman IM, Hawley S, Nelson PS, Coleman R, Wang YC, Corey E, Lange PH, Higano CS, Vessella RL (2008) Differential expression of angiogenesis associated genes in prostate cancer bone, liver and lymph node metastases. Clin Exp Metastasis 25:377–388
53. Azab F, Azab AK, Maiso P, Calimeri T, Flores LM, Liu Y, Quang P, Roccaro AM, Sacco A, Ngo HT, Zhang Y, Morgan B, Carrasco R, Ghobrial IM (2011) Eph-B2/ephrin-B2 interaction plays a major role in the adhesion and proliferation of Waldenstrom's macroglobulinemia. Clin Cancer Res 18(1):91–104
54. Trinidad EM, Ballesteros M, Zuloaga J, Zapata A, Onso-Colmenar LM (2009) An impaired transendothelial migration potential of chronic lymphocytic leukemia (CLL) cells can be linked to ephrin-A4 expression. Blood 114:5081–5090
55. Carvalho RF, Beutler M, Marler KJ, Knoll B, Becker-Barroso E, Heintzmann R, Ng T, Drescher U (2006) Silencing of EphA3 through a cis interaction with ephrinA5. Nat Neurosci 9:322–330
56. Hornberger MR, Dutting D, Ciossek T, Yamada T, Handwerker C, Lang S, Weth F, Huf J, Wessel R, Logan C, Tanaka H, Drescher U (1999) Modulation of EphA receptor function by coexpressed ephrinA ligands on retinal ganglion cell axons. Neuron 22:731–742
57. Antion MD, Christie LA, Bond AM, Dalva MB, Contractor A (2010) Ephrin-B3 regulates glutamate receptor signaling at hippocampal synapses. Mol Cell Neurosci 45:378–388
58. Kao TJ, Kania A (2011) Ephrin-mediated cis-attenuation of Eph receptor signaling is essential for spinal motor axon guidance. Neuron 71:76–91
59. Falivelli G, Lisabeth EM, Rubio dT, Perez-Tenorio G, Tosato G, Salvucci O, Pasquale EB (2013) Attenuation of eph receptor kinase activation in cancer cells by coexpressed ephrin ligands. PLoS One 8:e81445
60. Marquardt T, Shirasaki R, Ghosh S, Andrews SE, Carter N, Hunter T, Pfaff SL (2005) Coexpressed EphA receptors and ephrin-A ligands mediate opposing actions on growth cone navigation from distinct membrane domains. Cell 121:127–139
61. Stein E, Lane AA, Cerretti DP, Schoecklmann HO, Schroff AD, Van Etten RL, Daniel TO (1998) Eph receptors discriminate specific ligand oligomers to determine alternative signaling complexes, attachment, and assembly responses. Genes Dev 12:667–678
62. Gao PP, Yue Y, Cerretti DP, Dreyfus C, Zhou R (1999) Ephrin-dependent growth and pruning of hippocampal axons. Proc Natl Acad Sci U S A 96:4073–4077
63. Holmberg J, Clarke DL, Frisen J (2000) Regulation of repulsion versus adhesion by different splice forms of an Eph receptor. Nature 408:203–206
64. Hindges R, McLaughlin T, Genoud N, Henkemeyer M, O'Leary D (2002) EphB forward signaling controls directional branch extension and arborization required for dorsal-ventral retinotopic mapping. Neuron 35:475–487
65. Mann F, Ray S, Harris W, Holt C (2002) Topographic mapping in dorsoventral axis of the Xenopus retinotectal system depends on signaling through ephrin-B ligands. Neuron 35:461–473
66. McLaughlin T, Hindges R, Yates PA, O'Leary DD (2003) Bifunctional action of ephrin-B1 as a repellent and attractant to control bidirectional branch extension in dorsal-ventral retinotopic mapping. Development 130:2407–2418
67. Hansen MJ, Dallal GE, Flanagan JG (2004) Retinal axon response to ephrin-as shows a graded, concentration-dependent transition from growth promotion to inhibition. Neuron 42:717–730
68. Noren NK, Pasquale EB (2007) Paradoxes of the EphB4 receptor in cancer. Cancer Res 67:3994–3997
69. Rutkowski R, Mertens-Walker I, Lisle JE, Herington AC, Stephenson SA (2011) Evidence for a dual function of EphB4 as tumor promoter and suppressor regulated by the absence or presence of the ephrin-B2 ligand. Int J Cancer 422(3):433–42
70. Fang WB, Brantley-Sieders DM, Parker MA, Reith AD, Chen J (2005) A kinase-dependent role for EphA2 receptor in promoting tumor growth and metastasis. Oncogene 24:7859–7868

71. Zelinski DP, Zantek ND, Stewart JC, Irizarry AR, Kinch MS (2001) EphA2 overexpression causes tumorigenesis of mammary epithelial cells. Cancer Res 61:2301–2306
72. Duxbury MS, Ito H, Zinner MJ, Ashley SW, Whang EE (2004) EphA2: a determinant of malignant cellular behavior and a potential therapeutic target in pancreatic adenocarcinoma. Oncogene 23:1448–1456
73. Thaker PH, Deavers M, Celestino J, Thornton A, Fletcher MS, Landen CN, Kinch MS, Kiener PA, Sood AK (2004) EphA2 expression is associated with aggressive features in ovarian carcinoma. Clin Cancer Res 10:5145–5150
74. Miao H, Burnett E, Kinch M, Simon E, Wang B (2000) Activation of EphA2 kinase suppresses integrin function and causes focal-adhesion-kinase dephosphorylation. Nat Cell Biol 2:62–69
75. Miao H, Wei BR, Peehl DM, Li Q, Alexandrou T, Schelling JR, Rhim JS, Sedor JR, Burnett E, Wang B (2001) Activation of EphA receptor tyrosine kinase inhibits the Ras/MAPK pathway. Nat Cell Biol 3:527–530
76. Dohn M, Jiang J, Chen X (2001) Receptor tyrosine kinase EphA2 is regulated by p53-family proteins and induces apoptosis. Oncogene 20:6503–6515
77. Guo H, Miao H, Gerber L, Singh J, Denning MF, Gilliam AC, Wang B (2006) Disruption of EphA2 receptor tyrosine kinase leads to increased susceptibility to carcinogenesis in mouse skin. Cancer Res 66:7050–7058
78. Rutkowski R, Mertens-Walker I, Lisle JE, Herington AC, Stephenson SA (2012) Evidence for a dual function of EphB4 as tumor promoter and suppressor regulated by the absence or presence of the ephrin-B2 ligand. Int J Cancer 131:E614–E624
79. Kumar SR, Scehnet JS, Ley EJ, Singh J, Krasnoperov V, Liu R, Manchanda PK, Ladner RD, Hawes D, Weaver FA, Beart RW, Singh G, Nguyen C, Kahn M, Gill PS (2009) Preferential induction of EphB4 over EphB2 and its implication in colorectal cancer progression. Cancer Res 69:3736–3745
80. Heroult M, Schaffner F, Pfaff D, Prahst C, Kirmse R, Kutschera S, Riedel M, Ludwig T, Vajkoczy P, Graeser R, Augustin HG (2010) EphB4 promotes site-specific metastatic tumor cell dissemination by interacting with endothelial cell-expressed ephrinb2. Mol Cancer Res 8:1297–1309
81. Miao H, Li DQ, Mukherjee A, Guo H, Petty A, Cutter J, Basilion JP, Sedor J, Wu J, Danielpour D, Sloan AE, Cohen ML, Wang B (2009) EphA2 mediates ligand-dependent inhibition and ligand-independent promotion of cell migration and invasion via a reciprocal regulatory loop with Akt. Cancer Cell 16:9–20
82. Matsuoka H, Obama H, Kelly ML, Matsui T, Nakamoto M (2005) Biphasic functions of the kinase-defective Ephb6 receptor in cell adhesion and migration. J Biol Chem 280:29355–29363
83. Lackmann M, Mann RJ, Kravets L, Smith FM, Bucci TA, Maxwell KF, Howlett GJ, Olsson JE, Vanden Bos T, Cerretti DP, Boyd AW (1997) Ligand for EPH-related kinase (LERK) 7 is the preferred high affinity ligand for the HEK receptor. J Biol Chem 272:16521–16530
84. Winslow JW, Moran P, Valverde J, Shih A, Yuan JQ, Wong SC, Tsai SP, Goddard A, Henzel WJ, Hefti F (1995) Cloning of AL-1, a ligand for an Eph-related tyrosine kinase receptor involved in axon bundle formation. Neuron 14:973–981
85. Davis S, Gale NW, Aldrich TH, Maisonpierre PC, Lhotak V, Pawson T, Goldfarb M, Yancopoulos GD (1994) Ligands for EPH-related receptor tyrosine kinases that require membrane attachment or clustering for activity. Science 266:816–819
86. Batlle E, Henderson JT, Beghtel H, van den Born MM, Sancho E, Huls G, Meeldijk J, Robertson J, van de Wetering M, Pawson T, Clevers H (2002) Beta-catenin and TCF mediate cell positioning in the intestinal epithelium by controlling the expression of EphB/ephrinB. Cell 111:251–263
87. Macrae M, Neve RM, Rodriguez-Viciana P, Haqq C, Yeh J, Chen C, Gray JW, McCormick F (2005) A conditional feedback loop regulates Ras activity through EphA2. Cancer Cell 8:111–118

88. Ji XD, Li G, Feng YX, Zhao JS, Li JJ, Sun ZJ, Shi S, Deng YZ, Xu JF, Zhu YQ, Koeffler HP, Tong XJ, Xie D (2011) EphB3 is overexpressed in non-small-cell lung cancer and promotes tumor metastasis by enhancing cell survival and migration. Cancer Res 71:1156–1166
89. Hafner C, Schmitz G, Meyer S, Bataille F, Hau P, Langmann T, Dietmaier W, Landthaler M, Vogt T (2004) Differential gene expression of Eph receptors and ephrins in benign human tissues and cancers. Clin Chem 50:490–499
90. Cooke JE, Kemp HA, Moens CB (2005) EphA4 is required for cell adhesion and rhombomere-boundary formation in the zebrafish. Curr Biol 15:536–542
91. Tawadros T, Brown MD, Hart CA, Clarke NW (2012) Ligand-independent activation of EphA2 by arachidonic acid induces metastasis-like behaviour in prostate cancer cells. Br J Cancer 107:1737–1744
92. Petty A, Myshkin E, Qin H, Guo H, Miao H, Tochtrop GP, Hsieh JT, Page P, Liu L, Lindner DJ, Acharya C, MacKerell AD, Ficker E, Song J, Wang B (2012) A small molecule agonist of EphA2 receptor tyrosine kinase inhibits tumor cell migration in vitro and prostate cancer metastasis in vivo. PLoS One 7:e42120
93. Aslam MI, Abraham J, Mansoor A, Druker BJ, Tyner JW, Keller C (2014) PDGFRbeta reverses EphB4 signaling in alveolar rhabdomyosarcoma. Proc Natl Acad Sci U S A 111:6383–6388
94. Hiramoto-Yamaki N, Takeuchi S, Ueda S, Harada K, Fujimoto S, Negishi M, Katoh H (2010) Ephexin4 and EphA2 mediate cell migration through a RhoG-dependent mechanism. J Cell Biol 190:461–477
95. Bartley TD, Hunt RW, Welcher AA, Boyle WJ, Parker VP, Lindberg RA, Lu HS, Colombero AM, Elliott RL, Guthrie BA (1994) B61 is a ligand for the ECK receptor protein-tyrosine kinase. Nature 368:558–560
96. Alford SC, Bazowski J, Lorimer H, Elowe S, Howard PL (2007) Tissue transglutaminase clusters soluble A-type ephrins into functionally active high molecular weight oligomers. Exp Cell Res 313:4170–4179
97. Hattori M, Osterfield M, Flanagan JG (2000) Regulated cleavage of a contact-mediated axon repellent. Science 289:1360–1365
98. Beauchamp A, Lively MO, Mintz A, Gibo D, Wykosky J, Debinski W (2012) EphrinA1 is released in three forms from cancer cells by matrix metalloproteases. Mol Cell Biol 32:3253–3264
99. Janes PW, Saha N, Barton WA, Kolev MV, Wimmer-Kleikamp SH, Nievergall E, Blobel CP, Himanen JP, Lackmann M, Nikolov DB (2005) Adam meets Eph: an ADAM substrate recognition module acts as a molecular switch for ephrin cleavage in trans. Cell 123:291–304
100. Georgakopoulos A, Litterst C, Ghersi E, Baki L, Xu C, Serban G, Robakis NK (2006) Metalloproteinase/Presenilin1 processing of ephrinB regulates EphB-induced Src phosphorylation and signaling. EMBO J 25:1242–1252
101. Pascall JC, Brown KD (2004) Intramembrane cleavage of ephrinB3 by the human rhomboid family protease, RHBDL2. Biochem Biophys Res Commun 317:244–252
102. Tomita T, Tanaka S, Morohashi Y, Iwatsubo T (2006) Presenilin-dependent intramembrane cleavage of ephrin-B1. Mol Neurodegener 1:2
103. Litterst C, Georgakopoulos A, Shioi J, Ghersi E, Wisniewski T, Wang R, Ludwig A, Robakis NK (2007) Ligand binding and calcium influx induce distinct ectodomain/gamma-secretase-processing pathways of EphB2 receptor. J Biol Chem 282:16155–16163
104. Xu J, Litterst C, Georgakopoulos A, Zaganas I, Robakis NK (2009) Peptide EphB2/CTF2 generated by the gamma-secretase processing of EphB2 receptor promotes tyrosine phosphorylation and cell surface localization of N-methyl-D-aspartate receptors. J Biol Chem 284:27220–27228
105. Lin KT, Sloniowski S, Ethell DW, Ethell IM (2008) Ephrin-B2-induced cleavage of EphB2 receptor is mediated by matrix metalloproteinases to trigger cell repulsion. J Biol Chem 283:28969–28979
106. Gatto G, Morales D, Kania A, Klein R (2014) EphA4 receptor shedding regulates spinal motor axon guidance. Curr Biol 24:2355–2365

107. Inoue E, Deguchi-Tawarada M, Togawa A, Matsui C, Arita K, Katahira-Tayama S, Sato T, Yamauchi E, Oda Y, Takai Y (2009) Synaptic activity prompts gamma-secretase-mediated cleavage of EphA4 and dendritic spine formation. J Cell Biol 185:551–564
108. Eriksson O, Ramstrom M, Hornaeus K, Bergquist J, Mokhtari D, Siegbahn A (2014) The eph tyrosine kinase receptors ephb2 and epha2 are novel proteolytic substrates of tissue factor/coagulation factor viia. J Biol Chem 289:32379–32391
109. Wykosky J, Palma E, Gibo DM, Ringler S, Turner CP, Debinski W (2008) Soluble monomeric EphrinA1 is released from tumor cells and is a functional ligand for the EphA2 receptor. Oncogene 27:7260–7273
110. Alford S, Watson-Hurthig A, Scott N, Carette A, Lorimer H, Bazowski J, Howard PL (2010) Soluble ephrin a1 is necessary for the growth of HeLa and SK-BR3 cells. Cancer Cell Int 10:41
111. Ieguchi K, Tomita T, Omori T, Komatsu A, Deguchi A, Masuda J, Duffy SL, Coulthard MG, Boyd A, Maru Y (2014) ADAM12-cleaved ephrin-A1 contributes to lung metastasis. Oncogene 33:2179–2190
112. Nikolov DB, Xu K, Himanen JP (2013) Eph/ephrin recognition and the role of Eph/ephrin clusters in signaling initiation. Biochim Biophys Acta 1834:2160–2165
113. Lackmann M, Oates AC, Dottori M, Smith FM, Do C, Power M, Kravets L, Boyd AW (1998) Distinct subdomains of the EphA3 receptor mediate ligand binding and receptor dimerization. J Biol Chem 273:20228–20237
114. Himanen JP, Saha N, Nikolov DB (2007) Cell-cell signaling via Eph receptors and ephrins. Curr Opin Cell Biol 19:534–542
115. Himanen JP, Yermekbayeva L, Janes PW, Walker JR, Xu K, Atapattu L, Rajashankar KR, Mensinga A, Lackmann M, Nikolov DB, Dhe-Paganon S (2010) Architecture of Eph receptor clusters. Proc Natl Acad Sci U S A 107:10860–10865
116. Wimmer-Kleikamp SH, Janes PW, Squire A, Bastiaens PI, Lackmann M (2004) Recruitment of Eph receptors into signaling clusters does not require ephrin contact. J Cell Biol 164:661–666
117. Seiradake E, Harlos K, Sutton G, Aricescu AR, Jones EY (2010) An extracellular steric seeding mechanism for Eph-ephrin signaling platform assembly. Nat Struct Mol Biol 17:398–402
118. Seiradake E, Schaupp A, del Toro RD, Kaufmann R, Mitakidis N, Harlos K, Aricescu AR, Klein R, Jones EY (2013) Structurally encoded intraclass differences in EphA clusters drive distinct cell responses. Nat Struct Mol Biol 20:958–964
119. Schaupp A, Sabet O, Dudanova I, Ponserre M, Bastiaens P, Klein R (2014) The composition of EphB2 clusters determines the strength in the cellular repulsion response. J Cell Biol 204:409–422
120. Janes PW, Griesshaber B, Atapattu L, Nievergall E, Hii LL, Mensinga A, Chheang C, Day BW, Boyd AW, Bastiaens PI, Jorgensen C, Pawson T, Lackmann M (2011) Eph receptor function is modulated by heterooligomerization of A and B type Eph receptors. J Cell Biol 195:1033–1045
121. Komurov K, Padron D, Cheng T, Roth M, Rosenblatt KP, White MA (2010) Comprehensive mapping of the human kinome to epidermal growth factor receptor signaling. J Biol Chem 285:21134–21142
122. Shao Z, Zhang WF, Chen XM, Shang ZJ (2008) Expression of EphA2 and VEGF in squamous cell carcinoma of the tongue: correlation with the angiogenesis and clinical outcome. Oral Oncol 44:1110–1117
123. Fukai J, Yokote H, Yamanaka R, Arao T, Nishio K, Itakura T (2008) EphA4 promotes cell proliferation and migration through a novel EphA4-FGFR1 signaling pathway in the human glioma U251 cell line. Mol Cancer Ther 7:2768–2778
124. Salvucci O, de la Luz SM, Martina JA, McCormick PJ, Tosato G (2006) EphB2 and EphB4 receptors forward signaling promotes SDF-1-induced endothelial cell chemotaxis and branching remodeling. Blood 108:2914–2922

125. Dalva MB, Takasu MA, Lin MZ, Shamah SM, Hu L, Gale NW, Greenberg ME (2000) EphB receptors interact with NMDA receptors and regulate excitatory synapse formation. Cell 103:945–956
126. Nolt MJ, Lin Y, Hruska M, Murphy J, Sheffler-Colins SI, Kayser MS, Passer J, Bennett MV, Zukin RS, Dalva MB (2011) EphB controls NMDA receptor function and synaptic targeting in a subunit-specific manner. J Neurosci 31:5353–5364
127. White JM (2003) ADAMs: modulators of cell-cell and cell-matrix interactions. Curr Opin Cell Biol 15:598–606
128. Yumoto N, Wakatsuki S, Kurisaki T, Hara Y, Osumi N, Frisen J, Sehara-Fujisawa A (3) Meltrin beta/ADAM19 interacting with EphA4 in developing neural cells participates in formation of the neuromuscular junction. PLoS One e3322
129. Wei S, Xu G, Bridges LC, Williams P, White JM, DeSimone DW (2010) ADAM13 induces cranial neural crest by cleaving class B Ephrins and regulating Wnt signaling. Dev Cell 19:345–352
130. Solanas G, Cortina C, Sevillano M, Batlle E (2011) Cleavage of E-cadherin by ADAM10 mediates epithelial cell sorting downstream of EphB signalling. Nat Cell Biol 13:1100–1107
131. Nakada M, Niska JA, Miyamori H, McDonough WS, Wu J, Sato H, Berens ME (2004) The phosphorylation of EphB2 receptor regulates migration and invasion of human glioma cells. Cancer Res 64:3179–3185
132. Tanaka M, Kamata R, Takigahira M, Yanagihara K, Sakai R (2007) Phosphorylation of ephrin-B1 regulates dissemination of gastric scirrhous carcinoma. Am J Pathol 171:68–78
133. Huynh-Do U, Vindis C, Liu H, Cerretti DP, McGrew JT, Enriquez M, Chen J, Daniel TO (2002) Ephrin-B1 transduces signals to activate integrin-mediated migration, attachment and angiogenesis. J Cell Sci 115:3073–3081
134. Hu T, Shi G, Larose L, Rivera GM, Mayer BJ, Zhou R (2009) Regulation of process retraction and cell migration by EphA3 is mediated by the adaptor protein Nck1. Biochemistry 48:6369–6378
135. Ellis C, Kasmi F, Ganju P, Walls E, Panayotou G, Reith AD (1996) A juxtamembrane autophosphorylation site in the Eph family receptor tyrosine kinase, Sek, mediates high affinity interaction with p59fyn. Oncogene 12:1727–1736
136. Zisch AH, Kalo MS, Chong LD, Pasquale EB (1998) Complex formation between EphB2 and Src requires phosphorylation of tyrosine 611 in the EphB2 juxtamembrane region. Oncogene 16:2657–2670
137. Knoll B, Drescher U (2004) Src family kinases are involved in EphA receptor-mediated retinal axon guidance. J Neurosci 24:6248–6257
138. Zou JX, Wang B, Kalo MS, Zisch AH, Pasquale EB, Ruoslahti E (1999) An Eph receptor regulates integrin activity through R-Ras. Proc Natl Acad Sci U S A 96:13813–13818
139. Choi S, Park S (1999) Phosphorylation at Tyr-838 in the kinase domain of EphA8 modulates Fyn binding to the Tyr-615 site by enhancing tyrosine kinase activity. Oncogene 18:5413–5422
140. Hock B, Bohme B, Karn T, Feller S, Rubsamen-Waigmann H, Strebhardt K (1998) Tyrosine-614, the major autophosphorylation site of the receptor tyrosine kinase HEK2, functions as multi-docking site for SH2-domain mediated interactions. Oncogene 17:255–260
141. Vindis C, Cerretti DP, Daniel TO, Huynh-Do U (2003) EphB1 recruits c-Src and p52Shc to activate MAPK/ERK and promote chemotaxis. J Cell Biol 162:661–671
142. Kandouz M, Haidara K, Zhao J, Brisson ML, Batist G (2010) The EphB2 tumor suppressor induces autophagic cell death via concomitant activation of the ERK1/2 and PI3K pathways. Cell Cycle 9:398–407
143. Maddigan A, Truitt L, Arsenault R, Freywald T, Allonby O, Dean J, Narendran A, Xiang J, Weng A, Napper S, Freywald A (2011) EphB receptors trigger Akt activation and suppress Fas receptor-induced apoptosis in malignant T lymphocytes. J Immunol 187:5983–5994

144. Stahl S, Branca RM, Efazat G, Ruzzene M, Zhivotovsky B, Lewensohn R, Viktorsson K, Lehtio J (2011) Phosphoproteomic profiling of NSCLC cells reveals that ephrin B3 regulates pro-survival signaling through Akt1-mediated phosphorylation of the EphA2 receptor. J Proteome Res 10:2566–2578
145. Yang NY, Fernandez C, Richter M, Xiao Z, Valencia F, Tice DA, Pasquale EB (2011) Crosstalk of the EphA2 receptor with a serine/threonine phosphatase suppresses the Akt-mTORC1 pathway in cancer cells. Cell Signal 23:201–212
146. Fritsche-Guenther R, Noske A, Ungethum U, Kuban RJ, Schlag PM, Tunn PU, Karle J, Krenn V, Dietel M, Sers C (2010) De novo expression of EphA2 in osteosarcoma modulates activation of the mitogenic signalling pathway. Histopathology 57:836–850
147. Brantley-Sieders DM, Zhuang G, Hicks D, Fang WB, Hwang Y, Cates JM, Coffman K, Jackson D, Bruckheimer E, Muraoka-Cook RS, Chen J (2008) The receptor tyrosine kinase EphA2 promotes mammary adenocarcinoma tumorigenesis and metastatic progression in mice by amplifying ErbB2 signaling. J Clin Invest 118:64–78
148. Larsen AB, Stockhausen MT, Poulsen HS (2010) Cell adhesion and EGFR activation regulate EphA2 expression in cancer. Cell Signal 22:636–644
149. Larsen AB, Pedersen MW, Stockhausen MT, Grandal MV, Van DB, Poulsen HS (2007) Activation of the EGFR gene target EphA2 inhibits epidermal growth factor-induced cancer cell motility. Mol Cancer Res 5:283–293
150. Poliakov A, Cotrina ML, Pasini A, Wilkinson DG (2008) Regulation of EphB2 activation and cell repulsion by feedback control of the MAPK pathway. J Cell Biol 183:933–947
151. Gurniak CB, Berg LJ (1996) A new member of the Eph family of receptors that lacks protein tyrosine kinase activity. Oncogene 13:777–786
152. Matsuoka H, Iwata N, Ito M, Shimoyama M, Nagata A, Chihara K, Takai S, Matsui T (1997) Expression of a kinase-defective Eph-like receptor in the normal human brain. Biochem Biophys Res Commun 235:487–492
153. Freywald A, Sharfe N, Roifman CM (2002) The kinase-null EphB6 receptor undergoes transphosphorylation in a complex with EphB1. J Biol Chem 277:3823–3828
154. Gu C, Park S (2001) The EphA8 receptor regulates integrin activity through p110gamma phosphatidylinositol-3 kinase in a tyrosine kinase activity-independent manner. Mol Cell Biol 21:4579–4597
155. Huynh-Do U, Stein E, Lane AA, Liu H, Cerretti DP, Daniel TO (1999) Surface densities of ephrin-B1 determine EphB1-coupled activation of cell attachment through alphavbeta3 and alpha5beta1 integrins. EMBO J 18:2165–2173
156. Taddei ML, Parri M, Angelucci A, Onnis B, Bianchini F, Giannoni E, Raugei G, Calorini L, Rucci N, Teti A, Bologna M, Chiarugi P (2009) Kinase-dependent and -independent roles of EphA2 in the regulation of prostate cancer invasion and metastasis. Am J Pathol 174:1492–1503
157. Miao H, Strebhardt K, Pasquale EB, Shen TL, Guan JL, Wang B (2005) Inhibition of integrin-mediated cell adhesion but not directional cell migration requires catalytic activity of EphB3 receptor tyrosine kinase. Role of Rho family small GTPases. J Biol Chem 280:923–932
158. Kandouz M, Batist G (2010) Gap junctions and connexins as therapeutic targets in cancer. Expert Opin Ther Targets 14:681–692
159. Mellitzer G, Xu Q, Wilkinson DG (1999) Eph receptors and ephrins restrict cell intermingling and communication. Nature 400:77–81
160. Jones TL, Chong LD, Kim J, Xu RH, Kung HF, Daar IO (1998) Loss of cell adhesion in Xenopus laevis embryos mediated by the cytoplasmic domain of XLerk, an erythropoietin-producing hepatocellular ligand. Proc Natl Acad Sci U S A 95:576–581
161. Paul DL, Yu K, Bruzzone R, Gimlich RL, Goodenough DA (1995) Expression of a dominant negative inhibitor of intercellular communication in the early Xenopus embryo causes delamination and extrusion of cells. Development 121:371–381
162. Davy A, Bush JO, Soriano P (2006) Inhibition of gap junction communication at ectopic Eph/ephrin boundaries underlies craniofrontonasal syndrome. PLoS Biol 4:e315

163. Ishii M, Mueller I, Nakajima T, Pasquale EB, Ogawa K (2011) EphB signaling inhibits gap junctional intercellular communication and synchronized contraction in cultured cardiomyocytes. Basic Res Cardiol 106:1057–1068
164. Zhou N, Zhao WD, Liu DX, Liang Y, Fang WG, Li B, Chen YH (2011) Inactivation of EphA2 promotes tight junction formation and impairs angiogenesis in brain endothelial cells. Microvasc Res 82:113–121
165. Tanaka M, Kamata R, Sakai R (2005) EphA2 phosphorylates the cytoplasmic tail of Claudin-4 and mediates paracellular permeability. J Biol Chem 280:42375–42382
166. Shang X, Lin X, Howell SB (2014) Claudin-4 controls the receptor tyrosine kinase EphA2 pro-oncogenic switch through ss-catenin. Cell Commun Signal 12:59
167. Sukka-Ganesh B, Mohammed KA, Kaye F, Goldberg EP, Nasreen N (2012) Ephrin-A1 inhibits NSCLC tumor growth via induction of Cdx-2 a tumor suppressor gene. BMC Cancer 12:309
168. Dravis C, Henkemeyer M (2011) Ephrin-B reverse signaling controls septation events at the embryonic midline through separate tyrosine phosphorylation-independent signaling avenues. Dev Biol 355:138–151
169. Genet G, Guilbeau-Frugier C, Honton B, Dague E, Schneider MD, Coatrieux C, Calise D, Cardin C, Nieto C, Payre B, Dubroca C, Marck P, Heymes C, Dubrac A, Arvanitis D, Despas F, Altie MF, Seguelas MH, Delisle MB, Davy A, Senard JM, Pathak A, Gales C (2012) Ephrin-B1 is a novel specific component of the lateral membrane of the cardiomyocyte and is essential for the stability of cardiac tissue architecture cohesion. Circ Res 110:688–700
170. Tanaka M, Kamata R, Sakai R (2005) Phosphorylation of ephrin-B1 via the interaction with claudin following cell-cell contact formation. EMBO J 24:3700–3711
171. Joberty G, Petersen C, Gao L, Macara IG (2000) The cell-polarity protein Par6 links Par3 and atypical protein kinase C to Cdc42. Nat Cell Biol 2:531–539
172. Wang Q, Margolis B (2007) Apical junctional complexes and cell polarity. Kidney Int 72:1448–1458
173. Shin K, Fogg VC, Margolis B (2006) Tight junctions and cell polarity. Annu Rev Cell Dev Biol 22:207–235
174. Lee HS, Daar IO (2009) EphrinB reverse signaling in cell-cell adhesion: is it just par for the course? Cell Adh Migr 3:250–255
175. Lee HS, Nishanian TG, Mood K, Bong YS, Daar IO (2008) EphrinB1 controls cell-cell junctions through the Par polarity complex. Nat Cell Biol 10:979–986
176. Baum B, Georgiou M (2011) Dynamics of adherens junctions in epithelial establishment, maintenance, and remodeling. J Cell Biol 192:907–917
177. Takeichi M (2014) Dynamic contacts: rearranging adherens junctions to drive epithelial remodelling. Nat Rev Mol Cell Biol 15:397–410
178. Fagotto F, Rohani N, Touret AS, Li R (2013) A molecular base for cell sorting at embryonic boundaries: contact inhibition of cadherin adhesion by ephrin/ Eph-dependent contractility. Dev Cell 27:72–87
179. Orsulic S, Kemler R (2000) Expression of Eph receptors and ephrins is differentially regulated by E-cadherin. J Cell Sci 113(Pt 10):1793–1802
180. Parrinello S, Napoli I, Ribeiro S, Wingfield DP, Fedorova M, Parkinson DB, Doddrell RD, Nakayama M, Adams RH, Lloyd AC (2010) EphB signaling directs peripheral nerve regeneration through Sox2-dependent Schwann cell sorting. Cell 143:145–155
181. Kasemeier-Kulesa JC, Bradley R, Pasquale EB, Lefcort F, Kulesa PM (2006) Eph/ephrins and N-cadherin coordinate to control the pattern of sympathetic ganglia. Development 133:4839–4847
182. Glazier JA, Zhang Y, Swat M, Zaitlen B, Schnell S (2008) Coordinated action of N-CAM, N-cadherin, EphA4, and ephrinB2 translates genetic prepatterns into structure during somitogenesis in chick. Curr Top Dev Biol 81:205–247
183. Cooper MA, Son AI, Komlos D, Sun Y, Kleiman NJ, Zhou R (2008) Loss of ephrin-A5 function disrupts lens fiber cell packing and leads to cataract. Proc Natl Acad Sci U S A 105:16620–16625

184. Winning RS, Wyman TL, Walker GK (2001) EphA4 activity causes cell shape change and a loss of cell polarity in Xenopus laevis embryos. Differentiation 68:126–132
185. Miura K, Nam JM, Kojima C, Mochizuki N, Sabe H (2009) EphA2 engages Git1 to suppress Arf6 activity modulating epithelial cell-cell contacts. Mol Biol Cell 20:1949–1959
186. Son J, Park MS, Park I, Lee HK, Lee SH, Kang B, Min BH, Ryoo J, Lee S, Bae JS, Kim SH, Park MJ, Lee HS (2014) Pick1 modulates ephrinB1-induced junctional disassembly through an association with ephrinB1. Biochem Biophys Res Commun 450:659–665
187. Cheng C, Gong X (2011) Diverse roles of Eph/ephrin signaling in the mouse lens. PLoS One 6, e28147
188. Cheng C, Ansari MM, Cooper JA, Gong X (2013) EphA2 and Src regulate equatorial cell morphogenesis during lens development. Development 140:4237–4245
189. Nasreen N, Khodayari N, Sriram PS, Patel J, Mohammed KA (2014) Tobacco smoke induces epithelial barrier dysfunction via receptor EphA2 signaling. Am J Physiol Cell Physiol 306:C1154–C1166
190. Saito T, Masuda N, Miyazaki T, Kanoh K, Suzuki H, Shimura T, Asao T, Kuwano H (2004) Expression of EphA2 and E-cadherin in colorectal cancer: correlation with cancer metastasis. Oncol Rep 11:605–611
191. Hou F, Yuan W, Huang J, Qian L, Chen Z, Ge J, Wu S, Chen J, Wang J, Chen Z (2012) Overexpression of EphA2 correlates with epithelial-mesenchymal transition-related proteins in gastric cancer and their prognostic importance for postoperative patients. Med Oncol 29:2691–2700
192. Huang J, Xiao D, Li G, Ma J, Chen P, Yuan W, Hou F, Ge J, Zhong M, Tang Y, Xia X, Chen Z (2014) EphA2 promotes epithelial-mesenchymal transition through the Wnt/beta-catenin pathway in gastric cancer cells. Oncogene 33:2737–2747
193. Liu C, Huang H, Wang C, Kong Y, Zhang H (2014) Involvement of ephrin receptor A4 in pancreatic cancer cell motility and invasion. Oncol Lett 7:2165–2169
194. Zantek ND, Azimi M, Fedor-Chaiken M, Wang B, Brackenbury R, Kinch MS (1999) E-cadherin regulates the function of the EphA2 receptor tyrosine kinase. Cell Growth Differ 10:629–638
195. Chiu ST, Chang KJ, Ting CH, Shen HC, Li H, Hsieh FJ (2009) Over-expression of EphB3 enhances cell-cell contacts and suppresses tumor growth in HT-29 human colon cancer cells. Carcinogenesis 30:1475–1486
196. Batlle E, Bacani J, Begthel H, Jonkheer S, Gregorieff A, van de Born M, Malats N, Sancho E, Boon E, Pawson T, Gallinger S, Pals S, Clevers H (2005) EphB receptor activity suppresses colorectal cancer progression. Nature 435:1126–1130
197. Cortina C, Palomo-Ponce S, Iglesias M, Fernandez-Masip JL, Vivancos A, Whissell G, Huma M, Peiro N, Gallego L, Jonkheer S, Davy A, Lloreta J, Sancho E, Batlle E (2007) EphB-ephrin-B interactions suppress colorectal cancer progression by compartmentalizing tumor cells. Nat Genet 39:1376–1383
198. Schauer MC, Stoecklein NH, Theisen J, Kropil F, Baldus S, Hoelscher A, Feith M, Bolke E, Matuschek C, Budach W, Knoefel WT (2012) The simultaneous expression of both ephrin B3 receptor and E-cadherin in Barrett's adenocarcinoma is associated with favorable clinical staging. Eur J Med Res 17:10
199. Fang WB, Ireton RC, Zhuang G, Takahashi T, Reynolds A, Chen J (2008) Overexpression of EPHA2 receptor destabilizes adherens junctions via a RhoA-dependent mechanism. J Cell Sci 121:358–368
200. Potla L, Boghaert ER, Armellino D, Frost P, Damle NK (2002) Reduced expression of EphrinA1 (EFNA1) inhibits three-dimensional growth of HT29 colon carcinoma cells. Cancer Lett 175:187–195
201. Hess AR, Seftor EA, Gruman LM, Kinch MS, Seftor RE, Hendrix MJ (2006) VE-cadherin regulates EphA2 in aggressive melanoma cells through a novel signaling pathway: implications for vasculogenic mimicry. Cancer Biol Ther 5:228–233

202. Guo JQ, Zheng QH, Chen H, Chen L, Xu JB, Chen MY, Lu D, Wang ZH, Tong HF, Lin S (2014) Ginsenoside Rg3 inhibition of vasculogenic mimicry in pancreatic cancer through downregulation of VEcadherin/EphA2/MMP9/MMP2 expression. Int J Oncol 45:1065–1072
203. Kobielak A, Fuchs E (2004) Alpha-catenin: at the junction of intercellular adhesion and actin dynamics. Nat Rev Mol Cell Biol 5:614–625
204. Vasioukhin V (2012) Adherens junctions and cancer. Subcell Biochem 60:379–414
205. Dusek RL, Attardi LD (2011) Desmosomes: new perpetrators in tumour suppression. Nat Rev Cancer 11:317–323
206. Lin S, Gordon K, Kaplan N, Getsios S (2010) Ligand targeting of EphA2 enhances keratinocyte adhesion and differentiation via desmoglein 1. Mol Biol Cell 21:3902–3914
207. Garber K (2010) Of Ephs and ephrins: companies target guidance molecules in cancer. J Natl Cancer Inst 102:1692–1694
208. Nasreen N, Mohammed KA, Antony VB (2006) Silencing the receptor EphA2 suppresses the growth and haptotaxis of malignant mesothelioma cells. Cancer 107:2425–2435
209. Brantley DM, Cheng N, Thompson EJ, Lin Q, Brekken RA, Thorpe PE, Muraoka RS, Cerretti DP, Pozzi A, Jackson D, Lin C, Chen J (2002) Soluble Eph A receptors inhibit tumor angiogenesis and progression in vivo. Oncogene 21:7011–7026
210. Martiny-Baron G, Korff T, Schaffner F, Esser N, Eggstein S, Marme D, Augustin HG (2004) Inhibition of tumor growth and angiogenesis by soluble EphB4. Neoplasia 6:248–257
211. Scehnet JS, Ley EJ, Krasnoperov V, Liu R, Manchanda PK, Sjoberg E, Kostecke AP, Gupta S, Kumar SR, Gill PS (2009) The role of ephs, ephrins, and growth factors in Kaposi sarcoma and implications of ephrinB2 blockade. Blood 113:254–263
212. Djokovic D, Trindade A, Gigante J, Badenes M, Silva L, Liu R, Li X, Gong M, Krasnoperov V, Gill PS, Duarte A (2010) Combination of Dll4/Notch and Ephrin-B2/EphB4 targeted therapy is highly effective in disrupting tumor angiogenesis. BMC Cancer 10:641
213. Tanaka M, Kamata R, Yanagihara K, Sakai R (2010) Suppression of gastric cancer dissemination by ephrin-B1-derived peptide. Cancer Sci 101:87–93
214. Ansuini H, Meola A, Gunes Z, Paradisi V, Pezzanera M, Acali S, Santini C, Luzzago A, Mori F, Lazzaro D, Ciliberto G, Nicosia A, La MN, Vitelli A (2009) Anti-epha2 antibodies with distinct in vitro properties have equal in vivo efficacy in pancreatic cancer. J Oncol 2009:951917
215. Landen CN Jr, Lu C, Han LY, Coffman KT, Bruckheimer E, Halder J, Mangala LS, Merritt WM, Lin YG, Gao C, Schmandt R, Kamat AA, Li Y, Thaker P, Gershenson DM, Parikh NU, Gallick GE, Kinch MS, Sood AK (2006) Efficacy and antivascular effects of EphA2 reduction with an agonistic antibody in ovarian cancer. J Natl Cancer Inst 98:1558–1570
216. Krasnoperov V, Kumar SR, Ley E, Li X, Scehnet J, Liu R, Zozulya S, Gill PS (2010) Novel EphB4 monoclonal antibodies modulate angiogenesis and inhibit tumor growth. Am J Pathol 176:2029–2038
217. Martiny-Baron G, Holzer P, Billy E, Schnell C, Brueggen J, Ferretti M, Schmiedeberg N, Wood JM, Furet P, Imbach P (2010) The small molecule specific EphB4 kinase inhibitor NVP-BHG712 inhibits VEGF driven angiogenesis. Angiogenesis 13:259–267

# Chapter 11
# Computational Approaches to Modeling of Molecular Interactions in Multicellular Systems

**Kakajan Komurov**

**Abstract** Cells in metazoans co-exist as multicellular communities, where coordinated interactions within and between cells determine the phenotype of the cell community as a whole. With the recent technological advances in cell sorting and genome-wide molecular profiling techniques, it is now possible to perform unbiased systems analyses of heterogeneous cell populations within multicellular systems. Functional analyses of such multi-layered high throughput data is greatly facilitated by computational tools to build coherent models of molecular interactions in multicellular systems. I will discuss different computational techniques of pathway-level analyses of genomic data, and the recent efforts of their extension to the analyses of heterogeneous cell populations.

**Keywords** Multicellular networks • Cancer systems biology • Network biology

## 11.1 Multicellular Systems Biology

Systems biology is an interdisciplinary field of biology that treats biological networks as complex systems, and studies their emergent properties that are not manifest at the level of their individual components. Such "holistic" studies of biological systems became feasible with the advancement in the 'omics technologies that allowed for large-scale and high-throughput collection of data that is required for these analyses. Some of the earliest pioneering studies were carried out in yeast, a unicellular organism, where first collections of massive transcriptomic, genetic and protein-protein interaction data were carried out at a genome-scale [1–4]. Computational analyses of these data have yielded important insight into the global organizational features of intracellular biological networks, such as their topology, dynamics, sensitivity and robustness [5–8] (Box 11.1).

K. Komurov (✉)
Division of Experimental Hematology and Cancer Biology, Cincinnati Children's Hospital Medical Center, Cincinnati, OH 45229, USA
e-mail: Kakajan.Komurov@cchmc.org



M. Kandouz (ed.), *Intercellular Communication in Cancer*,
DOI 10.1007/978-94-017-7380-5_11

**Box 11.1 Network Analysis in Systems Biology**
Biomolecular interactions can be represented in the form of binary relationships in the form of a network [43]. Representation of biological systems as such binary networks has greatly facilitated their analyses by adopting tools from other quantitative disciplines, such as computer science and physics, to study their properties [43, 44]. Some of the pioneering studies in early 2000s suggested that biological networks possessed functional properties analogous to man-made systems often studied in physics, social and computer sciences [5, 6, 8, 44, 45]. For example, it was reported that a feature of complex networks often studied in social networks, called *centrality*, which measures the relative *importance* of nodes in a network, also correlated with biological *importance*, or, *essentiality* of genes in yeast [5]. More precisely, proteins that had more central positions in the biological network based on the number of their protein-protein interactions were also more likely to lead to cell death when knocked out from yeast [5]. More studies followed that showed how certain connectivity patterns of modules (tightly connected multi-component functional units in the network) increased the robustness of the biological system to random attacks (i.e. environmental mutations), and how expression patterns of genes within and between modules allowed for efficient control of global cell behavior [7, 46–48]. These initial systems-level computational analyses pioneered the use of network theory tools to study cellular systems.

Despite a common realization of the importance of multicellular interactions in normal physiology and disease, the field of systems biology has mainly focused on intracellular networks, perhaps owing to technical difficulties in obtaining mass-scale data from multicellular systems. As such, earlier computational studies of global properties of multicellular systems (population modeling) have been limiting and have mainly focused on select few-component processes [9–11]. Nevertheless, these studies have yielded potentially important, though preliminary, insights into how inter-coupled cell dynamics can give rise to cell population-level properties. For example, intercellular synchronization of glycolytic oscillations is a well-recognized phenomenon in yeast, and has been the subject of several theoretical studies [11, 12]. In the yeast population, the feedback loops due to glycolytic products (lactate, ATP and NADH) that freely diffuse between cells have been found to be crucial mediators of intercellular synchronization of glycolytic oscillations [11–13]. In mammals, synchronized multicellular oscillations are most prevalent in the circadian rhythm oscillations, particularly in the neurons of the suprachiasmatic nucleus (SCN) [14, 15]. Here, a concerted intra- and inter-cellular signaling by gamma-aminobutyric acid (GABA), Calcium and vasoactive intestinal polypeptide (VIP) was found to modulate the amplitude and synchrony of circadian oscillations in SCN neurons [14, 16]. These studies made use of dynamic simulation approaches to model the signaling dynamics of multicellular populations. These dynamic modeling techniques, such as using ordinary differential equations (ODE), are ideally suited for simulations of such smaller few-component systems, although their use in larger systems (e.g. many-component pathways) within a multicellular context may be prohibitively computationally expensive.

Bonzanni et al. overcame this particular limitation of ODEs by using the Petri-Nets [17], a non-parametric approach to model network dynamics, which greatly simplifies the study of network dynamics, though at the expense of quantitative accuracy [18]. They used Petri-Net modeling to model the dynamics of interactions of transcriptional and signaling pathways in different cells in orchestrating patterned cell fate commitments in *C. elegans* vulval development. This analysis by Bonzanni et al. yielded several interesting insights into the non-cell autonomous regulation of cell fate commitments, including how a microRNA network in one cell can affect the differentiation fate of another. Although examples of computational studies of such few-component systems are numerous, genome-wide computational modeling analyses of multicellular systems have been rare, perhaps owing to the scarcity of genome-scale data from multicellular systems. In the remainder of this chapter, I will discuss a few notable studies that have undertaken computational multicellular systems approaches.

## 11.2 Computational Analyses of Multicellular Networks

### *11.2.1 Analyzing Inter-Tissue Networks*

One of the earliest of genome-wide multicellular computational studies were carried out using microarray gene expression data collected from multiple tissues of normal and obese mice by Eric Schadt's group [19]. The authors of this seminal study treated each tissue, particularly hypothalamus, adipose tissue and the liver, as separate interacting cells, and analyzed the patterns of co-expression of genes in one tissue with those in the other tissues, in normal and obese mice. Thus, they obtained a gene co-expression network that spanned three different tissues and connections between co-expressed genes within as well as between tissues. Strikingly, they found that inter-tissue gene expression correlations, which are expected to reflect functional information flow between tissues, were largely independent of within-tissue correlation patterns of genes in either of the tissues. This indicates that inter-tissue communication patterns between genes cannot be recapitulated by separate analyses of individual tissues, pointing to the emergence of complex patterns in multicellular systems. Several groups have conducted similar inter-tissue gene co-expression studies later in different contexts to study tumor microenvironments, organismal stress and late-onset Alzheimer's disease [20–22]. These studies demonstrated the significance of multi-tissue (i.e. multicellular) network approaches in the study of complex multicellular systems.

Bernard Palsson's group have undertaken a different approach by utilizing genome-wide metabolic reconstruction in conjunction with constraint-based modeling to analyze the patterns of metabolic interactions at a global scale between different cell types in brain [23]. Palsson group have pioneered the reconstruction of genome-wide metabolic networks from genome sequences of various model organisms [24], and recently of human (Recon 1) [25]. Such genome-wide metabolic reconstructions have proven to be a powerful tool to analyze the global properties of metabolic fluxes that contribute to normal physiology and pathology [26]. After integrating their metabolic network with gene and protein expression data in different brain cell types in normal individuals and patients with Alzheimer's disease, they uncovered significant suppression of several core metabolic pathways in multiple brain regions, and identified an extensive aberrant cholinergic coupling between different cell types that potentially contribute to Alzheimer's disease [23]. This study too was followed by several more that utilized a similar approach to study multicellular metabolic interactions in different disease contexts [27, 28]. This and other groups later further extended the genome-wide metabolic reconstruction approach to reconstruct the metabolic map of the whole organism (inter-organ network) [29, 30], which may be especially useful in understanding systemic metabolic disorders such as diabetes, or to predict organism-wide effects of an experimental drug.

### *11.2.2 Analyzing Tumor-Stroma Interactions*

Though relatively rare, several studies have undertaken genome-wide gene expression profiling of different cell types within a functional community to delineate global mechanisms of cell-cell interactions. For example, taking advantage of minimal cross-hybridization between mouse and human gene mRNAs on human and mouse Illumina microarray chips, respectively, some studies have successfully conducted large-scale gene expression profiling of tumor and stromal populations in different stages of cancer progression [31, 32]. Using such an approach, Cascone et al. performed an analysis of stromal and tumor cells in mice to study the role of tumor-stroma interactions in the acquisition of resistance to drugs and identified fibroblast growth factor (FGF) signaling between the stroma and tumor cells as a critical factor contributing to acquired drug resistance [31]. Still several other studies have used laser-capture microdissection to dissect tumor and stromal cell populations for microarray gene expression profiling [33–35]. Although analyses in the respective studies mainly focused on a handful of genes without the use of sophisticated computational tools, it is conceptually possible to build global models of cell-cell interactions from such data by extending genomic data-based computational network analysis techniques (Box 11.2) to a multicellular context, such as in the multicellular network analysis approach recently developed by us (see next).

### *11.2.3 Community-Wide Molecular Networks (CMN) Approach*

We recently extended our network analysis approach, NetWalk [36] (see Box 11.2), to the analyses of genomic data from multicellular systems [37]. This approach, Community-wide Molecular Network approach (CMN), treats the whole

**Box 11.2 Methods for Network Query from Experimental Data**

A particularly useful approach for functional analyses of genomic data is the use of so-called *a priori* molecular networks, i.e. collection of functional interactions based on the existing knowledge between proteins and other biological molecules. Comprehensive catalogues of functional interactions between proteins and small-molecules can be found in online resources such as HPRD [49], BIND [50], MINT [51] and BioGRID [52]. Analyses of genomic data within the context of *a priori* knowledge on molecular interactions have proven useful in generating high confidence hypotheses regarding underlying mechanistic relationships of cellular processes [18, 43, 53–60]. Although such knowledge-based modeling is limited by prior biological information, it helps put the new genome-wide data within the context of what is already known to gain insight into the molecular mechanisms that are highlighted by the data. In knowledge-based analyses, prior knowledge is usually represented by networks of binary interactions between biological entities (e.g. genes, proteins, metabolites) [43], which are derived from previous literature or various prediction methods. An important challenge in network-based analyses of data is integration of experimental data with prior knowledge interactions for the retrieval of most relevant biomolecular networks.

Despite popularity of network-based analyses of genomic data, there is no standard way, although multiple computational methods and software have been developed for this purpose [61]. Retrieval of most relevant biological networks/pathways associated with the upper or lower end of the data distribution is not a trivial task, mainly because members of a biological pathway do not usually have similar data values (e.g. gene expression change), which necessitates the use of various computational algorithms for finding such networks of genes [53, 55, 57, 62–65]. Almost all of the existing network-based data analysis methods are so-called list-based network building methods. These methods use a pre-defined gene list of interest (seed genes) as seeds for iterative network building based on connectivity of non-seed genes with the seed genes. Most software tools, including commercial ones (IPA [53]), use this method. Yet others use an enrichment analysis to score pre-defined pathways for enrichment for the seed genes (e.g. MetaCore [66], BiologicalNetworks [62]). Since these methods use a seed gene list for their analyses, one of the main limitations of this approach is that it can miss potentially important information contained in the sub-threshold regions of the distribution.

To overcome the limitations of list-based network building approaches, we developed a data-biased random walk method, NetWalk [36]. In contrast to list-based methods, NetWalk is a random-walk-based method that overlays the whole data distribution onto the network, and iteratively scores every interaction in the network for their relevance to the experimental data based on simultaneous assessment of the local connectivity and the local data distribution. In this way, NetWalk can account for the whole data distribution

(continued)

**Box 11.2** (continued)
without requiring data cutoffs. Also, unlike other methods, NetWalk output is not restricted to a number of static networks of interest, but is a distribution of relevance scores assigned to each interaction in the network (Edge Flux values) that can be interactively queried for visualizations of networks of interest of desired size and composition. In addition, Edge Flux values can be used in traditional statistical analyses for comparative network analyses in the form of Edge Flux tables. For example, it is possible to conduct comparative analyses of active networks between several conditions, just like with regular gene expression data, which is a unique advantage of NetWalk results in the form of EF Tables. Such comparative analyses of networks between experimental samples cannot be done with any of the existing tools. We have shown that networks generated from the Edge Flux values of NetWalk run perform better based on (1) coherence with the input experimental data (results of the analysis should be highly concordant with the input), (2) ability to control for topological bias (e.g. due to historically more studied proteins) and (3) ability to score coherence of experimental data with local network connectivity [64, 67] (e.g. inter-connected nodes with similar experimental data values should score high).

NetWalk and other network analysis methods have been used extensively to analyze genomic datasets to extract context-specific meaningful relationships between genes [43, 61, 68].

multicellular system as a single molecular network. While most of the interactions in a CMN are intracellular, inter-cellular interactions are mediated by secreted and cell surface proteins of different cells. By combining multiple cellular networks into a single CMN, we can now apply conventional computational approaches to perform integrative analyses of intra- and inter-cellular molecular pathways that mediate the multicellular behavior. In particular, in the case when genome-wide gene expression data can be obtained from the different cell populations within the CMN, CMN-based NetWalk analysis of such data can provide valuable insight into the molecular interactions within and between cell types that contribute to the studied multicellular phenotype.

In our original study, we used the genome-wide gene expression data from Cascone et al. [31], who collected gene expression measurements from stromal and tumor tissues of bevacizumab-resistant lung cancer xenografts in mice. In this study, they injected human tumor cells that are bevacizumab-sensitive into nude mice to form tumors, and continuously treated them with bevacizumab till the tumors in mice became resistant. To understand the role of tumor-stroma interactions in such acquired resistance to targeted therapy, the authors then obtained gene expression data from tumor and stromal cell populations of tumors before and after the acquisition of drug resistance. However, the authors in that study only performed a candidate-based analysis, and focused on the induction of Fibroblast Growth Factor 2 (FGF2) and EGFR by stromal cells as a possible mechanism of the tumor resistance to bevacizumab. However, re-analyses of the data using our CMN-based NetWalk approach revealed an extensive network of intra- and inter-cellular interactions, involving signaling as well as metabolic pathways that potentially contribute to drug resistance of tumor cells. A particularly intriguing finding was a potential symbiotic relationship between tumor and stromal cells, where bevacizumab-induced nutrient stress, and consequently cell death, in the tumor cells were rescued by highly specific activation of gluconeogenic pathways in stromal cells. Interestingly, such symbiotic relationships between tumor and stromal cells, particularly in the form of metabolic product exchange, have been reported in the literature [38–41]. As tumor-stromal interactions are now being recognized as the fundamental elements contributing to tumorigenesis [42], more studies are likely to follow to make use of a similar multicellular systems approach to study tumor-stroma molecular interactions.

## 11.3 Conclusion

The field of multicellular systems biology is in its infancy, as genome-wide approaches, which have revolutionized cell-based biology, have been rarely applied to study heterogeneous multicellular communities. Although this can be largely attributed to the difficulty of identifying and separating different cell types of a multicellular system or lack of proper *in vitro* multicellular models, recent technological developments in FACS and microfluidics and *in vitro* organotypic cultures may spur a new interest in multicellular systems biology. Moreover, it is now even possible to perform single-cell transcriptomics to derive transcriptional profiles of every individual cell within a cell population. Therefore, the coming years will probably witness a burst in multicellular systems biology studies, which will spark the development of standardized computational methods and software for the analyses of multicellular systems.

## References

1. Uetz P et al (2000) A comprehensive analysis of protein-protein interactions in Saccharomyces cerevisiae. Nature 403(6770):623–627
2. Tong AH et al (2004) Global mapping of the yeast genetic interaction network. Science 303(5659):808–813
3. Krogan NJ et al (2006) Global landscape of protein complexes in the yeast Saccharomyces cerevisiae. Nature 440(7084):637–643
4. Hughes TR et al (2000) Functional discovery via a compendium of expression profiles. Cell 102(1):109–126
5. Jeong H et al (2001) Lethality and centrality in protein networks. Nature 411(6833):41–42

6. Ravasz E et al (2002) Hierarchical organization of modularity in metabolic networks. Science 297(5586):1551–1555
7. Han JD et al (2004) Evidence for dynamically organized modularity in the yeast protein-protein interaction network. Nature 430(6995):88–93
8. Maslov S, Sneppen K (2002) Specificity and stability in topology of protein networks. Science 296(5569):910–913
9. De Monte S et al (2007) Dynamical quorum sensing: population density encoded in cellular dynamics. Proc Natl Acad Sci U S A 104(47):18377–18381
10. Henson MA (2003) Dynamic modeling of microbial cell populations. Curr Opin Biotechnol 14(5):460–467
11. Wolf J, Heinrich R (1997) Dynamics of two-component biochemical systems in interacting cells; synchronization and desynchronization of oscillations and multiple steady states. Biosystems 43(1):1–24
12. Wolf J, Heinrich R (2000) Effect of cellular interaction on glycolytic oscillations in yeast: a theoretical investigation. Biochem J 345(Pt 2):321–334
13. Richard P et al (1996) Acetaldehyde mediates the synchronization of sustained glycolytic oscillations in populations of yeast cells. Eur J Biochem 235(1–2):238–241
14. Henson MA (2013) Multicellular models of intercellular synchronization in circadian neural networks. Chaos Solitons Fractals 50:48–64
15. Honma S et al (2000) Synaptic communication of cellular oscillations in the rat suprachiasmatic neurons. Neurosci Lett 294(2):113–116
16. Vasalou C, Herzog ED, Henson MA (2011) Multicellular model for intercellular synchronization in circadian neural networks. Biophys J 101(1):12–20
17. Bonzanni N et al (2009) Executing multicellular differentiation: quantitative predictive modelling of C. elegans vulval development. Bioinformatics 25(16):2049–2056
18. Ruths D et al (2008) The signaling petri net-based simulator: a non-parametric strategy for characterizing the dynamics of cell-specific signaling networks. PLoS Comput Biol 4(2), e1000005
19. Dobrin R et al (2009) Multi-tissue coexpression networks reveal unexpected subnetworks associated with disease. Genome Biol 10(5):R55
20. Zhang B et al (2013) Integrated systems approach identifies genetic nodes and networks in late-onset alzheimer's disease. Cell 153(3):707–720
21. Xiang Y, Zhang J, Huang K (2013) Mining the tissue-tissue gene co-expression network for tumor microenvironment study and biomarker prediction. BMC Genomics 14(Suppl 5):S4
22. Droujinine IA, Perrimon N (2013) Defining the interorgan communication network: systemic coordination of organismal cellular processes under homeostasis and localized stress. Front Cell Infect Microbiol 3:82
23. Lewis NE et al (2010) Large-scale in silico modeling of metabolic interactions between cell types in the human brain. Nat Biotechnol 28(12):1279–1285
24. Conrad TM, Lewis NE, Palsson BO (2011) Microbial laboratory evolution in the era of genome-scale science. Mol Syst Biol 7:509
25. Thiele I et al (2013) A community-driven global reconstruction of human metabolism. Nat Biotechnol 31(5):419–425
26. Bordbar A et al (2014) Constraint-based models predict metabolic and associated cellular functions. Nat Rev Genet 15(2):107–120
27. Kumar A et al (2014) Multi-tissue computational modeling analyzes pathophysiology of type 2 diabetes in MKR mice. PLoS One 9(7), e102319
28. Vazquez A (2013) Metabolic states following accumulation of intracellular aggregates: implications for neurodegenerative diseases. PLoS One 8(5), e63822
29. Bordbar A et al (2011) A multi-tissue type genome-scale metabolic network for analysis of whole-body systems physiology. BMC Syst Biol 5:180
30. Cui X et al (2014) A computational approach to estimate interorgan metabolic transport in a mammal. PLoS One 9(6), e100963

31. Cascone T et al (2011) Upregulated stromal EGFR and vascular remodeling in mouse xenograft models of angiogenesis inhibitor-resistant human lung adenocarcinoma. J Clin Invest 121(4):1313–1328
32. Iorns E et al (2012) Simultaneous analysis of tumor and stromal gene expression profiles from xenograft models. Breast Cancer Res Treat 131(1):321–324
33. Ma XJ et al (2009) Gene expression profiling of the tumor microenvironment during breast cancer progression. Breast Cancer Res 11(1):R7
34. Gregg JL et al (2010) Analysis of gene expression in prostate cancer epithelial and interstitial stromal cells using laser capture microdissection. BMC Cancer 10:165
35. Boersma BJ et al (2008) A stromal gene signature associated with inflammatory breast cancer. Int J Cancer 122(6):1324–1332
36. Komurov K, White MA, Ram PT (2010) Use of data-biased random walks on graphs for the retrieval of context-specific networks from genomic data. PLoS Comput Biol 6(8), e1000889
37. Komurov K (2012) Modeling community-wide molecular networks of multicellular systems. Bioinformatics 28(5):694–700
38. Pavlides S et al (2009) The reverse Warburg effect: aerobic glycolysis in cancer associated fibroblasts and the tumor stroma. Cell Cycle 8(23):3984–4001
39. Koukourakis MI et al (2006) Comparison of metabolic pathways between cancer cells and stromal cells in colorectal carcinomas: a metabolic survival role for tumor-associated stroma. Cancer Res 66(2):632–637
40. Colegio OR et al (2014) Functional polarization of tumour-associated macrophages by tumour-derived lactic acid. Nature 13(7519):559–63
41. Sonveaux P et al (2008) Targeting lactate-fueled respiration selectively kills hypoxic tumor cells in mice. J Clin Invest 118(12):3930–3942
42. Hanahan D, Coussens LM (2012) Accessories to the crime: functions of cells recruited to the tumor microenvironment. Cancer Cell 21(3):309–322
43. Barabasi AL, Oltvai ZN (2004) Network biology: understanding the cell's functional organization. Nat Rev Genet 5(2):101–113
44. Barzel B, Barabasi AL (2013) Universality in network dynamics. Nat Phys 9:673–681
45. Jeong H et al (2000) The large-scale organization of metabolic networks. Nature 407(6804):651–654
46. Komurov K, White M (2007) Revealing static and dynamic modular architecture of the eukaryotic protein interaction network. Mol Syst Biol 3:110
47. Fraser HB (2005) Modularity and evolutionary constraint on proteins. Nat Genet 37(4):351–352
48. Albert R, Jeong H, Barabasi AL (2000) Error and attack tolerance of complex networks. Nature 406(6794):378–382
49. Mishra GR et al (2006) Human protein reference database–2006 update. Nucleic Acids Res 34(Database issue):D411–D414
50. Bader GD et al (2001) BIND–the biomolecular interaction network database. Nucleic Acids Res 29(1):242–245
51. Chatr-aryamontri A et al (2007) MINT: the molecular interaction database. Nucleic Acids Res 35(Database issue):D572–D574
52. Stark C et al (2006) BioGRID: a general repository for interaction datasets. Nucleic Acids Res 34(Database issue):D535–D539
53. Calvano SE et al (2005) A network-based analysis of systemic inflammation in humans. Nature 437(7061):1032–1037
54. Chuang HY et al (2007) Network-based classification of breast cancer metastasis. Mol Syst Biol 3:140
55. Nikolsky Y, Nikolskaya T, Bugrim A (2005) Biological networks and analysis of experimental data in drug discovery. Drug Discov Today 10(9):653–662
56. Segal E et al (2003) Module networks: identifying regulatory modules and their condition-specific regulators from gene expression data. Nat Genet 34(2):166–176

57. Tomlins SA et al (2007) Integrative molecular concept modeling of prostate cancer progression. Nat Genet 39(1):41–51
58. Muller FJ et al (2008) Regulatory networks define phenotypic classes of human stem cell lines. Nature 455(7211):401–405
59. Ulitsky I, Shamir R (2009) Identifying functional modules using expression profiles and confidence-scored protein interactions. Bioinformatics 25(9):1158–1164
60. Pujana MA et al (2007) Network modeling links breast cancer susceptibility and centrosome dysfunction. Nat Genet 39(11):1338–1349
61. Mitra K et al (2013) Integrative approaches for finding modular structure in biological networks. Nat Rev Genet 14(10):719–732
62. Baitaluk M et al (2006) Biological networks: visualization and analysis tool for systems biology. Nucleic Acids Res 34:W466–W471 (Web Server issue)
63. Hu Z et al (2005) VisANT: data-integrating visual framework for biological networks and modules. Nucleic Acids Res 33:W352–W357 (Web Server issue)
64. Komurov K, White MA, Ram PT (2010) Use of data-biased random walks on graphs for the retrieval of context-specific networks from genomic data. PLoS Comput Biol 6(8), e1000889. doi:10.1371/journal.pcbi
65. Smoot ME et al (2011) Cytoscape 2.8: new features for data integration and network visualization. Bioinformatics 27(3):431–432
66. Ekins S et al (2006) Algorithms for network analysis in systems-ADME/Tox using the MetaCore and MetaDrug platforms. Xenobiotica 36(10–11):877–901
67. Komurov K (2011) Modeling community-wide molecular networks of multicellular systems. Bioinformatics 28(5):694–700. doi:10.1093/bioinformatics/btr718
68. Komurov K et al (2012) The glucose-deprivation network counteracts lapatinib-induced toxicity in resistant ErbB2-positive breast cancer cells. Mol Syst Biol 8:596

# Chapter 12
# Contact Normalization or Escape from the Matrix

**Harini Krishnan and Gary S. Goldberg**

**Abstract** Cancer is a complex process that involves interactions between numerous cell types. In many cases, tumor cell expansion is prevented by other cells in the microenvironment. The growth and morphology of genetically transformed cells can be normalized by junctional communication with surrounding nontransformed cells. Tumor cells need to overcome this process, called "contact normalization", before they can realize their malignant and metastatic potential. Here, we describe some fundamental aspects that underlie contact normalization, and how this information can be used to develop innovative ways to detect and treat many forms of cancer.

**Keywords** Cancer • Intercellular junctions • Cadherins • Connexins • Integrins • Contact normalization • Podoplanin

## 12.1 Introduction

Cancer is a leading cause of death that kills millions of people each year. There were 8.2 million cancer deaths worldwide in 2012. Cancer mortality is predicted to increase to 13 million each year, an increase of about 60 %, within the next two decades [1].

Over 32 million people were living with cancer within 5 years of diagnosis in 2012. In general, patients who survive 5 years after diagnosis are called "survivors" since the death rates of these patients are similar to that of the general population. However, death rates remain higher for many cancer survivors than the general population [2, 3].

---

Supported in part by funding from the New Jersey Health Foundation, the SOM Graduate School of Biomedical Sciences, and the Northarvest Bean Growers Association.

H. Krishnan • G.S. Goldberg (✉)
Graduate School of Biomedical Sciences and Department of Molecular Biology, School of Osteopathic Medicine, Rowan University, Stratford, NJ 08084, USA
e-mail: gary.goldberg@rowan.edu

 
M. Kandouz (ed.), *Intercellular Communication in Cancer*,
DOI 10.1007/978-94-017-7380-5_12

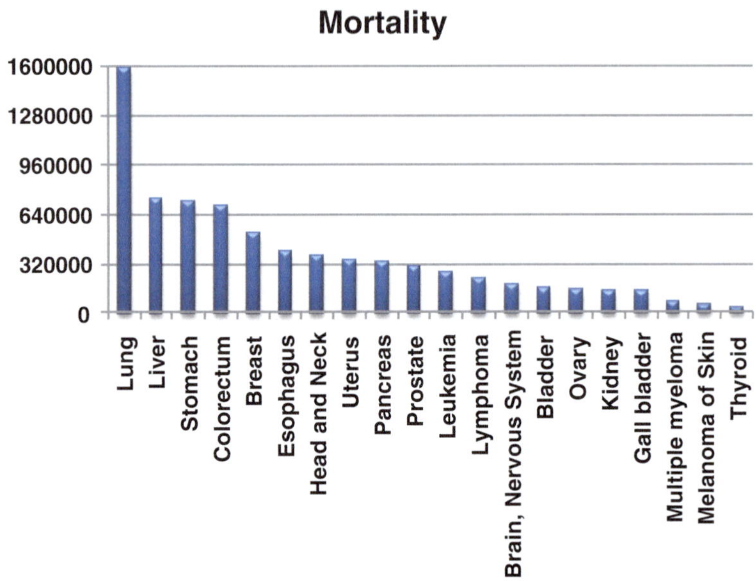


**Fig. 12.1 Mortality of some common cancers**. Worldwide estimated mortalities of all sexes and ages are shown for 2012 [1]. Head and Neck cancer includes cancers of the lip/oral cavity, nasopahrynx, other pharynx, and larynx. Uterine cancer includes cancers of the cervix uteri and corpus uteri. Lymphoma includes Hodgkin and non-Hodgkin lymphoma

As shown in Figs. 12.1 and 12.2, differences in mortality rate and 5 year survival rate between cancer types can be due to a combination of factors including genetic and biological differences in tumors, rate of disease progression, availability and access to screening and treatment, resistance to chemotherapy, lack of targeted therapies, stage of diagnosis of the disease, overall health and diet, as well as impact of the environment. For example, breast cancer has a 5 year survival rate of about 20 % [1, 4], while metastatic melanoma, pancreatic, hematopoietic, lymphoid, and brain cancers have much lower 5 year survival rates (Figs. 12.2 and 12.3). Some cancers are particularly aggressive and extremely challenging to treat. For example, the 5 year survival rate of pancreatic cancer is below 1 % [5].

Cancer arises from the clonal expansion of a single transformed cell. Transformation of normal cells to cancer cells can be triggered by carcinogens and spontaneous or hereditary genetic mutations. Cancer development is a multistep process that involves the interplay of many characteristics including evading growth suppression, sustaining pro-growth signals, resisting cell death, escaping immune destruction, inducing angiogenesis, activating invasion and metastasis, promoting inflammation, enabling limitless replication, increasing mutations and genomic instability, and reprogramming cellular energetics [6].

Cancer occurs when the homeostatic balance between cell growth and cell death pathways are derailed. This balance is normally controlled by proto-oncogenes and

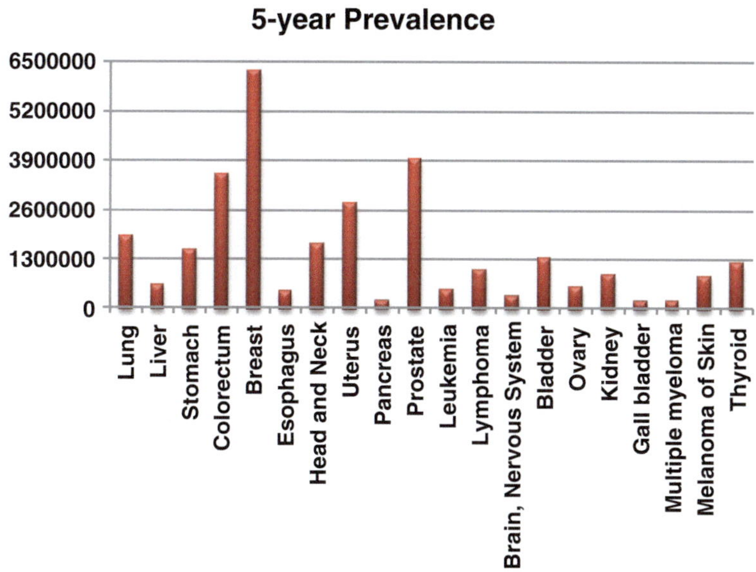


**Fig. 12.2 Five year prevalence of common cancers**. Worldwide 5 year survival rates among adults are shown for 2012 [1, 4]. Head and Neck cancer includes cancers of the lip/oral cavity, nasopahrynx, other pharynx and larynx. Uterine cancer includes cancers of the cervix uteri and corpus uteri. Lymphoma includes Hodgkin and non-Hodgkin lymphoma

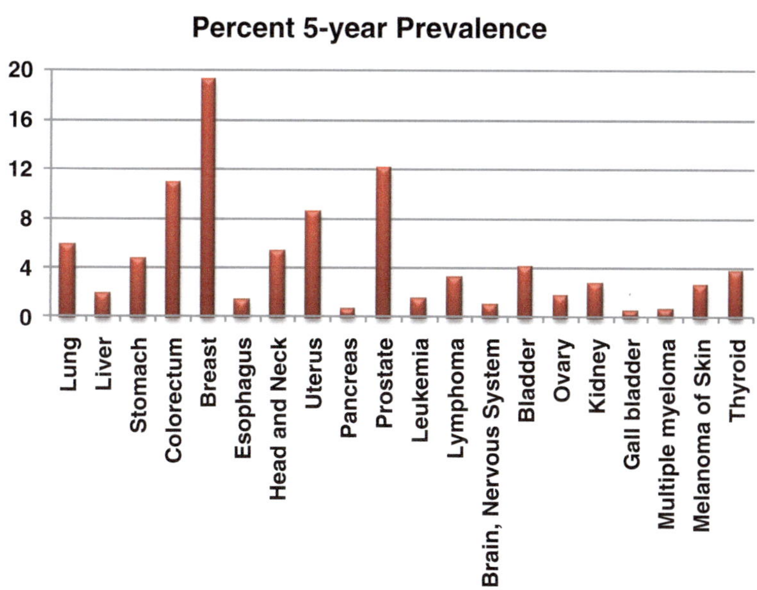


**Fig. 12.3 Percent 5 year prevalence of common cancers**. Worldwide percent 5 year prevalence among adults are shown for 2012 [1, 4]. Head and Neck cancer includes cancers of the lip/oral cavity, nasopahrynx, other pharynx and larynx. Uterine cancer includes cancers of the cervix uteri and corpus uteri. Lymphoma includes Hodgkin and non-Hodgkin lymphoma

tumor suppressors. Some principal oncogenes and tumor suppressor genes, their general functions, and major cancers implicated are listed in Tables 12.1 and 12.2.

## 12.2 Oncogenes

Alterations from proto-oncogenes to oncogenes are generally dominant; a change in one allele is usually sufficient to elicit a phenotype. Oncogenes can promote cell proliferation and motility. Mutations, gene amplification, chromosomal translocation, and epigenetic events can all lead to oncogene activation [6].

The Ras gene provides a classic example of oncogenic mutations that underlie some major types of cancer. The Ras gene family, including H-Ras, K-Ras, and N-Ras, encodes a GTPase that activates the mitogen-activated protein kinase (MAPK) pathway to promote cell proliferation and survival. Point mutations in Ras genes, especially in codons 12, 13, and 61, are common in tumorigenesis [13]. These mutations allow GTP binding to induce constitutive Ras activation [14]. Mutations the K-Ras locus have been found in nearly 30 % of all human tumors, and about 95 % of pancreatic adenocarcinomas. Many leukemias and lymphomas have mutations in N-Ras, while H-Ras mutations are found in many bladder cancers [13].

In contrast to point mutations found in Ras, gene amplification mutations leading to over expression are the most common type of aberration found in the Myc oncogene. These are especially common in solid tumors like breast cancer, neuroblastoma, and melanoma [11, 12]. The Myc transcription factor regulates multiple biological functions. Myc amplification can lead to many deleterious effects including genomic instability and several phases of tumor progression [71].

The Philadelphia (Ph) chromosome was the first chromosomal lesion to be associated with a malignant phenotype: chronic myelogenous leukemia (CML). The Ph chromosome results from a reciprocal translocation between chromosome 9 and 22. This translocation creates a highly expressed BCR-ABL fusion protein. Constitutive ABL tyrosine kinase activity drives uncontrolled expansion of the myeloid population of cells [20].

In addition to mutational changes in DNA sequence, epigenetics can also activate oncogenes. These events include DNA methylation and histone modifications that control gene expression [72]. For example, the Src tyrosine kinase regulates pleiotropic functions including cell proliferation, adhesion, migration, and angiogenesis [23]. While, activating mutations of the Src oncogene have been reported in colon cancer [8], overexpression and activation of the Src kinase occurs in many cancers including colorectal, breast, and pancreatic cancer [7, 9, 73]. Src can be inactivated by the transmembrane adapter protein CBP (Csk binding protein).

**Table 12.1** Some oncogenes, their general function, and cancers implicated

| Oncogene | General function | Major cancers implicated | References |
|---|---|---|---|
| *SRC* | Non-receptor tyrosine protein kinase | Colorectal cancer, breast cancer, pancreatic cancer | [7–9] |
| *MYC* | Nuclear phosphoprotein transcription factor | Burkitt's lymphoma, multiple myeloma, neuroblastoma, breast cancer, melanoma | [10–12] |
| *RAS* | Small GTPase | Pancreatic adenocarcinoma, haematopoietic and lymphoid cancer, colorectal cancer, melanoma | [13–15] |
| *BRAF* | Serine/threonine protein kinase | Metastatic melanoma, papillary thyroid cancer, colon cancer | [16–19] |
| *ABL* | Non-receptor tyrosine protein kinase | Leukemia (*BCR-ABL* fusion) | [20] |
| *ALK* | Receptor tyrosine protein kinase | Anaplastic large cell lymphoma, inflammatory myofibroblastic tumor, neuroblastoma, Non-small cell lung cancer | [21] |
| *AKT* | Serine/threonine protein kinase | Multiple myeloma, Anaplastic large cell lymphoma, ovarian cancer, thyroid carcinoma | [22, 23] |
| *cKIT* | Receptor tyrosine protein kinase | Small cell lung cancer, gastrointestinal stromal tumor, malignant melanoma, acute myeloid leukemia | [24] |
| *PI3K* | Phosphatidylinositol lipid kinase | Breast cancer, endometrial cancer, liver cancer, colorectal cancer | [25, 26] |
| *WNT* | Secreted lipid-modified signaling glycoprotein | Colorectal cancer, breast cancer, ovarian cancer | [27–29] |
| *CTNNB1* (β-Catenin) | Regulates cell-cell adhesion, transactivation of genes | Colorectal cancer, endometrial cancer, liver cancer, glioblastoma, melanoma | [27, 30–32] |
| *CCND1 (Cyclin D1)* | Drives G1/S transition of cell cycle | Mantle cell lymphoma, breast cancer, Non-small cell lung cancer, head and neck cancer, melanoma | [33, 34] |
| *HER2* | Receptor tyrosine protein kinase | Breast cancer, ovarian cancer, gastric cancer, prostate cancer | [35] |
| *BCL2* | Anti-apoptotic protein | Acute lymphoblastic leukemia, non-Hodgkins lymphoma, lung adenocarcinoma, renal carcinoma | [36, 37] |
| *NOTCH1* | Transmembrane receptor, transactivation of genes | T-cell acute lymphoblastic leukemias and lymphomas, breast cancer, melanoma | [38, 39] |

**Table 12.2** Some tumor suppressors, their general function, and cancers implicated

| Tumor suppressor genes | General function | Major cancers implicated | References |
|---|---|---|---|
| *RB* | Suppression of G1/S cell cycle transition, recruitment of chromatin remodeling enzymes | Retinoblastoma, prostate cancer, osteosarcoma, small cell lung cancer | [40, 41] |
| *TP53* | Cell cycle arrest at G1/S, maintains genome integrity, apoptosis activation | Li Fraumeni syndrome, lung cancer, ovarian cancer, head and neck cancer, colorectal cancer, breast cancer | [42–44] |
| *PTEN* | Phosphoinositide phosphatase, inhibits AKT/PKB pathway | Prostate cancer, glioblastoma, breast cancer, uterine and endometrial cancer | [43, 45] |
| *BRCA1/2* | DNA repair | Breast cancer, ovarian cancer, male breast cancer, prostate cancer | [46] |
| *APC* | Regulates cell-cell adhesion and Wnt signaling via β-catenin degradation | Familial adenomatous polyposis, colorectal cancer | [47, 48] |
| *ATM* | Phosphatidylinositol 3-kinase-related kinase, DNA damage response and repair, activates cell cycle checkpoints, redox sensor | Hematopoietic and lymphoid cancers, breast cancer | [49] |
| *FOXP1* | Transcription factor | Colon cancer, prostate cancer, breast cancer | [50–52] |
| *SWI/SNF* | Nucleosome remodeling complex | Malignant rhabdoid tumors, ovarian cancer, hepatocellular carcinoma | [53–55] |
| *CDKN2A* | Inhibition of cyclin dependent kinase 2A, inhibits G1/S cell cycle progression | Pancreatic cancer, melanoma | [56, 57] |
| *VHL* | E3 ubiquitin ligase for degradation of hypoxia inducible factor (HIF) | Kidney renal cell cancer, sporadic hemangioblastomas, familial pheochromocytoma | [43, 58, 59] |
| *CREBBP* | Transcription co-activation, intrinsic histone acetyltransferase activity | Acute lymphoblastic leukaemia, acute myeloid leukemia | [60–62] |
| *LKB1* | Serine/threonine protein kinase | Non-small cell lung cancer, Peutz–Jeghers syndrome associated gastrointestinal and breast cancer | [63, 64] |
| *NF1* | GTPase-activating protein (GAP), inactivates RAS | Neuroblastoma, Neurofibromatosis 1 associated sarcoma, glioma, astrocytoma and malignant schwannoma | [65, 66] |

(continued)

**Table 12.2** (continued)

| Tumor suppressor genes | General function | Major cancers implicated | References |
|---|---|---|---|
| *NF2* | Membrane-cytoskeleton scaffolding protein | Malignant mesothelioma, Neurofibromatosis 2 associated bilateral vestibular schwannoma and meningioma | [67, 68] |
| *NOTCH1* | Transmembrane receptor, transactivation of genes | Hepatocellular carcinoma, head and neck cancer, keratinocyte cancer | [69, 70] |

MAPK/PI3K pathways down regulate CBP by epigenetic histone modifications which, in turn, promote the transforming ability of Src [74].

## 12.3 Tumor Suppressor Genes

In contrast to oncogenes which are activated to promote cancer, tumor suppressor genes are inactivated to promote cancer. Tumor suppressor genes can be inactivated by DNA mutations or epigenetic modifications including promoter methylation. Tumor suppressor genes were originally found as "recessive oncogenes". Mutant tumor suppressor genes are usually recessive in that both alleles are inactivated in most cancers [6].

Retinoblastoma (Rb) is the first tumor suppressor to be discovered. The multifunctional Rb protein controls cell proliferation to prevent cancer progression. Rb frequently undergoes loss of heterozygosity in some common forms of cancer including retinoblastoma, prostate [41], and breast cancer [75].

The p53 gene is an extremely important tumor suppressor. It acts as acts as a "guardian of the genome" to prevent the generation of genetic mutations. Most p53 mutations are single base substitutions. In fact, about 90 % of p53 mutations occur between codons 125 and 300 that encode the DNA binding domain of the protein. Most tumors carry somatic mutations in the p53 gene. Germline p53 mutations cause Li-Fraumeni syndrome, which causes many cancers including sarcoma, breast cancer, and brain tumors [76].

In addition to DNA mutations, epigenetic events can also inhibit tumor suppressor gene expression to promote cancer. For example, the promoter region of the PTEN phosphatase gene is methylated in many ovarian cancers. Likewise, the BRCA1 "caretaker" gene promoter is often methylated in breast cancers [72]. Some genes can function both as oncogenes and tumor suppressor genes depending on their cellular context. For example, the NOTCH1 transmembrane receptor can promote leukemia and lymphoma progression, while inhibiting solid cancer progression of transformed hepatocytes keratinocytes (Tables 12.1 and 12.2) [70].

Cancer cells are highly heterogeneous. They carry multiple mutations of oncogenes and tumor suppressor genes. Metastatic cancers tend to possess more mutations than their primary counterparts. Thus, cells from the primary tumor acquire additional mutations that confer invasive and survival capabilities during their progression to malignancy. These malignant cancers are characterized by clonal expansion and dissemination of cells from the primary tumor to distant organs in the body by the process of metastasis [6].

## 12.4 Cancer Metastasis

Nearly all cancer deaths result from metastases. These cancers can present insurmountable challenges for surgery, radiation, and chemotherapy treatments. Cancer metastasis from primary tumors involves the orchestration of signaling events that control cell adhesion, anchorage, migration, and invasion. Tumor cells can migrate from primary sites into nearby blood or lymphatic vessels through which they can travel to and colonize distant tissues [6].

Metastasis is considered a rather a late event during tumor progression. However, there is increasing evidence that cancer cells can disseminate early in the tumorigenic process, in parallel with the development of primary tumors [77, 78]. Many factors including anatomical architecture, integrin affinities, and growth factor expression in new microenvironments dictate how specific cancers will disseminate throughout the body [79]. Table 12.3 lists some common sites of metastasis of prominent cancers.

**Table 12.3** Some common cancers and prominent sites of metastasis

| Cancer | Major sites of metastasis | References |
|---|---|---|
| Lung | Adrenal gland, bone, brain, liver, other lung | [80] |
| Liver | Lung, portal vein, portal lymph node | [81] |
| Stomach | Liver, lung, peritoneal cavity | [82] |
| Colorectum | Regional lymph nodes, liver, lung, peritoneal cavity | [82] |
| Breast | Bone, brain, liver, lung | [83] |
| Head and neck | Lymph nodes of the neck, salivary glands, lung | [84, 85] |
| Uterus and cervix | Vagina, peritoneal cavity, pelvic lymph nodes | [86] |
| Pancreas | Liver, peritoneal cavity, lung, bone, stomach, intestines | [87] |
| Prostate | Bone, lung, liver, adrenal gland | [88] |
| Leukemia | Lymph nodes, spleen, major blood vessels, central nervous system | [89] |
| Bladder | Regional lymph nodes, bone, lung, liver | [90] |
| Ovary | Peritoneal cavity, omentum, fallopian tube | [91] |
| Kidney | Adrenal gland, bone, brain, liver, lung | [92] |
| Melanoma of skin | Other areas of skin, subcutaneous tissue, bone, brain, liver, lung, skin/muscle | [93] |
| Thyroid | Regional lymph nodes, lung, bone, spine | [82, 92] |

Metastasis is achieved by sequential steps that are illustrated in Fig. 12.4. Each of these steps presents a challenge to tumor cell survival, making metastasis a rate limiting phenomenon. It is important to understand these processes since they also provide opportunities for evaluation and treatments as described below.

1. Tumor angiogenesis: Angiogenesis normally occurs during embryonic development, wound healing, and maintenance of the endometrium. Tumors hijack angiogenic pathways to promote their expansion. Tumor cells and stromal cells secrete angiogenic factors such as vascular endothelial growth factor A (VEGF-A), tumor necrosis factor alpha (TNFα), and basic fibroblast growth factor (bFGF) that signal through receptors in nearby vasculature. This triggers changes in αv integrins, VE cadherins, and matrix-metallo proteases (MMPs) to facilitate vascular endothelial migration, vessel sprouting, and neoangiogenesis in the tumors [6]. Angiogenesis inhibitors such as bevacizumab, sorafenib and sunitinib are currently being used to target cancers including glioblastoma, mammary carcinoma, hepatocarcinoma, and kidney cancers [94].
2. Separation of tumor cells from primary tumor mass: Loss of matrix anchorage and intercellular adhesion is needed for tumor cells to separate from their primary tumors. Matrix interactions are regulated by receptors including integrins, while intercellular adhesion is maintained by junctional proteins including cadherins and connexins. Consequently, alterations of these structures can be utilized as diagnostic biomarkers for many cancers [95].
3. Local invasion and migration: Dissemination of cells from the primary tumor to the vasculature requires extensive migration and invasion through the extracellular matrix and tumor stroma. Cell migration and invasion are mediated by growth factors, transmembrane receptors, oncogenic kinases, and proteases. For example, the Src tyrosine kinase induces the expression of a transmembrane receptor podoplanin (Pdpn) to induce tumor cell migration and invasion. Pdpn expression correlates with increases in matrix metalloprotease expression, activation of intracellular GTPases, and reorganization of the actin cytoskeleton. In many cases, there is a characteristic switch from epithelial to mesenchymal phenotype called epithelial mesenchymal transition. Epithelial to mesenchymal transition, activation of GTPases and proteases, and reorganization of the actin cytoskeleton are hallmarks of highly invasive tumors. Accordingly, chemotherapeutic agents are being developed to inhibit these proteases and cytoskeletal modulators [6].
4. Intravasation: Tumor cells that have migrated through the extracellular matrix can come into contact with blood vessels. These tumor cells can reach the lumen of blood vessels through the concerted action of cellular machinery that facilitates adhesion and invasion around vascular endothelium. In addition, tumor blood vessels are often fenestrated to allow passive shedding of tumor cells into the blood vasculature. Lymphatic vessels can also be penetrated due to a lack of tight junctions between lymphatic endothelial cells [96]. Chemotherapeutic agents can be used to inhibit tumor cell migration into these metastatic super-highways [6].

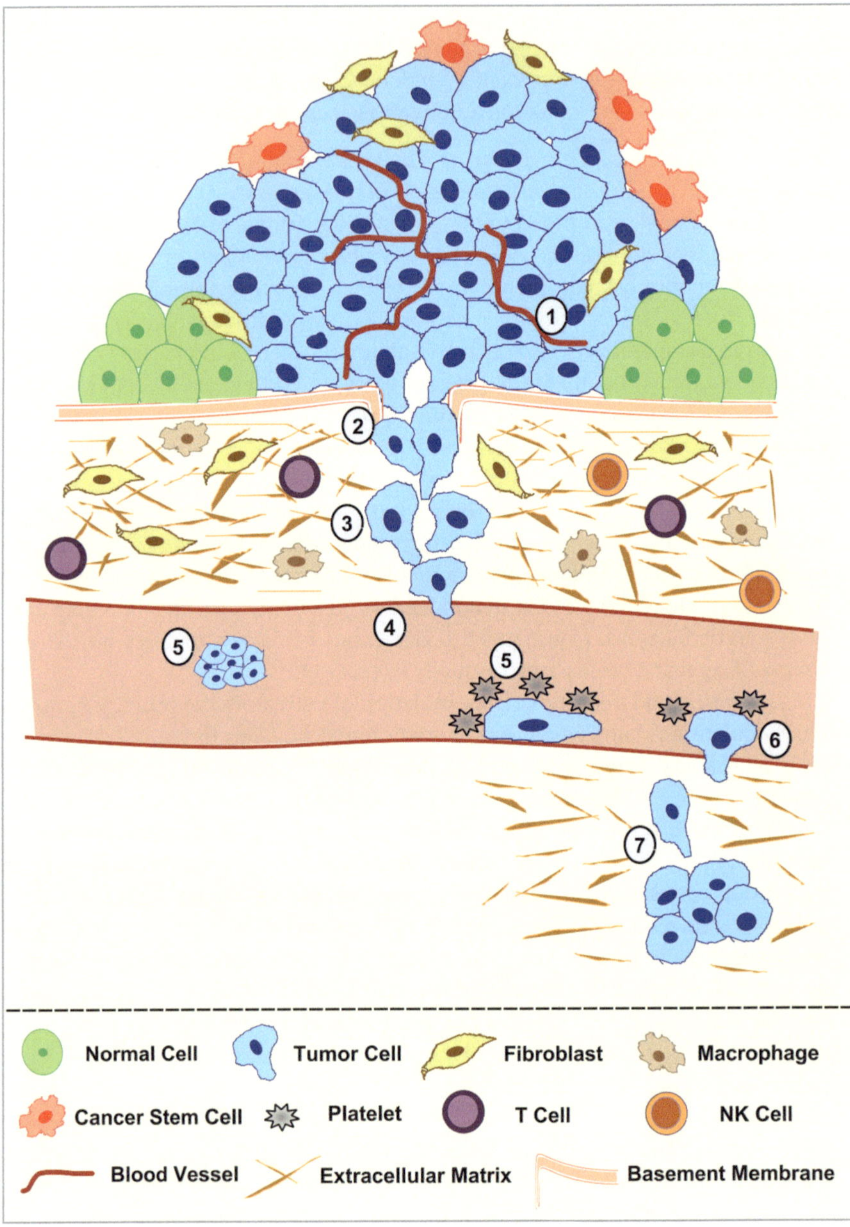


**Fig. 12.4 Stages of metastasis**. Tumor cells are surrounded by components that form the tumor microenvironment. These include normal cells, fibroblasts, extracellular matrix, blood vessels, and immune cells. Interactions between proliferating tumor cells and their microenvironment can promote cancer metastasis. Cancer metastasis is characterized by (*1*) tumor angiogenesis, (*2*) separation of tumor cells from the primary tumor mass, (*3*) local invasion and migration (*4*) intravasation, (*5*) survival in circulation, (*6*) extravasation, and (*7*) colonization

5. Survival in circulation: Anchorage independent growth is a major hallmark that differentiates cancer cells from their nontransformed precursors. Nonetheless, survival of tumor cells in circulation is rather challenging considering the hemodynamic shear forces involved and the potential for immune system destruction. Tumor cells circumvent this problem by travelling in clumps or by platelet aggregation. Platelet aggregation may help tumor cells escape immune system surveillance and promote metastasis [6]. For example, Pdpn expression in tumor cells induces platelet aggregation by CLEC-2 signaling and enhances tumor metastasis [97]. Agents that inhibit anchorage independent growth and tumor cell aggregation are being developed to combat several forms of metastatic cancer [6].
6. Extravasation: Adhesion of circulating tumor cells to vessel endothelium, and their subsequent extravasation out of the lumen, mimics aspects of leukocyte movement during inflammation. Extravasation of tumor cells involves interactions between mucins and selectins, followed by changes in integrins to stabilize contacts with the endothelium to promote transendothelial migration followed by extravasation. Cancer cells are known to upregulate mucin receptors that interact with selectins, and these can be targeted for chemotherapy. For example, antibodies that block interactions between E-selectin and PDPN on cancer cells with their respective binding partners can inhibit metastatic spread [98, 99].
7. Colonization: Colonization dictates life or death for circulating tumor cells. Tumor cells can undergo apoptosis if they do not find their new microenvironment permissive for growth and survival. Chemokines and growth factor signaling are needed for 'homing' of tumor cells to their new environment. Tumor cells can also accumulate further mutations that enable them to adapt to colonization sites [6].

Regardless of their journey, not all disseminated cells actively transform into macrometastasis. Many disseminated cancer cells remain dormant as micrometastasis. Patients can remain apparently cancer free for long periods of time before relapse. The tumor microenvironment can actively control tumor cell growth, migration, and metastasis thus contributing to cancer dormancy [100–102]. Mechanisms that underlie metastatic dormancy are not completely understood. However, elucidating the cellular and molecular basis for tumor dormancy should open new avenues for cancer detection and therapy.

## 12.5 Cancer Dormancy

Despite the fact that millions of people die of cancer every year, the majority of the population remains cancer free. Healthy individuals can carry asymptomatic microscopic tumors for an entire lifetime. Autopsies conducted in men and women who died of causes unrelated to cancer have revealed evidence of prostate and breast tumors, respectively [103, 104]. The frequency of these occult tumors far exceeds their actual incidence rates [103, 104]. Thus, individuals can carry latent microscopic tumors without developing cancer. For example, melanoma can remain dormant for several decades before changes in tissue equilibrium can trigger tumor progression [105]. Cancer dormancy can take place during any stage of tumor development including at the primary tumor, micrometastasis, secondary tumor, or after surgical and therapeutic interventions.

Minimal residual disease refers to a small number of cancer cells that remain dormant in the body after surgical or chemotherapeutic treatment. Although initially dormant, these cells can resume growth when carcinogenic signals are triggered. These cells are the major cause of cancer relapse in patients in remission for prolonged periods of time. Breast and prostate cancer patients who have been in cancer remission for many years can frequently suffer from sudden relapses. In fact, about 85 % of chronic myeloid leukemia patients in remission for several years after bone marrow transplant remain positive for Bcr-Abl fusion gene transcripts [106]. Evidence of minimal residual disease has been detected in the bone marrow and blood of individuals without clinical manifestation of cancers including mammary carcinoma, leukemia, lymphoma, pancreatic cancer, and melanoma [106].

As mentioned above, tumor cells interact with their extracellular matrix via integrins. Integrin signaling regulates fundamental aspects of cell growth, survival, differentiation, and migration. Accordingly, integrins can control the switch from cellular dormancy to a metastatic state [107]. For example, α5β1-integrin recognizes the urokinase plasminogen activator receptor (uPAR) to regulate cell proliferation and dormancy [108]. Integrins bind uPAR to activate the mitogenic extracellular regulated kinase (ERK) pathway and inhibit growth suppressive stress-activated protein kinase ($p38^{SAPK}$) signaling. When uPAR-integrin interactions are disrupted, p38 activity increases and ERK signaling decreases which forces cells into G0/G1 arrest. Consistent with these findings, low uPAR expression has been reported in dormant tumor cells [109]. In addition, α4-integrins in stromal macrophages interact with vascular cell adhesion protein (VCAM 1) on dormant breast cancer cells that have disseminated to the lung; this interaction induces PKB/AKT signaling to mediate cell dormant tumor cell survival [110]. Accordingly, MAPK and PKB/AKT inhibitors show promise as chemotherapeutic compounds [6].

Growing tumors are extremely dependent on blood vessels to deliver nutrients and oxygen for growth and expansion. Tumors that are unable to recruit blood vessels can become dormant. This process, called angiogenic dormancy, is illustrated in Fig. 12.5. Tumor suppressors such as p53 and p38 activate thrombospondin while inhibiting VEGF in order to suppress angiogenesis. Tumors can resume uncontrolled expansion when there is a switch from anti-angiogenic (thrombospondin, angiostatin, endostatin and vasostatin) to pro-angiogenic signals (VEGF-A, PDGF, bFGF) [106, 111]. For example, oncogenes such as ras and myc induce VEGF expression while suppressing thrombospondin expression to promote angiogenesis [106, 112]. As discussed below, the VEGF signaling pathway has emerged as a mechanism by which tumor cells awake from dormancy to break out of their microenvironment. Consequently, the VEGF pathway has also emerged as an important chemotherapeutic target.

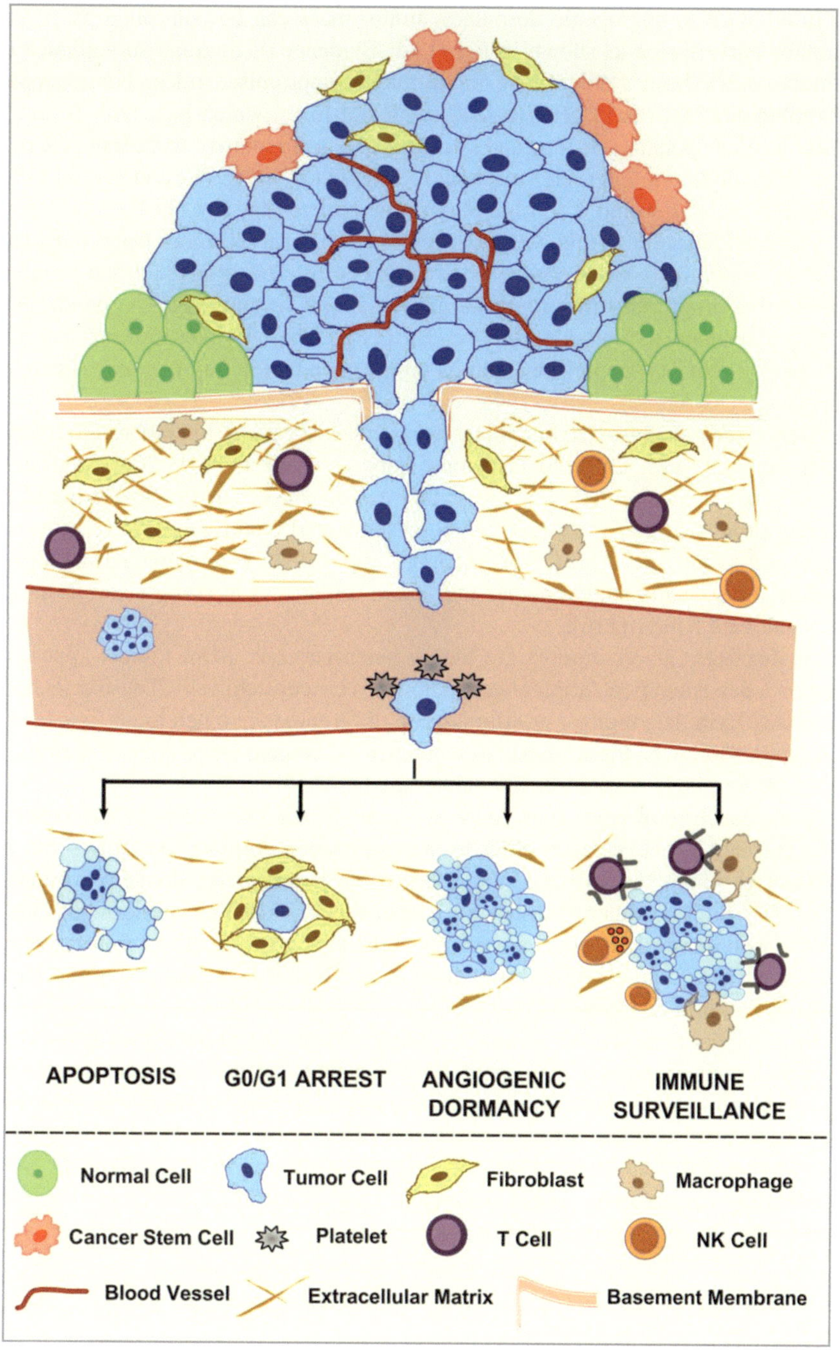


**Fig. 12.5 Tumor dormancy**. Cancers can be forced into dormancy during any stage of tumorigenesis. This figure illustrates tumor dormancy during metastasis of disseminated tumor cells. This process can manifest in a variety of ways including G0/G1 arrest of cell cycle of isolated tumor cells, and angiogenic suppression or immune surveillance of a group of tumor cells

In addition to angiogenic dormancy, tumor mass can be controlled by potent immune surveillance as shown in Fig. 12.5. Evidence of immune surveillance of tumors comes from studies done in immune compromised mice. For example, recombination activating gene (RAG2) deficient mice, which lack both B and T cells, develop spontaneous lung and intestinal adenocarcinomas. Mice that lack perforin, a cytolytic protein used by cytotoxic T lymphocytes (CTL) and natural killer cells (NK cells), develop B cell lymphomas with increasing age [113].

Other evidence for immune suppression of tumorigenesis comes from transplant patients. Individuals who receive kidney or bone marrow transplant and immune suppressive drugs can suffer donor origin melanoma, even though the donors had been in melanoma remission for decades [114]. These results indicate that the immune system can suppress melanoma progression by forcing them into a state of dormancy.

Tumor cells can also activate metastasic suppressor genes in response to microenvironmental cues. Over 20 metastatic suppressor genes have been identified. These genes encode proteins that can induce cell quiescence or apoptosis in order to prevent progression of metastasis when disseminated tumor cells extravasate into secondary organ sites [115]. For example, tumor cells expressing KISS1 (kisspeptins) form significantly smaller metastatic tumor foci in comparison to tumor cells that lack KISS1 [116].

As discussed above, tumors are highly heterogeneous. Most tumors appear to contain a persistent population of cells known as cancer stem cells. These cells have the ability to initiate tumors, proliferate, and differentiate into heterogeneous tumor cell types. Moreover, these stem cells can remain quiescent for prolonged periods of time and are highly resistant to chemotherapy [114].

Thus, cancer dormancy represents two sides of the same coin. While keeping cells dormant is beneficial in order to prevent cancer, dormant metastases can be resistant to chemotherapy and, therefore, carry potential for dangerous relapse [106]. Hence, it becomes essential for cancer therapy to eliminate dormant tumor cells as well those more active. For example, the eIF2 phosphatase inhibitor salubrinal can eliminate quiescent multiple myeloma cells that are resistant to the proteosome inhibitor bortezomib [117].

Most cancer therapies target proliferating tumor cells that have already escaped from their primary microenvironment. Extensive mechanistic understanding of how tumor cells interact with surrounding stroma is needed to more efficiently treat dormant cancer cells and cancer cells initial stages of tumorigenesis. Dormant and initiated cancer cells at early stages of tumorigenesis are in direct contact with surrounding normal cells. These interactions between normal and tumor cells are vital in the process of tumor growth and metastasis.

## 12.6 Contact Normalization

Normal cells usually stop dividing when they touch each other. This process, called contact growth inhibition, is mediated (at least in part) by cadherin junctions between cells. As mentioned above, cadherin junctions are often disrupted in cancer cells. Consequently, loss of contact growth inhibition is a major tumor cell hallmark [6].

However, although tumor cells do not exhibit contact inhibition by communicating with other tumor cells, they can be normalized by communicating with neighboring normal cells. This process, called contact normalization, is illustrated in Fig. 12.6. Although both processes are mediated by intercellular communication, contact normalization is quite different from contact growth inhibition. Contact inhibition is mediated by intercellular communication between tumor cells to control proliferation. In contrast, contact normalization is mediated by communication between tumor cells and normal cells to control cell morphology and phenotype in addition to proliferation [23, 100, 102, 118]. Indeed, nontransformed cells can rescue intercellular junctional communication (e.g. by cadherins and gap junctions) with adjacent transformed cells [105, 119–121].

Cells transformed by a variety of agents including chemicals [122, 123], viruses [124, 125], and oncogenes [126, 127] can be contact normalized by surrounding normal cells. Early evidence for contact normalization comes from studies transformed cells cultured with nontransformed cells over 50 years ago. In these experiments, fibroblasts infected with tumorigenic viruses retained normal morphology, and did not form characteristic cancer foci, when surrounded by nontransformed fibroblasts [128–130].

In addition to contact normalization, cell competition can exist at the interface between normal and transformed cells [131]. This phenomenon was first discovered in *Drosophila* imaginal discs. Here, cell survival is based on the expression levels of specific proteins in comparison to neighboring cells [132]. Competitive interactions

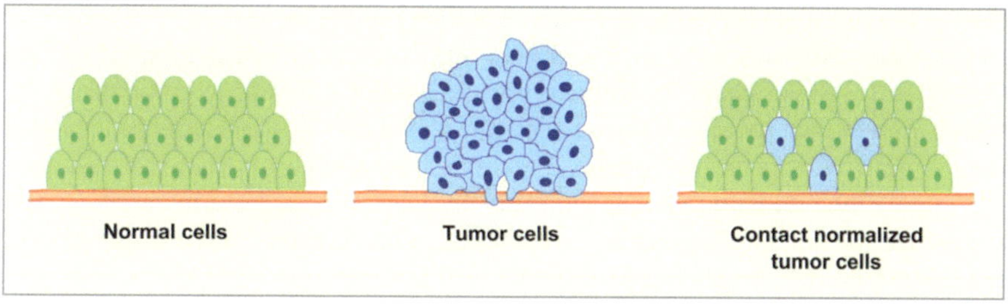


**Fig. 12.6 Contact normalization**. Normal cells are contact inhibited when they touch each other. In contrast, tumor cells lose the property of contact inhibition and grow into multilayered foci. However, normal cells can force adjacent tumor cells to assume a normal morphology and phenotype by the process of contact normalization. This process is particularly powerful during early stages of tumorigenesis. Tumor cells need to escape from contact normalization in order to break out of their microenvironment, grow, migrate, and metastasize

between normal and transformed epithelial cells result in apical extrusion and apoptosis [133–135]. For example, Src transformed MDCK cells cocultured with nontransformed MDCK cells are apically extruded from the monolayer with a concomitant increase in focal adhesion kinase (FAK) and myosin II activity. In contrast Src transformed MDCK cells cultured alone show no increase in FAK or myosin II, and are not apically extruded [134]. Thus, the presence of surrounding normal cells influences the signaling and behavior of transformed cells.

Early *in vivo* evidence of contact normalization comes from mice experiments in which normal mouse blastocysts were injected with highly malignant teratomas. Surprisingly, these chimeric blastocysts undergo normal embryonic development and give rise to offspring without tumors. In addition, the teratoma cells survive during development to form normal tissues and organs in these chimeric mice. Thus, malignant teratomas that are otherwise lethal in adult mice are clearly normalized in the microenvironment of developing blastocysts [136, 137].

It is now clear that genetic predisposition is not an absolute cause for cancer. Malignancies arise and progress from abnormal intercelluar interactions in the tumor microenvironment. Despite harboring tumorigenic mutations, tumor cells can be efficiently suppressed by neighboring normal cells [101, 120]. Particularly during early stages of neoplastic development, tumor cells are in direct contact with normal cells. This contact is mediated by cadherin and gap junctions. In general, cadherins and gap junctions promote contact normalization as illustrated in Fig. 12.7.

## 12.7 Cadherins

Cadherins belong to a superfamily of transmembrane glycoproteins that mediate intercellular recognition and adhesion in a calcium dependent manner. About 80 different cadherins including classical cadherins (E-cadherin, N-cadherin, VE-cadherin), desmosomal cadherins (desmoglein, desmocollin), and protocadherins ($\alpha$-, $\beta$-, $\gamma$-protocadherins) have been identified [100, 138].

Classical cadherins such as E-cadherin and N-cadherin are composed of five tandem extracellular cadherin domain repeats, a single pass transmembrane motif, and an intracellular domain. The extracellular region contains calcium binding sites located between successive domains that stabilize homophillic cadherin contacts. The intracellular domain interacts with catenins such as p120 catenin and $\beta$-catenin to regulate the actin cytoskeleton. These associations control fundamental aspects of cell morphology, growth, and migration [138].

As mentioned above, E-cadherin forms junctions that mediate contact inhibition of growth between epithelial cells [139]. Contact inhibited endothelial cells respond poorly to stimulation by VEGF. Endothelial cells express VE-cadherin (vascular endothelial cadherin) which prevents VEGFR-2 (VEGF receptor) phosphorylation in a $\beta$-catenin dependent manner to block VEGF induced proliferation [140]. Thus,

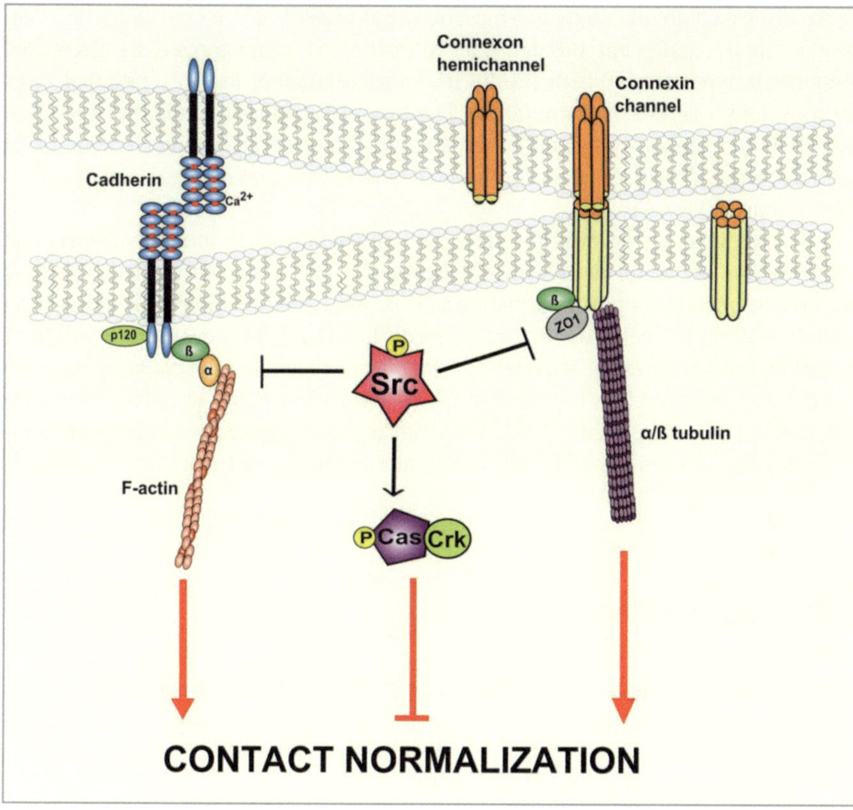


**Fig. 12.7 Intercellular communication and contact normalization**. Intercellular junctions such as cadherins and connexins promote contact normalization of tumor cells by neighboring normal cells. Intercellular cadherin interactions stabilize their intracellular links with catenins (p120, α, and β-catenin) to control the actin cytoskeleton and suppress cell proliferation. Individual connexins associate to form gap junction channels between cells. Connexins interact with β-catenin, zonula occludens, and tubulins to suppress cell growth. Cadherin and connexin interactions between normal and tumor cells promote contact normalization. SRC can phosphorylate cadherins, β-catenin, connexins, and CAS to disrupt intercellular junctional communication between normal and transformed cells in order to promote escape from contact normalization [23, 100]

contact inhibition stabilizes intercellular and cytoskeletal architecture to prevent cell growth and migration.

Growth factor and oncogenic signaling can disrupt cadherin mediated contact growth inhibition to promote cell growth and migration. For example, the Src tyrosine kinase phosphorylates cadherins, which causes β-catenin to dissociate from adherens junctions and translocate to the nucleus where it activates genes that promote cell proliferation and migration. Abnormal cadherin expression is found in several types of cancer, and is associated with increased invasion and metastasis [141].

As discussed above, epithelial-mesenchymal transition is a characteristic event seen in many malignant tumors. This transition is characterized by decreased E-cadherin expression, which results in dedifferentiation, loss of epithelial morphology, and increased cell migration. The loss of E-cadherin (and other epithelial markers) is accompanied by gain of mesenchymal markers including N-cadherin. In contrast to E-cadherin, N-cadherin can associate with signaling molecules that promote tumor cell motility [6].

Selective vascular permeability is essential for normal tissue homeostasis and extravasation of white blood cells. VE-cadherin can associate with the Vegfr-2 receptor to prevent its phosphorylation and subsequent signaling. This interaction inhibits endothelial cell growth and permeability [142]. However, upon VEGF-A signaling, the Src kinase is activated to phosphorylate and disrupt cadherin junction complexes composed of VE-cadherin, α-, β- and p120-catenin in order to increase endothelial cell permeability [143]. The resulting dysregulated vascular permeability facilitates transendothelial migration, intravasation, and extravasation during cancer metastasis [144].

## 12.8 Gap Junctions

Gap junctions form aqueous channels that connect the cytoplasm of adjacent cells. These channels are assembled by the connexin family of integral membrane proteins. Gap junctions allow adjacent cells to share intracellular signals such as metabolites and ions in a coordinated manner [145].

Connexins play an important role in cell growth control. Like cadherins, connexins can act as tumor suppressors, and gap junctions are often disrupted between tumor cells. For example, several cancers such as glioma and breast cancer show repressed Cx43 expression in comparison to corresponding normal cells. Moreover, restoration of Cx43 reverses the transformed phenotype of these cells [146, 147].

Studies have shown that gap junctional communication is not required for contact normalization [126]. This is consistent with previous reports of contact normalization in the absence of dye transfer between transformed and nontransformed cells [127]. However, while gap junctions are not absolutely required for contact normalization, gap junctional communication facilitates the ability of normal cells to control the growth of tumor cells [148, 149]. For example, inhibition of Cx43 expression decreases the ability of nontransformed cells to normalize transformed cells in coculture [148].

Similar to cadherin junctions, the Src kinase can phosphorylate Cx43 to disrupt gap junctional communication. In addition, several proteins downstream of Src including the Cas adaptor protein (see below) and MAPK can inhibit gap junctional communication [150–152]. Many tumor promoters also disrupt gap junctions in addition to cadherin junctions. Interestingly, cadherins are required for gap junction assembly and function [153].

## 12.9 Contact Normalization, Biomarkers, and Chemotherapeutic Targets

The Src tyrosine kinase regulates a variety of cellular functions including cell proliferation, differentiation, adhesion, migration, invasion, angiogenesis, and metabolism. Src transformed cells display characteristic neoplastic properties such as refractive morphology, uncontrolled proliferation, anchorage independence, growth factor independence, and increased motility. Furthermore, Src activation leads to the disruption of the actin cytoskeleton as well as matrix and intercellular adhesions. This disruption of the microenvironment and intercellular communication can lead to loss of contact growth inhibition by transformed cells [84].

Src plays an important role in tumor development and maintenance. Although rarely mutated, Src expression and kinase activity is frequently increased in many cancers as indicated in Table 12.1. Oncogenic Src kinase activation triggers several downstream effectors [23]. For instance, upon activation by integrins, Src phosphorylates focal adhesion proteins including Cas (Crk-associated substrate). Cas associates with Crk (CT10 regulator of kinase) to augment anchorage independent growth and migration. Src requires Cas to induce transformation [99]. Thus, Src, Cas, and Crk work together to regulate cell adhesion and migration associated with neoplastic transformation as shown in Fig. 12.7. However, molecular events underlying the effects of Src and Cas in tumorigenesis have not been completely elucidated.

While Src itself is an enticing chemotherapeutic target [154], substrates and downstream effectors of the Src kinase pathway that are needed to overcome contact normalization should serve as powerful biomarkers and chemotherapeutic targets. A layered culture system, illustrated in Fig. 12.8, has been employed to identify

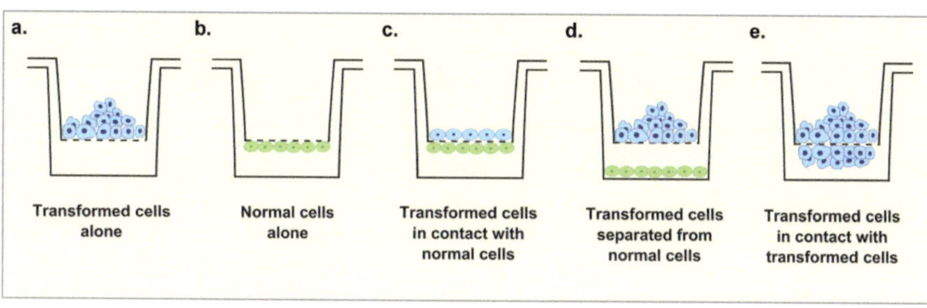


**Fig. 12.8 Layered culture system**. Transformed cells and nontransformed cells are plated on opposite sides of a porous membrane. The pores are large enough to allow intercellular junctional communication and contact normalization between the two populations, but small enough to present their migration and mixing. This system includes: (*a*) transformed cells alone plated on top of the membrane, (*b*) nontransformed cells alone at the bottom of the membrane, (*c*) transformed cells contact normalized by nontransformed cells, (*d*) transformed cells separated from nontransformed cells by 1 mm, (*e*) and transformed cells plated both on top and bottom of the membrane. The layered culture system can be used to identify genes affected by transformation and contact normalization [15, 126]

**Table 12.4** Prominent genes decreased by transformation and increased during contact normalization

| Genes decreased by *Src* transformation and increased during contact normalization | General function | Major cancers implicated | References |
|---|---|---|---|
| *FHL1* | Transcription coactivator | Breast cancer, gastric cancer, lung cancer, prostate cancer, brain tumour, renal cancer | [156–158] |
| *SDPR* | Serum deprivation response | Breast cancer, prostate cancer, kidney cancer | [157, 159] |
| *VCAM1* | Cell adhesion | Clear cell renal cancer | [160] |
| *SOX11* | Transcription factor | Ovarian cancer, hematopoietic malignancies | [161, 162] |
| *WNT5* | Secreted signaling protein of WNT family | Hematopoietic malignancies, colorectal cancer | [163, 164] |

these biomarkers and molecular targets [126, 155]. Comprehensive gene expression analysis can be used to identify genes in transformed cells that are affected by contact with nontransformed cells. These experiments reveal that while 10 % of the entire transcriptome is affected by oncogenic transformation, the expression of only 1 % of these genes (or about 0.1 % of the transcriptome) is reversed by contact normalization [15, 126]. In this paradigm, tumor promoters are identified as genes that are upregulated during transformation and down regulated during contact normalization. Conversely, tumor suppressors are identifies by genes that are down regulated during transformation and increased during contact normalization [99]. These studies reveal novel cancer biomarkers and functionally relevant chemotherapeutic targets, some of which are listed in Tables 12.4 and 12.5 and Fig. 12.10.

## 12.10 Contact Normalization and Tumor Suppressors

Several tumor suppressors have been found that are induced by contact normalization of transformed cells (see Table 12.4). These include four and a half LIM domains (Fhl1) and serum deprivation response protein (Sdpr) [126, 176]. While other tumor suppressors induced by contact normalization have been identified, Fhl1 and Sdpr are discussed here as examples.

## 12.11 Fhl1

Fhl1 belongs to a family of LIM-only proteins (cysteine rich zinc finger motifs named after their discovery in Lin-11, Isl-1, and Mec-3). As its name denotes, Fhl1 contains an N-terminal half LIM domain followed by four complete LIM domains. Fhl proteins can associate with intercellular junctions and transcription factors to

**Table 12.5** Prominent genes increased by transformation and decreased during contact normalization

| Genes increased by *Src* transformation and decreased during contact normalization | General function | Major cancers implicated | References |
|---|---|---|---|
| *1700012B09Rik* | Unannotated | No data available | [15] |
| *VEGFR2* | Transmembrane receptor involved in vasculogenesis, angiogenesis | Breast cancer, lung cancer, renal cancer, colorectal cancer | [165–168] |
| *PDPN* | Transmembrane receptor involved in platelet aggregation, cell migration, cell morphology | Mesothelioma, CNS tumors, head and neck cancer, melanoma, ovarian and testicular cancer | [169–173] |
| *TMEM163* | Transmembrane receptor that may bind zinc ions and recruit them to vesicular organelles | Papillary thyroid carcinoma, nodular lymphocyte-predominant Hodgkin lymphoma | [100, 174] |
| *PCDH7* | Cadherin family member involved in cell adhesion | Breast cancer, ovarian cancer | [175] |

control cellular processes including proliferation, differentiation, apoptosis, adhesion, migration, transcription, and signal transduction [177, 178].

Src utilizes Cas to block gap junctional communication and inhibit Fhl1 expression in order to promotes anchorage independent growth and migration. Ectopic expression of Fhl1 overrides the ability of Src transformed cells, breast cancer, and lung cancer cells to migrate and survive anchorage independence [156, 158, 176]. In fact, when Fhl1 expression is restored in lung cancer cells, they form significantly smaller tumors in mice in comparison to control lung cancer cells with low Fhl1 expression [158]. These data demonstrate a tumor suppressive role of Fhl1.

Fhl1 can shuttle between the cytoplasm and nucleus to regulate various cellular functions. The cooperative transcription functions of Fhl1 mimic the TGF-β pathway because of the involvement of SMADs. SMADs are intracellular proteins that transduce extracellular signals from TGF-β ligands to the nucleus where they activate transcription of genes involved in growth suppression. Fhl1 interaction with SMAD 2/3 triggers phosphorylation and nuclear accumulation of the Fhl1-SMAD 2/3/4 complex to affect transcription of growth regulatory genes including p21 and Myc. Evidently, Fhl1 can inhibit tumor cell growth through induction of p21 and down regulation of Myc [177].

Fhl1 expression is down regulated in malignancies including melanoma, breast, gastric, lung, prostate, ovarian, colon, thyroid, brain, renal, liver cancers [157, 158, 176]. Patients with low Fhl1 expression in gastric cancers have significantly lower survival rates than patients with high Fhl1 expression. Lower Fhl1 expression also correlates with increased serosal invasion and distant metastasis of gastric cancer [179].

Fhl1 is a novel tumor suppressor that is inhibited by the Src-Cas pathway. Induction of Fhl1 expression is associated with reversal of neoplastic phenotype. In addition, normal cells induce Fhl1 expression during contact normalization of transformed cells [176, 180, 181]. Clearly, loss of Fhl1 is associated with cancer progression and is, therefore, a clinically relevant cancer biomarker.

## 12.12 SDPR

Sdpr is a calcium-independent phospholipid-binding protein. SDPR is a substrate for protein kinase C (PKC) phosphorylation that induces caveolae formation in concert with caveolins [182, 183]. Caveolae play a significant role in signal transduction, endocytosis, and negative regulation of cell proliferation. Loss of Sdpr is associated with loss of membrane caveolae.

Sdpr was first characterized as a phosphatidylserine binding 68 kDa phosphoprotein from human platelets. Subsequent experiments found that Sdpr expression increases in serum starved cells during growth arrest. However, cells undergoing growth arrest by density dependent inhibition also show low levels of Sdpr expression. These data suggest that differential expression of Sdpr in contact inhibited and serum starved cells may be due to complex response pathways that involve intercellular communication [184].

As described above, Src utilizes Cas to suppress Fhl1 expression in order to promote anchorage independent cell growth and migration [176]. Expression of genes in transformed cells that express Cas can be compared with cells that do not express Cas to identify genes coordinately regulated by Fhl1, Src, and Cas. Interestingly, Fhl1 induces Sdpr expression. These data indicate that Sdpr acts downstream of Fhl1. Furthermore, Sdpr expression is suppressed in Src transformed cells but not in MEK transformed cells, indicating that Src inhibits Fhl1 and Sdpr independent of the MAPK pathway [157].

As with Fhl1, Sdpr expression is decreased by oncogenic Src kinase activity, but increased in transformed cells undergoing contact normalization by surrounding nontransformed cells [176, 180, 181]. In addition, Sdpr expression is suppressed in breast, prostate, and kidney cancer tissues in comparison to adjacent non-cancerous tissue [157, 159]. Thus, like Fhl1, Sdpr may serve as a useful biomarker for cancer progression.

## 12.13 Contact Normalization and Tumor Promoters

Tumor cells need to escape from contact normalization to proliferate, migrate, and metastasize. Genes that are elevated during transformation and inhibited during contact normalization are of great interest since they may be required for malignant growth and metastasis. These genes encode proteins that mediate events including glutamate transport (Tmem163), growth factor signaling (Vegfr2/Kdr), and cell migration (Pdpn) as shown in Table 12.5. These proteins are of particular interest since they can serve as chemotherapeutic targets.

## 12.14 Vegfr2

Vascular endothelial growth factors (VEGFs) are secreted homodimeric glycoproteins that play a critical role in the development of blood vascular system, lymphatic system, and wound healing. The mammalian VEGF family consists of five members: VEGF-A, VEGF-B, VEGF-C, VEGF-D, and placenta growth factor (PLGF). All VEGF ligands signal through three different VEGF tyrosine kinase receptors (Vegfr1-3). For example, VEGF-A signals through Vegfr1 and Vegfr2 to regulate hematopoiesis, cell migration, proliferation, and blood vessel formation [185].

Vegfr2, also known as KDR and FLK1, is a type III transmembrane kinase receptor. It is a 230 kDa protein that consists of an extracellular domain of about 760 amino acids, a single pass transmembrane domain of 25 amino acids, and a tyrosine kinase intracellular domain of about 570 amino acids. The extracellular region of Vegfr2 consists of seven immunoglobulin (Ig)-like domains, of which the second and third domains act as binding sites for VEGF-A. The Vegfr2 extracellular domain has multiple N-glycosylation sites which increase the molecular weight of the peptide from 150 to 230 kDa [185, 186].

The intracellular Vegfr2 region contains two kinase domains that are separated by an insert of about 70 amino acids, and a C-terminal tail. The intracellular region also contains multiple tyrosine residues that are phosphorylated to act as binding sites for several intracellular binding partners and signaling molecules. Vegfr2 associates with these other proteins to activate multiple signaling cascades that regulate various cellular functions. Vegfr2 also possesses serine phosphorylation and ubiquitination sites that regulate its signaling and turnover [185].

Vegfr2 is expressed by vascular and lymphatic endothelial cells. Homozygous null Vegfr2 mice die at E8.5-9.5 due to defective development of blood islands, endothelial cells, and hematopoietic cells, indicating that Vegfr2 is critical for vascular development [185]. Vegfr2 signaling is important for regulating endothelial cell proliferation, migration, survival, and permeability during vasculogenesis (development of new blood vessels) and angiogenesis (formation of blood vessels from pre-existing ones) in normal physiology, as well as pathological conditions including cancer [185, 187].

Vegfr2 receptor tyrosine activity is initiated by dimeric VEGF ligand binding. VEGF binding triggers Vegfr2 dimerization that leads to autophosphorylation of the receptor. Phosphorylated receptors then recruits several intracellular binding partners based on the presence of a Src homology-2 (SH2) domain. These binding partners include Src family kinases, Vegfr associated protein (VRAP), and phospholipase C (PLC). Engagement of these signaling molecules by Vegfr2 can further activate components of Src, MAPK, PI3K, MEK, and other pathways to promote the cellular functions listed above [187, 188].

Vegfr2 receptor tyrosine kinase activity is regulated by ligand availability. As described earlier, angiogenesis is critical for tumor growth and dissemination. Tumor angiogenesis is initiated in response to hypoxia, oncogenes, and growth factor signaling which increase ligand availability. For example, tumor cells release VEGF that activates Vegfr2 to stimulate vascular endothelial cell proliferation, migration, and permeability [6]. Integrin signaling can also promote tumor angiogenesis. In response to VEGF, integrin $\alpha_v\beta_3$ can associate with Vegfr2 to promote receptor activation. This receptor crosstalk can be facilitated by the activity of Src and small GTPases [188].

In addition to endothelial cells, many types of cancer cells, including glioma, lung, and breast cancer cells, can also express Vegfr2 [165, 166, 168, 189]. There is evidence that glioma stem cells may directly participate in tumor angiogenesis by transdifferentiating into vascular endothelial-like cells. Glioma cells can form blood channels to supply the growing tumor with nutrients and oxygen, a phenomenon known as vasculogenic mimicry. Vegfr2 expression maintains stemness of glioma cells to promote vasculogenic mimicry and sustain tumor growth [189]. Lung cancer cells that express VEGF and Vegfr2 create an autonomous autocrine signaling reaction to promote tumor cell growth and migration [165, 168]. Thus, VEGF and Vegfr2 can serve as functionally relevant cancer biomarkers and chemotherapeutic targets.

As described above, Src can phosphorylate adherence junctions to disrupt contact inhibition and promote escape from contact normalization. Vegfr2 is one of the top 3 genes that are increased by Src and suppressed by contact normalization [15]. This finding is consistent with the role of Vegfr2 as a tumor promoter. As mentioned earlier, VEGF-A signaling through Vegfr2 can activate Src mediated disruption of adherence junctions. During Src transformation and escape from contact normalization, Src induction of Vegfr2 may create a positive feedback loop to augment transformed cell proliferation, anchorage independent growth, and migration.

Angiogenesis can be inhibited by targeting VEGF ligands or Vegfr2. For example, bevacizumab (Avastin) is a humanized monoclonal antibody that forms a complex with VEGF-A to prevent receptor binding. Avastin was the first antiangiogenic drug to be approved by the FDA and is currently used to treat several forms of cancer. Other antiangiogenic drugs including sunitinib (Sutent) and sorafenib (Nexavar) are small molecule tyrosine kinase inhibitors that can be used to target Vegfr2 activity in tumor cells [185].

## 12.15 Podoplanin

Podoplanin (Pdpn), also known as aggrus, T1-α, OTS8, gp36, gp38, RT140, E11, and PA2.26, is a type-I transmembrane sialomucin glycoprotein. Pdpn was initially identified by as a 44 kDa protein with platelet aggregating properties in a highly metastatic clone of mouse colon adenocarcinoma cells (NL17) using a monoclonal antibody (8F11) [190]. This same antibody was used to identify Pdpn again as a

41 kDa protein in highly metastatic mouse melanoma cells (B16F10) [191]. Interestingly, this 8F11 antibody was able to block platelet aggregation *in vitro* [190, 191] and pulmonary metastasis of colon adenocarcinoma cells *in vivo* [192]. Pdpn was also independently discovered as a gene induced by 12-O-tetradecanoylphorbol-13-acetate (TPA) treatment and Ras transformation of mouse osteoblastic cells [192].

Pdpn is a transmembrane receptor with a large extracellular domain of about 130 amino acids, a single pass transmembrane domain, and a short intracellular domain of just 9 amino acids as shown in Fig. 12.9. The Pdpn extracellular domain is extensively *O*-glycosylated with sialic acid at serine and threonine residues [197]. Platelet

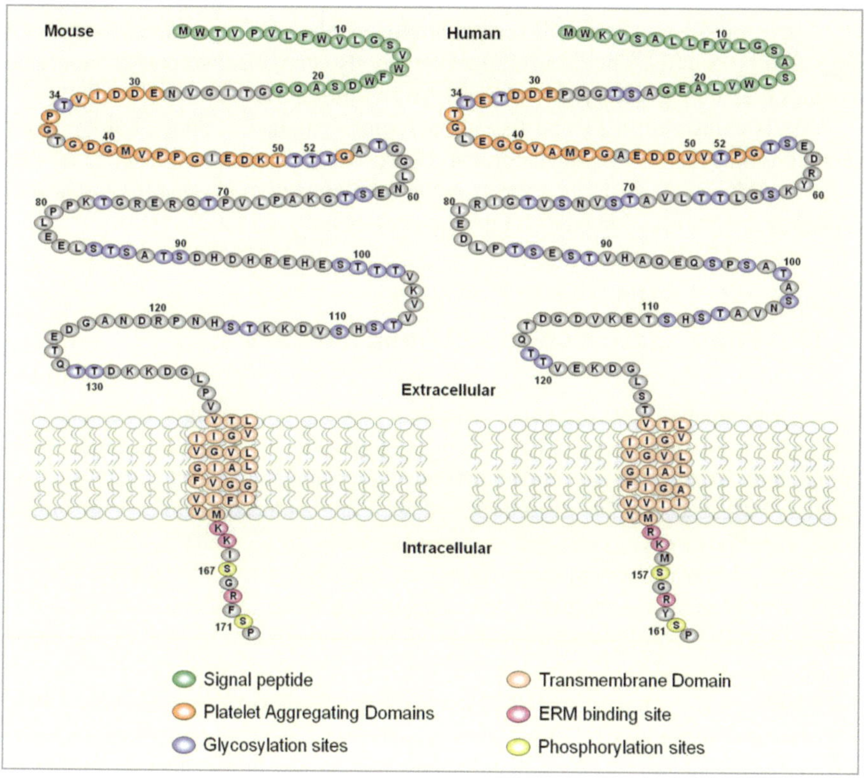


**Fig. 12.9 Podoplanin (Pdpn) topology**. Pdpn is a mucin-type glycoprotein containing a large extracellular domain, single pass transmembrane domain and a short intracellular domain. The extracellular domain contains three platelet aggregation (PLAG) domains (*orange circles*) and several serine and threonine residues that are potential *O*-glycosylation sites (*blue circles*) [193, 194]. The single pass hydrophobic transmembrane domain (*cream circles*) contains conserved GXXXG motif involved in receptor homodimerization [195]. The intracellular domain is only nine amino acid long and includes basic amino acids that act as binding sites for ERM family proteins (*pink circles*) as well as two conserved serine residues that can be phosphorylated to inhibit cell migration (*yellow circles*) [18, 196]. Mouse (172 aa) and human PDPN (162 aa) are shown as indicated

aggregating activity of Pdpn is mediated by three platelet aggregating (PLAG) domain repeats (EDXXVTPG) in the extracellular region [198]. Point mutations in these PLAG domains inhibit platelet aggregation and pulmonary metastasis [199]. Although, the predicted molecular weight of the Pdpn is about 20 kDa, extensive glycosylation increases its molecular weight to 40–55 kDa depending on cell types.

The Pdpn transmembrane domain is hydrophobic and contains GXXXG motifs that promote homodimerization and assembly of PDPN into membrane lipid rafts [195, 200]. Mutation of the transmembrane domain abrogates assembly of PDPN into lipid rafts and prevents subsequent induction of epithelial mesenchymal transition [195]. The transmembrane domain is followed by a short intracellular domain that is rich in basic amino acids. Although this intracellular tail contains only nine amino acids, it is critical for Pdpn function. For example, the Pdpn tail associates with ezrin proteins to promote reorganization of the actin cytoskeleton and cell migration [196, 201]. The intracellular domain also contains two conserved serine residues that are phosphorylated to decrease cell migration [18].

Pdpn is expressed in several tissues including lymphatics, lungs, kidney, heart and nervous system during development. However, its expression is more restricted in adults. Pdpn plays an important role in the development of lymphatic vessels, and is predominantly found in adult lymphatic endothelial cells [202].

## 12.16 Pdpn Expression and Binding Partners

During mammalian development, lymphatic vasculature sprouts from cardinal veins after the cardiovascular system is fully functional. Sprouting lymphatic endothelial cells express Pdpn, which binds to C-type lectin 2 receptor (CLEC2) on platelets to cause platelet aggregation. This Pdpn induced platelet aggregation closes junctions between cardinal veins and developing lymph sacs, thereby separating the lymphatic system from the blood vascular system [203, 204]. Pdpn is needed for lymphatic endothelial cell migration, adhesion, and capillary tube formation, and Pdpn null mice have severe defects in lymphatic system organization and function [202, 204, 205].

In addition to lymphatic endothelial cells, Pdpn is also expressed in type 1 lung alveolar epithelium [206]. Pdpn expression is associated with morphogenesis and differentiation of type I alveolar epithelial cells in the distal lung. Accordingly, Pdpn null mice show reduced and defective alveolar sacs, and die shortly after birth due to respiratory failure [207].

Pdpn expression in kidney podocytes is associated with the formation of interdigitating foot like processes that maintain selective filtration of the renal glomeruli. Pdpn down regulation results in flattening of foot processes and causes proteinuria [208, 209]. Interestingly, decreased Pdpn expression has been reported in nephrosis [208]. Pdpn expression is also found in osteocytes, osteoblasts, mesothelial cells, fibroblastic reticular cells, immune cells, parts of the nervous system such as spinal cord and choroid plexus, various constituents of the eye, and in parts of the heart during development [210, 211].

Pdpn plays a significant role in platelet aggregation, maintaining cell morphology, cell migration, invasion, and cancer progression. Pdpn does not contain any known enzymatic domains. Hence, Pdpn signaling is likely to be mediated by its binding partners. Pdpn binds to the endogenous lectins CLEC2 and galectin-8. Pdpn associates with CLEC2 on platelets to trigger a Src dependent phosphorylation of a cytoplasmic CLEC2 YXXL motif. This is followed by activation of spleen tyrosine kinase (SYK), which initiates a series of downstream signaling events that culminate with platelet aggregation involved in lymphatic development and metastatic dissemination of Pdpn expressing tumor cells [212]. Pdpn-CLEC2 signaling is also involved in immune responses such as dendritic cell migration and maintenance of high endothelial venule integrity during lymphocyte trafficking [213, 214]. Likewise, Pdpn binding to galectin-8 regulates lymphatic endothelial cell adhesion and migration [215]. Pdpn needs to be glycosylated to associate with CLEC-2 and galectin-8 [212, 215].

In addition to lectins, Pdpn also interacts with the cell surface glycoproteins CD9 and CD44. Pdpn interacts with the transmembrane regions 1 and 2 of a four pass tetraspanin membrane receptor CD9 to inhibit platelet aggregation and lung tumor cell metastasis [216]. Thus, Pdpn binding to CD9 elicits an opposing reaction to that of Pdpn and CLEC-2 interactions. Glycosylation dependent Pdpn binding to the standard isoform of CD44 regulates directional cell migration [217]. Coordinate expression of Pdpn and CD44 is correlated with epithelial mesenchymal transition, lamellipodia formation, and maintenance of cell adhesion within cancer cell nests [217, 218].

Other Pdpn ligands include the secreted chemokine SLC/CCL21 and heat shock protein HspA9. Pdpn-SLC/CCL21 complexes are found along the basolateral membrane of lymphatic endothelial cells and in the perivascular stroma, presumably to facilitate directed cell migration of CCR (receptor of SLC/CCL21) positive lymphocytes and dendritic cells during immune responses [219]. Pdpn binding to secreted HspA9 has been identified by mass spectrometry and immunoprecipitation analysis. Pdpn and HspA9 colocalize at the periphery of oral squamous cell carcinoma, suggesting a role for this interaction in tumor invasion [220].

As mentioned above, although the intracellular domain is just nine amino acids long, it interacts with intracellular signaling molecules to control cell migration, invasion, and lymphangiogenesis. The Pdpn intracellular domain contains the basic amino acids arginine and lysine that bind to ERM family proteins including ezrin and moesin. Pdpn associates with these proteins to activate RhoA GTPases, which induce ERM phosphorylation and stabilize the Pdpn-ERM complex [196]. Stable Pdpn-ERM complexes initiate Rho activation to induce reorganization of actin filaments in order to promote cell migration, invasion, and lymphatic capillary morphogenesis [196, 221, 222]. While Pdpn mediated single cell migration is accompanied by epithelial mesenchymal transition [196], collective cell migration appears to be independent of this transition [201].

As mentioned above, the Pdpn intracellular tail contains 2 conserved serine residues. These serine residues are reported to be putative protein kinase A (PKA) and protein kinase C (PKC) phosphorylation sites [223]. Recent studies indicate that Pdpn intracellular serines can be phosphorylated by PKA to inhibit cell migration

[18]. Additional work will be needed to elucidate the dynamics of Pdpn phosphorylation on cell behavior.

## 12.17 Pdpn Transcriptional Regulation

Pdpn expression can be regulated at the transcriptional level by multiple pathways that act downstream of tumor promoters such as TPA, Src, and Ras. The 5′ upstream region of the Pdpn gene lacks consensus TATA or CAAT sequences, but is GC rich and contains putative binding sites for various transcription factors [224]. This region contains four putative binding elements for the homeobox Prox1 transcription factor, which triggers transactivation of the Pdpn gene [225]. Prox1 induces lymphatic endothelial cell Pdpn expression to separate lymphatic and blood vascular systems [226]. In addition to Prox1, the calcineurin–NFATc1 signaling pathway has been implicated in controlling Pdpn expression during late lymphatic development [227].

The Pdpn promoter contains a TPA responsive element which is required for binding of the Fos transcription factor [228]. Accordingly, the Fos/AP-1 transcription factor induces Pdpn expression in skin cancer, glioma, and osteosarcoma cell lines [210]. The 5′ upstream regulatory region of the Pdpn gene also contains putative binding sites for Sp1, AP-4 and NF-1 transcription factors. For example, binding of Sp1 and Sp3 transcription factors to specific elements within the Pdpn promoter region has been shown to induce Pdpn expression in osteosarcoma cell lines [224].

Pdpn expression can be upregulated in response to cytokines and interleukins including TGF-β, IFN-γ, TNF-α, IL-6, IL-22, and IL-1β [229–231]. STAT-1/3 mediates Pdpn expression in keratinocytes induced by IFN-γ, while STAT-3 mediates Pdpn expression by IL-6 and IL-22, and TGF-β1 utilizes SMAD 2/3/4 and IMP-3 [230, 231]. As discussed above, the Src tyrosine kinase utilizes the Cas adaptor protein to induce Pdpn expression in transformed cells, and this effect is reversed during contact normalization [15].

In addition to regulation by tumor promoters and transcription factors, Pdpn expression can also be regulated epigenetically by DNA methylation and microRNAs. The effects of promoter methylation on Pdpn expression appear to be cell type specific. While DNA hypermethylation enhances Pdpn promoter activity in osteosarcoma cell lines, DNA hypermethylation is negatively correlated with PDPN transcript levels in glioma cell lines [224, 232]. The microRNAs miR-29b and miR-125a can directly target the 3′ untranslated region of Pdpn to inhibit its expression. This is evidenced in glioblastoma cell lines that show enhanced Pdpn expression with low levels of miR-29b and miR-125a. Restoration of these microRNAs inhibits proliferation and invasion of glioblastoma cells [233].

Pdpn can also be regulated at post-transcriptional and post-translational levels. Alternative polyadenylation during Pdpn transcription is reported to produce a 2.7 and 0.9 kb mRNA transcripts in cell lines and tissues. Alternative splicing may also produce variations of Pdpn that lack terminal tyrosine and serine residues in the c-terminal end [234]. In addition, PDPN can undergo proteolytic cleavage at various stages. Calpain mediated Pdpn peptide cleavage may regulate its localization and turnover [234]. In addition, sequential Pdpn proteolysis by metalloproteinase and presinilin-1/γ-secretase may release the Pdpn intracellular domain from the membrane to control its signaling abilities [235].

## 12.18 Pdpn in Cancer and Contact Normalization

Pdpn is typically used as a marker for lymphatics and lymphatic invasion in cancers [236]. However, many cancer cells express high levels of Pdpn including mammary carcinomas [201, 237–240], glioma [241], Melanoma [18, 242] and other skin cancers [243, 244], pancreatic carcinoma [245], lung cancer [246, 247], esophageal cancer [248], and oral cancer [249, 250]. In addition, Pdpn expression is associated with increased malignant progression and metastatic ability of these cancers. For example, Pdpn expression is notably increased in over 30 % of pre-malignant oral lesions and in over 60 % of oral cancers. In addition, over 70 % of primary oral cancers with cervical lymph node metastases express elevated levels of PDPN [251–255].

PDPN expression is predominantly found in the invasive front of tumors, consistent with its role in promoting tumor invasion [201, 223, 256, 257]. Pdpn expression is correlated with increased matrix metalloproteinase (MMP) activity that degrades extracellular matrix proteins to facilitate tumor invasion and metastasis [201, 222]. In addition to tumor cells, cancer associated fibroblasts also express Pdpn to promote tumor cell growth, migration, and metastasis [258]. For example, cancer associated fibroblasts express Pdpn to increase melanoma cell migration and survival [18]. Cancer associated fibroblast Pdpn expression also correlates with aggressive potential of melanoma and patient morbidity [242].

Pdpn has emerged as a functionally relevant biomarker and chemotherapeutic target for cancer cells that escape contact normalization in their microenvironment. Antibodies and lectins can be used to target Pdpn to combat tumor cell migration, invasion, and metastasis. As described above, the 8F11 antibody can target PDPN to inhibit platelet aggregation and lung metastasis of colon adenocarcinoma cells [259]. Likewise, the monoclonal NZ1 antibody generated against Pdpn PLAG domains inhibits platelet aggregation and glioma metastasis [260–262]. In addition to inhibiting platelet aggregation, neutralizing antibodies such as NZ1 can also induce antibody dependent cellular toxicity and complement dependent cytotoxicity to combat cancer progression [263, 264]. The anticancer drugs disulfiram and CARP1 functional mimetics can induce serine phosphorylation and ubiquitination of Pdpn to decrease cell migration and invasion of malignant pleural mesothelioma [265, 266]. Furthermore, a C-terminal fragment of the snake toxin aggretin can disrupt Pdpn-CLEC2 interactions to suppress platelet aggregation and pulmonary metastasis of hepatoma cells expressing Pdpn [267].

In addition to cytotoxic drugs and antibodies, lectins can be used to target Pdpn. For example, *Maackia amurensis* seed lectin (MASL) can be used to target Pdpn in order to inhibit tumor cell migration and survival *in vitro* and *in vivo* [171, 268]. This approach is encouraging since lectins can be taken orally to prevent and combat cancers that express Pdpn without notable side effects [171].

## 12.19 Conclusions

Contact normalization is an extremely powerful process that controls cancer cell growth and progression. Significant progress has been made to discover mechanisms that underlie this process. These studies have identified tumor promoters and suppressors that can be used as biomarkers and chemotherapeutic targets as shown in Fig. 12.10. However, much of this process remains to be elucidated. For example, Riken cDNA 1700012B09, Pdpn, Vegfr2, and Tmem163 rise to the top of the list of genes induced in transformed cells and suppressed by contact normalization [181]. However, while Pdpn and Vegfr2 have emerged as a prime cancer biomarkers and targets, 1700012B09Rik and Tmem163 remain relatively unexplored. More complete investigations of mechanisms that underlie contact normalization should illuminate promising avenues to detect and treat cancer.

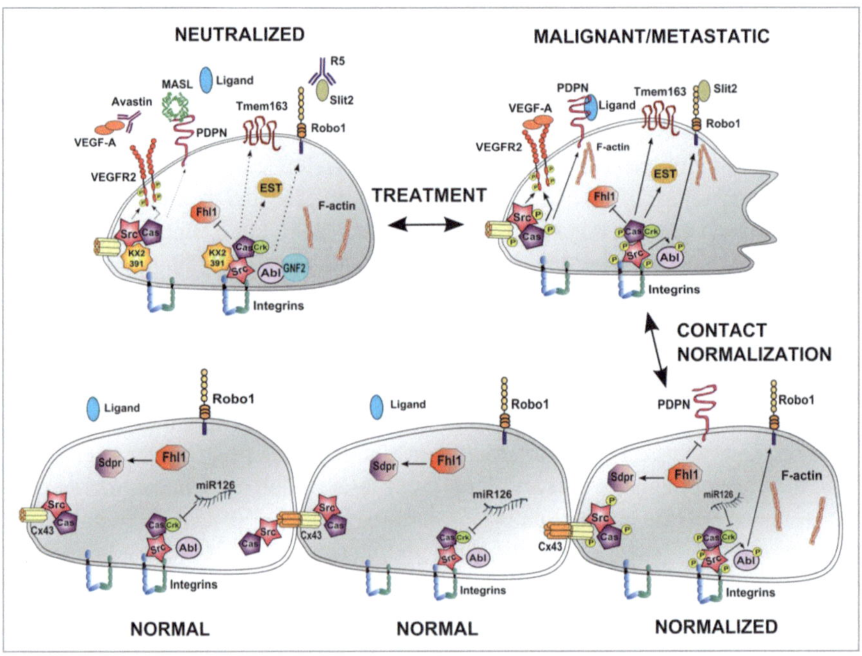


**Fig. 12.10 Contact normalization mechanisms and control of tumor cell migration.** Intercellular junctions enable normal cells to normalize the morphology and phenotype of adjacent tumor cells. Tumor promoters exemplified by Src activate downstream effectors such as Cas, Crk, and Abl that induce the expression of transmembrane receptors such as Pdpn, Vegfr2, Robo1, and Tmem163 to promote escape from contact normalization, migration, and metastasis of tumor cells. These receptors can be targeted to combat cancer progression. For example, the Avastin, MASL, and R5 monoclonal antibody can be used to target the Vegfr2 pathway, Pdpn, and Robo1, respectively. Kinase blockers exemplified by KX2-391 and GNF2 can be used to target oncogenic kinases including Src and Abl (*dotted arrow lines*). Multiple receptors and pathways may be targeted synergistically to more effectively combat cancer caused by tumor cells that escape contact normalization

## References


1. Ferlay J, Soerjomataram I, Ervik M, Dikshit R, Eser S, Mathers C, Rebelo M, Parkin DM, Forman D, Bray F (2013) GLOBOCAN 2012 v1.0, Cancer incidence and mortality worldwide: IARC CancerBase No. 11. http://globocan.iarc.fr
2. Bray F, Guerra Yi M, Parkin DM (2003) The comprehensive cancer monitoring programme in Europe. Eur J Public Health 13(3 Suppl):61–66
3. Klein CA (2011) Framework models of tumor dormancy from patient-derived observations. Curr Opin Genet Dev 21(1):42–49. doi:10.1016/j.gde.2010.10.011
4. Bray F, Ren JS, Masuyer E, Ferlay J (2013) Global estimates of cancer prevalence for 27 sites in the adult population in 2008. Int J Cancer 132(5):1133–1145. doi:10.1002/ijc.27711
5. Keleg S, Buchler P, Ludwig R, Buchler MW, Friess H (2003) Invasion and metastasis in pancreatic cancer. Mol Cancer 2:14
6. Hanahan D, Weinberg RA (2011) Hallmarks of cancer: the next generation. Cell 144(5):646–674. doi:10.1016/j.cell.2011.02.013
7. Aligayer H, Boyd DD, Heiss MM, Abdalla EK, Curley SA, Gallick GE (2002) Activation of Src kinase in primary colorectal carcinoma: an indicator of poor clinical prognosis. Cancer 94(2):344–351. doi:10.1002/cncr.10221
8. Irby RB, Mao W, Coppola D, Kang J, Loubeau JM, Trudeau W, Karl R, Fujita DJ, Jove R, Yeatman TJ (1999) Activating SRC mutation in a subset of advanced human colon cancers. Nat Genet 21(2):187–190. doi:10.1038/5971
9. Irby RB, Yeatman TJ (2000) Role of Src expression and activation in human cancer. Oncogene 19(49):5636–5642
10. Dang CV (2012) MYC on the path to cancer. Cell 149(1):22–35. doi:10.1016/j.cell.2012.03.003
11. Singhi AD, Cimino-Mathews A, Jenkins RB, Lan F, Fink SR, Nassar H, Vang R, Fetting JH, Hicks J, Sukumar S, De Marzo AM, Argani P (2012) MYC gene amplification is often acquired in lethal distant breast cancer metastases of unamplified primary tumors. Mod Pathol 25(3):378–387. doi:10.1038/modpathol.2011.171
12. Vita M, Henriksson M (2006) The Myc oncoprotein as a therapeutic target for human cancer. Semin Cancer Biol 16(4):318–330. doi:10.1016/j.semcancer.2006.07.015
13. Fernandez-Medarde A, Santos E (2011) Ras in cancer and developmental diseases. Genes Cancer 2(3):344–358. doi:10.1177/1947601911411084
14. Prior IA, Lewis PD, Mattos C (2012) A comprehensive survey of Ras mutations in cancer. Cancer Res 72(10):2457–2467. doi:10.1158/0008-5472.CAN-11-2612
15. Shen Y, Chen CS, Ichikawa H, Goldberg GS (2010) SRC induces podoplanin expression to promote cell migration. J Biol Chem 285(13):9649–9656. doi:10.1074/jbc.M109.047696
16. Davies H, Bignell GR, Cox C, Stephens P, Edkins S, Clegg S, Teague J, Woffendin H, Garnett MJ, Bottomley W, Davis N, Dicks E, Ewing R, Floyd Y, Gray K, Hall S, Hawes R, Hughes J, Kosmidou V, Menzies A, Mould C, Parker A, Stevens C, Watt S, Hooper S, Wilson R, Jayatilake H, Gusterson BA, Cooper C, Shipley J, Hargrave D, Pritchard-Jones K, Maitland N, Chenevix-Trench G, Riggins GJ, Bigner DD, Palmieri G, Cossu A, Flanagan A, Nicholson A, Ho JW, Leung SY, Yuen ST, Weber BL, Seigler HF, Darrow TL, Paterson H, Marais R, Marshall CJ, Wooster R, Stratton MR, Futreal PA (2002) Mutations of the BRAF gene in human cancer. Nature 417(6892):949–954. doi:10.1038/nature00766
17. Kimura ET, Nikiforova MN, Zhu Z, Knauf JA, Nikiforov YE, Fagin JA (2003) High prevalence of BRAF mutations in thyroid cancer: genetic evidence for constitutive activation of the RET/PTC-RAS-BRAF signaling pathway in papillary thyroid carcinoma. Cancer Res 63(7):1454–1457
18. Krishnan H, Ochoa-Alvarez JA, Shen Y, Nevel E, Lakshminarayanan M, Williams MC, Ramirez MI, Miller WT, Goldberg GS (2013) Serines in the intracellular tail of podoplanin (PDPN) regulate cell motility. J Biol Chem 288(17):12215–12221. doi:10.1074/jbc.C112.446823

19. Kumar R, Angelini S, Czene K, Sauroja I, Hahka-Kemppinen M, Pyrhonen S, Hemminki K (2003) BRAF mutations in metastatic melanoma: a possible association with clinical outcome. Clin Cancer Res 9(9):3362–3368
20. Maria Pérez-Caro IS-G (2007) BCR-ABL and human cancer. In: Srivastava R (ed) Apoptosis, cell signaling, and human diseases, vol I. Humana Press, pp 3–34. doi:10.1007/978-1-59745-200-7_1
21. Grande E, Bolos MV, Arriola E (2011) Targeting oncogenic ALK: a promising strategy for cancer treatment. Mol Cancer Ther 10(4):569–579. doi:10.1158/1535-7163.MCT-10-0615
22. Altomare DA, Testa JR (2005) Perturbations of the AKT signaling pathway in human cancer. Oncogene 24(50):7455–7464. doi:10.1038/sj.onc.1209085
23. Krishnan H, Miller WT, Goldberg GS (2012) SRC points the way to biomarkers and chemotherapeutic targets. Genes Cancer 3(5–6):426–435. doi:10.1177/1947601912458583
24. Lennartsson J, Ronnstrand L (2012) Stem cell factor receptor/c-Kit: from basic science to clinical implications. Physiol Rev 92(4):1619–1649. doi:10.1152/physrev.00046.2011
25. Karakas B, Bachman KE, Park BH (2006) Mutation of the PIK3CA oncogene in human cancers. Br J Cancer 94(4):455–459. doi:10.1038/sj.bjc.6602970
26. Yuan TL, Cantley LC (2008) PI3K pathway alterations in cancer: variations on a theme. Oncogene 27(41):5497–5510. doi:10.1038/onc.2008.245
27. Anastas JN, Moon RT (2013) WNT signalling pathways as therapeutic targets in cancer. Nat Rev Cancer 13(1):11–26. doi:10.1038/nrc3419
28. Benhaj K, Akcali KC, Ozturk M (2006) Redundant expression of canonical Wnt ligands in human breast cancer cell lines. Oncol Rep 15(3):701–707
29. Park JK, Song JH, He TC, Nam SW, Lee JY, Park WS (2009) Overexpression of Wnt-2 in colorectal cancers. Neoplasma 56(2):119–123
30. Morin PJ (1999) beta-catenin signaling and cancer. Bioessays 21(12):1021–1030, doi:10.1002/(SICI)1521-1878(199912)22:1<1021::AID-BIES6>3.0.CO;2-P
31. Polakis P (2000) Wnt signaling and cancer. Genes Dev 14(15):1837–1851
32. Polakis P (2012) Wnt signaling in cancer. Cold Spring Harb Perspect Biol 4(5). doi:10.1101/cshperspect.a008052
33. Diehl JA (2002) Cycling to cancer with cyclin D1. Cancer Biol Ther 1(3):226–231
34. Musgrove EA, Caldon CE, Barraclough J, Stone A, Sutherland RL (2011) Cyclin D as a therapeutic target in cancer. Nat Rev Cancer 11(8):558–572. doi:10.1038/nrc3090
35. Tai W, Mahato R, Cheng K (2010) The role of HER2 in cancer therapy and targeted drug delivery. J Control Release 146(3):264–275. doi:10.1016/j.jconrel.2010.04.009
36. Iqbal J, Neppalli VT, Wright G, Dave BJ, Horsman DE, Rosenwald A, Lynch J, Hans CP, Weisenburger DD, Greiner TC, Gascoyne RD, Campo E, Ott G, Muller-Hermelink HK, Delabie J, Jaffe ES, Grogan TM, Connors JM, Vose JM, Armitage JO, Staudt LM, Chan WC (2006) BCL2 expression is a prognostic marker for the activated B-cell-like type of diffuse large B-cell lymphoma. J Clin Oncol 24(6):961–968. doi:10.1200/JCO.2005.03.4264
37. Kirkin V, Joos S, Zornig M (2004) The role of Bcl-2 family members in tumorigenesis. Biochim Biophys Acta 1644(2–3):229–249. doi:10.1016/j.bbamcr.2003.08.009
38. Koch U, Radtke F (2010) Notch signaling in solid tumors. Curr Top Dev Biol 92:411–455. doi:10.1016/S0070-2153(10)92013-9
39. Palomero T, Ferrando A (2008) Oncogenic NOTCH1 control of MYC and PI3K: challenges and opportunities for anti-NOTCH1 therapy in T-cell acute lymphoblastic leukemias and lymphomas. Clin Cancer Res 14(17):5314–5317. doi:10.1158/1078-0432.CCR-07-4864
40. Gordon GM, Du W (2011) Targeting Rb inactivation in cancers by synthetic lethality. Am J Cancer Res 1(6):773–786
41. Sharma A, Yeow WS, Ertel A, Coleman I, Clegg N, Thangavel C, Morrissey C, Zhang X, Comstock CE, Witkiewicz AK, Gomella L, Knudsen ES, Nelson PS, Knudsen KE (2010) The retinoblastoma tumor suppressor controls androgen signaling and human prostate cancer progression. J Clin Invest 120(12):4478–4492. doi:10.1172/JCI44239

42. Gonzalez KD, Noltner KA, Buzin CH, Gu D, Wen-Fong CY, Nguyen VQ, Han JH, Lowstuter K, Longmate J, Sommer SS, Weitzel JN (2009) Beyond Li Fraumeni syndrome: clinical characteristics of families with p53 germline mutations. J Clin Oncol 27(8):1250–1256. doi:10.1200/JCO.2008.16.6959
43. Kandoth C, McLellan MD, Vandin F, Ye K, Niu B, Lu C, Xie M, Zhang Q, McMichael JF, Wyczalkowski MA, Leiserson MD, Miller CA, Welch JS, Walter MJ, Wendl MC, Ley TJ, Wilson RK, Raphael BJ, Ding L (2013) Mutational landscape and significance across 12 major cancer types. Nature 502(7471):333–339. doi:10.1038/nature12634
44. Muller PA, Vousden KH (2014) Mutant p53 in cancer: new functions and therapeutic opportunities. Cancer Cell 25(3):304–317. doi:10.1016/j.ccr.2014.01.021
45. Georgescu MM (2010) PTEN tumor suppressor network in PI3K-Akt pathway control. Genes Cancer 1(12):1170–1177. doi:10.1177/1947601911407325
46. Petrucelli N, Daly MB, Feldman GL (1993) BRCA1 and BRCA2 hereditary breast and ovarian cancer. In: Pagon RA, Adam MP, Ardinger HH et al (eds) GeneReviews. University of Washington, Seattle
47. Fearnhead NS, Britton MP, Bodmer WF (2001) The ABC of APC. Hum Mol Genet 10(7):721–733
48. Fodde R (2002) The APC gene in colorectal cancer. Eur J Cancer 38(7):867–871
49. Cremona CA, Behrens A (2014) ATM signalling and cancer. Oncogene 33(26):3351–3360. doi:10.1038/onc.2013.275
50. Banham AH, Beasley N, Campo E, Fernandez PL, Fidler C, Gatter K, Jones M, Mason DY, Prime JE, Trougouboff P, Wood K, Cordell JL (2001) The FOXP1 winged helix transcription factor is a novel candidate tumor suppressor gene on chromosome 3p. Cancer Res 61(24):8820–8829
51. Koon HB, Ippolito GC, Banham AH, Tucker PW (2007) FOXP1: a potential therapeutic target in cancer. Expert Opin Ther Targets 11(7):955–965. doi:10.1517/14728222.11.7.955
52. Krohn A, Seidel A, Burkhardt L, Bachmann F, Mader M, Grupp K, Eichenauer T, Becker A, Adam M, Graefen M, Huland H, Kurtz S, Steurer S, Tsourlakis MC, Minner S, Michl U, Schlomm T, Sauter G, Simon R, Sirma H (2013) Recurrent deletion of 3p13 targets multiple tumour suppressor genes and defines a distinct subgroup of aggressive ERG fusion-positive prostate cancers. J Pathol 231(1):130–141. doi:10.1002/path.4223
53. Li M, Zhao H, Zhang X, Wood LD, Anders RA, Choti MA, Pawlik TM, Daniel HD, Kannangai R, Offerhaus GJ, Velculescu VE, Wang L, Zhou S, Vogelstein B, Hruban RH, Papadopoulos N, Cai J, Torbenson MS, Kinzler KW (2011) Inactivating mutations of the chromatin remodeling gene ARID2 in hepatocellular carcinoma. Nat Genet 43(9):828–829. doi:10.1038/ng.903
54. Wang X, Haswell JR, Roberts CW (2014) Molecular pathways: SWI/SNF (BAF) complexes are frequently mutated in cancer – mechanisms and potential therapeutic insights. Clin Cancer Res 20(1):21–27. doi:10.1158/1078-0432.CCR-13-0280
55. Wiegand KC, Shah SP, Al-Agha OM, Zhao Y, Tse K, Zeng T, Senz J, McConechy MK, Anglesio MS, Kalloger SE, Yang W, Heravi-Moussavi A, Giuliany R, Chow C, Fee J, Zayed A, Prentice L, Melnyk N, Turashvili G, Delaney AD, Madore J, Yip S, McPherson AW, Ha G, Bell L, Fereday S, Tam A, Galletta L, Tonin PN, Provencher D, Miller D, Jones SJ, Moore RA, Morin GB, Oloumi A, Boyd N, Aparicio SA, Shih IM, Mes-Masson AM, Bowtell DD, Hirst M, Gilks B, Marra MA, Huntsman DG (2010) ARID1A mutations in endometriosis-associated ovarian carcinomas. N Engl J Med 363(16):1532–1543. doi:10.1056/NEJMoa1008433
56. Goldstein AM, Tucker MA (2001) Genetic epidemiology of cutaneous melanoma: a global perspective. Arch Dermatol 137(11):1493–1496
57. McWilliams RR, Wieben ED, Rabe KG, Pedersen KS, Wu Y, Sicotte H, Petersen GM (2011) Prevalence of CDKN2A mutations in pancreatic cancer patients: implications for genetic counseling. Eur J Hum Genet 19(4):472–478. doi:10.1038/ejhg.2010.198

58. Kaelin WG Jr (2004) The von Hippel-Lindau tumor suppressor gene and kidney cancer. Clin Cancer Res 10(18 Pt 2):6290S–6295S. doi:10.1158/1078-0432.CCR-sup-040025
59. Kim WY, Kaelin WG (2004) Role of VHL gene mutation in human cancer. J Clin Oncol 22(24):4991–5004. doi:10.1200/JCO.2004.05.061
60. Goodman RH, Smolik S (2000) CBP/p300 in cell growth, transformation, and development. Genes Dev 14(13):1553–1577
61. Iyer NG, Ozdag H, Caldas C (2004) p300/CBP and cancer. Oncogene 23(24):4225–4231. doi:10.1038/sj.onc.1207118
62. Mullighan CG, Zhang J, Kasper LH, Lerach S, Payne-Turner D, Phillips LA, Heatley SL, Holmfeldt L, Collins-Underwood JR, Ma J, Buetow KH, Pui CH, Baker SD, Brindle PK, Downing JR (2011) CREBBP mutations in relapsed acute lymphoblastic leukaemia. Nature 471(7337):235–239. doi:10.1038/nature09727
63. Korsse SE, Peppelenbosch MP, van Veelen W (2013) Targeting LKB1 signaling in cancer. Biochim Biophys Acta 1835(2):194–210. doi:10.1016/j.bbcan.2012.12.006
64. Shackelford DB, Abt E, Gerken L, Vasquez DS, Seki A, Leblanc M, Wei L, Fishbein MC, Czernin J, Mischel PS, Shaw RJ (2013) LKB1 inactivation dictates therapeutic response of non-small cell lung cancer to the metabolism drug phenformin. Cancer Cell 23(2):143–158. doi:10.1016/j.ccr.2012.12.008
65. Holzel M, Huang S, Koster J, Ora I, Lakeman A, Caron H, Nijkamp W, Xie J, Callens T, Asgharzadeh S, Seeger RC, Messiaen L, Versteeg R, Bernards R (2010) NF1 is a tumor suppressor in neuroblastoma that determines retinoic acid response and disease outcome. Cell 142(2):218–229. doi:10.1016/j.cell.2010.06.004
66. Korf BR (2000) Malignancy in neurofibromatosis type 1. Oncologist 5(6):477–485
67. Gutmann DH, Sherman L, Seftor L, Haipek C, Hoang Lu K, Hendrix M (1999) Increased expression of the NF2 tumor suppressor gene product, merlin, impairs cell motility, adhesion and spreading. Hum Mol Genet 8(2):267–275
68. Xiao GH, Gallagher R, Shetler J, Skele K, Altomare DA, Pestell RG, Jhanwar S, Testa JR (2005) The NF2 tumor suppressor gene product, merlin, inhibits cell proliferation and cell cycle progression by repressing cyclin D1 expression. Mol Cell Biol 25(6):2384–2394. doi:10.1128/MCB.25.6.2384-2394.2005
69. Lefort K, Mandinova A, Ostano P, Kolev V, Calpini V, Kolfschoten I, Devgan V, Lieb J, Raffoul W, Hohl D, Neel V, Garlick J, Chiorino G, Dotto GP (2007) Notch1 is a p53 target gene involved in human keratinocyte tumor suppression through negative regulation of ROCK1/2 and MRCKalpha kinases. Genes Dev 21(5):562–577. doi:10.1101/gad.1484707
70. Lobry C, Oh P, Aifantis I (2011) Oncogenic and tumor suppressor functions of Notch in cancer: it's NOTCH what you think. J Exp Med 208(10):1931–1935. doi:10.1084/jem.20111855
71. Wade M, Wahl GM (2006) c-Myc, genome instability, and tumorigenesis: the devil is in the details. Curr Top Microbiol Immunol 302:169–203
72. Kwon MJ, Shin YK (2011) Epigenetic regulation of cancer-associated genes in ovarian cancer. Int J Mol Sci 12(2):983–1008. doi:10.3390/ijms12020983
73. Han NM, Curley SA, Gallick GE (1996) Differential activation of pp 60(c-src) and pp62(c-yes) in human colorectal carcinoma liver metastases. Clin Cancer Res 2(8):1397–1404
74. Suzuki K, Oneyama C, Kimura H, Tajima S, Okada M (2011) Down-regulation of the tumor suppressor C-terminal Src kinase (Csk)-binding protein (Cbp)/PAG1 is mediated by epigenetic histone modifications via the mitogen-activated protein kinase (MAPK)/phosphatidylinositol 3-kinase (PI3K) pathway. J Biol Chem 286(18):15698–15706. doi:10.1074/jbc.M110.195362
75. Herschkowitz JI, He X, Fan C, Perou CM (2008) The functional loss of the retinoblastoma tumour suppressor is a common event in basal-like and luminal B breast carcinomas. Breast Cancer Res 10(5):R75. doi:10.1186/bcr2142

76. Olivier M, Hollstein M, Hainaut P (2010) TP53 mutations in human cancers: origins, consequences, and clinical use. Cold Spring Harb Perspect Biol 2(1):a001008. doi:10.1101/cshperspect.a001008
77. Husemann Y, Geigl JB, Schubert F, Musiani P, Meyer M, Burghart E, Forni G, Eils R, Fehm T, Riethmuller G, Klein CA (2008) Systemic spread is an early step in breast cancer. Cancer Cell 13(1):58–68. doi:10.1016/j.ccr.2007.12.003
78. Schardt JA, Meyer M, Hartmann CH, Schubert F, Schmidt-Kittler O, Fuhrmann C, Polzer B, Petronio M, Eils R, Klein CA (2005) Genomic analysis of single cytokeratin-positive cells from bone marrow reveals early mutational events in breast cancer. Cancer Cell 8(3):227–239. doi:10.1016/j.ccr.2005.08.003
79. Fidler IJ (2003) The pathogenesis of cancer metastasis: the 'seed and soil' hypothesis revisited. Nat Rev Cancer 3(6):453–458. doi:10.1038/nrc1098
80. Giugliano FM, Alberti D, Guida G, Palma GD, Iadanza L, Mormile M, Cammarota F, Montanino A, Fulciniti F, Ravo V, Muto P (2013) Non small-cell lung cancer with metastasis to thigh muscle and mandible: two case reports. J Med Case Rep 7(1):98. doi:10.1186/1752-1947-7-98
81. Lee YT, Geer DA (1987) Primary liver cancer: pattern of metastasis. J Surg Oncol 36(1):26–31
82. Leong SP, Cady B, Jablons DM, Garcia-Aguilar J, Reintgen D, Jakub J, Pendas S, Duhaime L, Cassell R, Gardner M, Giuliano R, Archie V, Calvin D, Mensha L, Shivers S, Cox C, Werner JA, Kitagawa Y, Kitajima M (2006) Clinical patterns of metastasis. Cancer Metastasis Rev 25(2):221–232. doi:10.1007/s10555-006-8502-8
83. Berman AT, Thukral AD, Hwang WT, Solin LJ, Vapiwala N (2013) Incidence and patterns of distant metastases for patients with early-stage breast cancer after breast conservation treatment. Clin Breast Cancer 13(2):88–94. doi:10.1016/j.clbc.2012.11.001
84. Betka J (2001) Distant metastases from lip and oral cavity cancer. ORL J Otorhinolaryngol Relat Spec 63(4):217–221. doi:10.1159/000055744
85. Noguti J, De Moura CF, De Jesus GP, Da Silva VH, Hossaka TA, Oshima CT, Ribeiro DA (2012) Metastasis from oral cancer: an overview. Cancer Genomics Proteomics 9(5):329–335
86. Colombo N, Preti E, Landoni F, Carinelli S, Colombo A, Marini C, Sessa C (2013) Endometrial cancer: ESMO Clinical Practice Guidelines for diagnosis, treatment and follow-up. Ann Oncol 24(Suppl 6):vi33–vi38. doi:10.1093/annonc/mdt353
87. Yachida S, Iacobuzio-Donahue CA (2009) The pathology and genetics of metastatic pancreatic cancer. Arch Pathol Lab Med 133(3):413–422. doi:10.1043/1543-2165-133.3.413
88. Bubendorf L, Schopfer A, Wagner U, Sauter G, Moch H, Willi N, Gasser TC, Mihatsch MJ (2000) Metastatic patterns of prostate cancer: an autopsy study of 1,589 patients. Hum Pathol 31(5):578–583
89. Viadana E, Bross ID, Pickren JW (1978) An autopsy study of the metastatic patterns of human leukemias. Oncology (Williston Park) 35(2):87–96
90. Shinagare AB, Ramaiya NH, Jagannathan JP, Fennessy FM, Taplin ME, Van den Abbeele AD (2011) Metastatic pattern of bladder cancer: correlation with the characteristics of the primary tumor. AJR Am J Roentgenol 196(1):117–122. doi:10.2214/AJR.10.5036
91. Lengyel E (2010) Ovarian cancer development and metastasis. Am J Pathol 177(3):1053–1064. doi:10.2353/ajpath.2010.100105
92. Ramadan S, Ugas MA, Berwick RJ, Notay M, Cho H, Jerjes W, Giannoudis PV (2012) Spinal metastasis in thyroid cancer. Head Neck Oncol 4:39. doi:10.1186/1758-3284-4-39
93. Bhatia S, Tykodi SS, Thompson JA (2009) Treatment of metastatic melanoma: an overview. Oncology (Williston Park) 23(6):488–496
94. Thompson Coon J, Hoyle M, Green C, Liu Z, Welch K, Moxham T, Stein K (2010) Bevacizumab, sorafenib tosylate, sunitinib and temsirolimus for renal cell carcinoma: a systematic review and economic evaluation. Health Technol Assess 14(2):1–184, iii-iv. doi:10.3310/hta14020

95. Brooks SA, Lomax-Browne HJ, Carter TM, Kinch CE, Hall DM (2010) Molecular interactions in cancer cell metastasis. Acta Histochem 112(1):3–25. doi:10.1016/j.acthis.2008.11.022
96. Bockhorn M, Jain RK, Munn LL (2007) Active versus passive mechanisms in metastasis: do cancer cells crawl into vessels, or are they pushed? Lancet Oncol 8(5):444–448. doi:10.1016/S1470-2045(07)70140-7
97. Takagi S, Sato S, Oh-hara T, Takami M, Koike S, Mishima Y, Hatake K, Fujita N (2013) Platelets promote tumor growth and metastasis via direct interaction between Aggrus/podoplanin and CLEC-2. PLoS One 8(8), e73609. doi:10.1371/journal.pone.0073609
98. Gout S, Tremblay PL, Huot J (2008) Selectins and selectin ligands in extravasation of cancer cells and organ selectivity of metastasis. Clin Exp Metastasis 25(4):335–344. doi:10.1007/s10585-007-9096-4
99. Krishnan H, Miller WT, Goldberg GS (2012) SRC points the way to biomarkers and chemotherapeutic targets. Genes Cancer 3(5–6):426–435
100. Jhon Alberto Ochoa-Alvarez CG, Krishnan H, Xiaoxuan Wu, Goldberg GS (2011) Contact normalization: mechanisms and pathways to biomarkers and chemotherapeutic targets. In: Extracellular and intracellular signaling. RSC Publishing, Cambridge. doi:10.1039/9781849733434-00105
101. Rubin H (2006) What keeps cells in tissues behaving normally in the face of myriad mutations? Bioessays 28(5):515–524. doi:10.1002/bies.20403
102. Rubin H (2008) Contact interactions between cells that suppress neoplastic development: can they also explain metastatic dormancy? Adv Cancer Res 100:159–202. doi:10.1016/S0065-230X(08)00006-7
103. Albertsen PC (2007) Commentary: occult prostate cancer – imposter or the real deal? Int J Epidemiol 36(2):281–282. doi:10.1093/ije/dym051
104. Nielsen M, Thomsen JL, Primdahl S, Dyreborg U, Andersen JA (1987) Breast cancer and atypia among young and middle-aged women: a study of 110 medicolegal autopsies. Br J Cancer 56(6):814–819
105. Klein G (2012) Tumor resistance. Oncoimmunology 1(8):1355–1359. doi:10.4161/onci.22194
106. Aguirre-Ghiso JA (2007) Models, mechanisms and clinical evidence for cancer dormancy. Nat Rev Cancer 7(11):834–846. doi:10.1038/nrc2256
107. Barkan D, Chambers AF (2011) beta1-integrin: a potential therapeutic target in the battle against cancer recurrence. Clin Cancer Res 17(23):7219–7223. doi:10.1158/1078-0432.CCR-11-0642
108. Aguirre-Ghiso JA, Liu D, Mignatti A, Kovalski K, Ossowski L (2001) Urokinase receptor and fibronectin regulate the ERK(MAPK) to p38(MAPK) activity ratios that determine carcinoma cell proliferation or dormancy in vivo. Mol Biol Cell 12(4):863–879
109. Allgayer H, Aguirre-Ghiso JA (2008) The urokinase receptor (u-PAR)--a link between tumor cell dormancy and minimal residual disease in bone marrow? APMIS 116(7–8):602–614. doi:10.1111/j.1600-0463.2008.00997.x
110. Giancotti FG (2013) Mechanisms governing metastatic dormancy and reactivation. Cell 155(4):750–764. doi:10.1016/j.cell.2013.10.029
111. Naumov GN, Bender E, Zurakowski D, Kang SY, Sampson D, Flynn E, Watnick RS, Straume O, Akslen LA, Folkman J, Almog N (2006) A model of human tumor dormancy: an angiogenic switch from the nonangiogenic phenotype. J Natl Cancer Inst 98(5):316–325. doi:10.1093/jnci/djj068
112. Paez D, Labonte MJ, Bohanes P, Zhang W, Benhanim L, Ning Y, Wakatsuki T, Loupakis F, Lenz HJ (2012) Cancer dormancy: a model of early dissemination and late cancer recurrence. Clin Cancer Res 18(3):645–653. doi:10.1158/1078-0432.CCR-11-2186
113. Swann JB, Smyth MJ (2007) Immune surveillance of tumors. J Clin Invest 117(5):1137–1146. doi:10.1172/JCI31405
114. Kleffel S, Schatton T (2013) Tumor dormancy and cancer stem cells: two sides of the same coin? Adv Exp Med Biol 734:145–179. doi:10.1007/978-1-4614-1445-2_8

115. Horak CE, Lee JH, Marshall JC, Shreeve SM, Steeg PS (2008) The role of metastasis suppressor genes in metastatic dormancy. APMIS 116(7–8):586–601. doi:10.1111/j.1600-0463.2008.01213.x
116. Nash KT, Phadke PA, Navenot JM, Hurst DR, Accavitti-Loper MA, Sztul E, Vaidya KS, Frost AR, Kappes JC, Peiper SC, Welch DR (2007) Requirement of KISS1 secretion for multiple organ metastasis suppression and maintenance of tumor dormancy. J Natl Cancer Inst 99(4):309–321. doi:10.1093/jnci/djk053
117. Schewe DM, Aguirre-Ghiso JA (2009) Inhibition of eIF2alpha dephosphorylation maximizes bortezomib efficiency and eliminates quiescent multiple myeloma cells surviving proteasome inhibitor therapy. Cancer Res 69(4):1545–1552. doi:10.1158/0008-5472.CAN-08-3858
118. Hogan C (2012) Impact of interactions between normal and transformed epithelial cells and the relevance to cancer. Cell Mol Life Sci 69(2):203–213. doi:10.1007/s00018-011-0806-3
119. Alt-Holland A, Zhang W, Margulis A, Garlick JA (2005) Microenvironmental control of premalignant disease: the role of intercellular adhesion in the progression of squamous cell carcinoma. Semin Cancer Biol 15(2):84–96. doi:10.1016/j.semcancer.2004.08.007
120. Bissell MJ, Hines WC (2011) Why don't we get more cancer? A proposed role of the microenvironment in restraining cancer progression. Nat Med 17(3):320–329. doi:10.1038/nm.2328
121. Soto AM, Sonnenschein C (2011) The tissue organization field theory of cancer: a testable replacement for the somatic mutation theory. Bioessays 33(5):332–340. doi:10.1002/bies.201100025
122. Hennings H, Robinson VA, Michael DM, Pettit GR, Jung R, Yuspa SH (1990) Development of an in vitro analogue of initiated mouse epidermis to study tumor promoters and antipromoters. Cancer Res 50(15):4794–4800
123. Mehta PP, Bertram JS, Loewenstein WR (1986) Growth inhibition of transformed cells correlates with their junctional communication with normal cells. Cell 44(1):187–196
124. Booth BW, Boulanger CA, Anderson LH, Smith GH (2011) The normal mammary microenvironment suppresses the tumorigenic phenotype of mouse mammary tumor virus-neu-transformed mammary tumor cells. Oncogene 30(6):679–689. doi:10.1038/onc.2010.439
125. Stoker MG (1967) Transfer of growth inhibition between normal and virus-transformed cells: autoradiographic studies using marked cells. J Cell Sci 2(3):293–304
126. Alexander DB, Ichikawa H, Bechberger JF, Valiunas V, Ohki M, Naus CC, Kunimoto T, Tsuda H, Miller WT, Goldberg GS (2004) Normal cells control the growth of neighboring transformed cells independent of gap junctional communication and SRC activity. Cancer Res 64(4):1347–1358
127. Martin W, Zempel G, Hulser D, Willecke K (1991) Growth inhibition of oncogene-transformed rat fibroblasts by cocultured normal cells: relevance of metabolic cooperation mediated by gap junctions. Cancer Res 51(19):5348–5351
128. Rubin H (1960) The suppression of morphological alterations in cells infected with Rous sarcoma virus. Virology 12:14–31
129. Stoker M (1964) Regulation of growth and orientation in hamster cells transformed by polyoma virus. Virology 24:165–174
130. Stoker MG, Shearer M, O'Neill C (1966) Growth inhibition of polyoma-transformed cells by contact with static normal fibroblasts. J Cell Sci 1(3):297–310
131. Hogan C, Kajita M, Lawrenson K, Fujita Y (2011) Interactions between normal and transformed epithelial cells: their contributions to tumourigenesis. Int J Biochem Cell Biol 43(4):496–503. doi:10.1016/j.biocel.2010.12.019
132. Moreno E (2008) Is cell competition relevant to cancer? Nat Rev Cancer 8(2):141–147. doi:10.1038/nrc2252
133. Hogan C, Dupre-Crochet S, Norman M, Kajita M, Zimmermann C, Pelling AE, Piddini E, Baena-Lopez LA, Vincent JP, Itoh Y, Hosoya H, Pichaud F, Fujita Y (2009) Characterization of the interface between normal and transformed epithelial cells. Nat Cell Biol 11(4):460–467. doi:10.1038/ncb1853

134. Kajita M, Hogan C, Harris AR, Dupre-Crochet S, Itasaki N, Kawakami K, Charras G, Tada M, Fujita Y (2010) Interaction with surrounding normal epithelial cells influences signalling pathways and behaviour of Src-transformed cells. J Cell Sci 123(Pt 2):171–180. doi:10.1242/jcs.057976
135. Vidal M, Larson DE, Cagan RL (2006) Csk-deficient boundary cells are eliminated from normal Drosophila epithelia by exclusion, migration, and apoptosis. Dev Cell 10(1):33–44. doi:10.1016/j.devcel.2005.11.007
136. Brinster RL (1974) The effect of cells transferred into the mouse blastocyst on subsequent development. J Exp Med 140(4):1049–1056
137. Mintz B, Illmensee K (1975) Normal genetically mosaic mice produced from malignant teratocarcinoma cells. Proc Natl Acad Sci U S A 72(9):3585–3589
138. Gumbiner BM (2005) Regulation of cadherin-mediated adhesion in morphogenesis. Nat Rev Mol Cell Biol 6(8):622–634. doi:10.1038/nrm1699
139. Polyak K, Kato JY, Solomon MJ, Sherr CJ, Massague J, Roberts JM, Koff A (1994) p27Kip1, a cyclin-Cdk inhibitor, links transforming growth factor-beta and contact inhibition to cell cycle arrest. Genes Dev 8(1):9–22
140. Grazia Lampugnani M, Zanetti A, Corada M, Takahashi T, Balconi G, Breviario F, Orsenigo F, Cattelino A, Kemler R, Daniel TO, Dejana E (2003) Contact inhibition of VEGF-induced proliferation requires vascular endothelial cadherin, beta-catenin, and the phosphatase DEP-1/CD148. J Cell Biol 161(4):793–804. doi:10.1083/jcb.200209019
141. Jeanes A, Gottardi CJ, Yap AS (2008) Cadherins and cancer: how does cadherin dysfunction promote tumor progression? Oncogene 27(55):6920–6929. doi:10.1038/onc.2008.343
142. Dejana E, Orsenigo F, Lampugnani MG (2008) The role of adherens junctions and VE-cadherin in the control of vascular permeability. J Cell Sci 121(Pt 13):2115–2122. doi:10.1242/jcs.017897
143. Koch S, Tugues S, Li X, Gualandi L, Claesson-Welsh L (2011) Signal transduction by vascular endothelial growth factor receptors. Biochem J 437(2):169–183. doi:10.1042/BJ20110301
144. Simoneau B, Houle F, Huot J (2012) Regulation of endothelial permeability and transendothelial migration of cancer cells by tropomyosin-1 phosphorylation. Vasc Cell 4(1):18. doi:10.1186/2045-824X-4-18
145. Goodenough DA, Goliger JA, Paul DL (1996) Connexins, connexons, and intercellular communication. Annu Rev Biochem 65:475–502
146. Hirschi KK, Xu CE, Tsukamoto T, Sager R (1996) Gap junction genes Cx26 and Cx43 individually suppress the cancer phenotype of human mammary carcinoma cells and restore differentiation potential. Cell Growth Differ 7(7):861–870
147. Huang RP, Fan Y, Hossain MZ, Peng A, Zeng ZL, Boynton AL (1998) Reversion of the neoplastic phenotype of human glioblastoma cells by connexin 43 (cx43). Cancer Res 58(22):5089–5096
148. Goldberg GS, Martyn KD, Lau AF (1994) A connexin 43 antisense vector reduces the ability of normal cells to inhibit the foci formation of transformed cells. Mol Carcinog 11(2):106–114
149. Zhu D, Kidder GM, Caveney S, Naus CC (1992) Growth retardation in glioma cells cocultured with cells overexpressing a gap junction protein. Proc Natl Acad Sci U S A 89(21):10218–10221
150. Pahujaa M, Anikin M, Goldberg GS (2007) Phosphorylation of connexin43 induced by Src: regulation of gap junctional communication between transformed cells. Exp Cell Res 313(20):4083–4090. doi:10.1016/j.yexcr.2007.09.010
151. Shen Y, Khusial PR, Li X, Ichikawa H, Moreno AP, Goldberg GS (2007) SRC utilizes Cas to block gap junctional communication mediated by connexin43. J Biol Chem 282(26):18914–18921. doi:10.1074/jbc.M608980200
152. Zhou L, Kasperek EM, Nicholson BJ (1999) Dissection of the molecular basis of pp60(v-src) induced gating of connexin 43 gap junction channels. J Cell Biol 144(5):1033–1045

153. Kamei J, Toyofuku T, Hori M (2003) Negative regulation of p21 by beta-catenin/TCF signaling: a novel mechanism by which cell adhesion molecules regulate cell proliferation. Biochem Biophys Res Commun 312(2):380–387
154. Aleshin A, Finn RS (2010) SRC: a century of science brought to the clinic. Neoplasia 12(8):599–607
155. Goldberg GS, Moreno AP, Lampe PD (2002) Gap junctions between cells expressing connexin 43 or 32 show inverse permselectivity to adenosine and ATP. J Biol Chem 277(39):36725–36730. doi:10.1074/jbc.M109797200
156. Ding L, Niu C, Zheng Y, Xiong Z, Liu Y, Lin J, Sun H, Huang K, Yang W, Li X, Ye Q (2011) FHL1 interacts with oestrogen receptors and regulates breast cancer cell growth. J Cell Mol Med 15(1):72–85. doi:10.1111/j.1582-4934.2009.00938.x
157. Li X, Jia Z, Shen Y, Ichikawa H, Jarvik J, Nagele RG, Goldberg GS (2008) Coordinate suppression of Sdpr and Fhl1 expression in tumors of the breast, kidney, and prostate. Cancer Sci 99(7):1326–1333. doi:10.1111/j.1349-7006.2008.00816.x
158. Niu C, Liang C, Guo J, Cheng L, Zhang H, Qin X, Zhang Q, Ding L, Yuan B, Xu X, Li J, Lin J, Ye Q (2012) Downregulation and growth inhibitory role of FHL1 in lung cancer. Int J Cancer 130(11):2549–2556. doi:10.1002/ijc.26259
159. Bai L, Deng X, Li Q, Wang M, An W, Deli A, Gao Z, Xie Y, Dai Y, Cong YS (2012) Downregulation of the cavin family proteins in breast cancer. J Cell Biochem 113(1):322–328. doi:10.1002/jcb.23358
160. Shioi K, Komiya A, Hattori K, Huang Y, Sano F, Murakami T, Nakaigawa N, Kishida T, Kubota Y, Nagashima Y, Yao M (2006) Vascular cell adhesion molecule 1 predicts cancer-free survival in clear cell renal carcinoma patients. Clin Cancer Res 12(24):7339–7346. doi:10.1158/1078-0432.CCR-06-1737
161. Gustavsson E, Sernbo S, Andersson E, Brennan DJ, Dictor M, Jerkeman M, Borrebaeck CA, Ek S (2010) SOX11 expression correlates to promoter methylation and regulates tumor growth in hematopoietic malignancies. Mol Cancer 9:187. doi:10.1186/1476-4598-9-187
162. Sernbo S, Gustavsson E, Brennan DJ, Gallagher WM, Rexhepaj E, Rydnert F, Jirstrom K, Borrebaeck CA, Ek S (2011) The tumour suppressor SOX11 is associated with improved survival among high grade epithelial ovarian cancers and is regulated by reversible promoter methylation. BMC Cancer 11:405. doi:10.1186/1471-2407-11-405
163. Liang H, Chen Q, Coles AH, Anderson SJ, Pihan G, Bradley A, Gerstein R, Jurecic R, Jones SN (2003) Wnt5a inhibits B cell proliferation and functions as a tumor suppressor in hematopoietic tissue. Cancer Cell 4(5):349–360
164. Ying J, Li H, Yu J, Ng KM, Poon FF, Wong SC, Chan AT, Sung JJ, Tao Q (2008) WNT5A exhibits tumor-suppressive activity through antagonizing the Wnt/beta-catenin signaling, and is frequently methylated in colorectal cancer. Clin Cancer Res 14(1):55–61. doi:10.1158/1078-0432.CCR-07-1644
165. Chatterjee S, Heukamp LC, Siobal M, Schottle J, Wieczorek C, Peifer M, Frasca D, Koker M, Konig K, Meder L, Rauh D, Buettner R, Wolf J, Brekken RA, Neumaier B, Christofori G, Thomas RK, Ullrich RT (2013) Tumor VEGF:VEGFR2 autocrine feed-forward loop triggers angiogenesis in lung cancer. J Clin Invest 123(4):1732–1740. doi:10.1172/JCI65385
166. Liang Y, Brekken RA, Hyder SM (2006) Vascular endothelial growth factor induces proliferation of breast cancer cells and inhibits the anti-proliferative activity of anti-hormones. Endocr Relat Cancer 13(3):905–919. doi:10.1677/erc.1.01221
167. Smith NR, Baker D, James NH, Ratcliffe K, Jenkins M, Ashton SE, Sproat G, Swann R, Gray N, Ryan A, Jurgensmeier JM, Womack C (2010) Vascular endothelial growth factor receptors VEGFR-2 and VEGFR-3 are localized primarily to the vasculature in human primary solid cancers. Clin Cancer Res 16(14):3548–3561. doi:10.1158/1078-0432.CCR-09-2797
168. Tanno S, Ohsaki Y, Nakanishi K, Toyoshima E, Kikuchi K (2004) Human small cell lung cancer cells express functional VEGF receptors, VEGFR-2 and VEGFR-3. Lung Cancer 46(1):11–19. doi:10.1016/j.lungcan.2004.03.006

169. Huttenlocher S, Seibold ND, Gebhard MP, Noack F, Thorns C, Hasselbacher K, Wollenberg B, Schild SE, Rades D (2014) Evaluation of the prognostic role of tumor cell podoplanin expression in locally advanced squamous cell carcinoma of the head and neck. Strahlenther Onkol. doi:10.1007/s00066-014-0694-1
170. Kimura N, Kimura I (2005) Podoplanin as a marker for mesothelioma. Pathol Int 55(2):83–86. doi:10.1111/j.1440-1827.2005.01791.x
171. Ochoa-Alvarez JA, Krishnan H, Shen Y, Acharya NK, Han M, McNulty DE, Hasegawa H, Hyodo T, Senga T, Geng JG, Kosciuk M, Shin SS, Goydos JS, Temiakov D, Nagele RG, Goldberg GS (2012) Plant lectin can target receptors containing sialic acid, exemplified by podoplanin, to inhibit transformed cell growth and migration. PLoS One 7(7), e41845. doi:10.1371/journal.pone.0041845
172. Raica M, Cimpean AM, Ribatti D (2008) The role of podoplanin in tumor progression and metastasis. Anticancer Res 28(5B):2997–3006
173. Shibahara J, Kashima T, Kikuchi Y, Kunita A, Fukayama M (2006) Podoplanin is expressed in subsets of tumors of the central nervous system. Virchows Arch 448(4):493–499. doi:10.1007/s00428-005-0133-x
174. Brune V, Tiacci E, Pfeil I, Doring C, Eckerle S, van Noesel CJ, Klapper W, Falini B, von Heydebreck A, Metzler D, Brauninger A, Hansmann ML, Kuppers R (2008) Origin and pathogenesis of nodular lymphocyte-predominant Hodgkin lymphoma as revealed by global gene expression analysis. J Exp Med 205(10):2251–2268. doi:10.1084/jem.20080809
175. Li AM, Tian AX, Zhang RX, Ge J, Sun X, Cao XC (2013) Protocadherin-7 induces bone metastasis of breast cancer. Biochem Biophys Res Commun 436(3):486–490. doi:10.1016/j.bbrc.2013.05.131
176. Shen Y, Jia Z, Nagele RG, Ichikawa H, Goldberg GS (2006) SRC uses Cas to suppress Fhl1 in order to promote nonanchored growth and migration of tumor cells. Cancer Res 66(3):1543–1552
177. Ding L, Wang Z, Yan J, Yang X, Liu A, Qiu W, Zhu J, Han J, Zhang H, Lin J, Cheng L, Qin X, Niu C, Yuan B, Wang X, Zhu C, Zhou Y, Li J, Song H, Huang C, Ye Q (2009) Human four-and-a-half LIM family members suppress tumor cell growth through a TGF-beta-like signaling pathway. J Clin Invest 119(2):349–361. doi:10.1172/JCI35930
178. Lin J, Ding L, Jin R, Zhang H, Cheng L, Qin X, Chai J, Ye Q (2009) Four and a half LIM domains 1 (FHL1) and receptor interacting protein of 140 kDa (RIP140) interact and cooperate in estrogen signaling. Int J Biochem Cell Biol 41(7):1613–1618. doi:10.1016/j.biocel.2009.02.007
179. Sakashita K, Mimori K, Tanaka F, Kamohara Y, Inoue H, Sawada T, Hirakawa K, Mori M (2008) Clinical significance of loss of Fhl1 expression in human gastric cancer. Ann Surg Oncol 15(8):2293–2300. doi:10.1245/s10434-008-9904-3
180. Li X, Jia Z, Shen Y, Ichikawa H, Jarvik J, Nagele RG, Goldberg GS (2008) Coordinate suppression of Sdpr and Fhl1 expression in tumors of the breast, kidney, and prostate. Cancer Sci 99(7):1326–1333
181. Shen Y, Chen CS, Ichikawa H, Goldberg GS (2010) SRC induces podoplanin expression to promote cell migration. J Biol Chem 285(13):9649–9656
182. Burgener R, Wolf M, Ganz T, Baggiolini M (1990) Purification and characterization of a major phosphatidylserine-binding phosphoprotein from human platelets. Biochem J 269(3):729–734
183. Hansen CG, Bright NA, Howard G, Nichols BJ (2009) SDPR induces membrane curvature and functions in the formation of caveolae. Nat Cell Biol 11(7):807–814. doi:10.1038/ncb1887
184. Gustincich S, Schneider C (1993) Serum deprivation response gene is induced by serum starvation but not by contact inhibition. Cell Growth Differ 4(9):753–760
185. Holmes K, Roberts OL, Thomas AM, Cross MJ (2007) Vascular endothelial growth factor receptor-2: structure, function, intracellular signalling and therapeutic inhibition. Cell Signal 19(10):2003–2012. doi:10.1016/j.cellsig.2007.05.013

186. Takahashi T, Shibuya M (1997) The 230 kDa mature form of KDR/Flk-1 (VEGF receptor-2) activates the PLC-gamma pathway and partially induces mitotic signals in NIH3T3 fibroblasts. Oncogene 14(17):2079–2089. doi:10.1038/sj.onc.1201047
187. Hoeben A, Landuyt B, Highley MS, Wildiers H, Van Oosterom AT, De Bruijn EA (2004) Vascular endothelial growth factor and angiogenesis. Pharmacol Rev 56(4):549–580. doi:10.1124/pr.56.4.3
188. Olsson AK, Dimberg A, Kreuger J, Claesson-Welsh L (2006) VEGF receptor signalling – in control of vascular function. Nat Rev Mol Cell Biol 7(5):359–371. doi:10.1038/nrm1911
189. Yao X, Ping Y, Liu Y, Chen K, Yoshimura T, Liu M, Gong W, Chen C, Niu Q, Guo D, Zhang X, Wang JM, Bian X (2013) Vascular endothelial growth factor receptor 2 (VEGFR-2) plays a key role in vasculogenic mimicry formation, neovascularization and tumor initiation by Glioma stem-like cells. PLoS One 8(3), e57188. doi:10.1371/journal.pone.0057188
190. Watanabe M, Okochi E, Sugimoto Y, Tsuruo T (1988) Identification of a platelet-aggregating factor of murine colon adenocarcinoma 26: Mr 44,000 membrane protein as determined by monoclonal antibodies. Cancer Res 48(22):6411–6416
191. Watanabe M, Sugimoto Y, Tsuruo T (1990) Expression of a Mr 41,000 glycoprotein associated with thrombin-independent platelet aggregation in high metastatic variants of murine B16 melanoma. Cancer Res 50(20):6657–6662
192. Nose K, Saito H, Kuroki T (1990) Isolation of a gene sequence induced later by tumor-promoting 12-O-tetradecanoylphorbol-13-acetate in mouse osteoblastic cells (MC3T3-E1) and expressed constitutively in ras-transformed cells. Cell Growth Differ 1(11):511–518
193. Kaneko MK, Kato Y, Kitano T, Osawa M (2006) Conservation of a platelet activating domain of Aggrus/podoplanin as a platelet aggregation-inducing factor. Gene 378:52–57
194. Steentoft C, Vakhrushev SY, Joshi HJ, Kong Y, Vester-Christensen MB, Schjoldager KT, Lavrsen K, Dabelsteen S, Pedersen NB, Marcos-Silva L, Gupta R, Bennett EP, Mandel U, Brunak S, Wandall HH, Levery SB, Clausen H (2013) Precision mapping of the human O-GalNAc glycoproteome through SimpleCell technology. EMBO J 32(10):1478–1488. doi:10.1038/emboj.2013.79
195. Fernandez-Munoz B, Yurrita MM, Martin-Villar E, Carrasco-Ramirez P, Megias D, Renart J, Quintanilla M (2011) The transmembrane domain of podoplanin is required for its association with lipid rafts and the induction of epithelial-mesenchymal transition. Int J Biochem Cell Biol 43(6):886–896. doi:10.1016/j.biocel.2011.02.010
196. Martin-Villar E, Megias D, Castel S, Yurrita MM, Vilaro S, Quintanilla M (2006) Podoplanin binds ERM proteins to activate RhoA and promote epithelial-mesenchymal transition. J Cell Sci 119(Pt 21):4541–4553
197. Kaneko MK, Kato Y, Kameyama A, Ito H, Kuno A, Hirabayashi J, Kubota T, Amano K, Chiba Y, Hasegawa Y, Sasagawa I, Mishima K, Narimatsu H (2007) Functional glycosylation of human podoplanin: glycan structure of platelet aggregation-inducing factor. FEBS Lett 581(2):331–336. doi:10.1016/j.febslet.2006.12.044
198. Kato Y, Fujita N, Kunita A, Sato S, Kaneko M, Osawa M, Tsuruo T (2003) Molecular identification of Aggrus/T1alpha as a platelet aggregation-inducing factor expressed in colorectal tumors. J Biol Chem 278(51):51599–51605. doi:10.1074/jbc.M309935200
199. Kunita A, Kashima TG, Morishita Y, Fukayama M, Kato Y, Tsuruo T, Fujita N (2007) The platelet aggregation-inducing factor aggrus/podoplanin promotes pulmonary metastasis. Am J Pathol 170(4):1337–1347. doi:10.2353/ajpath.2007.060790
200. Barth K, Blasche R, Kasper M (2010) T1alpha/podoplanin shows raft-associated distribution in mouse lung alveolar epithelial E10 cells. Cell Physiol Biochem 25(1):103–112. doi:10.1159/000272065
201. Wicki A, Lehembre F, Wick N, Hantusch B, Kerjaschki D, Christofori G (2006) Tumor invasion in the absence of epithelial-mesenchymal transition: podoplanin-mediated remodeling of the actin cytoskeleton. Cancer Cell 9(4):261–272. doi:10.1016/j.ccr.2006.03.010
202. Schacht V, Ramirez MI, Hong YK, Hirakawa S, Feng D, Harvey N, Williams M, Dvorak AM, Dvorak HF, Oliver G, Detmar M (2003) T1alpha/podoplanin deficiency disrupts normal

lymphatic vasculature formation and causes lymphedema. EMBO J 22(14):3546–3556. doi:10.1093/emboj/cdg342
203. Tammela T, Alitalo K (2010) Lymphangiogenesis: molecular mechanisms and future promise. Cell 140(4):460–476. doi:10.1016/j.cell.2010.01.045
204. Uhrin P, Zaujec J, Breuss JM, Olcaydu D, Chrenek P, Stockinger H, Fuertbauer E, Moser M, Haiko P, Fassler R, Alitalo K, Binder BR, Kerjaschki D (2010) Novel function for blood platelets and podoplanin in developmental separation of blood and lymphatic circulation. Blood 115(19):3997–4005. doi:10.1182/blood-2009-04-216069
205. Navarro A, Perez RE, Rezaiekhaligh MH, Mabry SM, Ekekezie II (2011) Polarized migration of lymphatic endothelial cells is critically dependent on podoplanin regulation of Cdc42. Am J Physiol Lung Cell Mol Physiol 300(1):L32–L42. doi:10.1152/ajplung.00171.2010
206. Williams MC, Cao Y, Hinds A, Rishi AK, Wetterwald A (1996) T1 alpha protein is developmentally regulated and expressed by alveolar type I cells, choroid plexus, and ciliary epithelia of adult rats. Am J Respir Cell Mol Biol 14(6):577–585. doi:10.1165/ajrcmb.14.6.8652186
207. Ramirez MI, Millien G, Hinds A, Cao Y, Seldin DC, Williams MC (2003) T1alpha, a lung type I cell differentiation gene, is required for normal lung cell proliferation and alveolus formation at birth. Dev Biol 256(1):61–72
208. Breiteneder-Geleff S, Matsui K, Soleiman A, Meraner P, Poczewski H, Kalt R, Schaffner G, Kerjaschki D (1997) Podoplanin, novel 43-kd membrane protein of glomerular epithelial cells, is down-regulated in puromycin nephrosis. Am J Pathol 151(4):1141–1152
209. Matsui K, Breitender-Geleff S, Soleiman A, Kowalski H, Kerjaschki D (1999) Podoplanin, a novel 43-kDa membrane protein, controls the shape of podocytes. Nephrol Dial Transplant 14(Suppl 1):9–11
210. Astarita JL, Acton SE, Turley SJ (2012) Podoplanin: emerging functions in development, the immune system, and cancer. Front Immunol 3:283. doi:10.3389/fimmu.2012.00283
211. Tomooka M, Kaji C, Kojima H, Sawa Y (2013) Distribution of podoplanin-expressing cells in the mouse nervous systems. Acta Histochem Cytochem 46(6):171–177. doi:10.1267/ahc.13035
212. Suzuki-Inoue K, Kato Y, Inoue O, Kaneko MK, Mishima K, Yatomi Y, Yamazaki Y, Narimatsu H, Ozaki Y (2007) Involvement of the snake toxin receptor CLEC-2, in podoplanin-mediated platelet activation, by cancer cells. J Biol Chem 282(36):25993–26001. doi:10.1074/jbc.M702327200
213. Acton SE, Astarita JL, Malhotra D, Lukacs-Kornek V, Franz B, Hess PR, Jakus Z, Kuligowski M, Fletcher AL, Elpek KG, Bellemare-Pelletier A, Sceats L, Reynoso ED, Gonzalez SF, Graham DB, Chang J, Peters A, Woodruff M, Kim YA, Swat W, Morita T, Kuchroo V, Carroll MC, Kahn ML, Wucherpfennig KW, Turley SJ (2012) Podoplanin-rich stromal networks induce dendritic cell motility via activation of the C-type lectin receptor CLEC-2. Immunity 37(2):276–289. doi:10.1016/j.immuni.2012.05.022
214. Herzog BH, Fu J, Wilson SJ, Hess PR, Sen A, McDaniel JM, Pan Y, Sheng M, Yago T, Silasi-Mansat R, McGee S, May F, Nieswandt B, Morris AJ, Lupu F, Coughlin SR, McEver RP, Chen H, Kahn ML, Xia L (2013) Podoplanin maintains high endothelial venule integrity by interacting with platelet CLEC-2. Nature 502(7469):105–109. doi:10.1038/nature12501
215. Cueni LN, Detmar M (2009) Galectin-8 interacts with podoplanin and modulates lymphatic endothelial cell functions. Exp Cell Res 315(10):1715–1723
216. Nakazawa Y, Sato S, Naito M, Kato Y, Mishima K, Arai H, Tsuruo T, Fujita N (2008) Tetraspanin family member CD9 inhibits Aggrus/podoplanin-induced platelet aggregation and suppresses pulmonary metastasis. Blood 112(5):1730–1739. doi:10.1182/blood-2007-11-124693
217. Martin-Villar E, Fernandez-Munoz B, Parsons M, Yurrita MM, Megias D, Perez-Gomez E, Jones GE, Quintanilla M (2010) Podoplanin associates with CD44 to promote directional cell migration. Mol Biol Cell 21(24):4387–4399. doi:10.1091/mbc.E10-06-0489

218. Tsuneki M, Yamazaki M, Maruyama S, Cheng J, Saku T (2013) Podoplanin-mediated cell adhesion through extracellular matrix in oral squamous cell carcinoma. Lab Invest 93(8):921–932. doi:10.1038/labinvest.2013.86
219. Kerjaschki D, Regele HM, Moosberger I, Nagy-Bojarski K, Watschinger B, Soleiman A, Birner P, Krieger S, Hovorka A, Silberhumer G, Laakkonen P, Petrova T, Langer B, Raab I (2004) Lymphatic neoangiogenesis in human kidney transplants is associated with immunologically active lymphocytic infiltrates. J Am Soc Nephrol 15(3):603–612
220. Tsuneki M, Maruyama S, Yamazaki M, Xu B, Essa A, Abe T, Babkair H, Cheng J, Yamamoto T, Saku T (2013) Extracellular heat shock protein A9 is a novel interaction partner of podoplanin in oral squamous cell carcinoma cells. Biochem Biophys Res Commun 434(1):124–130. doi:10.1016/j.bbrc.2013.03.057
221. Navarro A, Perez RE, Rezaiekhaligh M, Mabry SM, Ekekezie II (2008) T1alpha/podoplanin is essential for capillary morphogenesis in lymphatic endothelial cells. Am J Physiol Lung Cell Mol Physiol 295(4):L543–L551. doi:10.1152/ajplung.90262.2008
222. Wicki A, Christofori G (2007) The potential role of podoplanin in tumour invasion. Br J Cancer 96(1):1–5. doi:10.1038/sj.bjc.6603518
223. Martin-Villar E, Scholl FG, Gamallo C, Yurrita MM, Munoz-Guerra M, Cruces J, Quintanilla M (2005) Characterization of human PA2.26 antigen (T1alpha-2, podoplanin), a small membrane mucin induced in oral squamous cell carcinomas. Int J Cancer 113(6):899–910. doi:10.1002/ijc.20656
224. Hantusch B, Kalt R, Krieger S, Puri C, Kerjaschki D (2007) Sp1/Sp3 and DNA-methylation contribute to basal transcriptional activation of human podoplanin in MG63 versus Saos-2 osteoblastic cells. BMC Mol Biol 8:20. doi:10.1186/1471-2199-8-20
225. Pan Y, Wang WD, Yago T (2014) Transcriptional regulation of podoplanin expression by Prox1 in lymphatic endothelial cells. Microvasc Res 94C:96–102. doi:10.1016/j.mvr.2014.05.006
226. Hong YK, Harvey N, Noh YH, Schacht V, Hirakawa S, Detmar M, Oliver G (2002) Prox1 is a master control gene in the program specifying lymphatic endothelial cell fate. Dev Dyn 225(3):351–357. doi:10.1002/dvdy.10163
227. Kulkarni RM, Greenberg JM, Akeson AL (2009) NFATc1 regulates lymphatic endothelial development. Mech Dev 126(5-6):350–365. doi:10.1016/j.mod.2009.02.003
228. Durchdewald M, Guinea-Viniegra J, Haag D, Riehl A, Lichter P, Hahn M, Wagner EF, Angel P, Hess J (2008) Podoplanin is a novel fos target gene in skin carcinogenesis. Cancer Res 68(17):6877–6883. doi:10.1158/0008-5472.CAN-08-0299
229. Ekwall AK, Eisler T, Anderberg C, Jin C, Karlsson N, Brisslert M, Bokarewa MI (2011) The tumour-associated glycoprotein podoplanin is expressed in fibroblast-like synoviocytes of the hyperplastic synovial lining layer in rheumatoid arthritis. Arthritis Res Ther 13(2):R40. doi:10.1186/ar3274
230. Honma M, Minami-Hori M, Takahashi H, Iizuka H (2012) Podoplanin expression in wound and hyperproliferative psoriatic epidermis: regulation by TGF-beta and STAT-3 activating cytokines, IFN-gamma, IL-6, and IL-22. J Dermatol Sci 65(2):134–140. doi:10.1016/j.jdermsci.2011.11.011
231. Hwang YS, Xianglan Z, Park KK, Chung WY (2012) Functional invadopodia formation through stabilization of the PDPN transcript by IMP-3 and cancer-stromal crosstalk for PDPN expression. Carcinogenesis 33(11):2135–2146. doi:10.1093/carcin/bgs258
232. Peterziel H, Muller J, Danner A, Barbus S, Liu HK, Radlwimmer B, Pietsch T, Lichter P, Schutz G, Hess J, Angel P (2012) Expression of podoplanin in human astrocytic brain tumors is controlled by the PI3K-AKT-AP-1 signaling pathway and promoter methylation. Neuro Oncol 14(4):426–439. doi:10.1093/neuonc/nos055
233. Cortez MA, Nicoloso MS, Shimizu M, Rossi S, Gopisetty G, Molina JR, Carlotti C Jr, Tirapelli D, Neder L, Brassesco MS, Scrideli CA, Tone LG, Georgescu MM, Zhang W, Puduvalli V, Calin GA (2010) miR-29b and miR-125a regulate podoplanin and suppress

invasion in glioblastoma. Genes Chromosomes Cancer 49(11):981–990. doi:10.1002/gcc.20808
234. Martin-Villar E, Yurrita MM, Fernandez-Munoz B, Quintanilla M, Renart J (2009) Regulation of podoplanin/PA2.26 antigen expression in tumour cells. Involvement of calpain-mediated proteolysis. Int J Biochem Cell Biol 41(6):1421–1429. doi:10.1016/j.biocel.2008.12.010
235. Yurrita MM, Fernandez-Munoz B, Del Castillo G, Martin-Villar E, Renart J, Quintanilla M (2014) Podoplanin is a substrate of presenilin-1/gamma-secretase. Int J Biochem Cell Biol 46:68–75. doi:10.1016/j.biocel.2013.11.016
236. Breiteneder-Geleff S, Soleiman A, Kowalski H, Horvat R, Amann G, Kriehuber E, Diem K, Weninger W, Tschachler E, Alitalo K, Kerjaschki D (1999) Angiosarcomas express mixed endothelial phenotypes of blood and lymphatic capillaries: podoplanin as a specific marker for lymphatic endothelium. Am J Pathol 154(2):385–394. doi:10.1016/S0002-9440(10)65285-6
237. Kono T, Shimoda M, Takahashi M, Matsumoto K, Yoshimoto T, Mizutani M, Tabata C, Okoshi K, Wada H, Kubo H (2007) Immunohistochemical detection of the lymphatic marker podoplanin in diverse types of human cancer cells using a novel antibody. Int J Oncol 31(3):501–508
238. Niemiec JA, Adamczyk A, Ambicka A, Mucha-Malecka A, W MW, Rys J (2014) Triple-negative, basal marker-expressing, and high-grade breast carcinomas are characterized by high lymphatic vessel density and the expression of podoplanin in stromal fibroblasts. Appl Immunohistochem Mol Morphol 22(1):10–16. doi:10.1097/PAI.0b013e318286030d
239. Pula B, Wojnar A, Werynska B, Ambicka A, Kruczak A, Witkiewicz W, Ugorski M, Podhorska-Okolow M, Dziegiel P (2013) Impact of different tumour stroma assessment methods regarding podoplanin expression on clinical outcome in patients with invasive ductal breast carcinoma. Anticancer Res 33(4):1447–1455
240. Schoppmann SF, Berghoff A, Dinhof C, Jakesz R, Gnant M, Dubsky P, Jesch B, Heinzl H, Birner P (2012) Podoplanin-expressing cancer-associated fibroblasts are associated with poor prognosis in invasive breast cancer. Breast Cancer Res Treat 134(1):237–244. doi:10.1007/s10549-012-1984-x
241. Cortez MA, Nicoloso MS, Shimizu M, Rossi S, Gopisetty G, Molina JR, Carlotti C Jr, Tirapelli D, Neder L, Brassesco MS, Scrideli CA, Tone LG, Georgescu MM, Zhang W, Puduvalli V, Calin GA (2010) miR-29b and miR-125a regulate podoplanin and suppress invasion in glioblastoma. Genes Chromosomes Cancer 49:981–990
242. Kan S, Konishi E, Arita T, Ikemoto C, Takenaka H, Yanagisawa A, Katoh N, Asai J (2014) Podoplanin expression in cancer-associated fibroblasts predicts aggressive behavior in melanoma. J Cutan Pathol 41(7):561–567. doi:10.1111/cup.12322
243. Kawaguchi H, El Naggar AK, Papadimitrakopoulou V, Ren H, Fan YH, Feng L, Lee JJ, Kim E, Hong WK, Lippman SM, Mao L (2008) Podoplanin: a novel marker for oral cancer risk in patients with oral premalignancy. J Clin Oncol 26(3):354–360
244. Yuan P, Temam S, El Naggar A, Zhou X, Liu DD, Lee JJ, Mao L (2006) Overexpression of podoplanin in oral cancer and its association with poor clinical outcome. Cancer 107(3):563–569
245. Shindo K, Aishima S, Ohuchida K, Fujiwara K, Fujino M, Mizuuchi Y, Hattori M, Mizumoto K, Tanaka M, Oda Y (2013) Podoplanin expression in cancer-associated fibroblasts enhances tumor progression of invasive ductal carcinoma of the pancreas. Mol Cancer 12(1):168. doi:10.1186/1476-4598-12-168
246. Ito M, Ishii G, Nagai K, Maeda R, Nakano Y, Ochiai A (2012) Prognostic impact of cancer-associated stromal cells in patients with stage I lung adenocarcinoma. Chest 142(1):151–158. doi:10.1378/chest.11-2458
247. Ono S, Ishii G, Nagai K, Takuwa T, Yoshida J, Nishimura M, Hishida T, Aokage K, Fujii S, Ikeda N, Ochiai A (2013) Podoplanin-positive cancer-associated fibroblasts could have prognostic value independent of cancer cell phenotype in stage I lung squamous cell carcinoma:

usefulness of combining analysis of both cancer cell phenotype and cancer-associated fibroblast phenotype. Chest 143(4):963–970. doi:10.1378/chest.12-0913
248. Schoppmann SF, Jesch B, Riegler MF, Maroske F, Schwameis K, Jomrich G, Birner P (2013) Podoplanin expressing cancer associated fibroblasts are associated with unfavourable prognosis in adenocarcinoma of the esophagus. Clin Exp Metastasis 30(4):441–446. doi:10.1007/s10585-012-9549-2
249. Cirligeriu L, Cimpean AM, Raica M, Doros CI (2014) Dual role of podoplanin in oral cancer development. In Vivo 28(3):341–347
250. de Vicente JC, Rodrigo JP, Rodriguez-Santamarta T, Lequerica-Fernandez P, Allonca E, Garcia-Pedrero JM (2013) Podoplanin expression in oral leukoplakia: tumorigenic role. Oral Oncol 49(6):598–603. doi:10.1016/j.oraloncology.2013.02.008
251. Cueni LN, Hegyi I, Shin JW, Albinger-Hegyi A, Gruber S, Kunstfeld R, Moch H, Detmar M (2010) Tumor lymphangiogenesis and metastasis to lymph nodes induced by cancer cell expression of podoplanin. Am J Pathol 177(2):1004–1016
252. Funayama A, Cheng J, Maruyama S, Yamazaki M, Kobayashi T, Syafriadi M, Kundu S, Shingaki S, Saito C, Saku T (2011) Enhanced expression of podoplanin in oral carcinomas in situ and squamous cell carcinomas. Pathobiology 78(3):171–180
253. Huber GF, Fritzsche FR, Zullig L, Storz M, Graf N, Haerle K, Jochum W, Stoeckli SJ, Moch H (2011) Podoplanin expression correlates with sentinel lymph node metastasis in early squamous cell carcinomas of the oral cavity and oropharynx. Int J Cancer 129(6):1404–1409
254. Inoue H, Miyazaki Y, Kikuchi K, Yoshida N, Ide F, Ohmori Y, Tomomura A, Sakashita H, Kusama K (2012) Podoplanin promotes cell migration via the EGF-Src-Cas pathway in oral squamous cell carcinoma cell lines. J Oral Sci 54(3):241–250
255. Kreppel M, Drebber U, Wedemeyer I, Eich HT, Backhaus T, Zoller JE, Scheer M (2011) Podoplanin expression predicts prognosis in patients with oral squamous cell carcinoma treated with neoadjuvant radiochemotherapy. Oral Oncol 47(9):873–878
256. dos Santos Almeida A, Oliveira DT, Pereira MC, Faustino SE, Nonogaki S, Carvalho AL, Kowalski LP (2013) Podoplanin and VEGF-C immunoexpression in oral squamous cell carcinomas: prognostic significance. Anticancer Res 33(9):3969–3976
257. Nakashima Y, Yoshinaga K, Kitao H, Ando K, Kimura Y, Saeki H, Oki E, Morita M, Kakeji Y, Hirahashi M, Oda Y, Maehara Y (2013) Podoplanin is expressed at the invasive front of esophageal squamous cell carcinomas and is involved in collective cell invasion. Cancer Sci 104(12):1718–1725. doi:10.1111/cas.12286
258. Pula B, Witkiewicz W, Dziegiel P, Podhorska-Okolow M (2013) Significance of podoplanin expression in cancer-associated fibroblasts: a comprehensive review. Int J Oncol 42(6):1849–1857. doi:10.3892/ijo.2013.1887
259. Sugimoto Y, Watanabe M, Oh-hara T, Sato S, Isoe T, Tsuruo T (1991) Suppression of experimental lung colonization of a metastatic variant of murine colon adenocarcinoma 26 by a monoclonal antibody 8F11 inhibiting tumor cell-induced platelet aggregation. Cancer Res 51(3):921–925
260. Chandramohan V, Bao X, Kato Kaneko M, Kato Y, Keir ST, Szafranski SE, Kuan CT, Pastan IH, Bigner DD (2013) Recombinant anti-podoplanin (NZ-1) immunotoxin for the treatment of malignant brain tumors. Int J Cancer 132(10):2339–2348. doi:10.1002/ijc.27919
261. Kato Y, Kaneko MK, Kuno A, Uchiyama N, Amano K, Chiba Y, Hasegawa Y, Hirabayashi J, Narimatsu H, Mishima K, Osawa M (2006) Inhibition of tumor cell-induced platelet aggregation using a novel anti-podoplanin antibody reacting with its platelet-aggregation-stimulating domain. Biochem Biophys Res Commun 349(4):1301–1307. doi:10.1016/j.bbrc.2006.08.171
262. Kato Y, Vaidyanathan G, Kaneko MK, Mishima K, Srivastava N, Chandramohan V, Pegram C, Keir ST, Kuan CT, Bigner DD, Zalutsky MR (2010) Evaluation of anti-podoplanin rat monoclonal antibody NZ-1 for targeting malignant gliomas. Nucl Med Biol 37(7):785–794. doi:10.1016/j.nucmedbio.2010.03.010

263. Abe S, Morita Y, Kaneko MK, Hanibuchi M, Tsujimoto Y, Goto H, Kakiuchi S, Aono Y, Huang J, Sato S, Kishuku M, Taniguchi Y, Azuma M, Kawazoe K, Sekido Y, Yano S, Akiyama S, Sone S, Minakuchi K, Kato Y, Nishioka Y (2013) A novel targeting therapy of malignant mesothelioma using anti-podoplanin antibody. J Immunol 190(12):6239–6249. doi:10.4049/jimmunol.1300448
264. Kaneko MK, Kunita A, Abe S, Tsujimoto Y, Fukayama M, Goto K, Sawa Y, Nishioka Y, Kato Y (2012) Chimeric anti-podoplanin antibody suppresses tumor metastasis through neutralization and antibody-dependent cellular cytotoxicity. Cancer Sci 103(11):1913–1919. doi:10.1111/j.1349-7006.2012.02385.x
265. Cheriyan VT, Wang Y, Muthu M, Jamal S, Chen D, Yang H, Polin LA, Tarca AL, Pass HI, Dou QP, Sharma S, Wali A, Rishi AK (2014) Disulfiram suppresses growth of the malignant pleural mesothelioma cells in part by inducing apoptosis. PLoS One 9(4), e93711. doi:10.1371/journal.pone.0093711
266. Jamal S, Cheriyan VT, Muthu M, Munie S, Levi E, Ashour AE, Pass HI, Wali A, Singh M, Rishi AK (2014) CARP-1 functional mimetics are a novel class of small molecule inhibitors of malignant pleural mesothelioma cells. PLoS One 9(3), e89146. doi:10.1371/journal.pone.0089146
267. Chang CH, Chung CH, Hsu CC, Peng HC, Huang TF (2014) Inhibitory effects of polypeptides derived from a snake venom C-type lectin, aggretin, on tumor cell-induced platelet aggregation. J Thromb Haemost 12(4):540–549. doi:10.1111/jth.12519
268. Ochoa-Alvarez JA, Krishnan H, Pastorino JG, Nevel E, Kephart D, Lee JJ, Retzbach EP, Shen Y, Fatahzadeh M, Baredes S, Kalyoussef E, Honma M, Adelson ME, Kaneko MK, Kato Y, Young MA, Deluca-Rapone L, Shienbaum AJ, Yin K, Jensen LD, Goldberg GS (2015) Antibody and lectin target podoplanin to inhibit oral squamous carcinoma cell migration and viability by distinct mechanisms. Oncotarget 6(11):9045–9060

# Chapter 13
# Intercellular Communication, the Tumor Microenvironment, and Tumor Progression

**Mei-Fong Pang and Celeste M. Nelson**

**Abstract** Within a solid tumor, cancer cells interact with normal host cells as well as with insoluble and soluble factors, including extracellular matrix proteins and chemokines. Here, we survey the molecular mechanisms by which cancer cells interact with their surrounding microenvironment, with a particular focus on intercellular communication. This host-tumor crosstalk provides key signals that direct the growth, migration, and dissemination of cancer cells. Approaches that disrupt or coopt intercellular communication between the tumor and its host are exciting and potentially powerful anti-cancer therapeutic strategies.

**Keywords** Stroma • Microenvironment • Metastasis

## Abbreviations-activated

| | |
|---|---|
| CAF | Cancer-associated fibroblast |
| CAR | Coxsackie-adenovirus receptor |
| CC | Chemokine |
| CCR | Chemokine receptor |
| CSC | Cancerstem cell |
| ECM | Extracellular matrix |
| EGF | Epidermal growth factor |
| EMT | Epithelial-mesenchymal transition |
| GJ | Gap junction |
| GJIC | Gap junction intercellular communication |
| HCC | Hepatocellular carcinoma |
| HGF | Hepatocyte growth factor |
| ICAM | Intercellular adhesion molecule |

M.-F. Pang • C.M. Nelson (✉)
Departments of Chemical and Biological Engineering, Princeton University,
Princeton, NJ 08544, USA

Departments of Molecular Biology, Princeton University,
303 Hoyt Laboratory, William Street, Princeton, NJ 08544, USA
e-mail: celesten@princeton.edu

 
M. Kandouz (ed.), *Intercellular Communication in Cancer*,
DOI 10.1007/978-94-017-7380-5_13

| | |
|---|---|
| IL | Interleukin |
| LN | Lymph node |
| MSC | Mesenchymal stem cell |
| MVB | Multi-vesicular body |
| PDGF | Platelet-derived growth factor |
| SLN | Sentinel lymph node |
| TAK | Transforming growth factorkinase |
| siRNA | Short interfering RNA |
| TGF | Transforming growth factor |
| TNF | Tumor necrosis factor |
| VCAM | Vascular cell adhesion molecule |
| VEGF | Vascular endothelial growth factor |

## 13.1 Introduction

### *13.1.1 The Tumor Microenvironment*

Interactions between cells and their surrounding microenvironment are crucial to maintain functional tissue homeostasis. Defects in these interactions result in tissue pathology and disease progression. Solid tumors consist of cancer cells and the surrounding tumor microenvironment. The cellular components of the tumor microenvironment include fibroblasts, endothelial cells, and inflammatory cells. The physical and chemical components include the extracellular matrix (ECM) and diffusible molecules such as dissolved oxygen, cytokines, and growth factors [1–3] (Fig. 13.1). Cancer cells interact dynamically and co-evolve with this complex surrounding microenvironment [4]. Crosstalk between cancer cells and the tumor microenvironment can affect the behavior of cancer cells and promote tumor progression by conferring cancer cells with the ability to migrate, invade, and metastasize [1, 3].

## 13.2 Acquisition of Invasive and Metastatic Properties during Tumor Progression within the Tumor Microenvironment through Intercellular Communication

Several types of specialized intercellular junctional proteins maintain epithelial cell contacts with neighboring cells, including tight junctions, adherens junctions, and gap junctions [5]. Tight junctions are comprised of transmembrane proteins such as coxsackie adenovirus receptor (CAR) and occludin, which seal the space between

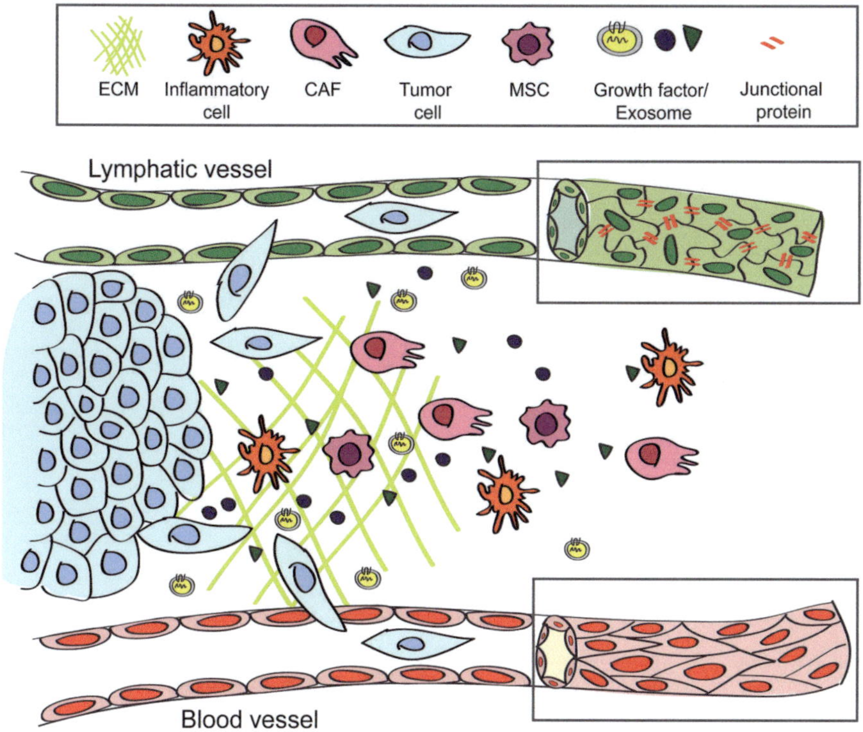


**Fig. 13.1 Thetumor microenvironment.** Intercellular interactions within the tumor microenvironment drive tumor progression to an invasive state. The physical components of the tumor microenvironment, including the ECM, can influence the metastatic properties of cancer cells. Intercellular communication between tumor cells and other cells including vascular endothelial cells, CAFs, and inflammatory cells directly or indirectly permits the acquisition of an invasive and migratory phenotype. For example, elevated levels of TGFβ secreted by stromal cells within the tumor microenvironment can induce EMT and CSC properties in cancer cells, leading to tumor dissemination and metastatic spread

neighboring cells and prevent the passage of molecules and ions. Adherens junctions form from complexes of E-cadherin that anchor cells to each other. Gap junctions (GJs) are communicating junctions made up of connexins that allow direct intercellular communication between neighboring cells [6]. Intercellular communication is important for the maintenance of homeostasis as well as tumor progression. Cancer cells use these various junctional complexes to transmit signals to their neighbors and respond collectively to different conditions [7]. Cancer cells can communicate with each other directly through these junctions [8, 9] or indirectly through paracrine signaling via the secretion of growth factors [10], chemokines, [11] and exosomes [12].

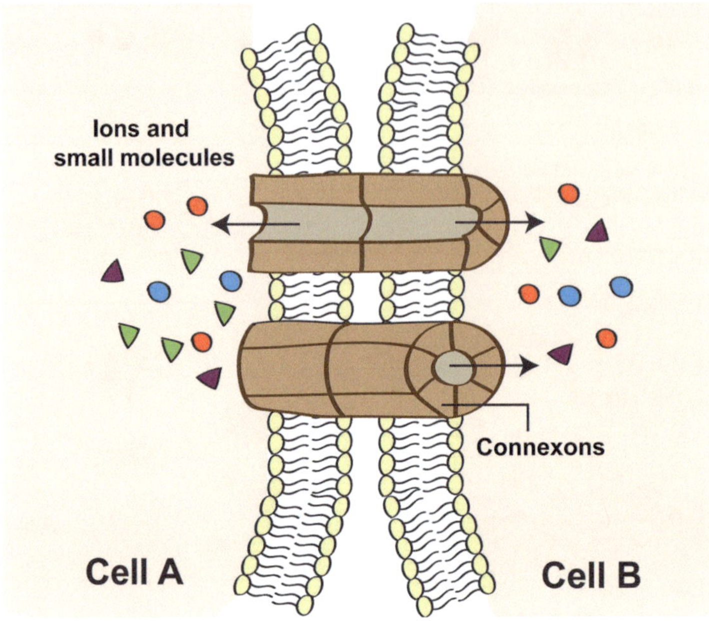


**Fig. 13.2 Gap junctions.** GJs are composed of connexons that each assemble from six subunits of GJ proteins, connexins. GJs connect the cytoplasm of adjacent cells and allow direct communication between cells through the passage of ions and small molecules

## *13.2.1 Direct Intercellular Communication Via Gap Junctions*

GJs are gated intercellular channels composed of connexons that each assemble from six connexins (Fig. 13.2). The opening of GJ channels allows the cell-to-cell transfer of small molecules ($M_r$ <1000) including ions [13], short interfering RNAs (siRNAs) [14, 15], and metabolites [16]. Gap junction intercellular communication (GJIC) thus allows for the direct communication between the cytoplasms of adjacent cells. GJIC can occur between the same type of cells (homospecific) or between different types of cells (heterospecific) [17, 18]. GJIC plays a critical role in maintaining cellular homeostasis by regulating processes including proliferation, differentiation, and barrier function [19–21]. Loss of GJs and altered localization of connexins are often observed in human cancers [9, 22–25]. Intercellular communication can be measured by passing the current from a microelectrode from one cell to an adjacent cell [26]. Using this method, Loewenstein and Kanno found that GJIC was reduced between cancer cells [26]. GJIC can also be measured using a dye-coupling assay in which fluorescent tracers are microinjected into individual cells and the transfer of dye into the adjacent cells is monitored and quantified [22]. Dye-coupling assays have revealed an absence of GJIC between cancer cells and the adjacent normal cellular compartment in primary human liver tumors. In these cells, connexin 32 (Cx32) localizes to the cytoplasm instead of at the plasma membrane

[22]. Similarly, immunoelectron microscopy analysis has revealed that squamous carcinoma cells have poorly developed GJs, with some of the connexins accumulating in the cytoplasm [27]. These studies suggest that aberrant localization of connexins might be one mechanism by which cancer cells inactivate GJIC. Conversely, exogenous expression of Cx32 into communication-deficient cancer cells can restore GJIC and attenuate tumor growth in vivo [9]. Consistent with these findings, Cx32-deficient mice are significantly more susceptible to developing hepatic tumors both spontaneously as well as when induced by treatment with diethylnitrosamine [28]. Similarly, Cx43-heterozygous mice are more prone to developing lung tumors in response to urethane administration than are wild-type mice [29]. These findings suggest a role for GJIC in tumorigenesis.

Furthermore, disrupting GJIC can alter metastatic potential. Loss of GJs is associated with enhanced migration capacity of cancer cells. Introducing the metastasis suppressor gene, breast cancer metastasis suppressor-1 (BRMS1), into highly metastatic breast cancer cells can restore GJIC to a level similar to that of normal breast tissue and reduce metastatic potential [30]. Similarly, ectopic expression of Cx32 increases intercellular adhesion between HeLa cells and suppresses their ability to migrate and invade [31]. Treatment with the GJIC inhibitor, oleamide, reverses the effect of Cx32 overexpression, suggesting that GJIC itself regulates the migratory and invasive phenotype of tumor cells [31]. The expression of Cx43 is downregulated in breast cancer cells compared to normal breast epithelial cells [30, 32]. Overexpression of Cx43 reduces breast cancer cell invasiveness by suppressing the expression of N-cadherin [33] and preventing tumor cell adhesion to endothelial cells [33]. Consistently, injection of Cx43-expressing breast cancer cells into immunocompromised mice results in smaller tumors and fewer lung metastases compared to control cells [34], suggesting that Cx43 alters the metastatic potential of breast cancer cells by regulating N-cadherin. Cx43-knockout mice have disrupted cell polarity and display defects in epithelial-mesenchymal transition (EMT) in the epicardium [35].

Heterospecific GJIC between tumor cells and other cell types within the tumor microenvironment may affect the metastatic potential of cancer cells [30]. It has been shown that the growth of transformed cells is inhibited where there is strong heterologous communication with the adjacent non-transformed cells. When the heterologous communication is blocked chemically, the growth inhibitory effect is abolished, suggesting an active role for GJIC in transmitting signals for growth control [36]. Electron microscopy analysis has revealed that leukocytes induce the expression of Cx43 and formation of GJs between themselves and endothelial cells during inflammation [37], suggesting that communication via GJIC may mediate leukocyte extravasation. Similarly, studies indicate that cancer cells and endothelial cells interact with each other to mediate cancer cell extravasation to distant sites [38–40]. Ectopic expression of Cx43 in breast cancer cells that lack GJIC promotes the establishment of functional heterologous communication with vascular endothelial cells and enhances transendothelial migration of the cancer cells [39]. Functional GJs are formed at the contact sites between cancer cells and endothelial cells where metabolic coupling occurs [40]. The modulation of GJIC between cancer cells and endothelial cells depends on the ability of cancer cells to adhere to the endothelium [40]. Breast cancer cells that overexpress Cx43 have an enhanced capacity to adhere to the pulmonary endothelium, whereas breast cancer cells overexpressing a dominant-negative mutant of Cx43 fail to adhere [41]. Elevated levels of Cx43 have also been found at contact sites between cancer cells and endothelial cells [40, 41].

Dye transfer assays have revealed that ectopic expression of Cx26 can induce the formation of heterologous GJs between melanoma cells and vascular endothelial cells in a coculture system [42]. The Cx26-expressing melanoma cells were also found to be more metastatic than mutant Cx26-expressing cells [42]. Consistent with these studies, Cx26 has been implicated in the metastasis of human cancers [43–46]. Immunohistochemical analysis shows a strong correlation between Cx26 expression and lymphatic vessel invasion [46]. Cx26 might potentiate lymphatic vessel invasion through the formation of heterologous GJIC between the cancer cells and lymphatics [46]. These data indicate that formation of heterologous GJIC between cancer cells and endothelial cells may play a role in cancer cell extravasation. Recently, bone marrow stromal cells were shown to transfer CXCL12-specific microRNAs (miRNAs) to breast cancer cells through GJIC [47]. Since reducing the levels of CXCL12 inhibits the proliferation of breast cancer cells and induces quiescence, this GJIC-mediated transmission of miRNA might contribute to breast cancer cell dormancy [47]. Direct intercellular communication involving GJs between cancer cells and other cell types within the tumor microenvironment can promote tumor progression, so therapeutic targeting of GJIC may inhibit the metastatic spread of cancers.

### 13.2.2 Indirect Intercellular Communication: Paracrine Signaling via Exosomes, Growth Factors, and Cytokines

Although cell-cell communication can occur in a contact-dependent manner via GJIC, cells can also communicate with distant target cells through the secretion of signaling molecules such as growth factors, cytokines, and exosomes [48]. The secreted molecules act as local mediators to regulate cell signaling within the surrounding microenvironment, in a process known as paracrine signaling. Exosomes are cup-shaped vesicles ranging from 50-100 nm in diameter [49]. These naturally occurring nanoparticles are generated from multivesicular bodies (MVBs). Fusion of MVBs with the plasma membrane leads to release of exosomes via exocytosis into the surrounding tumor microenvironment and circulation [49, 50–52] (Fig. 13.3). Accumulating evidence suggests that exosomes are important mediators of intercellular communication. Exosomes contain cargos of bioactive molecules including DNA fragments [53], mRNAs [54, 55], miRNAs [54–56], and proteins [57, 55], which can be transferred into recipient cells. Cancer cells [58–61] and cancer-associated fibroblasts (CAFs) [62] have been found to release exosomes that

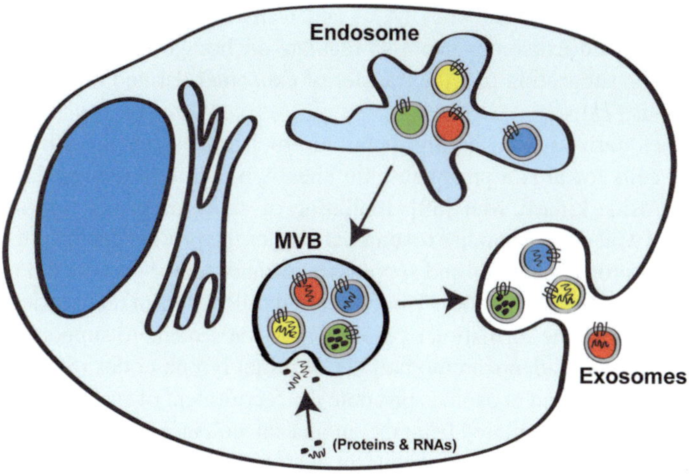


**Fig. 13.3 Mechanisms ofexosomebiogenesis.**Exosomes are derived from late endosomes called MVBs. Small molecules such as proteins and RNA are sorted in MVBs, and their fusion with the plasma membrane leads to secretion of exosomes via exocytosis

communicate with distant cell types to modulate immune evasion [63], angiogenesis [58], drug resistance [64], and formation of the pre-metastatic niche [60]. Higher levels of circulating exosomes are found in cancer patients than in healthy individuals [65, 66]. Cancer patients with high levels of exosomes have poorly differentiated tumors and a reduced rate of survival [67]. In vivo imaging of tumor-derived exosomes using GFP-labeled CD63, an exosome marker, in an orthotopic mouse model has revealed that breast cancer cells release exosomes into the tumor microenvironment and circulation. These tumor-derived exosomes are taken up by other cancer cells and CAFs in mice with breast cancer metastases, indicating that tumor-derived exosomes might be involved in promoting metastasis [52]. Membrane fusion assays and confocal microscopy analysis have revealed that the uptake of exosomes is mediated through direct fusion of the exosome membrane with the plasma membrane [68].

Double-stranded DNA (dsDNA) has been found in exosomes [69, 70] and whole genome sequencing using genomic DNA isolated from exosomes of cancer cell lines and serum from cancer patients has detected mutations in oncogenes and tumor suppressor genes including KRAS and p53 [69]. These findings suggest the possibility of using exosomes as a circulating biomarker to detect cancer. Microarray analysis of exosomes derived from mouse and human mast cell lines has revealed the presence of mRNAs and small RNAs such as miRNAs [54]. In vitro translation assays have demonstrated that these exosomal mRNAs are functional and can be translated into proteins [54]. Bioinformatics analysis using miRNA profiling of hepatocellular carcinoma (HCC) has identified a set of miRNAs that control the regulation of transforming growth factor-activated kinase-1 (TAK1) [56], which plays a crucial role in liver cancer [71]. Incubating cells with exosomes derived from HCC constitutively activates TAK1 and TAK1-associated signaling pathways [71]. HCC-derived exosomes can also regulate anchorage-independent growth in recipient cells, suggesting that the transfer of exosomal-derived miRNA can affect tumorigenesis [71].

Exosomes derived from highly metastatic melanomas can drive bone marrow progenitor cells towards a pro-metastatic phenotype through the regulation of the receptor tyrosine kinase, Met [60]. Inhibition of exosome production and down-regulation of Met expression in exosomes abolishes the pro-metastatic phenotype of bone marrow progenitor cells and suppresses primary tumor growth and metastasis in a xenograft mouse model [60]. Tumor-derived miRNA from metastatic adenocarcinoma can induce the formation of the pre-metastatic niche to support the growth of poorly metastatic adenocarcinoma [72]. Sentinel lymph nodes (SLNs) enriched with melanoma-derived exosomes promote the recruitment of melanoma cells [73]. In addition, exosomes isolated from melanoma can enhance the trafficking of melanoma cells towards lymphatic vessels in vivo by preconditioning the SLNs for tumor growth and tumor angiogenesis [73].

Bone marrow stromal cells secrete exosomes containing miRNA against CXCL12 to induce a quiescent state in breast cancer cells, indicating a role for stroma-derived exosomes in regulating breast cancer dormancy [47]. Evading immune destruction is one of the hallmarks of cancer [74] and cancer cells can communicate with immune cells to evade immune surveillance [63]. For example, exosomes derived from human prostate cancer cells induce the apoptosis of T cells in a Fas ligand (FasL)-dependent manner [63]. Blocking the Fas-FasL interaction using a FasL-neutralizing antibody prevents exosome-induced T cell apoptosis [63]. Exosomes secreted by chemoresistant breast cancer cells, which contain miRNAs that modulate the cell cycle and apoptosis, can also confer drug resistance to chemosensitive breast cancer cells [64]. For example, HER2-positive cancer cells release exosomes that can alter chemosensitivity to trastuzumab, a drug that targets the HER2 receptor [75]. HER2-positive exosomes abolish the antitumor activity of trastuzumab [75].

In addition to releasing exosomes to drive tumor progression, cancer cells and the surrounding stroma also communicate with each other through the secretion of various inflammatory cytokines, growth factors, and proteases to shape a metastasis-permissive tumor microenvironment [76]. Inflammatory cytokines and growth factors such as transforming growth factor-beta (TGFβ) [77], tumor necrosis factor-alpha (TNFα), epidermal growth factor (EGF) [78], hepatocyte growth factor (HGF) [79] and interleukins (IL) including IL-6 [80] and IL-8 [81] all induce EMT, a process in which polarized epithelial cells become highly migratory and invasive mesenchymal-like cells [82]. The activation of EMT inducing-transcription factors such as Snail, Slug, Twist, and Zeb alters gene expression, apical-basal polarity, and reduces cell-cell interactions [83, 84]. EMT is essential for tissue and organ generation during embryogenesis [85], and has also been found to be reactivated during pathological conditions such as fibrosis, wound healing, and cancer [86]. Although still somewhat controversial [87], accumulating evidence suggests that EMT facilitates tumor cell dissemination and metastasis during tumor progression [86]. The induction of EMT in cancer cells results in the acquisition of a metastatic phenotype [88]. Furthermore, recent studies suggest that the induction of EMT endows cells with cancer stem cell (CSC) properties [89]. These findings suggest the possibility of disrupting paracrine-mediated intercellular communication to inhibit tumor progression.

## 13.3 Tumor Cell Dissemination

### *13.3.1 Organ-Specific Metastasis: The Role of Chemokines*

Over a century ago, Stephen Paget examined the autopsy reports of 735 breast cancer cases and found that the distribution of metastasis was not random [90]. Paget suggested that tumor cells ("seed") can only grow and survive in appropriate microenvironments ("soil") [90]. Various types of cancers have specific patterns of metastasis [91] and it is now understood that the specificity of metastasis is mediated in part by specific molecules expressed by cancer cells and the resident host cells at metastatic sites. In particular, chemokines (CCs) and chemokine receptors (CCRs) mediate the circulation and homing of immune cells during inflammation and the immune response [92] (Fig. 13.4). CCs are small, secreted peptides that mediate the chemotactic migration of cells expressing appropriate CCRs [92], which are G protein-coupled receptors. For example, CCL21/CCR7 and CXCL12/CXCR4 are involved in the homing of dendritic cells to lymph nodes during inflammation [93]. Binding of CCs to CCRs activates signaling pathways that regulate cell polarity, migration, and adhesion [92]. Different cancer cells express different CCRs, suggesting that cells may respond specifically to chemotactic forces generated by CCs secreted by target destinations [94]. Therefore, cancer cells may exploit CC/CCR signaling during tumor dissemination [95]. In addition to mediating directional migration, CCs also stimulate the expression of cell adhesion molecules such as intercellular adhesion molecule-1 (ICAM-1) and vascular cell adhesion molecule-1 (VCAM-1) [96].

### *13.3.2 Role of Chemokines in Hematogenous Dissemination: Tumor-Vascular Endothelial Interactions*

Metastasis is a complex process that involves a defined series of events. Cancer cells must detach from the primary tumor site; intravasate into the blood or lymphatic vasculature; survive through the circulation; extravasate from the vasculature and colonize the distant site [97]. Intercellular communication between tumor cells and the stromal compartment plays an important role in facilitating hematogenous dissemination [98, 99]. Coculture of tumor cells with platelets and leukocytes can activate microvascular endothelial cells [100]. Activated microvascular endothelial cells produce elevated levels of the inflammatory CC, CCL5, to recruit inflammatory monocytes to shape a pro-metastatic microenvironment [100]. Administration

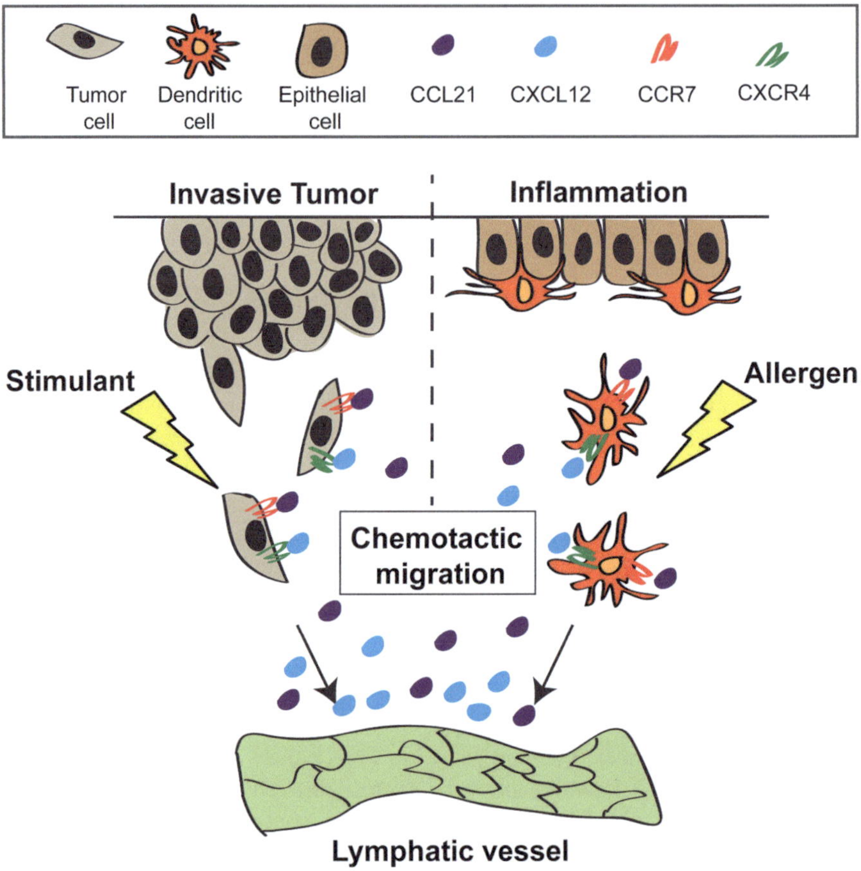


**Fig. 13.4 CC and CCR expression during lymphatic homing.** During inflammation, DCs are activated by allergens. Activated DCs upregulate the lymphatic homing receptors such as CCR7 and CXCR4. Induction of CCR7 and CXCR4 mediates the chemotactic migration of activated DCs towards the lymphatic vessels, which secrete specific ligands, CCL21 and CXCL12. Overexpression of CCR7 and CXCR4 has been implicated in LN metastasis in various human cancers. Therefore, it is thought that tumor cells hijack the DC route to metastasize towards the lymphatics

of CCL5 antagonist abolishes CCL5-dependent monocyte recruitment and strongly suppresses tumor metastasis in an experimental metastasis model, suggesting a pro-metastatic role for CCL5 [100]. Furthermore, components of the stromal compartment, such as mesenchymal stem cells (MSCs), can affect the phenotype of tumor cells through the secretion of CCL5. Binding of CCL5 to CCR5-expressing tumor cells significantly enhances metastasis and promotes tumor cell dissemination [101]. Microarray analysis of human breast cancer specimens shows an increase in the expression of CCL5 and its receptor, CCR5. Consistently, inhibiting CCR5 reduces breast cancer invasiveness in culture and decreases metastatic colonization in vivo [102].

CCs can also regulate physiological and pathological angiogenesis [103]. CXCL1, CXCL2, CXCL8 and CCL2 are among the pro-angiogenic CCs [104, 105]. CCL2 can induce the expression of vascular endothelial growth factor-A (VEGF-A) in prostate cancer cells, which suggests that CCL2 can promote tumor-induced angiogenesis indirectly [106]. Xenograft studies using CXCL8-expressing melanoma cells revealed that tumor growth, angiogenesis, and metastasis are significantly reduced in CXCR2-knockout mice, indicating the importance of CXCL8/CXCR2 signaling in regulating tumor-associated angiogenesis and dissemination [107].

### *13.3.3 Role of Chemokines in Lymphatic Dissemination: Tumor-Lymphatic Endothelial Interactions*

Histopathological analysis of human biopsies has often revealed the presence of tumor cells in close proximity to peritumoral lymphatic vessels [108]. Malignant cancer cells can induce tumor lymphangiogenesis through the secretion of lymphangiogenic factors including VEGF-A, VEGF-C, VEGF-D, and platelet-derived growth factor (PDGF)-BB for tumor dissemination [109, 110]. In addition to tumor lymphangiogenesis, CC-mediated interactions between cancer cells and lymphatic endothelial cells can also enhance cancer cell dissemination towards lymphatic vessels [109]. CC gradients generated by lymphatic vessels, including CCL21 and CXCL12, act as guidance cues to mediate the directional migration of cancer cells [111]. Lymphatic endothelial cells have unique, semi-open, and discontinuous 'button-like' junctions within the lymphatic vessels [112]. Tumor cells might have more access to lymphatic vessels due to this 'leaky' structure, and may use a similar mechanism of entry into the lymphatics as dendritic cells (DCs).

Several CC/CCR pairs, including CCL21/CCR7 and CXCL12/CXCR4, have been implicated in lymph node (LN) metastasis in various cancers [113–115]. Lymphatic vessels and secondary lymphoid organs secrete high levels of CCL21, which is the ligand for its cognate receptor, CCR7 [111, 116]. CCR7 is indispensable for lymphatic homing, as mature DCs in CCR7-deficient mice fail to migrate towards the draining lymph nodes [93]. Under normal conditions, DCs increase their expression of CCR7 upon activation and during maturation. Binding of lymphatic-secreted CCL21 to CCR7 mediates directional migration of DCs towards the lymphatics [111]. Recently, Shields and co-workers showed that metastatic tumor cells express CCR7 and secrete CCR7 ligands autologously [117]. Under the influence of autologous chemotaxis and interstitial flow caused by draining lymphatics, these tumor cells were guided to the lymphatics [117]. Ectopic expression of CCR7 in B16 melanoma cells results in a significant increase in the number of LN metastases in a xenograft mouse model [118]. LN metastasis in CCR7-expressing B16 melanoma mice is inhibited after treatment with anti-CCL21-blocking antibodies, indicating a role for the CCL21/CCR7 axis in lymphatic dissemination of cancer cells [118].

In addition to CCL21/CCR7, elevated levels of CXCL12/CXCR4 have been associated with LN metastasis [119–121]. CXCL12 is expressed by the high endothelial venules (HEV) and the T cell zones of the spleen and LNs [122]. Similar to CCL21, CXCL12 plays a crucial role in recruiting immune cells during the initiation of the immune response [123] (Fig. 13.4). T cells, B cells, and mature DCs express the receptor for CXCL12, CXCR4 [124, 125]. Activation of CXCR4 by CXCL12 increases the adhesion of tumor cells to endothelial cells by upregulating the expression of β1-integrin [126]. Tumor-associated lymphatic vessels can mediate targeted migration of CXCR4(+)/CD133(+) melanoma cells through the secretion of CXCL12 [127]. Inhibiting CXCR4 signaling using the specific inhibitor, AMD3100, reduces axillary LN metastasis, indicating a role for CXCL12/CXCR4 in LN metastasis [127]. Since both CCR7 and CXCR4 are lymphatic-homing receptors for DCs and have been closely linked with high incidence of LN metastasis [115, 128], it is plausible that metastatic tumor cells might exploit the DC trafficking route during inflammation by manipulating their chemotactic interaction with the lymphatics via CCs/CCRs to mediate lymphogenic dissemination.

## 13.4 Therapeutic Interventions Targeting Intercellular Communication

Increasing GJIC in cancers is beneficial to anti-cancer therapy [129]. Treatments that combine a GJ enhancer, quinolones (PQs), with cisplatin decrease mammary tumor growth compared to cisplatin treatment alone in a mouse xenograft model of breast cancer [130]. Histology analysis revealed that PQ treatment upregulates the expression of the GJ proteins, Cx43 and Cx26, in the tumor. Similarly, targeting paracrine signaling appears beneficial. The GE11 peptide can bind to EGF receptor with high specificity [131]. Engineered exosomes that contain GE11 allow for targeted delivery of let-7a miRNA to EGFR-positive breast cancer cells in a xenograft mouse model [132]. Systemic administration of GE11-positive exosomes loaded with let-7a miRNA suppressed tumor growth in vivo, suggesting the usefulness of exosomes to deliver miRNA as a therapeutic approach [132]. Engineered exosomes may also be used for targeted delivery of the chemotherapeutic drug, doxorubicin (Dox), to the tumor in a mouse model. Dox was encapsulated into exosomes engineered to express αv-integrin-specific iRGD peptide on the surface. Dox-containing iRGD exosomes were able to bind to αv-integrin-positive tumor cells and suppress tumor growth in tumor-bearing mice more efficiently than non-targeting control exosomes [133].

Chemotherapy-resistant breast cancer cells express high levels of TGFβ and acquire CSC-like properties [134]. The chemotherapeutic drug, paclitaxel, enriches the population of breast CSCs by upregulating TGFβ and IL-8 signaling. Co-administration of paclitaxel and the TGFβ type I receptor kinase inhibitor, LY2157299, into a xenograft mouse model inhibited tumor reestablishment, suggesting that this combination treatment blocks the development of chemoresistance and CSC expansion in breast cancer [135]. Another study showed that a combination treatment of CXCR4-specific inhibitor, AMD3100, and the chemotherapy drug, docetaxel, enhanced the antitumor effect of docetaxel in a tumor-bearing mouse model. These data reveal the possibility of using the CXCR4 inhibitor as a chemosensitizer in cancer treatment [136]. It is clear from these studies that targeting intercellular communication can attenuate tumor growth and metastatic spread. However, more studies need to be carried out to assess the usefulness of therapeutic interventions that target intercellular communication within the tumor microenvironment.

## 13.5 Conclusion

It is well established that intercellular communication between cancer cells and other cells within the surrounding tumor microenvironment is critical for tumorigenesis and tumor progression. Therefore, understanding the mechanisms underlying these interactions might lead to the development of novel therapeutics to block tumor progression and metastasis.

## References

1. Mueller MM, Fusenig NE (2004) Friends or foes – bipolar effects of the tumour stroma in cancer. Nat Rev Cancer 4(11):839–849
2. Korkaya H, Liu S, Wicha MS (2011) Breast cancer stem cells, cytokine networks, and the tumor microenvironment. J Clin Invest 121(10):3804–3809
3. Calorini L, Bianchini F (2010) Environmental control of invasiveness and metastatic dissemination of tumor cells: the role of tumor cell-host cell interactions. Cell Commun Signal 8:24
4. Polyak K, Haviv I, Campbell IG (2009) Co-evolution of tumor cells and their microenvironment. Trends Genet 25(1):30–38
5. Matter K, Aijaz S, Tsapara A, Balda MS (2005) Mammalian tight junctions in the regulation of epithelial differentiation and proliferation. Curr Opin Cell Biol 17(5):453–458
6. Evans WH, Martin PE (2002) Gap junctions: structure and function (review). Mol Membr Biol 19(2):121–136
7. Stains JP, Civitelli R (2005) Gap junctions regulate extracellular signal-regulated kinase signaling to affect gene transcription. Mol Biol Cell 16(1):64–72
8. Yamasaki H (1990) Gap junctional intercellular communication and carcinogenesis. Carcinogenesis 11(7):1051–1058
9. Eghbali B, Kessler JA, Reid LM, Roy C, Spray DC (1991) Involvement of gap junctions in tumorigenesis: transfection of tumor cells with connexin 32 cdna retards growth in vivo. Proc Natl Acad Sci U S A 88(23):10701–10705
10. Imagawa W, Pedchenko VK, Helber J, Zhang H (2002) Hormone/growth factor interactions mediating epithelial/stromal communication in mammary gland development and carcinogenesis. J Steroid Biochem Mol Biol 80(2):213–230

11. O'Hayre M, Salanga CL, Handel TM, Allen SJ (2008) Chemokines and cancer: migration, intracellular signalling and intercellular communication in the microenvironment. Biochem J 409(3):635–649
12. Camussi G, Deregibus MC, Bruno S, Cantaluppi V, Biancone L (2010) Exosomes/microvesicles as a mechanism of cell-to-cell communication. Kidney Int 78(9):838–848
13. Qu Y, Dahl G (2002) Function of the voltage gate of gap junction channels: selective exclusion of molecules. Proc Natl Acad Sci U S A 99(2):697–702
14. Wolvetang EJ, Pera MF, Zuckerman KS (2007) Gap junction mediated transport of shRNA between human embryonic stem cells. Biochem Biophys Res Commun 363(3):610–615
15. Valiunas V, Polosina YY, Miller H, Potapova IA, Valiuniene L, Doronin S, Mathias RT, Robinson RB, Rosen MR, Cohen IS, Brink PR (2005) Connexin-specific cell-to-cell transfer of short interfering rna by gap junctions. J Physiol 568(Pt 2):459–468
16. Chang Q, Tang W, Ahmad S, Zhou B, Lin X (2008) Gap junction mediated intercellular metabolite transfer in the cochlea is compromised in connexin30 null mice. PLoS One 3(12), e4088
17. Yamasaki H, Mesnil M, Omori Y, Mironov N, Krutovskikh V (1995) Intercellular communication and carcinogenesis. Mutat Res 333(1–2):181–188
18. Trosko JE, Ruch RJ (1998) Cell-cell communication in carcinogenesis. Front Biosci 3:d208–d236
19. Djalilian AR, McGaughey D, Patel S, Seo EY, Yang C, Cheng J, Tomic M, Sinha S, Ishida-Yamamoto A, Segre JA (2006) Connexin 26 regulates epidermal barrier and wound remodeling and promotes psoriasiform response. J Clin Invest 116(5):1243–1253
20. Langlois S, Maher AC, Manias JL, Shao Q, Kidder GM, Laird DW (2007) Connexin levels regulate keratinocyte differentiation in the epidermis. J Biol Chem 282(41):30171–30180
21. Maass K, Ghanem A, Kim JS, Saathoff M, Urschel S, Kirfel G, Grummer R, Kretz M, Lewalter T, Tiemann K, Winterhager E, Herzog V, Willecke K (2004) Defective epidermal barrier in neonatal mice lacking the c-terminal region of connexin43. Mol Biol Cell 15(10):4597–4608
22. Krutovskikh V, Mazzoleni G, Mironov N, Omori Y, Aguelon AM, Mesnil M, Berger F, Partensky C, Yamasaki H (1994) Altered homologous and heterologous gap-junctional intercellular communication in primary human liver tumors associated with aberrant protein localization but not gene mutation of connexin 32. Int J Cancer 56(1):87–94
23. Uchida Y, Matsuda K, Sasahara K, Kawabata H, Nishioka M (1995) Immunohistochemistry of gap junctions in normal and diseased gastric mucosa of humans. Gastroenterology 109(5):1492–1496
24. Tsai H, Werber J, Davia MO, Edelman M, Tanaka KE, Melman A, Christ GJ, Geliebter J (1996) Reduced connexin 43 expression in high grade, human prostatic adenocarcinoma cells. Biochem Biophys Res Commun 227(1):64–69
25. King TJ, Fukushima LH, Hieber AD, Shimabukuro KA, Sakr WA, Bertram JS (2000) Reduced levels of connexin43 in cervical dysplasia: inducible expression in a cervical carcinoma cell line decreases neoplastic potential with implications for tumor progression. Carcinogenesis 21(6):1097–1109
26. Loewenstein WR, Kanno Y (1966) Intercellular communication and the control of tissue growth: lack of communication between cancer cells. Nature 209(5029):1248–1249
27. Tada J, Hashimoto K (1997) Ultrastructural localization of gap junction protein connexin 43 in normal human skin, basal cell carcinoma, and squamous cell carcinoma. J Cutan Pathol 24(10):628–635
28. Temme A, Buchmann A, Gabriel HD, Nelles E, Schwarz M, Willecke K (1997) High incidence of spontaneous and chemically induced liver tumors in mice deficient for connexin32. Curr Biol 7(9):713–716
29. Avanzo JL, Mesnil M, Hernandez-Blazquez FJ, Mackowiak II, Mori CM, da Silva TC, Oloris SC, Garate AP, Massironi SM, Yamasaki H, Dagli ML (2004) Increased susceptibility to

urethane-induced lung tumors in mice with decreased expression of connexin43. Carcinogenesis 25(10):1973–1982
30. Saunders MM, Seraj MJ, Li Z, Zhou Z, Winter CR, Welch DR, Donahue HJ (2001) Breast cancer metastatic potential correlates with a breakdown in homospecific and heterospecific gap junctional intercellular communication. Cancer Res 61(5):1765–1767
31. Yang J, Liu B, Wang Q, Yuan D, Hong X, Yang Y, Tao L (2011) Connexin 32 and its derived homotypic gap junctional intercellular communication inhibit the migration and invasion of transfected hela cells via enhancement of intercellular adhesion. Mol Med Rep 4(5):971–979
32. McLachlan E, Shao Q, Laird DW (2007) Connexins and gap junctions in mammary gland development and breast cancer progression. J Membr Biol 218(1–3):107–121
33. Li Z, Zhou Z, Donahue HJ (2008) Alterations in cx43 and ob-cadherin affect breast cancer cell metastatic potential. Clin Exp Metastasis 25(3):265–272
34. Li Z, Zhou Z, Welch DR, Donahue HJ (2008) Expressing connexin 43 in breast cancer cells reduces their metastasis to lungs. Clin Exp Metastasis 25(8):893–901
35. Rhee DY, Zhao XQ, Francis RJ, Huang GY, Mably JD, Lo CW (2009) Connexin 43 regulates epicardial cell polarity and migration in coronary vascular development. Development 136(18):3185–3193
36. Mehta PP, Bertram JS, Loewenstein WR (1986) Growth inhibition of transformed cells correlates with their junctional communication with normal cells. Cell 44(1):187–196
37. Jara PI, Boric MP, Saez JC (1995) Leukocytes express connexin 43 after activation with lipopolysaccharide and appear to form gap junctions with endothelial cells after ischemia-reperfusion. Proc Natl Acad Sci U S A 92(15):7011–7015
38. El-Sabban ME, Pauli BU (1991) Cytoplasmic dye transfer between metastatic tumor cells and vascular endothelium. J Cell Biol 115(5):1375–1382
39. Pollmann MA, Shao Q, Laird DW, Sandig M (2005) Connexin 43 mediated gap junctional communication enhances breast tumor cell diapedesis in culture. Breast Cancer Res 7(4):R522–R534
40. El-Sabban ME, Pauli BU (1994) Adhesion-mediated gap junctional communication between lung-metastatic cancer cells and endothelium. Invasion Metastasis 14(1–6):164–176
41. Elzarrad MK, Haroon A, Willecke K, Dobrowolski R, Gillespie MN, Al-Mehdi AB (2008) Connexin-43 upregulation in micrometastases and tumor vasculature and its role in tumor cell attachment to pulmonary endothelium. BMC Med 6:20
42. Ito A, Katoh F, Kataoka TR, Okada M, Tsubota N, Asada H, Yoshikawa K, Maeda S, Kitamura Y, Yamasaki H, Nojima H (2000) A role for heterologous gap junctions between melanoma and endothelial cells in metastasis. J Clin Invest 105(9):1189–1197
43. Naoi Y, Miyoshi Y, Taguchi T, Kim SJ, Arai T, Maruyama N, Tamaki Y, Noguchi S (2008) Connexin26 expression is associated with aggressive phenotype in human papillary and follicular thyroid cancers. Cancer Lett 262(2):248–256
44. Ezumi K, Yamamoto H, Murata K, Higashiyama M, Damdinsuren B, Nakamura Y, Kyo N, Okami J, Ngan CY, Takemasa I, Ikeda M, Sekimoto M, Matsuura N, Nojima H, Monden M (2008) Aberrant expression of connexin 26 is associated with lung metastasis of colorectal cancer. Clin Cancer Res 14(3):677–684
45. Inose T, Kato H, Kimura H, Faried A, Tanaka N, Sakai M, Sano A, Sohda M, Nakajima M, Fukai Y, Miyazaki T, Masuda N, Fukuchi M, Kuwano H (2009) Correlation between connexin 26 expression and poor prognosis of esophageal squamous cell carcinoma. Ann Surg Oncol 16(6):1704–1710
46. Naoi Y, Miyoshi Y, Taguchi T, Kim SJ, Arai T, Tamaki Y, Noguchi S (2007) Connexin26 expression is associated with lymphatic vessel invasion and poor prognosis in human breast cancer. Breast Cancer Res Treat 106(1):11–17
47. Lim PK, Bliss SA, Patel SA, Taborga M, Dave MA, Gregory LA, Greco SJ, Bryan M, Patel PS, Rameshwar P (2011) Gap junction-mediated import of microrna from bone marrow stro-

mal cells can elicit cell cycle quiescence in breast cancer cells. Cancer Res 71(5):1550–1560
48. Ao M, Franco OE, Park D, Raman D, Williams K, Hayward SW (2007) Cross-talk between paracrine-acting cytokine and chemokine pathways promotes malignancy in benign human prostatic epithelium. Cancer Res 67(9):4244–4253
49. Thery C, Ostrowski M, Segura E (2009) Membrane vesicles as conveyors of immune responses. Nat Rev Immunol 9(8):581–593
50. Simons M, Raposo G (2009) Exosomes–vesicular carriers for intercellular communication. Curr Opin Cell Biol 21(4):575–581
51. Rabinowits G, Gercel-Taylor C, Day JM, Taylor DD, Kloecker GH (2009) Exosomal microrna: a diagnostic marker for lung cancer. Clin Lung Cancer 10(1):42–46
52. Suetsugu A, Honma K, Saji S, Moriwaki H, Ochiya T, Hoffman RM (2013) Imaging exosome transfer from breast cancer cells to stroma at metastatic sites in orthotopic nude-mouse models. Adv Drug Deliv Rev 65(3):383–390
53. Balaj L, Lessard R, Dai L, Cho YJ, Pomeroy SL, Breakefield XO, Skog J (2011) Tumour microvesicles contain retrotransposon elements and amplified oncogene sequences. Nat Commun 2:180
54. Valadi H, Ekstrom K, Bossios A, Sjostrand M, Lee JJ, Lotvall JO (2007) Exosome-mediated transfer of mrnas and micrornas is a novel mechanism of genetic exchange between cells. Nat Cell Biol 9(6):654–659
55. Xiao D, Ohlendorf J, Chen Y, Taylor DD, Rai SN, Waigel S, Zacharias W, Hao H, McMasters KM (2012) Identifying mrna, microrna and protein profiles of melanoma exosomes. PLoS One 7(10), e46874
56. Kogure T, Lin WL, Yan IK, Braconi C, Patel T (2011) Intercellular nanovesicle-mediated microrna transfer: a mechanism of environmental modulation of hepatocellular cancer cell growth. Hepatology 54(4):1237–1248
57. Staubach S, Razawi H, Hanisch FG (2009) Proteomics of muc1-containing lipid rafts from plasma membranes and exosomes of human breast carcinoma cells mcf-7. Proteomics 9(10):2820–2835
58. Hood JL, Pan H, Lanza GM, Wickline SA, Consortium for Translational Research in Advanced I, Nanomedicine (2009) Paracrine induction of endothelium by tumor exosomes. Lab Invest 89(11):1317–1328
59. Tadokoro H, Umezu T, Ohyashiki K, Hirano T, Ohyashiki JH (2013) Exosomes derived from hypoxic leukemia cells enhance tube formation in endothelial cells. J Biol Chem 288(48):34343–34351
60. Peinado H, Aleckovic M, Lavotshkin S, Matei I, Costa-Silva B, Moreno-Bueno G, Hergueta-Redondo M, Williams C, Garcia-Santos G, Ghajar C, Nitadori-Hoshino A, Hoffman C, Badal K, Garcia BA, Callahan MK, Yuan J, Martins VR, Skog J, Kaplan RN, Brady MS, Wolchok JD, Chapman PB, Kang Y, Bromberg J, Lyden D (2012) Melanoma exosomes educate bone marrow progenitor cells toward a pro-metastatic phenotype through met. Nat Med 18(6):883–891
61. Gu J, Qian H, Shen L, Zhang X, Zhu W, Huang L, Yan Y, Mao F, Zhao C, Shi Y, Xu W (2012) Gastric cancer exosomes trigger differentiation of umbilical cord derived mesenchymal stem cells to carcinoma-associated fibroblasts through tgf-beta/smad pathway. PLoS One 7(12), e52465
62. Luga V, Zhang L, Viloria-Petit AM, Ogunjimi AA, Inanlou MR, Chiu E, Buchanan M, Hosein AN, Basik M, Wrana JL (2012) Exosomes mediate stromal mobilization of autocrine wnt-pcp signaling in breast cancer cell migration. Cell 151(7):1542–1556
63. Abusamra AJ, Zhong Z, Zheng X, Li M, Ichim TE, Chin JL, Min WP (2005) Tumor exosomes expressing fas ligand mediate cd8+ t-cell apoptosis. Blood Cells Mol Dis 35(2):169–173

64. Chen WX, Liu XM, Lv MM, Chen L, Zhao JH, Zhong SL, Ji MH, Hu Q, Luo Z, Wu JZ, Tang JH (2014) Exosomes from drug-resistant breast cancer cells transmit chemoresistance by a horizontal transfer of micrornas. PLoS One 9(4), e95240
65. Taylor DD, Gercel-Taylor C (2008) Microrna signatures of tumor-derived exosomes as diagnostic biomarkers of ovarian cancer. Gynecol Oncol 110(1):13–21
66. Rosell R, Wei J, Taron M (2009) Circulating microrna signatures of tumor-derived exosomes for early diagnosis of non-small-cell lung cancer. Clin Lung Cancer 10(1):8–9
67. Silva J, Garcia V, Rodriguez M, Compte M, Cisneros E, Veguillas P, Garcia JM, Dominguez G, Campos-Martin Y, Cuevas J, Pena C, Herrera M, Diaz R, Mohammed N, Bonilla F (2012) Analysis of exosome release and its prognostic value in human colorectal cancer. Genes Chromosomes Cancer 51(4):409–418
68. Parolini I, Federici C, Raggi C, Lugini L, Palleschi S, De Milito A, Coscia C, Iessi E, Logozzi M, Molinari A, Colone M, Tatti M, Sargiacomo M, Fais S (2009) Microenvironmental ph is a key factor for exosome traffic in tumor cells. J Biol Chem 284(49):34211–34222
69. Kahlert C, Melo SA, Protopopov A, Tang J, Seth S, Koch M, Zhang J, Weitz J, Chin L, Futreal A, Kalluri R (2014) Identification of double-stranded genomic DNA spanning all chromosomes with mutated kras and p53 DNA in the serum exosomes of patients with pancreatic cancer. J Biol Chem 289(7):3869–3875
70. Thakur BK, Zhang H, Becker A, Matei I, Huang Y, Costa-Silva B, Zheng Y, Hoshino A, Brazier H, Xiang J, Williams C, Rodriguez-Barrueco R, Silva JM, Zhang W, Hearn S, Elemento O, Paknejad N, Manova-Todorova K, Welte K, Bromberg J, Peinado H, Lyden D (2014) Double-stranded DNA in exosomes: a novel biomarker in cancer detection. Cell Res 24:766–769. doi:10.1038/cr.2014.44
71. Bettermann K, Vucur M, Haybaeck J, Koppe C, Janssen J, Heymann F, Weber A, Weiskirchen R, Liedtke C, Gassler N, Muller M, de Vos R, Wolf MJ, Boege Y, Seleznik GM, Zeller N, Erny D, Fuchs T, Zoller S, Cairo S, Buendia MA, Prinz M, Akira S, Tacke F, Heikenwalder M, Trautwein C, Luedde T (2010) Tak1 suppresses a nemo-dependent but nf-kappab-independent pathway to liver cancer. Cancer Cell 17(5):481–496
72. Rana S, Malinowska K, Zoller M (2013) Exosomal tumor microrna modulates premetastatic organ cells. Neoplasia 15(3):281–295
73. Hood JL, San RS, Wickline SA (2011) Exosomes released by melanoma cells prepare sentinel lymph nodes for tumor metastasis. Cancer Res 71(11):3792–3801
74. Hanahan D, Weinberg RA (2011) Hallmarks of cancer: the next generation. Cell 144(5):646–674
75. Ciravolo V, Huber V, Ghedini GC, Venturelli E, Bianchi F, Campiglio M, Morelli D, Villa A, Della Mina P, Menard S, Filipazzi P, Rivoltini L, Tagliabue E, Pupa SM (2012) Potential role of her2-overexpressing exosomes in countering trastuzumab-based therapy. J Cell Physiol 227(2):658–667
76. Rasanen K, Vaheri A (2010) Activation of fibroblasts in cancer stroma. Exp Cell Res 316(17):2713–2722
77. Xu J, Lamouille S, Derynck R (2009) Tgf-beta-induced epithelial to mesenchymal transition. Cell Res 19(2):156–172
78. Lo HW, Hsu SC, Xia W, Cao X, Shih JY, Wei Y, Abbruzzese JL, Hortobagyi GN, Hung MC (2007) Epidermal growth factor receptor cooperates with signal transducer and activator of transcription 3 to induce epithelial-mesenchymal transition in cancer cells via up-regulation of twist gene expression. Cancer Res 67(19):9066–9076
79. Elliott BE, Hung WL, Boag AH, Tuck AB (2002) The role of hepatocyte growth factor (scatter factor) in epithelial-mesenchymal transition and breast cancer. Can J Physiol Pharmacol 80(2):91–102
80. Sullivan NJ, Sasser AK, Axel AE, Vesuna F, Raman V, Ramirez N, Oberyszyn TM, Hall BM (2009) Interleukin-6 induces an epithelial-mesenchymal transition phenotype in human breast cancer cells. Oncogene 28(33):2940–2947

81. Fernando RI, Castillo MD, Litzinger M, Hamilton DH, Palena C (2011) Il-8 signaling plays a critical role in the epithelial-mesenchymal transition of human carcinoma cells. Cancer Res 71(15):5296–5306
82. Nieto MA, Cano A (2012) The epithelial-mesenchymal transition under control: global programs to regulate epithelial plasticity. Semin Cancer Biol 22(5–6):361–368
83. Thiery JP, Acloque H, Huang RY, Nieto MA (2009) Epithelial-mesenchymal transitions in development and disease. Cell 139(5):871–890
84. Fuxe J, Vincent T, Garcia de Herreros A (2010) Transcriptional crosstalk between tgf-beta and stem cell pathways in tumor cell invasion: role of emt promoting smad complexes. Cell Cycle 9(12):2363–2374
85. Radisky DC (2005) Epithelial-mesenchymal transition. J Cell Sci 118(Pt 19):4325–4326
86. Kalluri R, Weinberg RA (2009) The basics of epithelial-mesenchymal transition. J Clin Invest 119(6):1420–1428
87. De Craene B, Berx G (2013) Regulatory networks defining emt during cancer initiation and progression. Nat Rev Cancer 13(2):97–110
88. Akhurst RJ, Derynck R (2001) Tgf-beta signaling in cancer–a double-edged sword. Trends Cell Biol 11(11):S44–S51
89. Mani SA, Guo W, Liao MJ, Eaton EN, Ayyanan A, Zhou AY, Brooks M, Reinhard F, Zhang CC, Shipitsin M, Campbell LL, Polyak K, Brisken C, Yang J, Weinberg RA (2008) The epithelial-mesenchymal transition generates cells with properties of stem cells. Cell 133(4):704–715
90. Paget S (1989) The distribution of secondary growths in cancer of the breast. 1889. Cancer Metastasis Rev 8(2):98–101
91. Balkwill F (2004) Cancer and the chemokine network. Nat Rev Cancer 4(7):540–550
92. Zlotnik A, Yoshie O (2012) The chemokine superfamily revisited. Immunity 36(5):705–716
93. Forster R, Schubel A, Breitfeld D, Kremmer E, Renner-Muller I, Wolf E, Lipp M (1999) Ccr7 coordinates the primary immune response by establishing functional microenvironments in secondary lymphoid organs. Cell 99(1):23–33
94. Muller A, Homey B, Soto H, Ge N, Catron D, Buchanan ME, McClanahan T, Murphy E, Yuan W, Wagner SN, Barrera JL, Mohar A, Verastegui E, Zlotnik A (2001) Involvement of chemokine receptors in breast cancer metastasis. Nature 410(6824):50–56
95. Zlotnik A (2006) Involvement of chemokine receptors in organ-specific metastasis. Contrib Microbiol 13:191–199
96. Constantin G, Majeed M, Giagulli C, Piccio L, Kim JY, Butcher EC, Laudanna C (2000) Chemokines trigger immediate beta2 integrin affinity and mobility changes: differential regulation and roles in lymphocyte arrest under flow. Immunity 13(6):759–769
97. Gupta GP, Massague J (2006) Cancer metastasis: building a framework. Cell 127(4):679–695
98. Fidler IJ (2003) The pathogenesis of cancer metastasis: the 'seed and soil' hypothesis revisited. Nat Rev Cancer 3(6):453–458
99. Witz IP (2008) The selectin-selectin ligand axis in tumor progression. Cancer Metastasis Rev 27(1):19–30
100. Laubli H, Spanaus KS, Borsig L (2009) Selectin-mediated activation of endothelial cells induces expression of ccl5 and promotes metastasis through recruitment of monocytes. Blood 114(20):4583–4591
101. Karnoub AE, Dash AB, Vo AP, Sullivan A, Brooks MW, Bell GW, Richardson AL, Polyak K, Tubo R, Weinberg RA (2007) Mesenchymal stem cells within tumour stroma promote breast cancer metastasis. Nature 449(7162):557–563
102. Velasco-Velazquez M, Jiao X, De La Fuente M, Pestell TG, Ertel A, Lisanti MP, Pestell RG (2012) Ccr5 antagonist blocks metastasis of basal breast cancer cells. Cancer Res 72(15):3839–3850
103. Keeley EC, Mehrad B, Strieter RM (2011) Chemokines as mediators of tumor angiogenesis and neovascularization. Exp Cell Res 317(5):685–690

104. Addison CL, Daniel TO, Burdick MD, Liu H, Ehlert JE, Xue YY, Buechi L, Walz A, Richmond A, Strieter RM (2000) The cxc chemokine receptor 2, cxcr2, is the putative receptor for elr + cxc chemokine-induced angiogenic activity. J Immunol 165(9):5269–5277
105. Salcedo R, Ponce ML, Young HA, Wasserman K, Ward JM, Kleinman HK, Oppenheim JJ, Murphy WJ (2000) Human endothelial cells express ccr2 and respond to mcp-1: direct role of mcp-1 in angiogenesis and tumor progression. Blood 96(1):34–40
106. Li X, Loberg R, Liao J, Ying C, Snyder LA, Pienta KJ, McCauley LK (2009) A destructive cascade mediated by ccl2 facilitates prostate cancer growth in bone. Cancer Res 69(4):1685–1692
107. Singh S, Varney M, Singh RK (2009) Host cxcr2-dependent regulation of melanoma growth, angiogenesis, and experimental lung metastasis. Cancer Res 69(2):411–415
108. Nathanson SD (2003) Insights into the mechanisms of lymph node metastasis. Cancer 98(2):413–423
109. Skobe M, Hawighorst T, Jackson DG, Prevo R, Janes L, Velasco P, Riccardi L, Alitalo K, Claffey K, Detmar M (2001) Induction of tumor lymphangiogenesis by vegf-c promotes breast cancer metastasis. Nat Med 7(2):192–198
110. Oh SJ, Jeltsch MM, Birkenhager R, McCarthy JE, Weich HA, Christ B, Alitalo K, Wilting J (1997) Vegf and vegf-c: specific induction of angiogenesis and lymphangiogenesis in the differentiated avian chorioallantoic membrane. Dev Biol 188(1):96–109
111. Saeki H, Moore AM, Brown MJ, Hwang ST (1999) Cutting edge: secondary lymphoid-tissue chemokine (slc) and cc chemokine receptor 7 (ccr7) participate in the emigration pathway of mature dendritic cells from the skin to regional lymph nodes. J Immunol 162(5):2472–2475
112. Baluk P, Fuxe J, Hashizume H, Romano T, Lashnits E, Butz S, Vestweber D, Corada M, Molendini C, Dejana E, McDonald DM (2007) Functionally specialized junctions between endothelial cells of lymphatic vessels. J Exp Med 204(10):2349–2362
113. Hopken UE, Foss HD, Meyer D, Hinz M, Leder K, Stein H, Lipp M (2002) Up-regulation of the chemokine receptor ccr7 in classical but not in lymphocyte-predominant hodgkin disease correlates with distinct dissemination of neoplastic cells in lymphoid organs. Blood 99(4):1109–1116
114. Liu F, Lang R, Wei J, Fan Y, Cui L, Gu F, Guo X, Pringle GA, Zhang X, Fu L (2009) Increased expression of sdf-1/cxcr4 is associated with lymph node metastasis of invasive micropapillary carcinoma of the breast. Histopathology 54(6):741–750
115. Ding Y, Shimada Y, Maeda M, Kawabe A, Kaganoi J, Komoto I, Hashimoto Y, Miyake M, Hashida H, Imamura M (2003) Association of cc chemokine receptor 7 with lymph node metastasis of esophageal squamous cell carcinoma. Clin Cancer Res 9(9):3406–3412
116. Manzo A, Bugatti S, Caporali R, Prevo R, Jackson DG, Uguccioni M, Buckley CD, Montecucco C, Pitzalis C (2007) Ccl21 expression pattern of human secondary lymphoid organ stroma is conserved in inflammatory lesions with lymphoid neogenesis. Am J Pathol 171(5):1549–1562
117. Shields JD, Fleury ME, Yong C, Tomei AA, Randolph GJ, Swartz MA (2007) Autologous chemotaxis as a mechanism of tumor cell homing to lymphatics via interstitial flow and autocrine ccr7 signaling. Cancer Cell 11(6):526–538
118. Wiley HE, Gonzalez EB, Maki W, Wu MT, Hwang ST (2001) Expression of cc chemokine receptor-7 and regional lymph node metastasis of b16 murine melanoma. J Natl Cancer Inst 93(21):1638–1643
119. Uchida D, Onoue T, Begum NM, Kuribayashi N, Tomizuka Y, Tamatani T, Nagai H, Miyamoto Y (2009) Vesnarinone downregulates cxcr4 expression via upregulation of kruppel-like factor 2 in oral cancer cells. Mol Cancer 8:62
120. Mirisola V, Zuccarino A, Bachmeier BE, Sormani MP, Falter J, Nerlich A, Pfeffer U (2009) Cxcl12/sdf1 expression by breast cancers is an independent prognostic marker of disease-free and overall survival. Eur J Cancer 45(14):2579–2587

121. Salmaggi A, Maderna E, Calatozzolo C, Gaviani P, Canazza A, Milanesi I, Silvani A, DiMeco F, Carbone A, Pollo B (2009) Cxcl12, cxcr4 and cxcr7 expression in brain metastases. Cancer Biol Ther 8(17):1608–1614
122. Salogni L, Musso T, Bosisio D, Mirolo M, Jala VR, Haribabu B, Locati M, Sozzani S (2009) Activin a induces dendritic cell migration through the polarized release of cxc chemokine ligands 12 and 14. Blood 113(23):5848–5856
123. Sozzani S, Allavena P, D'Amico G, Luini W, Bianchi G, Kataura M, Imai T, Yoshie O, Bonecchi R, Mantovani A (1998) Differential regulation of chemokine receptors during dendritic cell maturation: a model for their trafficking properties. J Immunol 161(3):1083–1086
124. Xu Q, Yuan X, Xu M, McLafferty F, Hu J, Lee BS, Liu G, Zeng Z, Black KL, Yu JS (2009) Chemokine cxc receptor 4--mediated glioma tumor tracking by bone marrow--derived neural progenitor/stem cells. Mol Cancer Ther 8(9):2746–2753
125. Sugiyama T, Kohara H, Noda M, Nagasawa T (2006) Maintenance of the hematopoietic stem cell pool by cxcl12-cxcr4 chemokine signaling in bone marrow stromal cell niches. Immunity 25(6):977–988
126. Cardones AR, Murakami T, Hwang ST (2003) Cxcr4 enhances adhesion of b16 tumor cells to endothelial cells in vitro and in vivo via beta (1) integrin. Cancer Res 63(20):6751–6757
127. Kim M, Koh YJ, Kim KE, Koh BI, Nam DH, Alitalo K, Kim I, Koh GY (2010) Cxcr4 signaling regulates metastasis of chemoresistant melanoma cells by a lymphatic metastatic niche. Cancer Res 70(24):10411–10421
128. Arigami T, Natsugoe S, Uenosono Y, Yanagita S, Arima H, Hirata M, Ishigami S, Aikou T (2009) Ccr7 and cxcr4 expression predicts lymph node status including micrometastasis in gastric cancer. Int J Oncol 35(1):19–24
129. Tanaka M, Grossman HB (2001) Connexin 26 gene therapy of human bladder cancer: induction of growth suppression, apoptosis, and synergy with cisplatin. Hum Gene Ther 12(18):2225–2236
130. Shishido SN, Nguyen TA (2012) Gap junction enhancer increases efficacy of cisplatin to attenuate mammary tumor growth. PLoS One 7(9), e44963
131. Li Z, Zhao R, Wu X, Sun Y, Yao M, Li J, Xu Y, Gu J (2005) Identification and characterization of a novel peptide ligand of epidermal growth factor receptor for targeted delivery of therapeutics. FASEB J 19(14):1978–1985
132. Ohno S, Takanashi M, Sudo K, Ueda S, Ishikawa A, Matsuyama N, Fujita K, Mizutani T, Ohgi T, Ochiya T, Gotoh N, Kuroda M (2013) Systemically injected exosomes targeted to egfr deliver antitumor microrna to breast cancer cells. Mol Ther 21(1):185–191
133. Tian Y, Li S, Song J, Ji T, Zhu M, Anderson GJ, Wei J, Nie G (2014) A doxorubicin delivery platform using engineered natural membrane vesicle exosomes for targeted tumor therapy. Biomaterials 35(7):2383–2390
134. Shipitsin M, Campbell LL, Argani P, Weremowicz S, Bloushtain-Qimron N, Yao J, Nikolskaya T, Serebryiskaya T, Beroukhim R, Hu M, Halushka MK, Sukumar S, Parker LM, Anderson KS, Harris LN, Garber JE, Richardson AL, Schnitt SJ, Nikolsky Y, Gelman RS, Polyak K (2007) Molecular definition of breast tumor heterogeneity. Cancer Cell 11(3):259–273
135. Bhola NE, Balko JM, Dugger TC, Kuba MG, Sanchez V, Sanders M, Stanford J, Cook RS, Arteaga CL (2013) Tgf-beta inhibition enhances chemotherapy action against triple-negative breast cancer. J Clin Invest 123(3):1348–1358
136. Domanska UM, Timmer-Bosscha H, Nagengast WB, Oude Munnink TH, Kruizinga RC, Ananias HJ, Kliphuis NM, Huls G, De Vries EG, de Jong IJ, Walenkamp AM (2012) Cxcr4 inhibition with amd3100 sensitizes prostate cancer to docetaxel chemotherapy. Neoplasia 14(8):709–718

# Index

**A**
Actin cytoskeleton, 57, 59, 65, 71, 99, 101, 111, 152, 155, 170, 171, 214, 259, 272, 305, 312, 313, 315, 322
Adherens junctions, 2, 4, 56, 57, 68, 171, 175, 202, 256, 270, 313, 344, 345
Adhesion, cell-cell, 2, 3, 5, 16, 56, 59, 60, 63, 64, 70, 72, 79, 82, 85, 87, 92–94, 97, 107, 151–162, 170, 171, 178, 271, 273, 347, 351
Adhesion, cell-matrix, 68, 155, 157–158, 160, 161, 271
Angiogenesis, 18, 81, 99, 108, 111, 158, 159, 175, 179, 193, 208, 229, 234, 258, 259, 261, 274, 298, 300, 305, 306, 308, 315, 317, 319, 320, 349, 350, 353
Arp2/3 complex, 72, 153, 154

**B**
Bi-directional signaling, 69, 257, 275
Biomarkers, 44, 45, 104, 195, 204, 206, 208, 212–215, 305, 315–316, 318, 320, 325, 326, 349
Breast cancer, 2, 3, 5, 6, 8, 11, 15, 16, 18–20, 47, 82, 84, 92, 99, 105, 107, 152, 156, 162, 193–195, 204, 205, 259, 264, 272, 298, 300–303, 308, 314, 316, 317, 320, 347–352, 354
Breast development, 2, 298, 355
Bystander effect, 19, 34, 35

**C**
Cadherins, 5–7, 12, 16, 18, 56, 57, 59–65, 68, 70–88, 90–94, 96–98, 104, 105, 107, 108, 110–112, 151–162, 170–180, 193, 213, 214, 229, 233, 261, 270–273, 305, 311–314, 317, 347
Cancer dormancy, 307, 308, 310, 350
Cancer stem cells, 31, 86, 108, 110, 160, 176, 231, 310, 351
Carbenoxolone, 35, 47
Catenin, 5, 14, 18, 34, 57, 59–66, 70–75, 77, 78, 80, 82, 85, 86, 89–91, 93–95, 111, 112, 152–156, 159, 160, 170–173, 176–180, 261, 270, 272, 301, 312–314
α-Catenin, 5, 14, 18, 57, 59, 61–63, 71, 72, 75, 77, 86, 152–155, 170, 171, 178, 261
β-Catenin, 5, 6, 14, 17, 19, 34, 35, 57, 59, 61–64, 66, 70, 72, 75–82, 84, 86, 89–91, 93, 95, 100, 103, 108, 111, 112, 152–154, 156, 159, 160, 170–173, 176, 178, 179, 193, 194, 214, 271, 272, 302, 312, 313
CDC42, 65, 67, 70–73, 90, 213, 215, 216
Cell junctions, 2, 4, 6, 7, 12, 19, 56, 76–78, 80, 85, 86, 96, 98, 99, 101, 104, 151, 153, 155, 158, 171, 173, 256, 270, 272
Chemokines, 96, 202, 228, 234, 235, 307, 323, 345, 351–354
Chemoresistance, 204–206, 219, 355
Chemotherapeutic targets, 308, 315, 319, 320, 325, 326
Cholinergic systems, 58, 100–101
*Cis* interaction, 262


M. Kandouz (ed.), *Intercellular Communication in Cancer*,
DOI 10.1007/978-94-017-7380-5

Claudins, 4–6, 12, 76, 111, 178, 269
Clustering, receptor, 266, 267
Colorectal cancer, 81, 82, 86, 89–94, 103–107, 109, 110, 195, 259, 260, 272, 301, 302, 316, 317
Community-wide molecular networks (CMN), 290–293
Computational analysis, 287–293
Computational modeling, 158, 160, 162, 289
Connexin-43, 34, 46, 202
Connexins, 6, 8–9, 11–19, 33–37, 46, 48, 178, 202, 211, 269, 305, 313, 314, 345, 346
Contact inhibition of locomotion, 261
Contact Normalization, 311–320, 324–326
CRUMB and SCRIB complexes, 67–68
Cx26, 8, 10, 11, 13–16, 18, 19, 34, 348, 354
CXCL12, 229, 234, 235, 348, 350–354
CXCR4, 230, 234, 235, 267, 351–355
Cytokines, 32, 73, 88, 89, 94, 98, 102, 106, 175, 193, 194, 201, 202, 218, 228, 229, 232, 324, 344, 348–351
Cytoplasmic extensions, 202, 213

**D**

Desmosomes, 2, 4, 5, 56, 83, 175, 273
Differentiation, 2, 4–7, 10–15, 17, 20, 31, 34, 44, 56, 57, 63, 64, 66, 68–70, 82, 85, 87–89, 92, 93, 95, 99, 102, 110, 111, 155, 158, 175, 193, 203, 231, 260, 263, 273, 275, 289, 308, 315, 317, 322, 346
Dissemination, 157, 176, 263, 274, 304, 305, 320, 323, 345, 350–354
DNMT, 83, 105–106, 112
Dual functions, 86, 263

**E**

E6/E7, 178–180
E-cadherins, 5, 7, 19, 35, 56, 57, 59–65, 67, 69–72, 74–88, 90–99, 102, 104–107, 111, 112, 151–160, 162, 170–176, 178–180, 193, 212–214, 270–272, 312, 314, 345
Endocytosis, 57, 65, 74–75, 78–79, 88, 97–99, 102, 158, 193, 318
Eph receptors, 256, 257, 260, 262–265, 267, 269
Ephrin ligand, 256, 257, 262–265, 272
Epigenetic regulation, 82, 84
Epithelial-mesenchymal transition, 170, 174, 271, 305, 314, 323, 347, 350
Epstein–Barr virus (EBV), 170, 179, 180, 194
ErbB-2, 178, 179
Exosomes, 48, 49, 190–196, 201, 202, 204, 206–209, 212, 218, 345, 348–351, 354
Extracellular matrix, 2, 32, 56, 68, 78, 80, 160, 175, 193, 261, 268, 305, 306, 308, 325, 344

**F**

Fhl1, 316–318
Focal adhesion kinase (FAK), 69, 76, 94, 97, 99, 101, 102, 110, 157, 259, 263, 312
Fusion, cell-cell, 227–235

**G**

Gap Junctional Intercellular Communication, 16–18, 46
Gap junctions, 2, 4, 10, 15, 17, 19, 33–35, 46, 47, 57, 175, 202, 203, 211, 214, 218, 269, 311–314, 317, 344–346
Glioblastoma, 30, 43–46, 204, 205, 208, 261, 265, 267, 301, 302, 305, 324
Growth factors, 2, 6, 34, 35, 45, 62, 63, 72, 80, 87, 94, 175, 178, 180, 232, 260, 264, 265, 267, 274, 290, 304, 305, 307, 313, 315, 319, 320, 344, 345, 348–350, 353

**H**

HDAC, 74–75, 83, 106–107, 112
Hematogenous dissemination, 176, 351
Hemichannels, 6, 8–10, 15, 33, 36, 46
Heterologous communication, 347
Heterotypic interactions, 80, 152, 261
Homotypic interactions, 261
Human papillomavirus (HPV), 170, 178–180

**I**

Immune systems, 32, 98, 103, 110, 194, 196, 228, 232, 307, 310
Inflammation, 56, 71, 88–89, 95, 99, 100, 102, 106, 111, 173, 210, 217, 228–230, 233, 298, 307, 347, 351, 352, 354
Integrins, 68, 69, 80, 87, 97, 103, 229, 234, 268, 305, 307, 308, 315
Inter-tissue networks, 289–290
Intestinal epithelial homeostasis, 55

## K

Kaiso, transcriptional repressor, 65
Kinase, 7, 44, 63, 67, 69, 72–77, 80, 88, 92, 97, 99, 107–110, 158, 171–173, 195, 210, 216, 256, 257, 259, 261, 263–265, 267–271, 273–275, 299–302, 305, 308, 313–315, 318–320, 323, 324, 326, 349, 350, 354
Kinase-independent functions, 268

## L

Leukocytes, 228–231, 234, 235, 307, 347, 351
Ligand-independence, 93, 263–264, 268
Lymphatic dissemination, 353
Lymphatic vasculature, 322, 351

## M

Mammary tumorigenesis, 273
Mechanoresponse, 154
Mesenchymal stem cells, 36, 47, 155, 228, 229, 352
Metabolic pathways, 216, 290, 293
Metastasis, 5, 7, 11, 16–18, 34, 44, 46, 81, 82, 86, 90, 92–96, 104, 105, 109, 111, 170, 171, 174–176, 178–180, 192–195, 207, 215, 216, 227, 228, 233–235, 259–261, 263, 265, 268, 270, 272, 274, 298, 304–307, 309, 310, 313, 314, 317, 318, 321–323, 325, 326, 347–355
Microbiota, intestinal, 101–103
MicroRNA, 44–49, 74–75, 192, 206–207, 289
Microtubules, 8, 64, 66, 67, 209, 213, 215
Migration, cell, 12, 17, 34, 45, 46, 48, 56, 65, 72, 78, 80, 82, 85, 90, 91, 95–99, 101, 102, 104, 108, 109, 152, 154, 157, 159, 170, 172, 194, 259, 263, 264, 267, 268, 271, 305, 313, 314, 317–320, 322, 323, 325, 347, 351, 353, 354
Multicellular interactions, 288
Multicellular networks, 290

## N

Nectins, 7, 57, 67, 70–71
Neural progenitor cells (NPCs), 32, 34
Non-stem tumor cell, 31

## O

1-Octanol, 35, 36
Oncogenes, 16, 45, 89, 171, 172, 175, 179, 180, 263, 298–304, 308, 311, 320, 349
Oral cancer, 169, 170, 173, 174, 176–180, 325
Organ-specific metastasis, 351

## P

p120ctn, 57, 61–66, 71, 72, 75, 77, 79, 80, 86, 87, 90, 91, 98, 99, 157
Papillomavirus, human, 170
PAR proteins, 67, 92
Pdpn, 305, 307, 319–326
Phosphatase, 69, 73–78, 111, 273, 302, 303, 310
Phosphorylation, 10, 16, 61–63, 67, 70, 73–80, 82, 88, 89, 92, 98, 99, 153, 154, 171, 173, 178, 259, 263, 264, 267–269, 271, 272, 312, 314, 317–319, 323, 325
Podoplanin, 305, 320–322
Polarity, 4, 6–8, 20, 56, 62, 65–68, 78, 92, 94, 97, 152, 171, 174, 175, 193, 269, 271, 347, 350, 351
Protein kinase C (PKC), 16, 99, 214, 270, 318, 323
Proteolysis, 74–75, 79–82, 87, 157, 265, 325

## R

Response, immune and inflammatory, 88, 351

## S

Serum deprivation response protein (SDPR), 316, 318
Shedding, 74–75, 80–81, 95, 108, 232, 264–265, 272, 305
Signaling, 4, 6, 9, 11, 17, 20, 35, 56, 60–66, 69, 70, 73, 75, 81, 85, 87, 91, 94, 98–100, 102, 157, 158, 161, 170–173, 175–177, 192–195, 201, 204, 205, 208, 213, 215, 218, 229, 260–261, 265–268, 298, 305, 308, 314, 317–319, 345, 347–351
Slug/snail, 72
Small GTPases, 65, 70–72, 175, 320
Soluble forms, 264, 265
Stem cells, 36, 44, 45, 48, 56, 85–87, 95, 103, 110, 155, 158, 160, 176, 204, 228, 231, 234, 235, 260, 272, 310, 320
Stress, 5, 57, 94, 98–103, 171, 172, 196, 203, 210, 211, 216, 229, 270, 289, 293
Stromal cells, 6, 12, 96, 175, 195, 204, 206, 208, 211, 212, 218, 219, 228, 233, 261, 290, 292, 293, 305, 345, 348, 350
Systems biology, 287, 288, 293

**T**
Therapeutic targets, 104, 108, 110, 157, 195, 206, 216, 348
Therapy, 8, 19, 30, 31, 34–37, 43, 46, 103–112, 202, 204, 215–216, 292, 298, 307, 310, 354
Tight junctions, 2, 4, 56, 57, 67, 94, 269, 270, 305, 344
Transcriptional regulation, 74–75, 88
Trans-interactions, 70, 79, 262
Tumor-lymphatic endothelial interactions, 353
Tumor microenvironments, 19, 32, 80, 96, 100, 193, 195, 201–203, 206–208, 212, 218, 219, 260, 289, 306, 307, 312, 344, 345, 347, 348, 350, 355
Tumor-stroma interactions, 160, 201, 290, 292
Tumor-stromal interactions, 204, 293
Tumor suppressor genes, 16, 45, 46, 105, 213, 270, 299, 302–304, 349
Tumor-vascular endothelial interactions, 351–353
Tunneling nanotubes, 4, 202, 205, 218

**V**
Vegfr, 314, 319
Vimentin, 95, 175, 193, 213, 215, 272
Vitamin D, 110–112

**W**
Wnt, 5, 34, 35, 61, 63–66, 72, 81, 82, 85–87, 89, 90, 93–95, 100, 111, 153, 170–172, 175, 178, 179, 193, 272, 302

MIX
Papier aus verantwortungsvollen Quellen
Paper from responsible sources
FSC® C105338

If you have any concerns about our products,
you can contact us on
**ProductSafety@springernature.com**

In case Publisher is established outside the EU,
the EU authorized representative is:
**Springer Nature Customer Service Center GmbH**
**Europaplatz 3, 69115 Heidelberg, Germany**

Printed by Libri Plureos GmbH
in Hamburg, Germany